Modern High Temperature Science

To

LEO BREWER

From His Friends
Students, and Colleagues
on the Occasion
of His Sixty-Fifth
Birthday

June 13, 1984

MODERN HIGH TEMPERATURE SCIENCE

A Collection of Research Papers from Scientists, Post-Doctoral Associates, and Colleagues of

PROFESSOR LEO BREWER

in celebration of his 65th birthday

Edited by

John L. Margrave

Rice University, Houston, Texas

Humana Press • Clifton, New Jersey

Special Volume

Leo Brewer Special Festschrift Volume

*For papers with multiple authorship, the asterisk identifies the author to whom inquiries and reprint requests should be addressed.

Preface

It is a great pleasure to have the opportunity to honor our distinguished colleague, Professor Leo Brewer, on the occasion of his sixty-fifth birthday, with this special volume of *High Temperature Science.*

Leo and his wife, Rose, are personal friends of several generations of students and postdoctoral researchers at the University of California at Berkeley. Their concern and understanding has been important to many of us over the past forty years.

Each paper in this volume has at least one author who was a graduate student or a postdoctoral researcher in Leo's laboratory at Berkeley. The variety of topics is indicative of the wide-ranging science done by Brewer-ites and by Leo Brewer himself. He has personally participated in the resolution of many of the classical problems of high-temperature science—from the heat of sublimation of graphite to the dissociation energy of nitrogen to the prediction of binary and ternary phase diagrams. He and his students have made major contributions to atomic and molecular spectroscopy. He has made significant contributions to the development of efficient systems for energy conversion and to ceramics. In addition to his research activities, Leo Brewer has been a long-time participant in the dynamic undergraduate teaching program of the Berkeley Chemistry Department. He has provided crucial insight for students involved in those career-shaping experiences that one endures while acquiring the basics of inorganic, organic, and physical chemistry with that interwoven common bond of thermodynamics.

Equally important to many of us, Leo has been a willing listener, a personal advisor, and an almost limitless source of ideas and imaginative approaches to difficult experimental problems for hundreds of scientists and engineers with real-world assignments.

The task of assembling these manuscripts has been amazingly easy because of the spontaneity and cooperation of the many authors. Most of the effort has been that of our publisher and his staff at the Humana Press, Tom Lanigan and Fran Lipton, and of my secretaries, Donna Montez and Margaret Griffin,and I appreciate their meticulous work. I commend the authors and congratulate Professor and Mrs. Brewer on this happy occasion.

John L. Margrave, *Editor*
Houston, Texas

0018–1536/84/1700–00ix$02.00

Leo Brewer

Versatile Chemist

PAUL W. GILLES

Department of Chemistry, University of Kansas, Lawrence, Kansas, 66045-2112, USA

Universally regarded as the founder of modern high temperature chemistry, Leo Brewer is the subject of these comments. My first impression of Leo, gained during the summer of 1943 when I was a beginning graduate student, was that of a young, vigorous, bright scientist who was always in a hurry. He ran up steps two or three at a time, seemed never to get tired, worked on some large secret project, and never said very much. Now, I know him better. His vigor and intelligence seem undiminished. He still runs out-of-doors. He still works on large, important problems and projects. He still does not talk meaninglessly. My remarks to follow include some comments about his teaching, his research, his service, his recognition, and his personality. But first, the biographical data.

The biography is short. He was born in St. Louis, June 13, 1919 and soon moved to California. He received a BS degree in chemistry from the California Institute of Technology in 1940. He then went to The University of California, Berkeley and has remained there. He received his PhD in chemistry in 1943. He joined the Manhattan Project following his graduate work and joined the faculty in 1946. He and Rose Strugo were married in 1945 and they have 3 children, Beth, Roger, and Gail, and two grandchildren.

His own description of some of his educational background was presented at the American Chemical Society's Symposium memorializing G. N. Lewis (*1*).

THE TEACHER

Brewer has been a teacher his entire professional life. He teaches college freshmen, and he teaches the mature scientist. He used his thermodynamic knowledge to help modify the freshman chemistry experiments in order to satisfy changing environmental and legal requirements. Into the coursework of other undergraduate courses he has brought the results of his research activities. He has approached the teaching of thermodynamics at the undergraduate level in new ways. And, of course, his work with Pitzer in revising Lewis and Randall (2) has been a great service to students of chemistry. One high-temperature chemist who knew Brewer as his undergraduate teacher described him recently by saying that Brewer unequivocally was the best teacher he had ever had.

Brewer works with graduate students in a way that appears to have undergone little change in forty years. Some recent PhD students have described the way he guided their work, and these descriptions are in agreement with my recollection of the way that he led those of us who were among his earliest graduate students. We knew then, and current students know now, that they are very much on their own. Brewer allows his students much freedom. He is not demanding. His questions seem to be phrased so as to inspire thought and action, and he is always available. He is not an empire builder for himself. His research group has never been large. He has not routinely had a senior postdoctoral research associate; rather, he handles the students himself. He and his wife usually arrange a social event once or twice a year with the currently available coworkers, and these are described as "pleasant and relaxed." One recent PhD summarized his experience as a "happy and productive" one. Perhaps by his own example, or perhaps by the questions he asks, he emphasizes the importance of breadth of experience and knowledge. A mature scientist emphasized recently that Brewer really listens at scientific meetings and frequently has meaningful comments even though the work is far removed from his own.

Some years ago I prepared a genealogical chart for Brewer and his students. To prepare such a chart now is almost out of the question. He has educated thirty-eight PhD students and twenty-one postdoctoral research associates. Of the postdoctoral research associates, fourteen hold or have held faculty positions, two are in prestigious positions in national laboratories—one in the United States and one in the United Kingdom—and three are in research institutes. Of his graduate students, some nine occupy faculty positions, two have very responsible positions in Washington, and nine are in national labs. His scientific descendents extend at least four generations deep, the longest chain possibly being Brewer, Gilles, Margrave, Grimley, and Muenow.

The PhD students who have worked with him include: Robert A. Berg, George D. Brabson, Richard G. Brewer, Garry Bullard, Chin-An Chang, Shih-Ger Chang, Ara Chutjian, Paul T. Cunningham, Russell K. Edwards, Guy R. B. Elliott, John L. Engelke, Paul W. Gilles, David A. Goodman, David W. Green, Lucy G. Hagan, Robert H. Hauge, William T. Hicks, Helen L. Johansen, James S. Kane, Baldwin A. King, Drucilla

Davis Knutsen, Oscar H. Krikorian, Norman L. Lofgren, Donald F. Mastick, Richard F. Porter, Alan W. Searcy, Martin D. Shetlar, David Solan, Fred E. Stafford, Joel B. Tellinghuisen, Lieselotte K. Templeton, Sandor Trajmar, John Ling-Fai Wang, Robert M. Walsh, Paul R. Wengert, Earl F. Worden, Jr., and Bea-Jane Lin Yang. The postdoctoral research associates who have worked with him include Lester Andrews, Elizabeth B. Brackett, M. S. Chandrasekhariah, Pramod D. Desai, Jean Drowart, Walt W. Duley, Niels Engel, Haakon Haraldsen, Barton L. Houseman, C. Goeffrey James, Dan Jensen, Robert H. Lamoreaux, John L. Margrave, C. Beat Meyer, Ketil Motzfeldt, Leon F. Phillips, Lewis E. J. Roberts, Gerd M. Rosenblatt, Walter W. Smeltzer, Gollakota R. Somayajulu, and Karl Wieland. The thirty-eighth PhD student is John K. M. Gibson, who recently finished his work.

THE SCIENTIST

Brewer has previously described his attitude about the choice of research problems (3):

> I would contend that the relevance of an experiment can be measured in terms of how widely the result of the experiment will be felt. The significance of the experiment should not be judged by the practical value of the immediate results, but by the contribution to improved understanding of the behavior of all materials. The most fruitful design of an experiment to test a model of bonding, for example, might require measurements on a rare material that might have little prospects of practical use; nevertheless, the measurements on this useless material could help develop an understanding that could point the way to the design of extremely practical materials. Developmental research aimed at information required for an immediate practical goal can be very costly without the fundamental understanding that can pinpoint the fruitful directions. A small advance in understanding can repay its costs manyfold in redirection of numerous developmental projects.

Brewer's research is described in his own words as follows (4):

> High temperature chemistry is characterized by the occurrence of unusual species and phases that are often unstable at conventional temperatures. Because of the difficulty of carrying out measurements under conditions where it is often difficult to contain the materials and to avoid contamination, it is important to design experiments to yield information that can be used with predictive models. In this manner, one can often calculate chemical behavior under conditions where measurements had not been made or would not be practical.
>
> In the past, our research has aimed mostly at improving our understanding of the behavior of gases at high temperatures and of refractory containment materials. At present our main thrust is aimed at improving our understanding of the thermodynamics of metallic alloys. From many alloys we have a good understanding of the interactions and

> have quantitative predictive models which allow prediction of the behavior of unstudied alloys . . . . Our main thrust at present is to characterize the very stable intermetallic compounds that result from the reaction of transition metals from the left-hand side of the periodic table, which do not have enough electrons to use all of their bonding orbitals, with transition metals from the right-hand side, which have so many valence electrons that they are paired in nonbonding orbitals.

The emphasis he places on the broad and extensive knowledge that enables a scientist to change the focus of activities is perhaps derived from his own research experience. His research activities began in his undergraduate days when he studied the chemistry of organic unsaturated ketones. His graduate dissertation was on the kinetics of organic reactions. Then, when he joined the Manhattan Project he entered a totally different area and started the new field of high temperature chemistry. He initiated work on the synthesis of new inorganic refractory materials for the containment of electropositive liquid metals, critical evaluations of thermodynamic data, gas spectroscopic investigations of vaporization studies, gaseous transport studies, and the use of liquids as catalysts for vaporization studies. Early in his career he enunciated what is sometimes called Brewer's rule: in a saturated vapor an increase in temperature causes minor species to become more important, and, hence, the vapor becomes more complex. As his own statement suggests, he has emphasized throughout his work the creation and use of models to suggest crucial experiments that, when carried out, can help further refine the model. He has an uncanny ability to use these models and some few data to draw wide-ranging conclusions; for example, about the properties of the titanium–oxygen system, with which the writer is acquainted, and the vapor pressure of uranium.

He has always done much of his research himself. In order to work out phase diagrams he uses a hand calculator for which he has developed many programs (*5*). To study the thermodynamic properties of the strongly bonded intermetallic phases that are among his current interests, he and his associates have used several techniques, including mass spectrometrically measured pressures of the metallic gaseous atoms, high-temperature solid-state electrolytic EMF cells, and gas equilibration studies.

His current work appears to be directed toward expanding what is called the Brewer-Engel theory. Building on the ideas of Hume-Rothery, Lewis, Pauling, and Engel he has incorporated the notion of electron promotion and has developed a generalized acid-base theory relating to intermetallic compound formation (*1*). That Lewis's generalization of acids and bases should be further generalized in Berkeley is appropriate. In his work he has clarified the close connection between spectroscopically measured electronic states of gaseous atoms and the thermodynamic properties of metals and alloys. Thus, he has utilized his broad knowledge to join two scientific fields seemingly far apart, that concerning isolated gaseous atoms and that concerning dense solid materials.

Recently he has become interested in sulfur chemistry and sulfur dioxide abatement from coal-fired power plants. He organized and contributed to a DOE workshop in 1981 and edited its proceedings (*6*). He has provided a firm thermodynamic basis for future work in sulfur abatement (*7*).

THE SERVANT

His service to the scientific world lies in many areas. He has been on the editorial board of *High Temperature Science* since its inception, and he was instrumental in its beginning. Singlehandedly he has compiled Part II of the Bibliography on the High-Temperature Chemistry and Physics of Materials since its inception. He was instrumental in founding the Materials and Molecular Research Division of the Lawrence Berkeley Laboratory and led it for many years. At the request of the Atomic Energy Commission and its successors, the Energy Research and Development Administration and the Department of Energy, he has worked on numerous committees, most recently the DOE Selection Committee for the Fermi Award and the DOE Council for Materials Sciences. He has served for other Washington-based agencies as well. He recently served as a Master for the court in a technologicaly based law suit. When a new textook in physical chemistry came out a few years ago, many of us who used it resolved never to use it again because of its many errors in the thermodynamics chapters. But Brewer took the effort to educate the author; and the improved subsequent edition clearly shows his influence. He has participated in the activities of the International Union of Pure and Applied Chemistry and the International Atomic Energy Agency. Under the auspices of the latter, he has prepared a comprehensive, critical compilation of the properties of molybdenum (*8*). He was instrumental in founding the National Academy of Sciences National Research Council Committee on High Temperature Chemistry. He organized the first Gordon Research Conference on High Temperature Chemistry in 1960, and has been a frequent attendee, several times a post-banquet speaker, and always a senior advisor.

THE RECOGNITION

His awards include the Hume-Rothery Award of the Metallurgical Society of the American Institute of Mining, Metallurgical and Petroleum Engineers in 1982, the Palladium Medal of the Electrochemical Society in 1971, the E. O. Lawrence Award in 1961, and the Baekeland Award of the New Jersey Section of the American Chemical Society in 1953. He has been a member of the National Academy of Sciences since 1959.

THE PERSON

Brewer is a private person, a gentle and patient man, a helpful individual, and an optimistic and courageous human who leads a disciplined life. He rarely discusses his own affairs or his thoughts. He helps others by mentioning to them professional opportunities, by commenting on their publications, often typing the comments himself, by listening to and counseling on their personal problems, by writing or speaking in support of them, and by giving advice when it is sought. Rarely does he become angry, but he rises vigorously to defend his ideas if he is challenged. His principal hobby is horticulture. Rose, his wife, has been described as the one who provides the ballast—the supportive, continuous, and unwavering stability—for his life. Her splendid perception of and perspective on human and social issues brings to their life a depth of meaning. She is a tremendous help for him. One can sense their close and warm relationship and their love for their children and now the grandchildren.

At the time he was awarded the Palladium Medal, I wrote about Leo in a somewhat more personal and humorous vein (*9*). Those comments remain appropriate, but I shall not repeat them.

One person who has worked with him extensively summarized some aspects of his human qualities in terms of his integrity, his lack of arrogance, his treating people as if they are intelligent, his notion that each person is important, and his democratic ideals. The same person suggested four salient attributes of his scientific work: his independent creativity, as shown by his work in many fields; his penetrating perceptions of principles and their importance, and the associated avoidance of unproductive side-tracks and details; his continued productivity over four decades; and the enormous extent and breadth of his knowledge.

ACKNOWLEDGMENTS

The writer wishes to acknowledge helpful comments from H. F. Franzen, Frank T. Greene, David A. Goodman, James S. Kane, Glen F. Kessinger, Karen Krushwitz, John L. Margrave, C. Beat Meyer, Paul C. Nordine, Gerd M. Rosenblatt, Alan W. Searcy, Robert I. Sheldon, and Joel B. Tellinghuisen.

REFERENCES

1. Brewer, L., The Generalized Lewis Acid-Base Theory: Surprising Recent Developments, paper presented at American Chemical Society G. N. Lewis Symposium at Las Vegas, NV, March 30, 1982, *J. Chem. Educ.* **61,** 101 (1984).
2. Lewis, G. N., Randall, M., Pitzer, K. S., and Brewer, L., *Thermodynamics*, 2nd Edition, McGraw-Hill, New York, 1961.
3. Brewer, L., *J. Electrochem. Soc.* **119,** 7C (1972).

4. Brewer, L., High Temperature Thermodynamics, Annual Report for 1982, Materials and Molecular Research Division, Lawrence Berkeley Laboratory, University of California, Berkeley, CA, LBL-15150, 1983, p. 103.
5. Brewer, L., HP-41C Calculator Programs for Fitting of Data by Analytical Functions. Lawrence Berkeley Laboratory Report LBL-15346 (1982). Also, Estimation of Thermodynamic Data and Phase Diagrams Using HP-65 Calculator Programs, LBL-4994, 1981.
6. Brewer, L., ed., *Proceedings of Flue-Gas Desulfurization Conference,* Morgantown, WV, Nov. 6–7, 1980, CONF-801176, 1981, 363 pp., and author of paper No. 16, pp. 303–355; also available as Lawrence Berkeley Laboratory Report, LBL-11758.
7. Brewer, L., Thermodynamic Values for Desulfurization Processes. in *Flue Gas Desulfurization,* edited by Hudson, J. L., and Rochelle, G. T., American Chemical Society Symposium Series, No. 188, 1982, pp. 1–39. Also available as LBL-13846.
8. Brewer, L., *Atomic Energy Review, Special Issue No. 7, Molybdenum,* International Atomic Energy Agency, Vienna 1982.
9. Gilles, P. W., Leo Brewer, 1971 Palladium Medalist, in *J. Electrochem. Soc.* **119,** 5C (1972).

0018–1536/84/1700–0001$06.00

The Responsibility of High Temperature Scientists

LEO BREWER

Materials and Molecular Research Division, Lawrence Berkeley Laboratory, and Department of Chemistry, University of California, Berkeley, California 94720

Received November 22, 1983; Accepted, December 31, 1983

When I was first introduced to high temperature research on joining the Manhattan Project in January 1943, I was surprised by the novelty and unexpected complexity of high temperature systems. At room temperature, one expects a mixture of aluminum metal with an aluminum halide or aluminum oxide to be restricted to the zero and 3+ oxidation states. The predictions of chemical behavior upon heating, based on room temperature behavior, are grossly in error because of neglect of the additional oxidation states that become increasingly important as the temperature is increased. In high temperature aluminum–oxygen systems, for example, the trivalent gaseous species has not even been reported, but Al_2O, AlO, Al_2O_2, and AlO_2 gases have been established. For gaseous systems in equilibrium with condensed phases, one can demonstrate that the gaseous phase will become more and more complicated as the temperature is increased both with respect to the number of species and with respect to the complexity of the molecules.

This general behavior of high temperature systems requires a very broad outlook on the part of investigators of these systems. One must consider a wide variety of species to gain an adequate understanding of high temperature behavior. High temperature research is difficult. One not only has to consider the possibility of many unusual species, but the experimental conditions are difficult to control. A common difficulty is the matter of containment without excessive contamination. Even the problem of adequately characterizing the temperature can be a difficult job. The complexity of the systems and the difficulty of making reliable measurements set severe limits on the amount of information that can be obtained by direct experiment. But present-day technology requires a great deal of information about high temperature behavior to prepare materials by high temperature processes and to improve the efficiency of energy utilization and other high temperature operations.

The difficulty of obtaining the critically needed information stimulates the development of various theoretical models that can extend the knowledge of chemical behavior under less extreme conditions to conditions that are not readily accessible experimentally. Generally, advances in high temperature science require data obtained by classical techniques as well as experiments at the novel frontiers of science (*1*). Because of the difficulty of making measurements under extreme conditions, one almost always has to combine high temperature measurements with measurements taken under milder conditions and apply thermodynamic and other tools to extrapolate to the conditions of interest. For example, calorimetric determinations of heat capacities and enthalpies of formation and X-ray determinations of densities provide indispensible data for extrapolation to extreme conditions in the practical applications of high temperature science. Even these relatively easy classical measurements are available for only a small fraction of the systems of interest. Thus it becomes essential to develop various predictive models that can extend the available information to systems which have not been studied adequately. These models allow each reliable measurement to be multiplied in value many fold in providing yet unmeasured values.

The lack of needed information is a serious problem for all disciplines (2), but it is a particularly critical problem for high temperature scientists. Even information that may be available is not always readily accessible. I wish to discuss the responsibility of scientists in the matter of making information more readily available. It is impractical for most scientists and engineers to retrieve, critically evaluate, and put into usable form any significant fraction of the information they need. The retrieval of information from the literature is often a serious bottleneck. It is not only the tedious and time-consuming work of finding papers dispersed in many journals and published in many languages that stands in the way of utilizing the exponentially growing mass of information. A more serious obstacle is the need for a broad experience in the field to be able to evaluate critically the reliability of the measurements. A large fraction of the data in the primary literature is unreliable. This is particularly true of high temperature literature because of the difficulty of obtaining measurements under extreme conditions. It requires a great deal of experience to extract reliable information from the literature. Most people are not familiar enough with the experimental or theoretical techniques to be able to critically evaluate the various often conflicting values. Reliable data are even more important when no measurements exist for the compounds of interest and one must use various predictive correlation models to obtain the needed information. If the input data for the model are in error, the predictions can be greatly in error. As an example of the need for predictive models to detect serious systematic errors, in the review of the thermodynamic properties of molybdenum and its binary systems (3), it was necessary to reject all experimental determinations of the thermodynamic properties of gaseous molybdenum iodides since the values were far outside the range of uncertainty of bonding models. The reported values must have been for oxyiodides formed by interaction with the silica containers used in the experiments. The pure binary iodides could not have attained the reported partial pressures.

Critically evaluated data compilations are particularly important in high temperature research because of the intricacies of high temperature experimentation and the inherent complexity of high temperature systems. It is most important to be able to call upon the experts who are most capable of making the necessary choices among the reported values to provide critically evaluated compilations of recommended values. There is a tremendous economic saving if each user of data need not take the time to search the literature independently. There is an even greater saving if the user is directed to the most reliable data and is not misled by the erroneous values. The wrong choice between divergent values could make the difference between success and failure of a proposed process. Each of us has to critically evaluate data of interest to our particular subfield. It is important to make the results of our evaluations generally available to others who would not be as expert in our subfield to arrive at a reliable evaluation.

Sometimes, no one person has the breadth of experience to properly evaluate the interrelation between data from different fields. I would like to illustrate the above remarks about the need for critically evaluated compilations with the example of a recent compilation that required collaboration of considerable number of atomic spectroscopists. For calculations I planned to do, I needed accurate Gibbs energies of the gasous atoms. In reviewing the sources of currently tabulated values, I found that most values were calculated using the classical atomic energy levels tabulated by Charlotte Moore (4), an extremely useful source of observed levels, but for some of the transition metals, and most of the lanthanides and actinides, there are unobserved levels that lie low enough to contribute appreciably even at temperatures below 3000 K. Hultgren et al. (5) did try to update the observed levels and the JANAF Tables (6) did estimate the energies and multiplicities of missing levels for the calculations of their tabulations. Gurvich et al. (7) have also included estimated energy levels in their calculations. However, there are a considerable number of gaseous atoms that have not been completely treated. As I have had to estimate the missing levels and carry out the calculations of the thermodynamic properties for my own use, I have made the extra effort to put the results in a form that would be usable by others. Analytical equations will be published for all elements. I discuss the limitations to the accuracy for all elements, but I am presenting here the energy levels used only for the actinides from lawrencium to protoactinium and values of $-(G^\circ - H]^\circ_{298})/RT$ from 298 to 3000 K for lawrencium to francium.

The lanthanides and actinides are particularly difficult because of the large number of electronic configurations involving *f, d, p,* and *s* atomic orbitals that result in many low-lying energy levels, often with high multiplicity, and generally a high density of states. Because of configuration interaction, it is often difficult to assign levels to specific configurations. In addition, LS assignments are often not meaningful. However, the total number of levels and their *J* values are fixed. More than a decade ago, when I needed to know the energies of the various configurations of the lanthanides and actinides, there were few characterized levels. But it was possible (*8, 9*) with the Engel correlations between crystal structure and electronic configurations and chemical bonding models to use the enthalpies of sublimation of the metals to calculate the energies at which each configuration became important. I was then able to use various spectroscopic models proposed by Racah (*10*), Camus (*11*), Fred (*12*), and Nir (*13*) to fill in missing data. Nugent and Vander Sluis (*14*) have presented calculations that confirm these models. I am not an expert in this field, but I received enthusiastic cooperation and help from the experts. My previous calculations were not aimed at fixing the energies of each distinct level, as is now required for the thermodynamic calculations for the gaseous atoms. Again, I have had excellent help from the spectroscopic experts. Jean Blaise, in particular, was very helpful with my previous compilation as well as with the present one in providing me with literature sources and unpublished results. I received valuable help from J. Conway, H. Crosswhite, R. Engleman Jr., M. S. Fred, B. Pulliam, K. Rajnak, and E. F. Worden, as indicated in the literature citations given below. This collaboration has made possible thermodynamic calculations of high accuracy.

In addition to uncertainty caused by unobserved electronic levels, the uncertainty of the translational contribution (*15*) to $-(G^\circ - H_{298})/RT$ is $2.5(\Delta k/k) + 1.5(\Delta M/M)$, where Δk and ΔM are the uncertainties of the Boltzman constant and of the atomic weight, respectively. The uncertainties arising from Planck's constant or Avogadro's number are much smaller. The generally accepted (*16*) value of $\Delta k/k$ has been of the order of 4×10^{-5}, although recent measurements (*17*) indicate that it may have been reduced to the order of 10^{-5}. The values of the constants used are $N_0 = 6.022095 \times 10^{23}$ mol^{-1} for Avogadro's number, $h = 6.62618 \times 10^{-34}$ *J s* for Planck's constant, and $k = 1.38066 \times 10^{-23}$ J K^{-1} for the Boltzmann constant. These values yield for the translational contribution,

$$-(G^\circ - H^\circ_{298})/RT = 3/2 \ln M + 5/2 [\ln T + 298.15/T] - 3.6649$$

The atomic weights and their uncertainties are given by the IUPAC Commission on Atomic Weights (*18*). For many of the elements $\Delta M/M$ is less than 10^{-5} and the uncertainty of the translational contribution is 0.0001 because of Δk or possibly a factor of 4 lower if the accuracy of the recent measurements is confirmed. Elements of this group include He, Be, N, F, Na, Al, Si, P, K, Sc, V, Mn, Co, As, Br, Rb, Y, Nb, Rh, Ag, I, Cs, La, Pr, Tb, Ho, Tm, Ta, Re, Au, Tl, Bi, Ra, Ac, Th, Pa, U, Np, Pu, and the short lived elements such as Tc, Pm, Po, At, Rn, Fr, Am, Cm, Bk, Cf, Es, Fm, Md, No, and Lr, for which the atomic weight used is that of the longest-lived isotope. If thermodynamic data are needed for another isotope of atomic weight M', add 1.5 ln (M'/M) to the $-(G° - H°_{298})/RT$ values tabulated below for atomic weight M. The atomic weights of the isotopes can be accurately calculated from the Δ values given in meV by Lederer and Shirley (*19*) by adding $1.07356 \times 10^{-3}\Delta$ to the integral mass values. Thus for over half the elements, it would be desirable to reduce the uncertainty of the electronic contribution to $-(G° - H°_{298})/RT$ to below 0.0001. If ΔQ is the contribution that missing levels could make to the electronic partition function, $-(G° - H°_{298})/RT$ would be low by an error of $\Delta Q/Q$. Missing levels with a total degeneracy of 10 and an energy of 20000 cm^{-1} would yield at 3000 K

$$\Delta Q = 10\ e^{-1.4388(20000)/3000} = 7 \times 10^{-4}$$

If Q were around 7, the error would be 0.0001.

Generally, missing levels above 20,000 cm^{-1} need not be considered for temperature up to 3000 K unless Q should be close to unity and the density of states is very high just above 20,000 cm^{-1}. For the lanthanides and actinides that have high densities of states, the density is also high at lower energies and Q is very large. In such instances, even energy levels above 18,000 cm^{-1} can be neglected. Thirty-two of the elements, Li, B, S, Ca, Ti, Ni, Zn, Ga, Ge, Se, Kr, Sr, Zr, Mo, Ru, Sn, Sb, Te, Xe, Nd, Sm, Gd, Dy, Er, Yb, Hf, W, Os, Ir, Pt, Hg, and Pb, have $\Delta M/M$ uncertainties greater than 10^{-4} and would increase the uncertainty beyond the limit set by the Boltzmann constant. Although the atomic weight contribution to the uncertainty in $-(G° - H°_{298})/RT$ would be close to 0.001 for Li, Ge, Se, Sm, Os, and Pb, the uncertainty would exceed 0.001 only for B. For these elements with considerable atomic weight uncertainties, the inclusion of all levels below 20,000 cm^{-1} can be relaxed somewhat.

The low-lying energy levels needed for accurate calculations of $-(G° - H°_{298})/RT$ are tabulated for the heavy actinides starting with Lr. LS nomenclature is used generally for identification of the states, although JJ designations are sometimes given. In many instances of extensive mixing, the assignment of LS or configuration designations is not very meaningful and there are contradictions between the various lists. No matter how they are labeled, the number of levels for each J value is fixed and the LS labeling is the most convenient way of keeping track of whether any levels are missing. For purposes of accurate thermodynamic calculations, it is not necessary to have correct assignments as long as *J* values can be given for all levels, observed and still unobserved, up to at least 17,000 cm^{-1}. In all instances the *J* value is given for calculation of the degeneracy. As discussed above, the levels need not be given to as high an energy when the electronic partition function is large. The electronic partition function was combined with the translation contribution calculated from the atomic weight that is listed below with each element to calculate $-(G^\circ - H^\circ_{298})/RT$ using the standard statistical mechanical equations (*15*). For No, the electronic contribution is so small that the heat capacity is constant over a wide temperature range and an exact analytical equation can be tabulated. For the other elements, the calculated values were fit to a quartic polynomial for the two temperature ranges, 298–1000 and 1000–3000 K. The analytical equation for 298–1000 was fit exactly at 298.15 K. However, there will be deviations between the values calculated from the summed partition function and those calculated from the fitted polynomial that can amount to more than 0.001 at some temperatures. For those who might want higher accuracy, the exact values to three or four places beyond the decimal point are tabulated for 298.15, 1000, 2000, and 3000K. If higher accuracy is desired for other temperatures, the energy levels tabulated below can be summed to yield values with an uncertainty as low as 0.0001 for many of the elements. All values given have been left in the dimensionless form, divided by *R*, as given by the statistical mechanical equations. Since the principal interest is in the calculation of equilibrium constants, which requires $\Delta G^\circ/RT$, there is no point to multiplying by *R* and then having to divide again later.

Lr I *Levels, Complete to 25,000 cm^{-1}*, M = *260.1053*

The values in double parentheses were estimated by procedures used by Brewer (*8, 9*).

	s^2p		ds^2
2*P* 1/2	0.	2*D* 3/2	((8000.)) ± 2000
3/2	((4000.)) ± 500	5/2	((13000.)) ± 2000
	dsp		
4*F* 3/2	((27500.)) ± 2000		

No I *Levels, complete to 24,000 cm^{-1}*, M = *259.101*

The values in double parentheses were estimated by procedures used by Brewer (*8, 9*).

	s^2		*sp*
1S0	0.	3*P*0	((19500.)) ± 1000
		1	((21000.)) ± 1000
		2	((24000.)) ± 2000

Md I *Levels, complete to 19,000 cm^{-1}, M = 258.0985*

The values in double parentheses were estimated by procedures used by Brewer (*8*, *9*).

$f^{13}s^2$		$f^{13}sp$	
2*F* 7/2	0.	4*F* 7/2 (7/2, 0)	((19000.)) ± 1000
5/2	((14000.)) ± 1000.		

Fm I *Levels, complete to 18,500 cm^{-1}, M = 257.095*

Values for $f^{12}s^2$ were calculated by Crosswhite and Fred (*20*) using parameters of Goodman et al. (*2*). The $f^{12}sp$ value was estimated by the procedures used by Brewer (*8*, *9*).

$f^{12}s^2$		$f^{12}sp$	
3*H*6	0.	5*H*6(6,0)	((18500.)) ± 1000
5	12708.		
3*F*4	3238		
3	16802.		
2	12330.		
1*G*4	16103.		

Es I *Levels, complete to 19,300 cm^{-1}, M = 254.088*

The observed levels are from Worden et al. (*22*) and Blaise et al. (*23*) plus unpublished observations of ^4F 9/2 and ^4I 13/2 by J. Blaise and J.-F. Wyart. The values in parentheses are values calculated by Crosswhite. (10/19/81).

$f^{11}s^2$		$f^{10}ds^2$	
4*I* 15/2	0.	6*I* 15/2	17802.89
13/2	10244.30	17/2	19209.15
11/2	((17916.))	$f^{11}sp$	
9/2	((16428.))		
4*H* 11/2	8759.25	6*I* 17/2	19367.93
4*F* 9/2	7894.51		
7/2	((18827.))		
5/2	((16523.))		
2*P* 3/2	((11519.))		

Cf I *Levels, complete to 19,000 cm^{-1}, M = 251.080*

The observed levels are from Blaise et al. (*23*) and from a 5/31/83 list of observed levels provided by J. G. Conway. The f^{10} *sp* $J = 7$ and f^9ds^2 $J = 6$ levels are from a 8/18/83 list provided by J. Blaise. The values in double parentheses are values estimated assuming a regular variation with *J*.

	$f^{10}s^2$	
5I8	0.	
7	9078.145	
6	11074.39	
5	15846.145	
4	15375.47	
5F5	8516.38	
4	13965.75	
3	((16000.))	± 500
2	((18000.))	± 1000
5G6	16820.39	
5	((17500.))	± 500
4	18002.22	
3	((19000.))	± 1000
3P2	10589.25	
1	((13000.))	± 1000
0	((16000.))	± 1000

	$f^{10}sp$
7I9	18483.10
8	17459.215
7	19322.22

	f^9ds^2
7H8	16909.535
7	17308.175
6	19055.82

Bk I Levels, complete to 19,000 cm^{-1}, M = 247.0703

The observed levels are from Conway et al. (*24*), Worden and Conway (*25*), and Blaise et al. (*23*). Values in single parentheses are observed levels for which a more complete assignment has been estimated. Values in double parentheses are unobserved levels which were estimated assuming a regular variation with *J*.

		f^9s^2	
6H	15/2	0.	
	13/2	6530.72	
	11/2	9535.13	
	9/2	10605.57	
	7/2	12067.63	
	5/2	((13000.))	± 1000
6F	11/2	5416.69	
	9/2	5757.44	
	7/2	10985.83	
	5/2	12568.38	
	3/2	((15000.))	± 1000
	1/2	((17000.))	± 1000
4G	11/2	19013.78	

		f^8ds^2
8H	17/2	15290.142
	15/2	20569.745
	13/2	17532.10
	11/2	15987.97
	9/2	(16541.45)
	7/2	(18624.117)
	5/2	(16276.96)
	3/2	(17484.35)
8G	15/2	10587.345
	13/2	9141.115
	11/2	13769.315
	9/2	13439.485
	7/2	13191.915
	5/2	13931.00
	3/2	15626.86
	1/2	16145.84
8F	13/2	16038.00
	11/2	(17559.16)
	9/2	(19177.22)

		f^9ds
8H	17/2	17182.483
	15/2	17547.622
	13/2	18717.833
	11/2	20878.52

		f^8s^2p
(6,½)	11/2	17777.808
	13/2	17993.628

		f^9sp
(15/2,0)	15/2	*16913.77*
(15/2,1)	17/2	*17665.98*
	15/2	*19334.71*
	13/2	*19179.21*

		f^8ds^2
8*D*	11/2	9300.585
	9/2	10735.955
	7/2	14920.166
	5/2	(19197.97)
6*H*	15/2	16039.84
	13/2	(18540.75)
	11/2	(18733.63)

Cm I Levels, complete to 18,000 cm^{-1}, M = 247.070

The levels reported by Worden and Conway (*26*) have been modified by removal of three false levels (Worden and Conway, June 1983) and by removal of the odd level at 18,002.9 cm^{-1} and addition of two odd and four even levels between 15,300 and 19,600 cm^{-1} by J. Blaise (Aug. 1983).

f^7ds^2

		7*D*1	10133.857		
9*D*2	0.	2	9671.692		
3	302.153	3	9458.053	5*F*3	(17971.43)
4	815.655	4	8958.447	4	(17280.02)
5	1764.268	5	9064.88	5	19089.345
6	3809.358				

f^7d^2s

		9*F*1	15300.445		
11*F*2	10144.927	2	15546.646		
3	10484.864	3	15924.576	11*P*4	15719.057
4	10971.171	4	16516.293	5	16915.927
5	11641.681	5	17463.053	6	17047.49
6	12534.983	6	19059.385		
7	13720.28				
8	15110.02				

f^8sp

(6,0) 6	17656.657
(6,1) 7	18009.48
6	20197.521
5	19741.312

f^8s^2

7*F*6	1214.203	3*H*6	15302.54		
5	5136.52			5*H*5	19584.139
4	4877.61	5*D*4	14521.027	5*H*4	18491.582
3	7208.827				
2	7521.123	3*P*2	18945.13		
1	8696.685				
0	8887.255				

f^9s^2p

		7P2	16998.503
9P3	9263.374	3	16645.951
4	9784.543	4	16480.937
5	15721.679		

f^8ds

9G8	17656.156
7	16932.75
6	17036.202
5	18060.13

f^7dsp

11F2	15252.71
3	15677.75
4	16314.782
5	17315.732
6	18865.047

Am I Levels, complete to 18,000 cm^{-1}, M = 243.0614

Values in single parentheses are calculated values from Blaise et al. (23). The f^7sp and the f^7ds 10*D* 9/2 and 11/2 levels are those assigned by Fred and Tomkins (27). They also reported the J = 5/2 level at 17,858 cm^{-1}. The remaining observed levels are unpublished results of Bruce Pulliam (28). Values in double parentheses are estimated values assuming regular variation with J.

f^7s^2

8S	7/2	0.
6P	7/2	(14258.) ± 500
	5/2	(16639.) ± 500
	3/2	(18294.) ± 500

f^7ds

10D	5/2	((14000.)) ± 300
	7/2	14507.
	9/2	15135.
	11/2	15765.
	13/2	14539.
8D	3/2	((14000.)) ± 1000
	5/2	((14300.)) ± 500
	7/2	14752.
	9/2	((15500.)) ± 1000
	11/2	((16500.)) ± 1000
6D	9/2	((14900.)) ± 1000
	7/2	15018.
	5/2	((15200.)) ± 1000
	3/2	((15500.)) ± 1000
	1/2	((16000.)) ± 1000

f^6ds^2

8H	3/2	10684.
	5/2	12974.
	7/2	15273.
	9/2	17390.
8G	1/2	14944.
	3/2	13293.
	5/2	15860.
	7/2	((16900.)) ± 500
	9/2	((18000.)) ± 1000
8F	1/2	17671.
	3/2	16277.
	5/2	17858.

f^7sp

10P	7/2	15608.
	9/2	16512.

Pu I Levels, complete to 20,000 cm^{-1}, M = 244.0642

The observed levels are from the Aug. 1983 list of M. Fred and J. Blaise.

f^5d^2s

9L4	14912.011	9K3	16532.104	9I2	19558.257
5	16520.962	4	18046.108		
6	18602.505	5	20043.465		

f^5s^2p

3	17987.917
2	18718.882

f^6s^2

7F0	0.
1	2203.606
2	4299.659
3	6144.515
4	7774.653
5	9179.262
6	10238.473
5L6	16604.786
5H3	18652.287
4	19307.447
5	19317.922
5G2	17776.483
5D0	9772.532
1	13677.903
2	17305.142

f^6ds

9H1	13528.246
2	14341.947
3	15745.648
4	17323.291
5	18856.461
9D2	16276.332
3	17765.281
4	19779.507
9G0	17336.413
1	16776.530
2	18538.782
9F1	19337.431
7H2	17572.608
3	19343.298
1	18627.281

f^6sp

9G0	15449.472					5D0	19594.767
1	17045.776	7G1	19959.027	9F1	18578.669		
2	19074.292						

f^5ds^2

	$f^5(^6H)$		$f^5(^6H)$
7K4	6313.866		
5	9386.801	5K5	16155.109
6	12351.522	6	19776.904
7	15074.958		
8	17554.704		
9	19872.154		
7I3	9724.351		
4	12159.465	5I4	18591.122
5	14292.176		
6	16888.909		
7	19236.116		

	$f^5(^6H)$		$f^5(^6F)$				
7*H*2	8768.939		14692.549				
3	11840.715		14737.788				
4	14025.007		14853.317				
5	17911.977		16834.379				
6	20521.579		18672.411				
					$f^5(^6H)$		
7*G*1	10486.922		14763.705		15406.760		
2	13517.647		15249.635		16734.151	5*G*2	19203.415
3	15424.387		16595.109		18147.975		
4	17081.874		18963.921				
5	18897.584						
7*F*0	11747.245		16202.112				
1	12177.963		15856.888				
2	12322.613		17615.482				
3	13726.661		19426.512				
4	16304.260						
5	19059.958						
5*I*4	19865.603	7*D*1	17500.977	5*D*0	18181.485		
		2	19281.917				
		7*P*2	18346.917				
		3	19496.402				

Np I Levels, complete to 16,000cm^{-1}, M = 237.0482

M. Fred et al. (*29*) have presented 460 levels of Np I. They have continued work on the complicated spectra and the present tabulation is based on a July 21, 1983 list from J. Blaise. Some assignments are in question and are followed by "?". Values in parentheses are observed values for which a complete assignment had not been made. Values in double parentheses are calculated (Feb. 4, 1977) values of unobserved levels that are expected to be below 16,000 cm^{-1}.

f^4ds^2

		$f^4(^5I)$						
6*L*	11/2	0.						
	13/2	3502.855	4*L*	13/2	9955.11	4*L*	13/2	(18405.475)
	15/2	6903.44		15/2	13781.465	4*M*	15/2	16866.87
	17/2	10020.99						
	19/2	12930.155						
	21/2	16051.925						
6*K*	9/2	2033.965						
	11/2	5185.015	4*K*	11/2	12988.025			
	13/2	7792.035		13/2	14805.14			
	15/2	10502.85						
	17/2	12948.015						
	19/2	15319.605			$f^4(^5G)$			
6*I*	7/2	3450.995			12714.11			
	9/2	6643.51			12846.685	4*I*	9/2	15892.12?
	11/2	8706.57, 9694.045			((16480.))		11/2	14885.81
	13/2	11382.815					13/2	17554.13
	15/2	13049.990						
	17/2	14870.46						$f^4(^5F)$
6*H*	5/2	7015.155			14067.315			9524.395
	7/2	8278.305			13734.265			9507.645
	9/2	11954.36			(17418.42)			((12500.))
	11/2	14030.845			(16046.85)			(15686.62)
	13/2	16400.18						((15860.))
	15/2?	((17700.))						19638.0 ?

Term	J	$f^4(^5I)$	$f^4(^5G)$	$f^4(^5F)$	$f^4(^5S)$
4*H*	7/2	6892.085,((15700.))		18077.065	
	9/2	7871.875,14613.2		16905.21	
	11/2	(11560.13)			
	13/2	(14005.22)			
6*G*	3/2	6474.18	15668.74	((11180.))	
	5/2	11231.62	16831.865	15224.33	
	7/2	10892.125	(17881.635)	13024.87	
	9/2	11329.09		((15730.))	
4*G*	5/2	4615.67,10769.42		((13120.))	
	7/2	14669.005		((15980.))	
	9/2	((15500.))±500			
	11/2	((16528.))			
6*F*	1/2		((10960.))	((12750.))	
	3/2		16342.075	12630.305	14346.06
	5/2		(16190.23)	14569.485	
	7/2			(17881.635)	
6*D*	1/2		((13750.))	((15620.))	((16030.))
	3/2		((18040.))	(16764.905)	((13790.))
	5/2				((13730.))
	7/2			(16435.34)	((13540.))
	9/2			((16410.))	((16860.))
6*P*	3/2			((14610.))	
	5/2			((16870.))	

f^4d^2s

Term	J		Term	J	
8*M*	11/2	7112.43			
	13/2	8950.625	6*M*	13/2	13706.47
	15/2	11279.675		15/2	(16432.67)
	17/2	13794.985			
	19/2	16418.975			
	21/2	19016.935			

8*L*	9/2	96869.64							
	11/2	11478.22	6*L*	11/2					
	13/2	13488.305		13/2	(19081.19)				
	15/2	15622.57?							
	17/2	17726.34							
	19/2	19974.82							
8*K*	7/2	11819.76			14829.405				
	9/2	13548.11			(17134.945)				
	11/2	15099.34			(18930.31)				
	13/2	16741.96			18899.60				
	15/2	18822.565			$f^4(^5G)$	$f^4(^5F)$			
8*I*	5/2	12852.36			18158.8	15926.23			
	7/2	13872.36			(18077.065)	(16949.69)			
	9/2	14933.54				((17000.))	4*I*	9/2	(17597.17)
	11/2	16371.64				(18920.31)		11/2	(17439.58)
	13/2	18003.525						13/2	17554.13
8*H*	3/2	13816.48				17695.19			
	5/2	14629.28				18334.815			
	7/2	(15684.95)				(18657.275)			
	9/2	(15917.375)							
8*G*	1/2	15315.73							
	3/2	15423.005							
	5/2	(15497.22)							

		f^5sp			
8*I*	5/2	18654.895			
8*H*	3/2	18816.525			
	5/2	(18851.885)			
		f^5s^2			
6*H*	5/2	2831.14			
	7/2	5456.135	4*G*	7/2	16798.125
	9/2	7805.795	4*I*	9/2	16154.315
	11/2	9854.96	4*K*	11/2	16810.745
	13/2	11647.785	4*L*	13/2	17255.18
	15/2	((14500.))±1000			
		f^5ds			
8*K*	7/2	13384.205			
	9/2	14721.665	6*K*	9/2	17663.14
	11/2	((16000.))			
8*I*	5/2	((16000.))			
	7/2	17662.38			
8*H*	3/2	((16000.))			
	5/2	17248.64			

		f^4s^2P
(4,½)	9/2	11940.075
	7/2	12692.765
(4,½)	11/2	15743.235
	9/2	16507.73
	7/2	18388.14
	5/2	18194.40

		f^4dsp			
8*M*	11/2	14338.88			
	13/2	16334.01	6*M*	13/2	17007.70
	15/2	19827.885			
8*L*	9/2	16459.34			
	11/2	17859.20			
8*K*	7/2	18117.345			
	9/2	19646.625			

U I *Levels, complete to 18,000 cm^{-1}, M = 238.029*

Blaise and Radziemski (*30*) published a list of the levels of U *I*. Since then, Blaise has been correcting and revising the list. One of these revised lists was used by Crosswhite in the preparation of a Gmelin Handbook chapter (*31*). Another such revised list has been combined with the results of unpublished calculations by Kathie Rajnak, Hank Crosswhite, and Hannah Crosswhite, together with assignments based on new isotope shifts by Rolf Engleman, Jr. as a February 1983 list at Los Alamos. These lists were combined with recent revisions from J. Blaise (7/13/83) to obtain the following list.

When an energy value is given with no parentheses, configuration and *LS* labels have been provided with no significant contradictions between the various lists. When an energy value is given with a single parenthesis following the number, a complete assignment has been provided in at least one list, but there has been some contradiction in one of the other lists and a choice had to be made. When an energy is given with parentheses before and after the number, none of the lists gave a complete designation of configuration and *LS* and the assignment was made on the basis of being at the expected energy. When an energy is given with double parentheses before and after the number, the level has not been observed and the value given has been calculated by Rajnak, Crosswhite, and Crosswhite.

	f^3ds^2				
5*L*6	0.				
7	3800.830	3*L*7	10069.177		
8	7645.646	8	14501.808		
9	11308.154	9	18511.121		
10	14845.328				
5*K*5	620.323				
6	4275.707	3*K*6	7005.533		
7	7326.118	7	11677.038)		
8	10685.788	8	17540.416		
9	13535.187				
5*I*4	4453.419			5*I*4	11403.464)
5	7864.204	3*I*5	5762.079	5	11290.266
6	10987.587)	6	10288.617)	6	(13361.397)
7	13567.948)	7	14790.990	7	((17010.))
8	14842.801			8	(18621.395)
5*H*3	3868.486				
4	8133.291)	3*H*4	5991.374)		
5	11968.650)	5	10254.999)		
6	14174.346	6	12910.508		
7	(17928.964)				

5*G*2	7191.682						
3	7103.922	3*G* 3	(13936.705)				
4	10557.037	4	(12627.560)				
5	13876.403	5	(14562.354)				
6	(16847.021)						
5*H*3	8878.548			5*H*3	13433.173		
4	12362.448			4	(16047.611)		
5	14344.522			5	(18214.588)		
6	(15435.243)			6	((17640.))		
7	(18256.474)			7	((18870.))		
5*G*2	8856.992			5*G*2	11973.589	5*G*2	((15190.))
3	11788.927	3*G*3	(14774.232)	3	11943.945	3	((15880.))
4	(16588.910)	4	15986.814	4	(15026.778)	4	(16699.374)
5	(16154.522)			5	(18359.990)		
6	(16376.313)			6	(18111.753)		
5*F*1	12107.542			5*F*1	((16220.))		
2	13951.627	3*F*2	10708.604	2	17507.479		
3	10540.266	3	(15169.851)	3	(17909.858)		
4	((14440.))	4	(17302.664)	4	(17533.315)		
5	(15799.241)						
5*D*0	((13710.))			5*D*0	15025.079		
1	10103.439	3*D*1	((14830.))	1	(13826.896)	3*D*1	17222.983
2	15347.918	2	((15890.))	2	((18260.))	2	((18480.))
3	(14576.689)						
4	12884.796						
		3*P*0	5988.059			3*P*0	((17830.))
5*P*1	13719.061	1	((17630.))				
2	(17098.984)	2	((15590.))				
3	(17091.703)						

The 79 levels below 17,000 cm^{-1} are used individually in calculating the thermodynamic properties. The 13 levels between 17,000 and 18,000 cm^{-1} with a total multiplicity of $g = 109$ are used with an average energy of 17,500 cm^{-1}. Their contribution to $-(G^\circ - H^\circ_{298})/RT$ at 3000 K is only 2×10^{-5}, which could be neglected. Higher levels are even less important.

	f^3d^2s		
7*M*6	6249.029		
7	8118.632	5*M*7	12826.317
8	10347.345	8	16244.488
9	13127.926		
10	16040.501		

*7L*5	10081.030						
6	11457.312	*5L*6	14543.775				
7	13346.911	7	15712.862				
8	15458.493						
9	17882.937						
*7K*4	10208.488			*7K*4	14411.306		
5	11633.164	*5K*5	(16602.245)	5	14970.524)	*5K*5	(16825.752)
6	13402.537	6	(18005.947)	6	(17102.856)		
7	15353.800						
8	17428.302						
*7I*3	10819.934	*7I*3	(14488.035)				
4	11558.696	4	14274.369				
5	13632.064						
6	(15804.313)						
7	16766.126						
*7H*2	14191.049						
3	14281.945	*5H*3	(15560.488)	*5H*3	(16575.364)		
4	15778.044	4	(16451.765)	4	(17217.089)		
5	15542.022	5	(17461.027)	5	(17589.426)		
6	(15906.340)						
*7G*1	15999.879						
2	16942.169	*5G*2	13149.733)				
3	(17261.703)	3	(15906.822)				
4	(18065.156)	4	(17799.319)				
		5	(17882.529)				
*7F*0	((16320.))						
1	((17730.))	*5F*1	((16690.))				
		2	(16746.582)				
		3	(16983.306)				

The 44 levels below 17,000 cm^{-1} are used individually. The 10 levels between 17,000 and 18,000 cm^{-1} with a total multiplicity of 110 are taken with an average energy of 17,500 cm^{-1} and contribute only 2×10^{-5} to $-(G^\circ - H^\circ_{298})/RT$ even at 3000 K.

f^4s^2		$f^2d^2s^2$	
*5I*4	7020.710	*5L*6	11502.623
5	10051.315	7	15631.855
6	12643.393		
7	14858.786	*5K*5	11613.977
8	16862.069	6	15638.366
*5G*2	14911.841	*5I*4	13710.331
3	15007.579	5	16294.016
4	(13825.400)		

5*F*1	((13910.))	5*H*3	15831.130
2	12204.403		
3	(15584.40)	5*G*2	15732.203
4	(17893.878)		

f^3s^2p			
5*K*5	13463.392	5*K*5	(17559.322)
6	16505.772	6	((17760.))
5*I*4	12035.632	5*I*4	(17468.213)
5	17908.172	5	17369.549
5*H*3	17968.723	5*G*2	((16390.))

The 22 levels of f^4s^2, $f^2d^2s^2$, and f^3s^2p below 17,000 cm^{-1} are used individually. The eight levels between 17,000 and 18,000 cm^{-1} with a total multiplicity of 80 add 1×10^{-5} to $-(G° - H°_{298})/RT$ at 3000 K.

f^3dsp			
7*M*6	14643.867	7*K*4	16121.933
7	16900.387	5	16929.758
7*L*5	15720.681		
6	17361.895		

The only f^2d^3s level below 18,000 cm^{-1} is 7*L*5, which is calculated at 16,930 cm^{-1}.

The f^4ds 7*L* levels at $J = 5$, 14,839.737 and $J = 6$, 16,195.359; 7*I*3, 17,154.805; and 5*L*6, (17070.469) cm^{-1} are the only levels of this configuration expected below 18000 cm^{-1}.

For f^3d^3, Rajnak, Crosswhite, and Crosswhite calculated five levels below 17,000 cm^{-1} with below 17,000 cm^{-1}, but their parameters were rather uncertain. The 7*M*6 level was estimated by Brewer (*8*) to be at 21,000 ± 4000 cm^{-1}. J. Blaise (7/83) guesses that it might be the level observed at 23,084 cm^{-1}. In the Gmelin list, the lowest d^3 level assigned by Crosswhite was $J = 4$ at 17,048.861. As a compromise, the only d^3 levels used were 7*M*6 at ((17,000)) and 7*K*4 at (17,048.861).

In summary, estimates of the lowest levels of each configuration of U *I* are accurate enough to limit possible contributions of levels below 21,000 cm^{-1} to nine configurations. Of these, there have been no assignments of the f^3d^3 configuration to any low energy levels, but it is unlikely that any lie below 18,000 cm^{-1}. Similarly, for f^2d^3s, calculations indicate only one possible level below 18,000 cm^{-1}, which is calculated to be close to 17,000 cm^{-1} and would not be of significance for thermodynamic calculations up to 3000 K. For the f^4ds and f^3s^2p configurations, all expected low levels have been observed and no levels of consequence are believed missing. All of the low levels of f^3dps have been observed, but there is an extra observed level at 16,929 that does not match the calculations. Thus, for the higher five configurations, the presently available information is complete enough so that there is no significant uncertainty in the thermodynamic calculations resulting from those configurations.

All of the low levels of the f^4s^2 configurations have been observed except for the 5*F*1 level, which is calculated to be at 13,910 cm^{-1} with an uncertainty of several hundred cm^{-1}. This is at a low enough energy that it would of some value to pin down the energy more accurately. The f^3d^2s configuration has been fit rather well. The only unobserved levels, 7*F*0, calculated at 16,320, and 5*F*1, calculated at 16,690, have low enough degeneracies and high enough energies to be of no concern. For f^3ds^2, there are five levels that are unobserved between 15,190 and 16,690 that do not contribute significantly to the uncertainty, but there are three levels: 5*D*0 at 13,710, 5*F*4 at 14,440, and 3*D*1 at 14,830 cm^{-1}, that are unobserved and that are the limitations to the accuracy of the thermodynamic calculations.

Pa I Levels, complete to 15,000 cm^{-1}, M = 231.0359

The levels below are from an 8/17/83 list provided by J. Blaise. Values in parentheses are levels that were unassigned. Values in double parentheses are estimated.

		f^2ds^2	$f^2(^3F)$			
4*K*	11/2	0.				
	13/2	3711.625		2*K*	13/2	10937.66
	15/2	7512.695			15/2	(13741.735)
	17/2	11198.27				
4*I*	9/2	825.415				
	11/2	4121.45		2*I*	11/2	4576.005
	13/2	7383.295			13/2	(13507.955)
	15/2	10403.44				
4*H*	7/2	2966.53	5149.465			
	9/2	6149.805	(8337.88)	2*H*	9/2	3323.86
	11/2	(9156.99)	(11261.16)		11/2	(8571.065)
	13/2	(13270.66)	(16278.69)			
4G	5/2	1618.325	6708.56			
	7/2	4713.87	7658.77	2G	7/2	(9213.61)
	9/2	7330.815	(8596.27)		9/2	(11583.98)
	11/2	(10049.876)	(13771.63)			
4*F*	3/2	(7028.30)				
	5/2	(8081.965)		2*F*	5/2	5938.65
	7/2	(9794.35)			7/2	(9521.755)
	9/2	(11743.42)				

f^2d^2s

6*L*	11/2	7000.29		6*H*		11409.22
	13/2	(8506.455)				(11808.81)
	15/2	10684.635				(12360.57)
	17/2	13037.125				(13345.935)
	19/2	((16000.))±2000				(17100.271)
6*K*	9/2	9762.485		6*G*		((12000.))±2000
	11/2	10620.15			5/2	(14650.25)
	13/2	11811.455			7/2	(12204.755)
	15/2	(15654.51)			9/2	(14715.24)
					11/2	(14781.85)
6*I*	7/2	8168.225	11981.73			
	9/2	9338.35	(12777.265)			
	11/2	11120.555	(13066.20)			
	13/2	(14381.715)				
	15/2	(18037.425)				

fd^2sp

6*I*	7/2	15061.15
	9/2	(17665.32)

fds^2p

4*I*	9/2	(13742.87)
	11/2	(15452.99)
	13/2	(17100.271)

f^2d^3

6*L*	11/2	((>19000.))

fd^2s^2

4*I*	9/2	2659.405
	11/2	6302.23
	13/2	(9248.93)
	15/2	((12000.)±1000
4*H*	7/2	1978.22
	9/2	5335.73
	11/2	8419.075
	13/2	11498.725
4*G*	5/2	(6425.83)
	7/2	(6951.275)
	9/2	(6956.065)
	11/2	(10057.11)

fd^3s

6*I*	7/2	(7585.025)
	9/2	(8583.09)
	11/2	(10816.55)
	13/2	(13765.29)

f^2dsp

6*L*	11/2	14393.41			
	13/2	15671.52			
6*K*	9/2	15722.38			
	11/2	17156.785	4*K*	11/2	15745.035
6*I*	7/2	16309.985			

		f^2s^2p
4*I*	9/2	(11444.705)
	11/2	(12784.62)
4*H*	7/2	(12118.75)
	9/2	(14209.36)
	11/2	(15745.035)

		f^3s^2
4*I*	9/2	13018.61
	11/2	(16495.29)
	13/2	19452.45)

	Odd				Even	
Unassigned *J* =	9/2	9718.755	9/2	13462.04	7/2	13452.785
	7/2	9824.51	7/2	13529.04	7/2	13769.609
	5/2	10145.20	11/2	14152.97	9/2	14227.13
	7/2	11106.62	5/2	14519.355	9/2	14285.85
	9/2	11485.925	5/2	14683.005	11/2	14324.175
	7/2	11682.055	11/2	14692.62	7/2	14629.104
	9/2	11923.885	7/2	14825.895	11/2	15141.64
	9/2	12346.805	9/2	14876.545	9/2	15301.549
	7/2	12782.14	9/2	15456.27	9/2	15539.78
	7/2	12911.26	7/2	15609.55	7/2	15644.065

Calculation of Thermodynamic Values

In calculating the values of $-(G° - H°_{298})/RT$, the range of electronic energy levels used depended upon the multiplicity of the ground state and low-lying states. When all of the levels in a 100 cm^{-1} interval, e.g., 18,000–19,000 cm^{-1}, did not contribute more than 0.0001 at 3000 K, any higher levels were not used. Some of the energy levels listed were too high in energy to be used, but they are listed to indicate that a given *J* value had been accounted for. The number of levels used and the accuracy of the resulting calculations are listed for each element.

Lr, No, and Md: No experimental spectroscopic values are available and all of the estimated values that are listed were used inthe calculations. For Lr, the uncertainty of the tabulated $-(G° - H°_{298})/RT$ values because of the uncertainty of the estimated energy levels increased from 0.0001 at 298 K, 0.007 at 1000 K, 0.04 at 2000K, to 0.06 at 3000 K. For No, the contribution of excited states is so small that the heat capacity is essentially constant over a wide temperature range and an exact equation for $-(G° - H°_{298})/RT$ is tabulated as well as the constants of the quartic polynomial obtained by a least square fit. The uncertainty of the calculated values is 0.0001 up to 3000 K. For Md, the heat capacity is not constant, but the uncertainty in the calculated values of $-(G^\circ - H^\circ_{298})/RT$ is 0.0001 even at 3000 K.

Fm: No experimental values are available, but the calculated energy levels are probably accurate to ± 10 cm^{-1}. All of the tabulated values were used and the uncertainty in the $-(G^\circ - H^\circ_{298})/RT$ values increased from 0.0001 at 298 and 1000 K to 0.0005 at 2000 K and to 0.001 at 3000 K.

Es: All of the tabulated levels were used. The unobserved levels are high enough in energy and have a low enough uncertainty so that the uncertainty of the calculated $-(G^\circ - H^\circ_{298})/RT$ values is 0.0001 up to 3000 K. Ward (*32*) has used the same levels as given here for the Third Law treatment of his vapor pressure data.

Cf: With a ground state multiplicity of 17 and estimated levels as low as 13,000 cm^{-1}, there is no reason to use more than the fourteen lowest observed levels plus the four lowest estimated values. The uncertainty of the $-(G^\circ - H^\circ_{298})/RT$ values was 0.0001 through 2000 K but increased to 0.0002 at 3000 K. Ward et al. (*33*) have calculated values from 298 to 1173 K for Cf^{249}. Their value of $-(G^\circ - H^\circ_{298})/RT$ at 298 K is 0.012 lower because of the difference in atomic weights, but the difference increases to almost 0.015 at their higher temperatures.

Bk: The calculations need only the forty-two levels below 18,800 cm^{-1}, of which three are estimated levels. The uncertainty of the $-(G^\circ - H^\circ_{298})/RT$ values was 0.0001 through 2000 K but increased to 0.0004 at 3000 K. Ward and Kleinschmidt (*34*) have calculated values from 298 to 2000 K for Bk^{249} based on the levels published by Conway et al. (*24*). Their $-(G^\circ - H^\circ_{298})/RT$ values were uniformly higher by 0.012 because of the difference in atomic weights.

Cm: The twenty-nine odd and twenty-six even levels below 19,000 cm^{-1} were used. The listing is believed to be complete to above 18,000 cm^{-1} and no estimated values were needed. The uncertainty of the $-(G^\circ - H^\circ_{298})/RT$ values was 0.0001 even at 3000 K. The values for Cm^{244} listed as $-(G^\circ - H^\circ_{298})/T$ values by Oetting et al. (*35*) were actually values of $-(G^\circ - H^\circ_0)/T$ and their $H_{298} - H_0$ values were mislabeled. After conversion to $-(G^\circ - H^\circ_{298})/RT$ and addition of 0.018 for the difference in atomic weights, their values are in agreement with the values given here.

Am: It is believed that the four lowest observed levels include all levels expected below 14,000 cm^{-1}. All of the higher twenty-nine levels listed were used. Even though only fifteen have been observed, the uncertainties of the estimated values do not have a significant effect on the uncertainty of the calculated $-(G^\circ - H^\circ_{298})/RT$ values which is 0.0001 up through 2000 K but which increased to 0.0005 at 3000 K. The $-(G^\circ - H^\circ_{298})/RT$ value calculated for Am^{241} by Oetting et al. (*35*) at 298 K is 0.0124 low because of the selection of a different atomic weight, but the difference increases with temperature becoming 0.0131 at 2000 K and 0.0295 at 3000 K because their calculations did not include the unobserved values.

Pu: The Pu levels expected below 20,000 cm^{-1} have all been observed, but $-(G^\circ - H^\circ_{298})/RT$ can be calculated up to 3000 K with an uncertainty of 0.0001 using the twenty-seven even and fifty odd levels below 19,3000 cm^{-1}. Oetting et al. (*35*) provide values for Pu^{239}. If 0.0311 is added to their values of $-(G^\circ - H^\circ_{298})/RT$ to convert to Pu^{244}, their values agree within 0.0001 through 2000 K.

Np: Ninety-two even observed levels below 17,134 cm^{-1} and seventeen odd observed levels below 17,007 cm^{-1} were used in the calculations, but there are sixteen unobserved levels of the ds^2 configuration expected below 16,000 cm^{-1} plus the f^5s^2 $6H15/2$ level expected at 14,500 cm^{-1}. Calculated values were provided by J. Blaise as indicated in the list of levels. The four lowest calculated levels are $6F$ 1/2 at 10,960, $6G$ 3/2 at 11,180, $6H$ 9/2 at 12,500, and $6F$ 1/2 at 12,750 cm^{-1}. In addition to use of the seventeen calculated values below 16,000 cm^{-1}, seven of the levels calculated to be above 16,000 cm^{-1} were used in the thermodynamic calculations. Since the list of calculated values is not complete above 16,000 cm^{-1} and in view of the uncertainty caused by the calculated energies for the much lower levels, there was no reason to use all of the calculated energy levels available between 16,000 and 17,000 cm^{-1}. If the calculated energies for the low lying levels have uncertainties of not over 100 cm^{-1}, the uncertainty of the $-(G^\circ - H^\circ_{298})/RT$ values is 0.0001 up to 3000 K. The $-(G^\circ - H^\circ_{298})/RT$ values of Oetting et al. (*35*) agree within 0.0001 through 1000 K, but are 0.0002 low at 2000 K and 0.009 low at 3000 K.

U: Uranium is unusual in that levels from eight configurations f^3ds^2, f^3d^2s, f^4s^2, $f^2d^2s^2$, f^3s^2p, f^3dsp, f^2d^3s, and f^4ds must be considered up to 17,000 cm^{-1} and additional levels from f^3d^3 are not much higher. A total of 140 levels have been observed below 17,000 cm^{-1} and calculated values are available for the 13 missing levels. Between 17,000 and 18,000 cm^{-1}, twenty-eight levels have been observed. Also calculated or estimated values have been provided for seven levels. Of the unobserved levels, the ones that might limit the accuracy of the thermodynamic calculations are f^4s^2 $^5F_1$ calculated to be at 13,910 cm^{-1} and the three levels of f^4d^2, $^5D_0$ at 13,710, $^5F_4$ at 14,440, and $^3D_1$ at 14,830 cm^{-1}. The electronic partition function is so large, varying from 13.55 at 298 K to 34.84 at 3000 K, and the uncertainties of these calculated values are not believed to be over 100 cm^{-1} so that the uncertainty of $-(G^\circ - H^\circ_{298})/RT$ is not over 0.0001 even at 3000 K. The values of $-(G^\circ - H^\circ_{298})/RT$ calculated by Oetting et al. (*35*) from the observed levels listed at Los Alamos in 1975 by D. W. Steinhaus thus do not differ frm those given here until 3000 K where the difference is 0.0004.

Pa: Oetting et al. (*35*) have calculated thermodynamic values based on observed levels tabulated by J. Blaise in 1975. The present calculations are based on the 8/17/83 list of Blaise. The seventy-three observed even levels below 16,278 cm^{-1} and the forty-four odd levels below 16,071 cm^{-1} are used together with two estimated levels at 12,000 cm^{-1} with a total degeneracy of 20. In view of the uncertainty of the estimated levels, there was no reason to use estimated or observed levels above 16,000 cm^{-1}. The present values of $-(G° - H°_{298})/RT$ agree within 0.0002 with those of Oetting et al. through 1000 K but are higher by 0.011 at 2000 K and by 0.047 at 3000 K. The values tabulated here are accurate to 0.0001 up to 2000 K, but the value at 3000 K is rounded to ±0.001 because of the uncertainty of the two estimated levels.

Th: Oetting et al. (*35*) have calculated thermodynamic values based on the observed levels tabulated by Zalubas (*36*) in 1968. Giacchetti and Blaise (*37*) found in 1970 two even levels $d^3s\ ^3P_0$ and $d^2sp\ ^5G_2$ at 14,247.3 and 14,465.2 cm^{-1}, respectively. The 1976 publication by Zalubas (*38*) lists all of the observed levels used here and they are not reproduced here.

The present calculations are based on the thirty-one even levels observed below 18,430 cm^{-1} and the thirty-two odd levels observed below 18,380 cm^{-1} plus an estimated value of 17,000 ± 1000 cm^{-1} for the $fds_2\ ^1F_3$ level. The uncertainty is 0.0001 up to 3000 K. The $-(G° - H°_{298})/RT$ values calculated by Oetting et al. (*35*) are 0.0002 larger through 2000 K and 0.0013 larger at 3000 K. It is not clear why there is this difference.

Ac: Hilsenrath et al. (*39*) have calculated thermodynamic values using the spectroscopic values tabulated by Moore (*4*). These values include the estimate of 10,000 cm^{-1} for the $s^2p\ ^2P$ level. No additions have been found in the literature since then. The Hilsenrath values were corrected for the now accurately known atomic weight. The accuracy is 0.0001 to 1000 K, but is only 0.001 at 3000 K.

Ra: Moore (*4*) has tabulated experimental values for all levels expected below 18,000 cm^{-1} and values of $-(G° - H°_{298})/RT$ are accurate to 0,0001 even at 3000 K. The values of Hilsenrath et al. (*39*) were corrected for the now accurately known atomic weight.

Fr: The low-lying levels of Fr can be estimated by comparison with the other alkali metals with reasonable accuracy. This has been done by Stull and Sinke (*40*). Their values of $-(G° - H°_{298})/RT$ have been used with correction of atomic weight. The accuracy is 0.0001 to 1000 K, but is poorer than 0.001 at 3000 K.

Equations for $-(G^\circ - H^\circ_{298})/RT$

$$-(G^\circ - H^\circ_{298})/RT = A + BT + CT^2 + DT^3 + ET^4$$

Species	T_{init}, K	T_{max}, K	A	10^3B	10^6C	10^9D	$10^{12}E$
$Lr^{260}(g)$	298	1000	22.867	−6.775	+18.9111	−17.3325	+5.725
	1000	3000	21.084	+2.964	− 0.8058	+ 0.1630	−0.0148
$No^{259}(g)$	298	3000	18.9149	+2.5[298.15/T − ln(298.15/T)]			
	298	1000	22.169	−6.7789	+18.9123	−17.3146	+5.7027
	1000	3000	20.291	+3.19674	− 0.97551	+ 0.18992	−0.015971
$Md^{258}(g)$	298	1000	24.242	−6.7789	+18.9123	−17.3146	+5.7027
	1000	3000	22.364	+3.1972	− 0.9758	+ 0.18992	−0.01595
$Fm^{257}(g)$	298	1000	24.719	−6.7577	+18.855	−17.257	+5.691
	1000	3000	22.889	+3.065	− 0.85370	+ 0.15860	−0.0130
$Es^{254}(g)$	298	1000	24.912	−6.779	+18.913	−17.315	+5.703
	1000	3000	23.023	+3.227	− 1.0036	+ 0.2000	−0.0168
$Cf^{251}(g)$	298	1000	24.955	−6.779	+18.9126	−17.315	+5.7029
	1000	3000	23.068	+3.219	− 0.9952	+ 0.1960	−0.01618
$Bk^{247}(g)$	298	1000	24.871	−6.785	+18.928	−17.335	+5.712
	1000	3000	22.975	+3.2593	− 1.0579	+ 0.23164	−0.02070
$Cm^{247}(g)$	298	1000	24.832	−9.6513	+26.7795	−24.4953	+8.0414
	1000	3000	22.140	+4.62787	− 1.56742	+ 0.32840	−0.028836
$Am^{243}(g)$	298	1000	24.152	−6.7789	+18.9123	−17.3146	+5.7027
	1000	3000	22.279	+3.1795	− 0.95463	+ 0.17877	−0.013744
$Pu^{244}(g)$	298	1000	22.062	−6.5835	+18.1833	−16.2706	+5.3403
	1000	3000	20.290	+2.73112	− 0.33953	+ 0.048993	−0.00493
$Np^{237}(g)$	298	1000	24.512	−6.689	+18.587	−16.864	+5.561
	1000	3000	22.724	+2.86064	− 0.57383	+ 0.100047	−0.00822
$U^{238}(g)$	298	1000	24.898	−7.919	+22.113	−20.358	+6.737
	1000	3000	22.797	+3.5721	− 1.1458	+ 0.2679	−0.02547
$Pa^{231}(g)$	298	1000	24.688	−7.8111	+21.4728	−19.3035	+6.3043
	1000	3000	22.491	+3.56557	− 0.85458	+ 0.155658	−0.013022
$Th^{232}(g)$	298	1000	23.602	−6.6925	+18.6712	−17.0661	+5.6558
	1000	3000	21.931	+2.62634	− 0.46310	+ 0.077875	−0.006917
$Ac^{227}(g)$	298	1000	23.359	−6.737	+18.604	−16.785	+5.49
	1000	3000	21.524	+2.945	− 0.5935	+0.0577	
$Ra^{226}(g)$	298	1000	21.978	−6.85	+19.01	−17.353	+5.70
	1000	3000	20.255	+2.801	− 0.6434	+ 0.070	
$Fr^{223}(g)$	298	2000	19.3831	+2.5 [298.15/T − ln(298.15/T)]			

Species	Values of $-(G^\circ - H^\circ_{298})$ at: 298.15K	1000K	2000K	3000K	$(H^\circ_{298} - H^\circ_0)/R$,
$Lr^{260}(g)$	22.1139	23.391	24.857	25.929	745.38
$No^{259}(g)$	21.4149	22.6857	24.0459	24.9354	745.38
$Md^{258}(g)$	23.4885	24.7593	26.1195	27.010	745.38
$Fm^{257}(g)$	23.9682	25.2455	26.6646	27.629	745.38
$Es^{254}(g)$	24.1582	25.4290	26.7933	27.7119	745.38
$Cf^{251}(g)$	24.2010	25.4717	26.8351	27.7520	745.38
$Bk^{247}(g)$	24.1162	25.3875	26.7842	27.810	745.38
$Cm^{247}(g)$	23.7495	25.5004	27.2925	28.4484	889.26
$Am^{243}(g)$	23.3985	24.6693	26.0302	26.9399	745.38
$Pu^{244}(g)$	21.3261	22.7259	24.7075	26.3515	745.60
$Np^{237}(g)$	23.7669	25.1023	26.8185	28.1779	745.51
$U^{238}(g)$	24.0159	25.4653	27.0934	28.3709	781.65
$Pa^{231}(g)$	23.8063	25.3451	27.2405	28.643	764.13
$Th^{232}(g)$	22.8588	24.1656	25.8442	27.1850	745.38
$Ac^{227}(g)$	22.6034	23.9326	25.501	26.578	745.50
$Ra^{226}(g)$	21.2101	22.4808	23.8418	24.7552	745.38
$Fr^{223}(g)$	21.8831	23.1539	24.517	25.428	745.38

ACKNOWLEDGMENTS

This work was supported by the Division of Materials Sciences, Office of Basic Energy Sciences, US Department of Energy under contract No. DE-AC03-76SF00098

REFERENCES

1. Rosenblatt, G. M., Chairman, *High Temperature Science, Future Needs and Anticipated Developments,* NRC Committee on High Temperature Science and Technology, National Academy of Sciences, Washington, DC, 1979.
2. Stockmayer, W. H., Chairman, *National Needs for Critically Evaluated Physical and Chemical Data.* NRC Committee on Data Needs (CODAN), National Academy of Sciences, Washington, 1978.
3. Brewer, L., and Lamoreaux, R. H., Thermochemical Properties, *Atomic Energy Review Special Issue No. 7, Molybdenum: Physico-chemical Properties of Its Compounds and Alloys,* International Atomic Energy Agency, Vienna, 1980, pp. 74–77.
4. Moore, C. E., *Atomic Energy Levels,* Vol. I (1949), Vol. II (1952), Vol. III (1958), NBS Circular 467, Supt. of Documents, US Govt. Printing Office, Washington, DC.
5. Hultgren, R., Desai, P. D., Hawkins, D. T., Gleiser, M., Kelley, K. K., and Wagman, D. D., *Selected Values of the Thermodynamic Properties of the Elements,* Am. Soc. for Metals, Metals Park, OH, 1973.

6. *JANAF Thermochemical Tables,* second ed., NSRDS-NBS-37, US Govt. Printing Office, Washington, DC, 1971, and supplements in *J. Phys. Chem. Ref. Data 3,* 311 (1974); **4,** 1 (1975); **7,** 793 (1978); **11,** 695 (1982).
7. Glushko, V. P., Gurvich, L. V., Bergman, G. A., Veitz, I. V., Medvedev, V. A., Khachkuruzov, G. A., and Yungman, V. S., *Thermodynamic Properties of Individual Substances,* High-Temperature Institute, State Institute of Applied Chemistry, National Academy of Sciences of the U.S.S.R., Moscow, Vol. I: compounds of O, H, F, Cl, Br, I, He, Ne, Ar, Kr, Rn, S, N, and P (1978); Vol. II: C, Si, Ge, Sn, Pb (1979); Vol. III: B, Al, Ga, In, Tl, Be, Mg, Ca, Sr, Ba (1981); Vol. IV: Cr, Mo, W, V, Nb, Ta, Ti, Zr, Hf, Sc, Y, La, Th, U, Pu, Li, Na, K, Rh, Cs (1982).
8. Brewer, L., *J. Opt. Soc. Am.* **61,** 1101 (1971).
9. Brewer, L., *J. Opt. Soc. Am.* **61,** 1666–81 (1971).
10. Racah, G., *Cours au College de France,* Paris, 1964, as reported in ref. (*11*) (see below).
11. Camus, P., *J. Phys. (France)* **27,** 717 (1966).
12. Fred, M., *Advan. Chem. Ser.* **71,** 180 (1967).
13. Nir, S., *J. Opt. Soc. Am.* **60,** 354 (1970).
14. Nugent, L. H., and Vander Sluis, K. L., *J. Opt. Soc. Am.* **61,** 1112 (1971).
15. Pitzer, K. S., and Brewer, L., 2nd Ed. of Lewis, G. N., and Randall, M., *Thermodynamics,* McGraw-Hill, New York, 1961.
16. Cohen, E. R., and Taylor, B. N., *J. Phys. Chem. Ref. Data* **2,** 663 (1973).
17. Colclough, A. R., Quinn, T. J., and Chandler, R. D. R., *Proc. Roy. Soc. London Ser. A* **368,** 125 (1979).
18. Holden, N. E., *Pure Appl. Chem.* **52,** 2349 (1980).
19. Lederer, C. M., Shirley, V. S., Browne, E., Dairike, J. M., and Doebler, R. E., *Tables of Isotopes,* 7th Ed., Wiley, New York, 1978. and Jan. 1982 preprint of Atomic Mass Tables for Nuclides by A. H. Wapstra, G. Audi, and K. Bos.
20. Crosswhite, H., and Fred, M. S., private communication, 6/30/83.
21. Goodman, L. S., Diamond, H., Stanton, H. E., and Fred, M. S., *Phys. Rev.* **A4,** 473 (1971).
22. Worden, E. F., Lougheed, R. W., Gutmacher, R. G., and Conway, J. G., *J. Opt. Soc. Am.* **64,** 77 (1974).
23. Blaise, J., Wyart, J.-F., Conway, J. G., and Worden, E. F., *Phys. Scripta* **22,** 224 (1980).
24. Conway, J. G., Worden, E. F., Blaise, J., Camus, P., and Vergès, J., *Spectrochim. Acta* **32B,** 101 (1977).
25. Worden, E. F., and Conway, J. G., *At. Data Nucl. Data Tables* **22,** 329 (1978).
26. Worden, E. F., and Conway, J. G., *J. Opt. Soc. Am.* **66,** 109 (1976).
27. Fred, M., and Tomkins, F. S., *J. Opt. Soc. Am.* **47,** 1076 (1957).
28. Pulliam, B., private communication.
29. Fred, M., Tomkins, F. S., Blaise, J. E., Camus, P., and Vergès, J., *J. Opt. Soc. Am.* **67,** 7 (1977).
30. Blaise, J., and Radzimski, L., *J. Opt. Soc. Am.* **66,** 644 (1976).
31. Crosswhite, H., *Gmelin Handbook No. 55, Uranium, Supplement Volume A5, Spectra,* Springer-Verlag, New York, 1982.
32. Ward, J. W., *J. Less-Common Metals* **93,** 279 (1983).
33. Ward, J. W., Kleinschmidt, P. D., Haire, R. G., and Brown, D., *ACS Symposium Series* **131,** Edelstein, N. M., (1980), pp. 199–220.
34. Ward, J. W., and Kleinschmidt, P. D., *J. Chem. Phys.* **77,** 1464 (1982).
35. Ooetting, F. L., Rand, M. H., and Ackerman, R. J., *The Chemical Thermodynamics of Actinides Elements and Compounds.* Part 1. *The Actinides Elements,* International Atomic Energy Agency, Vienna, 1976.

36. Zalubas, R., *J. Opt. Soc. Am.* **58,** 1195 (1968).
37. Giacchetti, A., and Blaise, J., European Group for Atomic Spectroscopy, Paper 57, 1970.
38. Zalubas, R., *J. Res. NBS* **80A,** 221 (1976).
39. Hilsenrath, J., Messina, C. G., and Evans, W. H., Air Force Weapons Laboratory Report AFWL TDR-64-44, *Tables of Ideal Gas Thermodynamic Functions,* 1964.
40. Stull, D. R., and Sinke, G. C., *Thermodynamic Properties of the Elements,* American Chemical Society, Washington, D.C. 1956.

0018–1536/84/1700–0031$04.40

Determination of the Dissociation Energies of Gaseous Iron Monoxide and Manganese Monoxide by the Mass Spectrometric Knudsen Cell Method

S. SMOES AND J. DROWART*

Laboratorium voor Fysische Chemie, Vrije Universiteit Brussel, Pleinlaan 2, B-1050 Brussels, Belgium

Received September 23, 1983; Accepted December 31, 1983

ABSTRACT

The dissociation energies $D_0^\circ(FeO) = 95.9 \pm 1.8$ and $D_0^\circ(MnO) = 88.1 \pm 1.8$ kcal mol^{-1} were determined by studying the metathesis reactions $FeO(g) + Se(g) = Fe(g) + SeO(g)$ and $MnO(g) + Fe(g) = Mn(g) + FeO(g)$ by the mass spectrometric Knudsen cell method in the respective temperature intervals, 1640–1810 K and 1660–2060 K. These results are critically compared with the literature data, which implied the need for their reevaluation and a review of the thermodynamic data for FeOH(g) and FeO_2(g). For the latter molecule, thermodynamic functions are presented.

Index Entries: Iron monoxide, dissociation energy of; manganese monoxide, dissociation energy of; mass spectrometry, Knudsen cell method for FeO and MnO; Knudsen cell mass spectrometry, of FeO and MnO; FeOH(g); FeO_2(g), thermodynamic functions for.

INTRODUCTION

In 1951, Brewer and Mastick (*1*) from total pressure measurements deduced upper limits for the dissociation energies of FeO(g) and MnO(g) and showed that the Birge-Sponer extrapolation method leads to D_0°

*Author to whom all correspondence and reprint requests should be addressed.

values (*2, 3*) that are too high. This remains the case when use is made of more recent and accurate values for the vibration frequencies and anharmonicity constants for these molecules. Subsequent determinations of dissociation energies by flame photometry both for FeO(g) (*4–6*) and for MnO(g) (*7, 8*) as well as by the mass spectrometric Knudsen cell method (*9–12*), again both for FeO(g) (*13–18*) and for MnO(g) (*13*) are in general agreement (*19*) with one another and with upper limits set by total pressure measurements or spectroscopic data (*2, 3*), but they show appreciable dispersion. For FeO(g), the latter is only in part caused by the varying assumed nature of the electronic ground state, which proved difficult to determine experimentally (*20–24*) or theoretically (*25–27*) and which was only recently established to be $X^5\Delta$ (*28*).

D_0°(FeO) and D_0°(MnO) were therefore redetermined when these molecules were observed under experimental conditions well-suited for their study during the measurement of the dissociation energy of the molecule SeO(g) (*29*) by the mass spectrometric Knudsen cell method. The results of these measurements are presented here, together with a critical re-evaluation of the literature data, which shows that in particular for FeO(g), the flame photometric determinations of the dissociation energy are influenced by the presence of other metal-bearing species, the gaseous hydroxide and the gaseous dioxide.

EXPERIMENTAL

Measurement of D_0°(FeO) and D_0°(MnO) was performed by studying oxygen metathesis in the gas phase generated by a mixture of Fe_2O_3(s), MnSe(s) (*30*), and EuS(s) (*31*), placed in alumina or tungsten Knudsen cells, whose effusion orifices were either channels, 1.04 mm in diameter and 2.0 mm long or knife-edged, 0.9 mm in diameter. Further experimental details and procedures were as described earlier (*11, 12, 32*).

Ions produced by electron impact from the neutrals in the molecular beam were Fe^+, FeO^+, Mn^+, MnO+, Se^+, Se_2^+, SeS^+, and SO+. The measured threshold energies, 9.9 + 0.3 eV for SeO^+, determined at 1644 K relative to IP(Se_2) = 8.88 ± 0.02 eV (*33*) and 7.7 ± 0.3 eV for MnO^+, measured at 1989 K relative to IP(FeO) = 8.71 ± 0.1 eV (*17, 18*), indicate that these ions are formed by direct ionization of the corresponding neutrals. The same applies to the atomic ions, whose appearance potentials agreed within experimental uncertainty with the spectroscopic values for the ionization potentials (*34*). After verifying the absence of dissociative ionization, the measurements reported were performed with 20 eV ionizing electrons. No search was made for FeO_2(g) or MnO_2(g) since the corresponding ions are masked by those from FeS(g) and MnS(g). For SeS(g) there was however no interference with SeO_2(g) because the latter molecule is entirely dissociated at the temperatures and oxygen activities involved in this study; this was carefully verified by isotopic distribution measurements.

RESULTS

The simultaneous observation of the species cited above makes it possible to consider several reactions, among which those are retained that could be studied over the largest temperature interval and hence provide the greatest accuracy in determining the reaction enthalpy ΔH_T° by the 2nd law of thermodynamics. These reactions are:

$$FeO(g) + Se(g) = Fe(g) + SeO(g) \qquad (1)$$
$$MnO(g) + Fe(g) = Mn(g) + FeO(g) \qquad (2)$$
$$SeO(g) + SeS(g) = Se_2(g) + SO(g) \qquad (3)$$

Isotopically corrected relative intensities are listed in Tables 1 and 2. The equilibrium constants also given there were calculated under the assumption that the ratios $A(MO)/A(M)$ of the instrumental sensitivities A, each proportional to the product of the ionization cross-section, of the extraction coefficient from the ion source, of the transmission coefficient through the mass analyzer, and of the multiplication factor of the detector, a secondary electron multiplier, are the same for M = Fe, Mn, or Se. This assumption is motivated by the symmetry of the reactions, and by the proximity of the various atomic and molecular masses.

The variation of log K with T^{-1}, represented in Fig. 1, is given by the least squares equations, in which the uncertainties are standard deviations:

$$\log K(1) = (1338 \pm 260)K/T + (0.078 \pm 0.150)$$
$$\log K(2) = (1910 \pm 77)K/T - (0.529 \pm 0.040)$$

in the temperature intervals 1644–1807 and 1644–2061 K, respectively. Log $K(3)$ is given, between 1308 and 1807 K, by (*29*):

$$\log K(3) = (2753 \pm 77)K/T - (0.376 \pm 0.049)$$

The reaction enthalpies ΔH_0° deduced from these relations, using the enthalpies for Fe(g), Mn(g), and Se(g) from ref. (*35*), for SeO(g) from ref. (*29*), and for FeO(g) and MnO(g) from the appendix, as well as the third law values are†:

$$\Delta H_0^\circ(1) = D_0^\circ(FeO) - D_0^\circ(SeO) = -5.7 \pm 1.7 \text{ kcal mol}^{-1} \qquad \text{(3rd law)}$$
$$= -5.6 \pm 2.4 \text{ kcal mol}^{-1} \qquad \text{(2nd law)}$$
$$\Delta H_0^\circ(2) = D_0^\circ(MnO) - D_0^\circ(FeO) = -7.8 \pm 1.8 \text{ kcal mol}^{-1} \qquad \text{(3rd law)}$$
$$= -8.9 \pm 0.9 \text{ kcal mol}^{-1} \qquad \text{(2nd law)}$$

Combined with $D_0^\circ(SeO) = 101.6 \pm 1.5$ kcal mol^{-1} (*29*), the dissociation energies derived therefrom and preference being given to the 3rd law results, are:

$$D_0^\circ(FeO) = 95.9 \pm 1.8 \text{ kcal mol}^{-1}$$
$$D_0^\circ(MnO) = 88.1 \pm 1.8 \text{ kcal mol}^{-1}$$

†Throughout this paper, 1 cal = 1 cal_{th} = 4.184J; 1 atm = 101325 Nm^{-2}; for the standard states p° = 1 atm.

TABLE 1
Reaction: $FeO(g) + Se(g) = Fe(g) + SeO(g)$
Relative Ion Intensities, Equilibrium Constants, and Reaction Enthalpies, ΔH_0°/kcal mol^{-1}

Exp.	T, K	I^+ (FeO)	I^+ (Se)/10^2	I^+ (Fe)/10^2	I^+ (SeO)	log K	$-\Delta H_0^\circ$
3	1644	0.164	11.6	2.51	4.83	0.807	5.62
	1644	0.164	11.8	2.45	4.93	0.802	5.58
	1707	0.480	20.0	6.37	8.95	0.770	5.56
	1707	0.480	19.1	6.37	8.65	0.783	5.67
	1740	0.934	29.0	12.3	13.9	0.800	5.92
	1740	0.934	25.9	12.3	12.7	0.808	5.98
	1740	1.032	10.1	12.1	5.43	0.802	5.94
	1774	2.75	13.53	23.6	8.62	0.739	5.56
	1807	4.55	6.17	37.7	4.27	0.759	5.85
	1807	4.51	4.19	36.9	2.85	0.745	5.73
	1807	4.38	2.85	36.4	1.82	0.726	5.57
	1807	4.38	2.53	36.4	1.58	0.717	5.50
4	1656	0.197	9.05	3.38	3.50	0.823	5.78
	1657	0.197	11.0	3.38	4.22	0.822	5.77
	1750	1.11	15.1	14.1	7.24	0.783	5.82
	1750	1.11	12.4	13.9	6.03	0.785	5.84
	1750	1.10	10.3	13.6	5.13	0.792	5.89
					Average:	5.74 ± 0.14	

TABLE 2
Reaction $MnO(g) + Fe(g) = Mn(g) + FeO(g)$
Relative Ion Intensities, Equilibrium Constants and Reaction Enthalpies, ΔH_0°/kcal mol^{-1}

Exp.	T, K	I^+ (MnO)	I^+ (Fe)/10^2	I^+ (Mn)/10^2	I^+ (FeO)	log K	$-\Delta H_0^\circ$
1	1668	0.300	6.21	7.50	1.05	0.624	7.95
2	1666	0.070	3.81	6.30	0.180	0.628	7.98
	1902	2.90	12.7	10.2	11.3	0.493	7.92
	2061	20.5	116	72.0	85.1	0.412	7.78
3	1644	0.090	2.51	5.80	0.164	0.619	7.81
	1740	0.525	12.1	23.0	1.03	0.572	7.87
	1807	1.495	37.7	41.3	4.55	0.523	7.75
4	1657	0.090	3.38	7.80	0.175	0.651	8.11
	1750	0.510	13.7	23.6	1.10	0.568	7.89
					Average of 58 points:		7.80 ± 0.17

The uncertainties ascribed to ΔH_0° and D_0° are based on twice the standard deviations, on an estimated maximum temperature uncertainty of 0.5%, and on the estimated uncertainties in the thermodynamic functions described in the appendix. For the third law results, an uncertainty

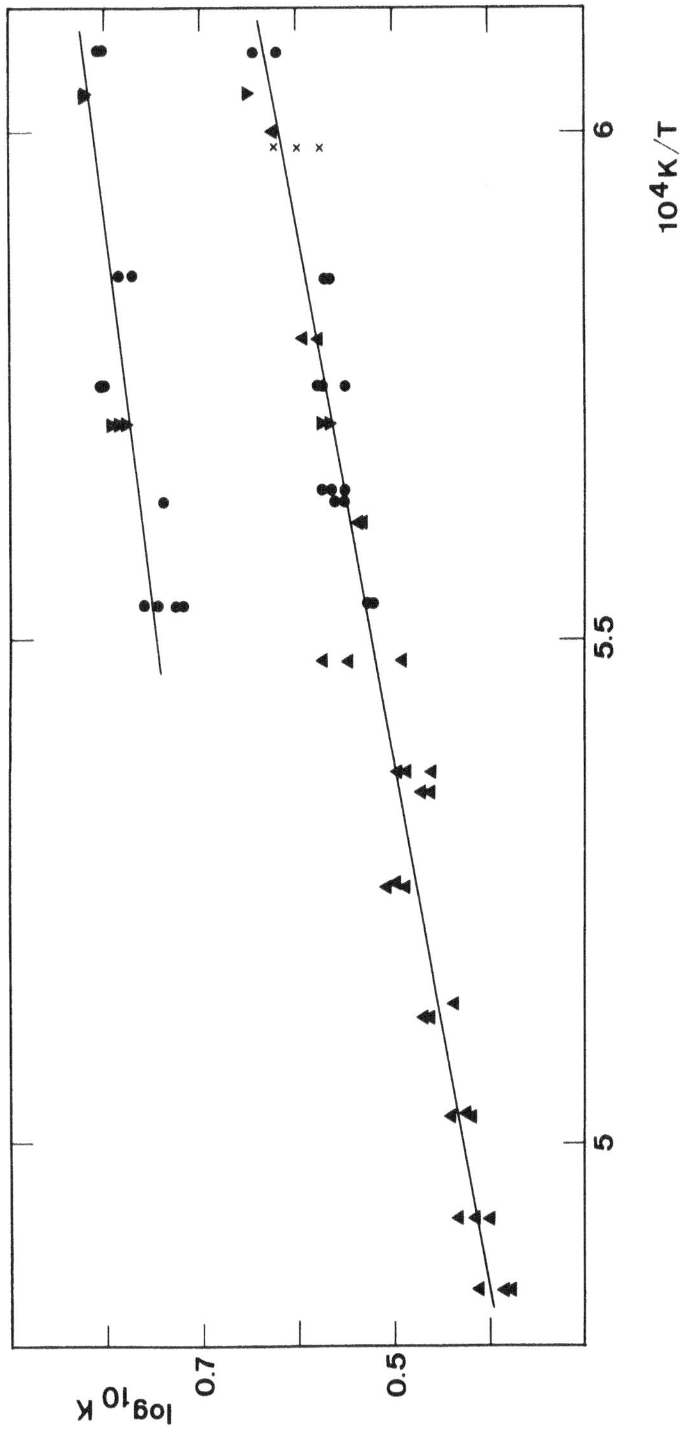

Fig. 1. Variation with reciprocal temperature of the equilibrium constants of the reactions: $FeO(g) + Se(g) = Fe(g) + SeO(g)$ (upper line) and $MnO(g) + Fe(g) = Mn(g) + FeO(g)$ (lower line); ×, Expt. 1; ▲, Expt. 2; ●, Expt. 3; ▼, Expt. 4.

of 50% in the relative sensitivies is further taken into account. Since the latter potential source of systematic error is the same for FeO(g), MnO(g), SeO(g), and SeS(g), indirectly involved in the results, it is included only once in the compounded uncertainty in D_0° for FeO(g), MnO(g), and SeO(g).

DISCUSSION

The results for both D_0°(FeO) and D_0°(MnO) show excellent consistency between the 2nd and 2rd law data, which, it is worth stressing, could, in the case of FeO, only be obtained by adopting a $X^5\Delta$ ground state in measurements in which the agreement between the two quasi-independent methods of determining thermodynamic information could be verified simultaneously for the three different reactions (1)–(3). As indicated in the experimental section, the atomic and molecular masses of reactants and products are close one to another, so that there is good reason to accept compensation between the several instrumental and molecular parameters intervening in the conversion of the observed experimental quantities, i.e., ionic currents or their ratios, into the quantities to be measured, i.e., partial pressures or their ratios and therefrom, equilibrium constants. In addition, for the molecule SeO(g), comparison with independent spectroscopic data also leads to excellent agreement (*29*). For these various reasons, the present data are considered to be characterized by good internal and external consistency; they should therefore be quite accurate.

Under these conditions, it appeared of interest critically to assess the literature results for the molecules FeO(g) and MnO(g). In so doing, the original data were first systematically recalculated with the same thermodynamic functions as those used in this work. For the flame spectrophotometric measurements, the incidence of interfering species was next taken into account.

D_0°(FeO)

In flame photometric measurements (*4, 5*), D_0°(FeO) was previously calculated by setting the FeO(g) concentration equal to the difference between the known total iron concentration and that of free Fe(g), calculated from atomic line intensities. Other FeX(g) iron-bearing species can thus cause overestimation of the FeO(g) concentration and, hence, of its dissociation energy by the amount $RT \ln [p^*(\text{FeO})/p(\text{FeO})]$. In this expression, p^* is the apparent pressure and p the true, but unknown, pressure, the ratio of these quantities being given by $p^*(\text{FeO})/p(\text{FeO}) = 1 + \Sigma[p(\text{FeX})/p(\text{FeO})]$. Corrections were therefore made for the two more likely interfering species, FeO_2(g) and FeOH(g).

The incidence of FeO_2(g) on D_0°(FeO) determined by flame photometry was calculated by combining the thermodynamic data for the reactions $FeO_2(g) + Fe(g) = 2FeO(g)$ (*17*), $FeO(g) + O(g) = Fe(g) + O_2(g)$, and $O_2(g) = 2\ O(g)$ (*36*) to obtain the equilibrium constant for the reaction

$$FeO(g) + O(g) = FeO_2(g) \tag{4}$$

which directly gives $p(FeO_2)/p(FeO)$ when $p(O)$ in the flame is known. For consistency, the primary data for $FeO_2(g)$, measured between 1718 and 1764 K (*17*), were recalculated with the thermodynamic fucntions given in the appendix. For the flame data in ref. (*5*), both the pressures labeled there "equilibrium" and "actual" were considered. As is shown in Table 3, $FeO_2(g)$ alone, on the basis of the above treatment, causes the apparent FeO(g) concentration in the particular flames of refs. (*4*) and (*5*) be overestimated by about 4–35% depending upon temperature and oxygen pressure and $D_0^\circ(FeO)$ to come out too high by up to 1.3 kcal mol^{-1}. If the stability of $FeO_2(g)$ were calculated by using $D_0^\circ(FeO)$ determined in this work, the correction in the FeO(g) concentration in these flames would be augmented by 15–25% and the correction for $D_0^\circ(FeO)$ would be increased by up to 1.6 kcal mol^{-1}.

To calculate the influence of FeOH(g) on $D_0^\circ(FeO)$, the reaction

$$FeO(g) + (½)H_2(g) = FeOH(g) \tag{5}$$

is retained since it again directly leads to the ratio $p(FeOH)/p(FeO)$ required to calculate $p^*(FeO)/p(FeO)$. The enthalpy difference associated with reaction (5) is, however, not well established since two alternate determinations (*6, 18*) of the stability of FeOH(g) yield quite different results. For consistency all necessary primary data were therfore taken within the same publication (see Appendix).

With ΔH_0° (5) = −36.0 kcal mol^{-1} (based on ref. *18*) and the Gibbs functions for FeOH(g) (*18*), H_2 (*36*), and FeO(g), one derives $p(FeOH)/p(FeO) = 0.04$ or 0.06 at 2412 K, depending upon whether the actual or the equilibrium $p(O)$ and $p(OH)$ pressures in ref. (*5*) are retained to calculate $p(H_2)$, for which either $\log p(H_2)/atm = -4.03$ or $\log p(H_2)/atm = -3.68$ is obtained. With ΔH_0° (5) = −43.9 kcal mol^{-1} (based on ref. 6), there results $p(FeOH)/p(FeO) = 0.21$ or 0.32 at 2412 K for the flame used in ref. (*5*), again depending upon whether the actual or the equilibrium pressure is retained.

The corrections in the FeO pressure in ref. (*5*) resulting from the presence of FeOH(g) and $FeO_2(g)$ are summarized in Table 3, which shows that these molecules jointly cause the uncorrected FeO(g) concentration to be about 1.2–1.5 times too high. The flame photometric Study 4 unfortunately may not be similarly corrected in a quantitative fashion for the presence of FeOH(g) because only the oxygen partial pressure is reported there. Consideration of typical flame compositions (*7*) and very approximate reconstruction of partial hydrogen pressures nevertheless suggest that FeOH(g) might also there represent between 10 and 50% of the apparent FeO(g) concentration and that, after correction for the pres-

TABLE 3
Incidence of FeO_2(g) and FeOH(g) on Flame Photometric Results for D_0° (FeO)/kcal mol^{-1}

Ref.	T, K	p(O)/atm × 10^3	p(OH)/atm × 10^3	$p(Fe)^a / p^*(FeO)$	$p(FeO_2) / p(FeO)$	$p(FeOH) / p(FeO)$	$p(Fe) / p(FeO)$	D_0°(FeO)
4	2431	0.5	[b]	1.44	0.035	[c]	≥1.49	≤100.4
				0.65			≥0.67	≤104.3
				2.83			≥2.93	≤ 97.2
				1.51			≥1.56	≤100.2
	2334	1.1	[b]	0.59	0.18	[c]	≥0.70	≤ 96.1
				0.32			≥0.38	≤ 99.0
				0.65			≥0.77	≤ 95.7
				0.23			≥0.27	≤100.6
	2240	0.8	[b]	0.74	0.34	[c]	≥0.99	≤ 92.0
				0.21			≥0.28	≤ 97.6
				Average and standard deviation				≤ 98.3 ± 3.5
5	2412	1.86[d]	2.4[d]	0.61	0.14	0.06[f]	0.73	96.8
						0.32[g]	0.89	95.9
		4.3[e]	3.7[e]	0.61	0.32	0.04[f]	0.83	92.1
						0.21[g]	0.93	91.6

[a] p^* (FeO) = apparent partial FeO pressure (see text).
[b] Not reported.
[c] Not calculable.
[d] Equilibrium pressures given in Table 1 of ref. (*5*);
[e] Actual pressure in Table 1 of ref. (*5*)
[f] Based on D_0° (Fe–OH) reported in ref. (*18*).
[g] Based on D_0° (Fe–OH) reported in ref. (*6*).

ence of FeO(g) as done in Table 3, a further reduction in D_0°(FeO) by 0.5–2 kcal mol^{-1} is probably justified. A major problem in the flame photometric measurements, however, seems to be the need to establish which of the reported "equilibrium" or "actual" pressures is the more appropriate one. Comparison of the systematically revised data in Table 4 indeed shows that the former leads to good consistency among all results, while in flame photometric publications, the pressure labeled "actual" is preferred.

The mass spectrometric values for D_0°(FeO) are only partly (*16, 17,* and the present datum) in mutual agreement, two (*14, 15*) being noticably higher and two (*13, 18*), appreciably lower. The results in agreement are all based on the study of metathesis reactions and are not discussed further.

D_0°(FeO) = 99.8 kcal mol^{-1} (*14*) is based (*19*) on the observation of $p(FeO)/p(Fe) \sim 0.2$ at 1600°C over a composition within the FeO(l) phase considered to be congruently vaporizing.

A minimum in the total vapor pressure within the FeO(l) phase leading to a constant composition $FeO_{1.116}$(l) at 1600°C, for which $p(O_2) = 1.66 \times 10^{-6}$ atm, has indeed been well documented (*37*). Calculation shows, however, that the congruency condition under Knudsen effusion, $\Sigma jp(Fe_iO_j)M^{-0.5}(Fe_iO_j)/\Sigma ip(Fe_iO_j)M^{-0.5}(Fe_iO_j) = 1.116$ is not satisfied when, at 1873 K, use is made of: the above pressure for molecular oxygen or the almost identical value 1.98×10^{-6} atm given in a comprehensive assessment for the entire Fe—O system (*38*); the corresponding pressure for atomic oxygen; the pressure for iron calculated from its values for the standard state (*35*) and its activity in the FeO(l) phase as a function of composition (*38*); $p(FeO) \approx 0.2\, p(Fe)$ (*14*); for completeness, $p(FeO_2)$, which is negligible, derived from the above data and those in the appendix (*39*). Perusal of the original publication (*37*) shows on the other hand that FeO(l) was contained in platinum crucibles for the experiments leading to univariant behavior for $FeO_{1.116}$(l) at 1600°C and that FeO(l) was observed to interact with platinum crucibles, particularly at the higher temperatures and lower oxygen pressures, eventually causing their partial melting. Therefore, and taking the activity data for the Fe–Pt system at 1880 K (*40*) into account, it is suggested that $FeO_{1.116}$(l) is not the congruently effusing composition in the Fe–O system, but a univariant composition in the Fe–O–Pt system. It is hence also suggested that $p/(FeO)/p(Fe) \approx 0.2$ at 1600°C does not correspond to $FeO_{1.116}$(l) with $p(O_2) \approx 2 \times 10^{-6}$ atm, but to a composition richer in oxygen and thus having a higher oxygen activity. On the basis of the presently accepted data (*38*) for the FeO(l) phase, it is further advanced that a congruently effusing stoichiometry within the FeO(l) phase in a truly inert container at 1600°C has a composition O/Fe < 1.116, with a ratio $p(FeO)/p(Fe) < 0.2$. This is in qualitative agreement with the observation (*17*) that the $p(O_2)/p(O)$ ratio increased by about a factor of five at 1756 K upon going from an alumina to a platinum cell initially containing

Fe_2O_3(s). It is hence concluded that D_0°(FeO) = 99.8 kcal mol^{-1} based on the data in refs. (*14*) and (*19*) is an upper limit to the actual value.

D_0°(FeO) = 99.5 kcal mol^{-1} (*15*) depends on pressure calibration from intensity ratios for atomic and molecular oxygen and use of the known equilibrium constant for the O_2(g) = 2 O(g) reaction to calculate the molecular oxygen pressure, but some difficulty in discriminating contributions to the O^+ intensity is noted (*15*). This can only lead to overestimation of the pressure for atomic oxygen and hence to underestimation of the O_2 pressure, which ultimately leads to a value for D_0°(FeO) that is too high. D_0°(FeO) = 99.5 kcal mol^{-1} (*15*) is therefore also an upper limit to the actual value.

For D_0°(FeO) = 93.5 kcal mol^{-1} (*13*), experimental details are not available, precluding a further analysis of this value.

D_0°(FeO) = 90.6 kcal mol^{-1} (*18*) results from measurements in which either H_2(g) or H_2O(g) were introduced over iron oxide in a molybdenum cell in which equilibrium was approached from both sides. The equilibrium constant for the reaction FeO(g) + H_2(g) = Fe(g) + H_2O(g) is the same in both cases, but its numerical value and thus D_0°(FeO) depend on the ratio of the estimated ionization cross-sections $\sigma(H_2O)/\sigma(H_2) \approx 1$ or ≈ 0.5. The former is preferred (*18*). Alternate estimation, assuming that 2 eV above the respective ionization potentials $\sigma(H_2O) \simeq \sigma(O) \simeq 0.14 \times 10^{-16}$ cm^2 (*41*) when the truncated united atom model is adopted would lead with $\sigma(H_2) \simeq 0.09 \times 10^{-16}$ cm^2 (*18*) to $\sigma(H_2O)/\sigma(H_2) \simeq 1.6$.

Relative intensities for H_2O^+ and H_2^+ reported at four temperatures (*18*) actually make it possible to consider the reaction H_2O(g) = H_2(g) + (½)O_2(g) and by comparison with oxygen pressures in the Fe–Mo–O system given in the appendix, to derive $A(H_2O)/A(H_2) = 1.5 \pm 0.7$ and hence $\sigma(H_2O)/\sigma(H_2) \simeq 1.5 \pm 0.7$ if there is compensation between the other instrumental factors for these two molecules. When the truncated united atom model is next used to estimate the ionization cross-section for FeO(g) at 2 eV above the ionization potential, $\sigma(\mathrm{FeO}) \simeq \sigma(\mathrm{Se}) \simeq 0.93 \times 10^{-16}$ cm^2 (*41*) is obtained. With $\sigma(\mathrm{Fe}) \simeq 2.05 \times 10^{-16}$ cm^2 (*41*), $A(\mathrm{FeO})/A(\mathrm{Fe}) \simeq \sigma(\mathrm{FeO})/\sigma(\mathrm{Fe}) \simeq 0.45$, D_0°(FeO) = 95.0 ± 3 kcal mol^{-1} is then calculated by either considering the original reaction (*18*) FeO(g) + H_2(g) = Fe(g) + H_2O(g) or FeO(g) = Fe(g) + (½)O_2(g) with the oxygen pressures in the Fe—Mo—O system. Within the uncertainty limits, D_0°(FeO) derived from ref. (*18*) is then in agreement with the other data. A relative ionization cross-section for FeO(g) lower than for Fe(g) is consistent with observations for several oxides and other binary molecules (*12*).

Turning to spectroscopic data, Birge Sponer extrapolation with the molecular constants cited in the appendix yields D_0°(FeO) $\simeq$ 118 kcal mol^{-1}, as before (*2, 3*) appreciably higher than the thermochemical results, but not high enough to correspond to convergence to the Fe(^3P) +

$O(^3P)$ states as could be the case if the $X^5\Delta$ state correlated with the latter (*20*).

In principle, a lower limit to $D_0^\circ(FeO)$ could be obtained from the higher rovibronic level in excited FeO*(g) observed in the chemiluminescence (*20*) resulting from reaction of O_3(g) or N_2O(g) at room temperature with Fe(g) cooled to about 700K: $D_0^\circ(FeO) \geq h\nu + D_0^\circ(O—X) + E(\text{products}) - E(\text{reactants})$, where $h\nu$ is the highest energy photon emitted by FeO*(g), D_0°(O—X) the dissociation energy of the O—O_2 bond, 24.2 kcal mol^{-1}, in O_3(g) (*36*) or of the O—N_2 bond, 38.6 kcal mol^{-1} in N_2O(g) (*36*), *E*(reactants), translational and internal energy transferred from the reactants O_3(g) or N_2O(g) and Fe(g) and *E*(products), translational energy carried by the products and internal energy present in either O_2(g) or N_2(g) formed in the process. In practice a number of problems arise. These are: absence of information on the energy transferred from the reactants, e.g., even at 700 K, especially the lower excited substates with $J = 3$ to O located at 416, 704, 888, and 978 cm^{-1} above the $J = 4$ component of the ^5D ground state of Fe(g) are still appreciably populated; absence of information on whether O_2(g) or N_2(g) are formed in their $v'' = 0$ or in their $v'' = 1$ vibrational level, the latter being conceivable if internal energy partitioning were favored upon dissociation of the Fe—O—X intermediate complex. Other probably more serious problems are that the spectra emitted by FeO*(g) in the region of the orange and blue systems of interest here are exceedingly complex and blended (*28*); that FeO*(g) produced by reaction with O_3(g) or N_2O(g) is known (*20*) to be rotationally hot; that only band heads for vibrational levels are available (*20*); that the transitions in the orange system are predominantly of the $\Delta\Omega = 0$ type (*28*) and that hence it should be known which Ω'' component of the $X^5\Delta$ ground state is reached in the more intense transitions.

If these various problems are ignored, the highest vibrational level reported in the orange system when Fe(g) reacts with N_2O(g), $v' = 2$ in the B state at 18,998 cm^{-1} yields $D_0^\circ(FeO) \geq 92.6$ kcal mol^{-1}. Likewise, the $v' = 2$ to $v'' = 2$ transition in the blue system, i.e., between 21,724 and 1733 cm^{-1} above $v'' = 0$ observed when Fe(g) reacts with O_3(g) leads to $D_0^\circ(FeO) \geq 91.3$ kcal mol^{-1}, whereas an edge in the spectrum at about 404 nm read from Fig. 3 in ref. (*20*) might correspond to $D_0^\circ(FeO) \geq$ 95.0 kcal mol^{-1}. The thermochemical data and the chemiluminescence are thus quite compatible, but because of the limitations enumerated, the latter cannot presently provide an accurate lower limit for $D_0^\circ(FeO)$.

In view of the above discussion, it is concluded that $D_0^\circ(FeO) = 95.9 \pm 1.8$ kcal mol^{-1} is consistent with all available data for the dissociation energy of the molecule FeO(g). Weighed or unweighed averages of all results in Table 4 lead to about the same value. $D_0^\circ(FeO) = 95.9 \pm 1.8$ kcal mol^{-1} is therefore proposed to be the best present estimate for this quantity.

TABLE 4
Dissociation Energy of FeO(g)/kcal mol^{-1} (Revised Literature Data)

Reaction studied	T, K	Method	Ref.	D_0° (FeO)[a]
FeO(g) = Fe(g) = O(g)	2240–2431	Flame	*4*	≤98.3 ± 11.5[a, b]
FeO(g) = Fe(g) + O(g)	2412	Flame	*5*	96.4 or 91.9[b]
FeO(g) + H(g) = Fe(g) + OH(g)	2035–2615	Flame	*6*	97.5 ± 5[c]
FeO(g) = Fe(g) + ½ O_2(g)	[d]	MS	*13*	~93.5 ± 5[e]
FeO(g) = Fe(g) + ½ O_2(g)	1873	MS	*14*	≤99.8 ± 5
FeO(g) = Fe(g) + ½ O_2(g)	1885–2150	MS	*15*	≤99.5 ± 3
FeO(g) + PO(g) = Fe(g) + PO_2(g)	1670	MS	*16*	96.8 ± 3
FeO(g) + O(g) = Fe(g) + O_2(g)	1639–1765	MS	*17*	95.2 ± 3
FeO(g) + H_2(g) = Fe(g) + H_2O(g)	1723–1879	MS	*18*	90.6 or 95.0 ± 3[f]
FeO(g) + Se(g) → SeO(g) + Fe(g)	1657–1807	MS	This work	95.9 ± 1.8

[a]The uncertainties are those cited in the original publication.
[b]See Table 4.
[c]2nd law value.
[d]Not reported.
[e]Calculated from the originally reported value, 90.3 kcal mol^{-1} (*13*), assuming that the Gibbs function used there for FeO(g) was, as for other transition metal oxides, taken from L. Brewer and M. S. Chandrasekharaiah, UCRL-8713 (1959) and that the measurements were performed in the vicinity of 1750 K.
[f]See text

D_0°(MnO)

Previous results, systematically recalculated, and the presently determined value for D_0°(MnO) are compared in Table 5.

The flame photometric datum (*7*) is based on relative intensity measurements for Mn(g) at two temperatures that were treated by an ingenious 3rd law procedure. Recalculation by the same iterative method as in the original publication, using however the thermodynamic functions for MnO(g) computed here, rather than more approximate ones (*7*), yields D_0°(MnO) = 88.0 ± 3 kcal mol^{-1} instead of the reported and usually cited value, D_0°(MnO) = 92 ± 9 kcal mol^{-1}.

The determination D_0°(MnO) = 96 ± 3 kcal mol^{-1} (*8*) based on the measurement of the variation of intensities for emission lines for both Mn(g) and MnO(g) in flames is appreciably higher than the other values. This can in all probability be ascribed to interference between thermal excitation and chemiluminescence, which already caused only the measurements for the hotter flames to be retained (*8*) when deriving D_0°(MnO). Even so, the observed temperature dependence for the intensity of the atomic Mn line at 4033 Å corresponds to an activation energy of 60 kcal mol^{-1} (*8*), there where the term values for states involved in the transition imply an excitation energy of 71 kcal mol^{-1}. No way to recalculate the data was found.

For the flame photometric experiments (5) involving determination of absolute pressures and third law evaluation of the results, D_0°(MnO) = 84.9 ± 2.3 kcal mol^{-1} calculated with the reported actual pressure is, as for FeO(g) discussed above, appreciably lower than D_0°(MnO) = 89.0 ± 2.3 kcal mol^{-1} calculated with the equilibrium oxygen pressure, which is in turn in agreement with D_0°(MnO) determined here.

In principle, the flame photometric results (*5, 7*) are upper limits for D_0°(MnO) and are presented as such in Table 5 because no correction is made for the possible presence of bound manganese species other than MnO(g) itself. The agreement with D_0°(MnO) measured here however suggests that their concentation is low.

For MnOH(g), this conclusion is consistent with the value 75 ± 3 kcal mol^{-1} reported (*8*) for the Mn—OH bond energy, although the latter figure may be uncertain since interference between chemiluminescence and thermal excitation is also noted for this molecule (*8*). Estimation of D_0°(Mn—OH) by comparison with D_0°(Fe—OH) = 83 ± 3 (*18*) or 92 ± 5 kcal mol^{-1} (*6*) yields D_0°(Mn—OH) = 75 ± 5 or 83 kcal mol^{-1}, where the latter would still not cause MnOH(g) to have appreciably interfered with the determination of partial pressures for MnO(g) in flames (*5, 7*).

For MnO_2(g), only estimates can be made. By comparison of the atomization energies of the gaseous monoxides and dioxides of the elements adjacent to Mn in the periodic table, $D_{0,at}^\circ(MnO_2)/D_0^\circ(MnO) \approx 2.1$ or D_0°(O—MnO) ≈ 97 kcal mol^{-1} is obtained. Analogy with $MnSe_2$(g) and MnSe(g) (*30*) yields D_0°(O—MnO) ≈ 105 kcal mol−1. With D_0°(O—

TABLE 5
Dissociation Energy of MnO(g)/kcal mol^{-1} Revised Literature Data)

Reaction Studied	*T*, K	Method	*Ref.*	D_0° (MnO)
MnO(g) = Mn(g) + O(g)	2110–2460	Flame	*7*	≤88 ± 9
MnO(g) = Mn(g) + O(g)	2412	Flame	*5*	≤89.0 ± 2.3[a]
				≤84.9 ± 2.3[b]
MnO(g) + H_2(g) = Mn(g) + H_2O(g)	2000–2500	Flame	*8*	96 ± 3[c]
MnO(g) = Mn(g) + (½)O_2(g)	[d]	MS	*13*	85.4 ± 4
MnO(g) + Fe(g) = Mn(g) + FeO(g)	1660–2060	MS	*This work*	88.1 ± 1.8

[a]*p*(O)/atm = (1.86 + 0.05) × 10^{-3}, i.e., equilibrium value (*5*).
[b]*p*(O)/atm = (4.3 ± 0.2) × 10^{-3}, i.e., actual value (*5*).
[c]2nd law value.
[d]Not reported.

MnO) = 101 kcal mol^{-1}, one calculates $p(MnO_2)/p(MnO) \approx 0.1$ at 2412 K for $p(O) = 1.9 \times 10^{-3}$ atm, which would imply decreasing $D_0^\circ(MnO)$ by about 0.5 kcal mol^{-1} for the flame photometric measurement (5).

For MnO(g), as for FeO(g), the Birge Sponer extrapolation method leads to values for the dissociation energy that are appreciably higher than the thermochemical result: D_0° = 104 kcal mol^{-1} for the ground state (2); 97 kcal mol^{-1} for the excited state (2). Total pressure measurements likewise yield much too high a value, $D_0^\circ(MnO) \leq 107$ kcal mol^{-1} (*1*), from which it was concluded that the partial pressure of MnO(g) is but a small fraction of the total decomposition pressure. These method thus cannot assist in further circumscribing the actual value for $D_0^\circ(MnO)$.

On the basis of the agreement between the second and third law results in the present determination and of their consistency with the different measurements discussed above, $D_0^\circ(MnO) = 88.1 \pm 1.8$ kcal mol^{-1} is hence retained as best present estimate of the dissociation energy of this molecule.

ACKNOWLEDGMENTS

This paper is respectfully dedicated to Prof. Leo Brewer on his 65th birthday.

The authors are grateful to Dr. H. Kleykamp (Kernforschungszentrum Karlsruhe, GFR) for communicating the results of DTA and measurements for the Fe—Mo—O system. They thank Dr. Ph. Spencer and Dr. K. Hack (T. H. Aachen, GFR) for discussions and independent calculations for the Fe—Mo—O system. They are indebted to the Fund for Joint Basic Research, Belgium and the National Fund for Scientific Research, Belgium for an equipment grant and support.

Appendix

THERMODYNAMIC FUNCTIONS

(a) FeO(g)

ϕ_T°, the Gibbs function and $(H_T^\circ - H_0^\circ)$, the heat content, given in Table 6, are calculated with the usual statistical-mechanical formulae. The same molecular constants, $\omega_e = 880.41$, $\omega_e x_e = 4.630$, $B_e = 0.51872$, $\alpha_e = 3.83 \times 10^{-3}$, and $D_e = 7.24 \times 10^{-7}$ cm^{-1} (*24, 28*) are used for the different components of the $X^5\Delta_i$ multiplet ground state (*28*) with a spin-

TABLE 6
Enthalpies (cal mol^{-1}) and Gibbs Functions (cal k^{-1} mol^{-1})

Specie	Function	Temperature, K					
		298.15	1600	1800	2000	2200	2400
FeO	$H^\circ_T - H^\circ_O$	2445	14428	16361	18325	20325	22361
	$\Phi^\circ_O(T)$	48.10	62.61	63.68	64.64	65.52	66.32
MnO	$H^\circ_T - H^\circ_O$	2118	13368	15171	16981	18797	20619
	$\Phi^\circ_O(T)$	49.30	62.27	63.26	64.15	64.96	65.71
SeO	$H^\circ_T - H^\circ_O$	2329	13656	15517	17402	19301	21237
	$\Phi^\circ_O(T)$	48.00	61.58	62.59	63.50	64.33	65.10
FeO_2	$H^\circ_T - H^\circ_O$	2452	20480	23418	26364	29294	32273
	$\Phi^\circ_O(T)$	51.63	69.70	71.22	72.60	73.86	75.02

orbit coupling constant A = -94.9 cm^{-1} (*28*). Ab initio calculations lead to either 3000 or 7000 cm^{-1} (*25, 26*) for the term value of the $^5\Sigma^+$ state to which photoelectron spectroscopy for FeO^-(g) assigns (*21*) an excitation energy of ca. 4000 cm^{-1} (*3b, 21*). The latter value is retained for the calculation of the thermodynamic functions presented here, which further take into account a $^5\phi$ state with multiplet components located between 9904 and 10,446 cm^{-1} (*28*), a $^5\Pi$ state near 10,200 cm^{-1} (*28*) and a $^3\Gamma$ state (*28*) that is tentatively placed here at 9000 cm^{-1}. Computation of the thermodynamic functions with and without the excited states shows that their contribution is in fact small in the temperature interval of interest here, in which uncertainties of 0.1 cal mol^{-1} K^{-1} and 0.5 kcal mol^{-1} are ascribed to ϕ_T° and $(H_T^\circ - H_0^\circ)$, respectively.

(b) FeO_2(g)

Infrared absorption spectra for FeO_2 prepared in rare gas matrices (*42*) are interpreted in terms of a cyclic isosceles C_{2v} structure with an apex angle of about 44°. This structure, however, corresponds to an excited state in a theoretical treatment (*43*) of the FeO_2 moiety in hemoglobin, where for the ground state a singlet bent but open (*136°*) configuration is preferred. Extensive ESR studies for MnO_2 (*44*) and for MnF_2 (*45*) in rare gas matrices in contrast lead to linear structures for both molecules, which possess respectively one less and one more electron than FeO_2 in orbitals that are not expected (*44, 46*) to determine the spatial structure. CrF_2, isoelectronic with FeO_2, is also considered to be linear (*47*). A linear symmetric structure is therefore retained here for FeO_2(g), which remains consistent with the observation of only two IR transitions and their pattern after isotopic substitution (*42*). By comparison with MnO_2 (. . . $\delta^2\sigma^1$, $^4\Sigma$) (*44*) and MnF_2 (. . . $\delta^2\pi^2\sigma^1$, $^6\Sigma$) (*45*), a configuration . . . $\delta^2\pi^1\sigma^1$ is assumed for FeO_2(g), leading to a $^5\Pi$ ground state. Spin-orbit interaction is neglected in the calculation of the thermodynamic functions, made with vibration frequencies ν_1 = 946 cm^{-1} (*42*), ν_2 = 517 cm^{-1} (*42*), and ν_3 = 650 cm^{-1}, estimated here and an interatomic distance r = 1.62 Å, assumed to be approximately the same in FeO_2 (g) as in FeO(g).

Treatment with these functions of the data (*17*) for the reaction Fe(g) + FeO_2(g) = 2FeO(g) studied between 1718 and 1764 K yields ΔH_0° = 10.5 kcal mol^{-1} with a standard deviation of 0.4 kcal mol^{-1}. Combined with $\Delta H_0^\circ = -22.7 \pm 0.2$ kcal mol^{-1} for the reaction Fe(g) + O_2(g) = FeO(g) + O(g) investigated in the same study, ΔH_0° = 105.7 kcal mol^{-1} is obtained for the reaction FeO_2(g) = FeO(g) + O(g).

(c) FeOH(g)

Addition of the reactions Fe(g) + OH(g) = FeO(g), ΔH_0° = 3.8 kcal mol^{-1} (recalculated from ref. 6 with the thermodynamic functions for FeO(g) computed here), Fe(g) + H_2O(g) = FeOH(g) + H(g), ΔH_0° = 26.3

kcal mol^{-1} (*6*) and (½)H_2(g) + OH(g) = H_2O(g), ΔH°_0 = −66.4 kcal mol^{-1} (*36*) yields ΔH°_0 = −43.9 kcal mol^{-1} for reaction (5), FeO(g) + (½)H_2(g) = FeOH(g).

Alternately, addition of the reaction FeO(g) + H_2(g) = FeOH(g) + H(g), ΔH°_0 = 15.6 kcal mol^{-1} [also recalculated with the thermodynamic functions for FeO(g) computed here] and H(g) = (½)H_2(g), ΔH°_0 = −51.6 kcal mol^{-1} (*36*) leads to ΔH°_0 (5) = −36.0 kcal mol^{-1} from ref. (*18*).

The numerical values of the Gibbs function for FeOH(g) used here are those reported in ref. (*18*).

(d) Comparison of Known and Estimated Pressures in the Fe–Mo–O System and Calculation of $A(H_2O)/A(H_2)$ Therefrom

In this section are compared, at 1747, 1772, 1787, and 1804 K: the oxygen pressures corresponding to ΔG° (formation, H_2O, g) (*36*) when $p(H_2O) = p(H_2)$; the oxygen pressures for the Mo(s) + MoO_2(s) system (*36*); the oxygen pressures for the Fe(s) + Fe_{1-y} O(l) system at the phase limit (*38*); tentative oxygen pressures for univariant domains within the Fe–Mo–O system suggested by comparison of the phase diagrams for the Fe–Mo (*40, 48, 49*) and Fe–O (*38*) systems at the temperatures of interest here and of the Fe–Mo–O system at other temperatures (*50–54*).

The mass flow upon passing H_2(g) or H_2O(g) over condensed phases in experiments such as in ref. (*18*) is indeed not sufficient to cause major modifications in the overall composition of the system. This implies that the $I(H_2O^+)/I(H_2^+)$ ratios determined mass spectrometrically (*18*) are a measure of the oxygen potential established by phase relations among the compounds present in the system. Since very similar $I(H_2O^+)/I(H_2{}^+)$ ratios were observed, upon introducing either H_2(g) or H_2O(g) (*18*), the system should either be steady-state but bi-variant, or univariant, the latter being the more probable. If the oxygen potential for the latter can be calculated, ΔG° (formation, H_2O, g) fixes $p(H_2O)/p(H_2)$, which in turn permits calculation of $A(H_2O)/A(H_2)$ = $[I(H_2O^+)]/p(H_2O)/p(H_2)$. In the presence of an excess of molybdenum provided by the cell in which Fe_2O_3(s) was initially placed (*18*), possible univariant domains in the Fe–Mo–O system at the temperatures of interest here could be: Mo(Fe)(s) + MoO_2(s) + Fe_2MoO_4(s) or Mo(Fe)(s) + FeMo(σ) + Fe_2MoO_4(s) above 1758 K, with in addition Mo(Fe)(s) + (Fe_2MoO_4(s) + $Fe_2Mo_3O_8$(s) below this temperature. Fe_2MoO_4(s) indeed melts congruently at about 1853 K (*55*). It co-exists with $Fe_2Mo_3O_8$(s) up to 1758 K, next with MoO_2(s) until 1833 K on the one side, with $Fe_{1-y}O$(s) until 1623 K on the other side, where it possibly is nonstoichiometric (*55*). Oxygen decomposition pressures, given in Table 7, were calculated with the following thermodynamic data: ΔG° (formation, FeMo, σ) = $[1.59 - 1.987 \times 10^{-3}T/K]$ kcal mol^{-1} (*48*) or ΔG° (formation, FeMo, σ) = $[15.527 - 19.437 \times 10^{-3}T/K + 5.363 \times 10^{-6}(T/K)^2]$ kcal mol^{-1} (*49*), and ΔG° (formation, Fe_2MoO_4,s) = $[(-258.533 + 63.743 \times 10^{-3}T/K \pm 0.65]$ kcal

TABLE 7
Comparison of Known and Estimated Oxygen Pressures in Possible Univariant Domains in the Fe–Mo–O System

System	$-\log [p(O_2)/\text{atm}]$			
	1747K	1772K	1787K	1804K
$Mo(s) + MoO_2(s)$ (*36*)	8.58	8.35	8.21	8.05
$Mo(Fe)(s) + MoO_2(s) + Fe_2MoO_4(s)$	~8.53[a]	~8.30	~8.16	~8.00
$Mo(Fe)(s) + Fe_2MoO_4(s) + FeMo(\sigma)$	~9.00	~8.77	~8.64	~8.49
$FeMo(\sigma) + Fe_2MoO_4(s) + Fe_xMo_yO_z$	~8.95	~8.72	~8.59	~8.44
$FeMo(\sigma) + Fe(Mo)(l)$[b] $+ FeO(l)$	~8.87	~8.65	~8.55	~8.40
$Fe(s) + FeO(l)$ (*38*)	9.12	8.92	8.81	8.66
$2H_2O(g) = 2H_2(g) + O_2(g)$ $[p(H_2O) = p(H_2)]$	8.97	8.76	8.64	8.51

[a]At 1747 K, $-\log [p(O_2)/\text{atm}] \approx 8.53$ is calculated also for the $Mo(Fe)(s)$ $[a(Mo) \simeq 0.85]$ $+ Fe_2MoO_4(s) + Fe_2Mo_3O_8(s)$ univariant domain, which ceases to exist above ca. 1758 K.

[b]In the Fe–Mo binary system, $FeMo(\sigma)$ coexists with $Fe_5Mo_3(R)$ until 1761 K.

mol^{-1} (*53*). The last equation, $\Delta G°$ (formation, MoO_2,s) (*36*) and $\Delta G°$ (formation, $Fe_2Mo_3O_8$,s) = [(−542.017 + 147.347 × $10^{-3}T$/K) ± 1.1] kcal mol^{-1} (*53*) are entirely consistent with disproportionation of $Fe_2Mo_3O_8$(s) around 1758 K. Activities of about 0.9 and 0.8, respectively, were retained for molybdenum in the Mo(Fe)(s) + MoO_2(s) + Fe_2MoO_4(s) and the Mo(Fe)(s) + FeMo(σ) + Fe_2MoO_4(s) systems. Oxygen pressures are also given for the more iron-rich domains, although these are not of direct importance here, where they provide lower limits for $p(O_2)$, however. In the FeMo(σ) + Fe(Mo)(l) + FeO(l) system, a(Fe) ~ 0.7 has been adopted to obtain $p(O_2)$ by interpolation of the oxygen pressures as a function of composition in $Fe_{1-y}O$(l) (*38*).

On the thermodynamic grounds, the oxygen pressures for the Mo(s) + FeMo(σ) + Fe_2MoO_4(s) system are retained for comparison with the $I(H_2O^+)/I(H_2^+)$ ratios (*18*), rather than those for the Mo(s) + MoO_2(s) + Fe_2MoO_4(s) system. The latter, however, cannot be entirely excluded, which explains the uncertainty limits given below for $A(H_2O)/A(H_2)$.

Subtraction of −2 log [$p(O_2)$/atm] for the system retained from −2 log $K(H_2O)$ (*36*) given in Table 7 gives 2 log[$p(H_2O)/p(H_2)$], whence follows $p(H_2O)/p(H_2)$ = 0.97, 0.99, 1.00, and 1.02, respectively, at 1747, 1772, 1787, and 1804 K, where $I(H_2O^+)/I(H_2^+)$ = 1.67, 1.51, 1.40, and 1.57, respectively (*18*). Therefrom results $A(H_2O)/A(H_2) \simeq \sigma(H_2O)/\sigma(H_2)$ = 1.5 (+0.3, −0.7).

The procedure followed obviously implies that treatment of the equilibrium FeO(g) = Fe(g) + (½)O_2(g) using $I(FeO^+)/I(Fe^+)$ given (*18*) at the same temperatures as $I(H_2O^+)/I(H_2^+)$, A(FeO)/A(Fe) estimated here and the oxygen pressures for the system Mo(s) + FeMo(σ) + Fe_2MoO_4(s) yields the same value for D_0°(FeO) as does the reaction studied (*18*) when use is made of the same A(FeO)/A(Fe) ratio as here and of $A(H_2O)/A(H_2)$ derived above.

(e) MnO(g)

For MnO(g), it is established (*56–58*) that the ground state is $X^6\Sigma^+$. The vibrational constants used in the present calculations are ω_e = 839.6 and $\omega_e x_e$ = 4.79 cm^{-1}; the rotational constants are B_e = 0.50325, α_e = 4.06 × 10^{-3} and D = 7.2 × 10^{-7} cm^{-1}. The uncertainty in the thermodynamic functions for MnO(g) is considered to be negligible in comparison with the other sources of inaccuracies in the calculation of D_0°(MnO).

REFERENCES

1. Brewer, L., and Mastick, D. F., *J. Chem. Phys.* **19**, 834 (1951).
2. Gaydon, A. G., *Dissociation Energies*, Chapman and Hall, London, 2nd Ed., 1953; 3rd Ed., 1968.
3. (a) Herzberg, G., *Spectra of Diatomic Molecules*, Van Nostrand, New York, 1950; (b) Huber, K. P., and Herzberg, G., Constants of Diatomic Molecules, Van Nostrand–Reinhold, New York, 1979.

4. Lagerqvist, A., and Huldt, L., Z. *Naturforsch.* **8A,** 493 (1953).
5. Zeegers, P. J. Th., Townsend, W. P., and Winefordner, J. D., Spectrochim. Acta **24B,** 243 (1968).
6. Jensen, D. E., and Jones, G. A., *J. Chem. Soc. Faraday Trans. I.,* **69,** 1448 (1973).
7. Huldt, L., and Lagerqvist, A., *Ark. Fys.* **3,** 525 (1951).
8. Padley, P. J., and Sugden, T. M., *Trans. Faraday Soc.* **55,** 2054 (1959).
9. Chupka, W. A., and Inghram, M. G., *J. Chem. Phys.* **21,** 371 (1953); *J. Phys. Chem.* **59,** 100 (1955).
10. Honig, R. E., *J. Chem. Phys.* **22,** 126 (1954).
11. Inghram, M. G., and Drowart, J., in *High Temperature Technology,* McGraw-Hill, New York, 1960, p. 219.
12. Drowart, J., and Goldfinger, P., *Angew. Chem., Intl. Ed.* **6,** 581 (1967).
13. Burns, R. P., cited by Cheetham, C. J. and Barrow, R. F., *Adv. High Temp. Chem.* **1,** 7 (1967).
14. Washburn, C. A., UCRL-10991, Lawrence Radiation Lab., Univ. of California, Berkeley, California, cited in ref. (*19*).
15. Balducci, G., De Maria, G., Guido, M., and Piacente, V., *J. Chem. Phys.* **55,** 2596 (1971).
16. Smoes, S., and Drowart, J., *Faraday Symposia of the Chemical Society No.* **8,** 139 (1973).
17. Hildenbrand, D. L., *Chem. Phys. Lett.* **34,** 352 (1975).
18. Murad, E., *J. Chem. Phys.* **73,** 1381 (1980).
19. Brewer, L., and Rosenblatt, G. M., *Adv. High Temp. Chem.* **2,** 1 (1969).
20. West, J. B., and Broida, H. P ., *J. Chem. Phys.* **62,** 2566 (1975).
21. Engelking, P. C., and Lineberger, W. C., *J. Chem. Phys.* **66,** 5054 (1977).
22. Green, D. W., Reedy, G. T. and Kay, J. G., *J. Mol. Spectr.* **78,** 217 (1979).
23. DeVore, T. C. and Gallaher, T. N., *J. Chem. Phys.* **70,** 4429 (1979).
24. Harris, S. M., and Barrow, R. F., *J. Mol. Spectr.* **84,** 334 (1980).
25. Bagus, P. S., and Preston, J. H. T., *J. Chem. Phys.* **59,** 2986 (1973).
26. Walsh, S. P., and Goddard, W. A., III, cited in ref. (*22*).
27. Michels, M. H., cited in ref. (*17*).
28. (a) Cheung, A. S-C, Gordon, R. M ., and Merer, A. J., *J. Mol. Spectr.* **87,** 289 (1981); (b) Cheung, A. S-C., Nelson, L., Lyyra, A. M ., Merer, A. J., and Taylor, A. W., *J. Mol. Spectr.* **95,** 213 (1982).
29. Smoes, S., and Drowart, J., *J., Chem. Soc. Faraday Trans.* **I,** in press.
30. Smoes, S., Pattje, W. R., and Drowart, J., *High Temp. Sci.* **10,** 109 (1978).
31. Smoes, S., Drowart, J., and Welter, J. M., *J. Chem. Thermodyn.* **9,** 275 (1977).
32. Drowart, J., Pattoret, A., and Smoes, S., *Proc. Brit. Ceram. Soc.* **8,** 67 (1967).
33. Berkowitz, J., and Chupka, W. A., *J. Chem. Phys.* **50,** 4245 (1969).
34. Rosenstock, H. M., Draxl, K., Steiner, B. W., and Herron, J. T., *J. Phys. Chem. Ref. Data Suppl. 1,* **6** (1977).
35. Hultgren, R., Desai, P. D., Hawkins, D. T., Kelley, K. K., and Wagman, D. D., *Selected Values of the Thermodynamic Properties of the Elements,* American Society for Metals, Metals Park, Ohio, 193.
36. JANAF Thermochemical Tables, 2nd ed., NSRDS-NBS 37 (1971).
37. Darken, L. S., and Gurry, R. W., *J. Am. Chem. Soc.* **68,** 798 (1946).
38. Spencer, P. J., and Kubaschewski, O., *Calphad* **2,** 147 (1978).
39. Omission of p(FeO) or use of p(FeO) calculated from the presently determined value for D_0°(FeO) with $p(O_2)$ from ref. (*38*) and p(Fe) calculated from the activity of iron in FeO(l) from ref. (*38*) and the standard state pressure from ref. (*35*) also show that $FeO_{1.116}$(l) does not effuse congruently at 1873 K. This implies that the possible inaccuracy in the ratio p(FeO)/p(Fe) $\approx$ 0.2

(*14*) does not influence the conclusions reached in this section with respect to the vaporization behavior of the condensed phase FeO(l), which depend only on the thermodynamic data for the latter (*37, 38*).

40. Hultgren, R., Desai, P. D., Hawkins, D. T., Gleiser, M., and Kelley, K. K., *Selected Values of the Thermodynamic Properties of Binary Alloys*, American Society for Metals, Metals Park, Ohio, 1973.
41. Mann, J. B. in *Recent Developments in Mass Spectrometry*, Ogata, K., and Hayakawa, P., eds., Tokyo University Press, Tokyo, 1970, p. 814.
42. Abramowitz, S., Acquista, N., and Levin, I. W., *Chem. Phys. Lett.* **50,** 423 (1977).
43. Goddard, W. A., III, and Olafson, B. D., *Proc. Natl. Acad. Sci. US* **72,** 2335 (1975).
44. Ferrante, R. F., Wilkerson, J. L., Graham, W. R. M., and Weltner Jr., W., *J. Chem. Phys.* **67,** 5904 (1977).
45. DeVore, T. C., Van Zee, R. J., and Weltner, Jr., W., *J. Chem. Phys.* **68,** 3522 (1978).
46. Hayes, E. F., *J. Phys. Chem.* **70,** 3740 (1966).
47. Hastie, J. W., Hauge, R. H., and Margrave, J. L., *Ann. Rev. Phys. Chem.* **21,** 475 (1970).
48. Brewer, L., and Lamoreaux, R. H., *Atom. Energ. Rev., Special Issue No. 7,* IAEA , Vienna (1980).
49. Nüssler, H. D., Hosler, T., and Kubaschewski, O., *Z. Metallkde* **71,** 396 (1980).
50. Monnier, G., *Bull. Soc. Chim. France* **298,** 1252 (1959).
51. Schmahl, N. G., and Dillenburg, H., *Z. Physik. Chem. N. F.* **77,** 113 (1972).
52. Rezukhina, T. N., and Kashina, T. A., *J. Chem. Thermodyn.* **10,** 279 (1978).
53. Kleykamp, H., and Schauer, V., *J. Less Comm. Metals* **81,** 229 (1981).
54. Rusakov, L. I., Novokhatskii, P. A., Lenev, L. M., and Savinskaya, A. A., *Dokl. Akad. Nauk SSSR* **161,** 410 (1965).
55. Kleykamp, H., private communication.
56. Das Sarma, J. M., *Z. Phys.* **157,** 98 (1959).
57. Pinchemel, B., and Schamps, J., *Can. J. Phys.* **53,** 431. (1975); *Chem. Phys.* **18,** 489 (1976).
58. Gordon, R. M., and Merer, A. J., *Can. J. Phys.* **58,** 642 (1980).

0018–1536/84/1700–0053$02.40

The Surface Diffusion of High Temperature Vapors in Porous Alumina

NATHAN S. JACOBSON,† ELIZABETH J. OPILA,‡ AND ALAN W. SEARCY*

Materials and Molecular Research Division, Lawrence Berkeley Laboratory, and Department of Materials Science and Mineral Engineering, University of California, Berkeley CA 94720

Received September 23, 1983; Accepted December 31, 1983

ABSTRACT

In porous alumina barriers ~1 mm thick, 45% porosity, and ~1.2 μm average pore diameter, transmission coefficients for NaCl, NaF, and LiF vapors are ~2.6, 6, and 12 times that of He, indicating that surface diffusion contributes to the transmission. For Zn vapor, transmission coefficients are 1.2–3.5 that of He through the ~1.2 μm pores and 18 times that of He through 0.14 μm pores. The temperature dependences of LiF and NaCl transmission by surface diffusion are close to those for Knudsen effusion, a result that indicates that the activated complex for surface diffusion is very weakly bound to the surface. Above their melting points, the transmissions of LiF and NaCl are irreproducible, but are usually markedly increased, perhaps because a liquid-like film partially penetrates the barrier.
Index Entries: Alumina, porous; Knudsen effusion; surface diffusion, at high temperatures; transport, in porous media; vapors, alkali halide; vapors, zinc; diffusion, of vapors in porous alumina.

INTRODUCTION

Equilibrium pressures derived from effusion studies usually assume that Knudsen flow is the only significant path of vapor transport through the orifice (*1*). This may not always be the case. Effusion studies are

*Author to whom all correspondence and reprint requests should be addressed.
†Present address: NASA-Lewis Research Center, Cleveland, Ohio.
‡Present address: Monolithic Memories, Inc., Sunnyvale, California.

sometimes thwarted by the "creep" or "bulk surface flow" of large amounts of material up the cell walls and out of the orifice (2,3). In this laboratory (4), this phenomenon has been observed with CrF_3, CoF_2, and FeF_2. It is a matter for concern that less obvious transport by surface diffusion may often contribute to effusion fluxes and introduce errors into vapor pressure measurements.

Winterbottom and Hirth (5–7) drew attention to this possibility and developed a theory for combined surface diffusion of a dilute monolayer of adsorbed vapor molecules and Knudsen flow through short cylindrical or conical orifices. To calculate from the model the surface diffusion contribution to the total flux, the surface diffusion coefficient and the energy of desorption are required. Generally, these data are unavailable. However, Winterbottom and Hirth used data for silver vapor on nickel and molybdenum surfaces to support their theoretical conclusion that the greatest surface diffusion contributions will be found with orifices of small diameters and small length-to-radius ratios or knife-edges. Ward et al. (*8,9*) have done computer simulations of Knudsen and surface diffusion through orifices and obtained results in general agreement with the model of Winterbottom and Hirth.

There has been little experimental data with which to compare this model. Boyer and Meadowcroft (*10*) have measured the free energy of vaporization of silver under conditions that Winterbottom and Hirth predict will promote surface diffusion. Use of a molybdenum cell lid with a small diameter knife-edge orifice yields an apparent $\Delta G = 227{,}600 + 85T$ J/mol. When corrected for surface diffusion, this value becomes much closer to the accepted free energy of $\Delta G = 266{,}100 + 110T$ J/mol. The difference between the apparent value of ΔG and the accepted value becomes much less for larger orifice diameters. Grimley et al. (*11*) have observed altered angular distributions for samarium vapor emerging from tantalum and boron nitride orifices, but they attribute this to bulk diffusion in the orifice walls.

The surface diffusion contributions to effusion modeled by Winterbottom and Hirth arise primarily from end effects, so that surface diffusion that obeys their model would be unimportant in long channels or in porous barriers, which can be viewed as a network of long irregularly shaped channels. However, extensive studies with barriers traversed by pores of ~10 nm cross-section at temperatures at or near room temperature show that for such small pores surface diffusion by gases can be a major path of transport (*12*). In a typical study, the transmission behavior of a gas under investigation is compared to that of He, for which adsorption and surface diffusion are negligible at room temperature. CO, for example, passes several times faster than He through a graphon plug, which has ~5 nm pores (*12*). The enhancement of CO flow is attributed to diffusion in a dilute monolayer adsorbed on the pore walls. Gilliland et al. (*13*) found that various hydrocarbons pass through Vycor up to 17 times faster than He does. They could not reconcile this result with submonolayer coverage and instead developed a theory based on a spreading film. At only a few hundred degrees centigrade, kT, where k is the Boltzmann constant and T is the absolute temperature, becomes large relative to the bonding energies of most gases to surfaces. Surface diffusion then becomes unimportant because most gas molecules desorb before they have moved a significant distance from the point of impact with the surface.

Mohazzabi and Searcy (*14*) have shown that porous barrier studies can provide insights into equilibration of monomer–dimer vapor mixtures in pressure gradients along surfaces. It seemed likely that barrier studies might provide useful information on surface diffusion of high temperature vapors if the bonding energy of the vapor molecules to the surface is large enough to be comparable to kT. To test this possibility, various high temperature vapors are passed through porous alumina. The diffusion rates are compared to Knudsen diffusion rates measured for He. Transmission probabilities are used to describe the diffusion. These are defined as the ratio of J_e, the molar flux of vapor per unit area that escapes from the plug, to J_o, the flux per unit area that strikes the plug. Transmission probabilities near those measured for He indicate that Knudsen flow is the primary mechanism of transport; larger transmission probabilities indicate the presence of an additional path of transport–presumably surface diffusion.

EXPERIMENTAL

Characterization of Porous Alumina

Two types of porous alumina were used–one with an average pore cross-section of 1.2 ± 0.2 μm and one with an average pore cross section of 0.14 ± 0.02 μm. Both were used as disks 0.95 ± 0.05 mm thick. The alumina of larger pore size was made at the Lawrence Berkeley Laboratory ceramics shop from an alumina greenware (Western Gold and Platinum, Belmont CA). This material was a mixture of ~1 μm cross-section alumina particles (99.5% pure) and a resin binder. The mixture was fired at about 1250°C for about 1 h. This process vaporized the resin binder and left a partially sintered porous product, called Alumina I.

The surface areas of five samples were measured with a BET apparatus. Different samples have similar porosities and pore sizes. The average area was 0.61 ± 0.11 m^2/g. The low surface areas rule out the possibility of a significant contribution of smaller pores to the measured porosity. A typical mercury porosimeter scan is shown in Fig. 1. Most of the penetration occurred at pressures corresponding to pore diameters of 1.1–0.6 μm. From the maximum penetration volume, the relative porosity was determined to be 0.45 ± 0.02.

The porous alumina with a smaller average pore diameter was prepared by isostatically pressing 0.6 μm alumina (Alcoa A16-SG) at 240

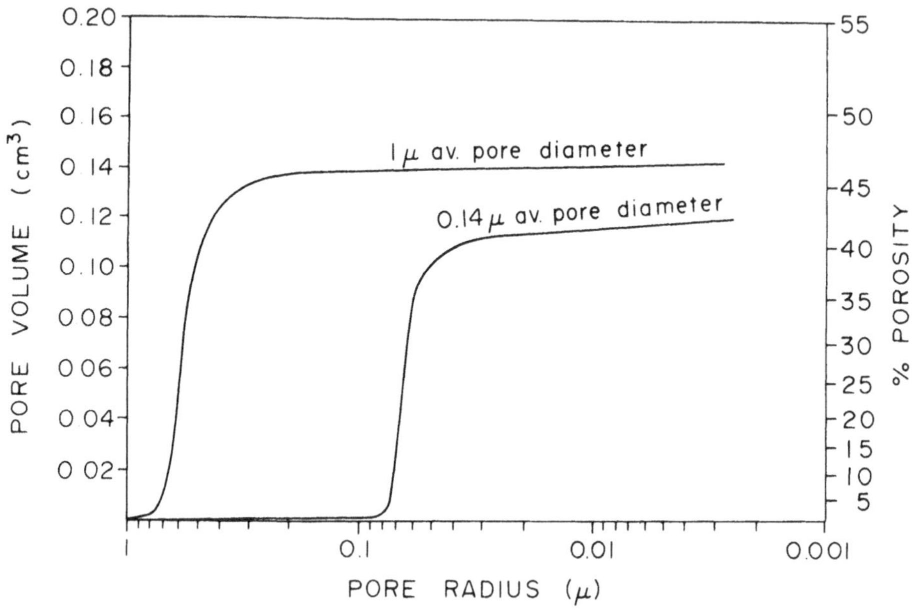

Fig. 1. Mercury porosimeter scans for both porous aluminas used.

MPa and firing for about 1 h at 1000°C. This material was also characterized with mercury porosimetry, BET surface area measurements, and scanning electron microscopy. A typical porosimeter plot is also shown in Fig. 1. Alumina II had an average pore diameter of 0.14 ± 0.01 μm, a porosity of 0.42 ± 0.01 and a surface area of 5.4 ± 0.1 m^2/g.

If a porous material is approximated as a series of parallel capillaries, its average pore diameter, d, can be estimated from the relation $d = 4\epsilon/S(\epsilon - 1)\,\rho$, where S is the surface area per gram, ϵ is the relative porosity, and ρ is the density (*15*). This method yielded $d = 1.2 \pm 0.2$ μm for Alumina I. Scanning electron microscope (SEM) pictures confirm that pore cross-sections are of the order of 1 μm, but show the pores to have very irregular shapes. These deviations from ideality reduce experimental transmission probabilities below those predicted for a bundle of parallel capillaries (*16*).

For a straight cylindrical tube with a length-to-radius ratio of (l/d) 100 or greater, the transmission probability is $(4/3)(d/l)$ (*17*). For a porous barrier, this becomes $(1/\tau)(4/3)(d/l)$, where τ, the tortuosity factor, is the ratio of transmission by Knudsen flow through ideal cylindrical orifices to transmission through a barrier of the same relative pore volume to surface area (*18*).

One apparatus was used for He and other gases, and a second was used for various high temperature vapors. The apparatus for gases is shown in Fig. 2. It consists of a gas bulb from which a gas can be leaked through a porous barrier. The rate of leakage is proportional to the transmission coefficient. The porous barrier is mounted on a smooth mullite

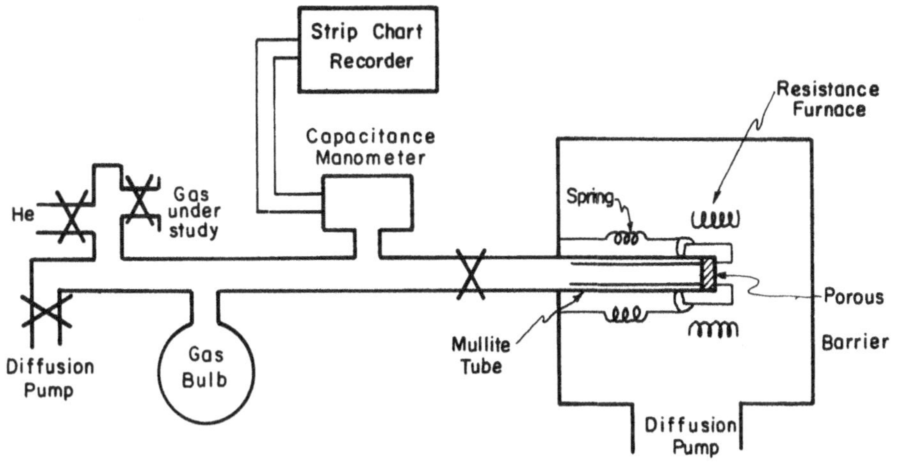

Fig. 2. Apparatus for measuring the transmission properties of helium.

tube and is held in place by a spring-loaded cap. This arrangement is designed to minimize escape other than through the porous barrier.

The rate of leakage from the gas bulb is the time derivative of the ideal gas law: $dn/dt = (V/RT_m)dP/dt$. Here V is the volume of the gas bulb and manifold, and T_m is the temperature of the gas in the manifold. The flow rate through the porous barrier is given by the Hertz-Knudsen-Langmuir equation, modified by W, the transmission probability:

$$J = (1/A)dn/dt = WP/(2\pi MRT_b)^{1/2}$$

where M is molecular weight of He and T_b is the temperature of He in the barrier. Equating and integrating these two expressions for J gives

$$W = [(2\pi MRT_b)^{1/2}V/(A_pRT_m)]\ (\Delta \ln P/\Delta t)$$

Here A_p is the exposed area of the barrier fronted by pores (i.e., the exposed area multiplied by the porosity) and $\Delta\ln P/\Delta t$ is the rate of leakage from the bulb. This quantity was calculated from the output of the capacitance manometer.

The transmission properties of the high temperature vapors were determined from weight losses. A diagram of the apparatus is shown in Fig. 3. Alumina cells could not be used for LiF because it penetrated through at grain boundaries. Free energy calculations indicate that the reaction of NaF(g) and Al_2O_3 to form Na_2O and AlF_3 is not favorable under these conditions. The formation of cryolite, Na_3AlF_6, does have a free energy change of −674 kJ/mol at 1100°K:

$$12\ NaF(g) + Al_2O_3(s) = 2\ Na_3AlF_6(s) + 3\ Na_2O(s)$$

However, this indicates that cryolite would form only if the pressure for NaF exceeds 200 Pa. Since the pressure in the cell was never greater than 1 Pa, this reaction cannot occur. There was also no physical evidence of a reaction occurring to any large extent. After an experiment, the barrier

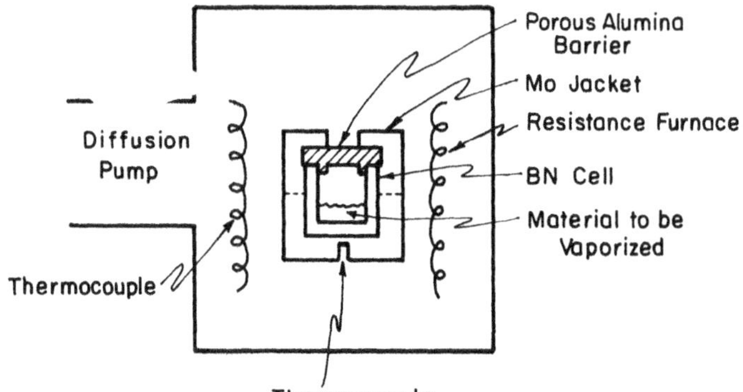

Fig. 3. Apparatus for measuring the transmission properties of high temperature vapors.

gained at most about 1 mg compared to a total weight loss of 5 mg or more of NaF. The used barrier showed only alumina X-ray diffraction peaks, and the remaining sample showed only sodium fluoride peaks. Similar calculations and examinations indicated no evidence for reactions of LiF or NaCl vapors with alumina. A barrier was sectioned and studied by Auger spectroscopy after LiF runs. A small quantity of fluorine was detected at the inner surface, but fluorine was undetectable in the interior of the barrier.

Thick-walled boron nitride cells showed no evidence of reaction with the vapors studied, had good mechanical strength, and could be ground to give a good fit between the cell body and alumina barrier. As in the apparatus for gases, the system is designed to minimize escape via routes other than through the barrier.

Zinc and three alkali halide salts–sodium chloride, sodium fluoride, and lithium fluoride–were examined. Reagent grade chemicals were used for each. Solids will establish essentially equilibrium pressures in the source cell if their vaporization coefficients are large relative to the transmission probability through the barrier (*1*). Zinc and the alkali halides have vaporization coefficients near unity (*19*), and barrier transmission coefficients were found to be $<10^{-2}$ in all measurements.

Exploratory heatings were made with the cell set at various heights in the furnace in order to find a position that gave no measurable condensation of vapor on the barrier. The cell, barrier, and material to be vaporized were initially weighed, loaded into the furnace, and rapidly brought to temperature. The temperature was maintained for a known time period (generally about 10 h). The cell assembly was then rapidly cooled and weighed again. Several measurements were taken for each solid. Equilibrium fluxes were determined from weight-loss measurements on cells with orifices. These measurements were used for temperature calibrations and to give the flux impinging on the bottom of the barrier (J_o). Weight-loss measurements on cells with porous barrier lids gave the flux emerging from the barrier (J_e). The transmission probabilities are

$W = J_e/J_o$. For NaCl, porous barriers such as those used here shift the monomer–dimer equilibrium to yield only traces of dimer (*14*). Similar behavior was found for NaF and LiF. Therefore, both orifice and porous barrier fluxes are expressed in terms of equivalent monomer units–$J_m + 2J_d$, where J_m is the monomer flux and J_d is the dimer flux.

The vacuum chamber used for these experiments was the Knudsen cell chamber of a Nuclide model HT-12-60 mass spectrometer. The temperature dependence of the effusing vapors could be determined using the mass spectrometer. In a mass spectrometer, ion intensity, I, and temperature, T, are related to flux by $J = gIT/\sigma$ (*20*). Here g is the machine constant and σ is the ionization cross section. Thus, when ion intensities are measured as a function of temperature, a plot of ln IT vs $1/T$ yields an enthalpy from the slope. For Knudsen effusion, this enthalpy is the enthalpy difference between the vapor and the condensed phase. If surface diffusion is the dominant path for vapor escape from a cell, the corresponding quantity calculated from the slope is the enthalpy difference between the transition state complex for surface diffusion and the condensed phase.

RESULTS AND DISCUSSION

Tortuosity factors have been found to have experimental values that range from 1 to 10. For the ~1 mm thick alumina barrier with 1 μm pores, these limiting values of τ give predicted transmission probabilities of 1.3×10^{-3} to 1.3×10^{-4}. Four runs with He yielded $W = (6.51 \pm 0.86) \times 10^{-4}$ or $\tau = 2.0$. Other noncondensible gases gave similar results; for Ar, N_2, O_2, and CO_2, $\tau = 1.9$; for SO_2, $\tau = 2.0$; for NH_3, $\tau = 2.1$; for CH_4, $\tau = 2.4$. Average deviations for repeated measurements with the same gas were 15% except for CH_4 for which the average deviations were 23%. Gases for which surface diffusion is important should show larger transmission probabilities than that of He. Within the scatter in the data, transmission of each of these gases can be accounted for by Knudsen flow alone.

Table 1 shows that transmission coefficients for the three alkali halides through the alumina barrier with 1.2 μm pores, unlike those for the gases, exceeded the transmission coefficient for He by factors of 2 or more. The increase in transmission coefficients over that for He presumably must be caused by contributions from surface diffusion.

Table 2 shows data obtained for zinc under three different sets of circumstances. Data reported under Ia are data collected under the same conditions as those reported for the alkali halides in Table 1. Data reported under heading Ib were made with the same apparatus and the same porous alumina, but later in time and with complete recalibrations of the apparatus for gas transmission and of that for studies of high temperature vapors. The He calibration is in reasonable agreement with that first obtained; 11 runs give an average transmission coefficient that is

TABLE 1
Transmission Probabilities for Helium and Alkali Halide Vapors[a]

Gas or vapor	Temp. K	No. of measure-ments	Av. incident flux, mol/cm²-s	Av. trans-mitted flux, mol/cm²-s	Av. trans-mission prob.,[b] W	W/W_{He}
He	295	4	—	—	$(6.51 \pm .86) \times 10^{-4}$	1.0
NaCl	1001	6	1.77×10^{-5}	3.00×10^{-8}	$(1.69 \pm .23) \times 10^{-3}$	2.6
NaF	1097	4	1.69×10^{-6}	6.61×10^{-9}	$(3.91 \pm .43) \times 10^{-3}$	6.0
LiF	1074	3	4.39×10^{-6}	3.49×10^{-8}	$(7.95 \pm .68) \times 10^{-3}$	12.2

[a]Uncertainties are standard deviations.

TABLE 2
Transmission Probabilities for Helium and Zinc Vapors[a]

Gas or vapor	Temp. K	No. of measure-ments	Av. incident flux, mol/cm²-s	Av. trans-mitted flux, mol/cm²-s	Av. trans-mission prob.,[b] W	W/W_{He}
			Alumina Ia with 1.2 ± 0.2 μm pores			
He	295	4	—	—	$(6.5 \pm 0.8) \times 10^{-4}$	1
Zn	630	5	7.62×10^{-6}	6.16×10^{-9}	$(8.09 \pm 1.45) \times 10^{-4}$	1.2
			Alumina Ib with 1.2 ± 0.2 μm pores			
He	295	11	—	—	$(8.0 \pm 1.3) \times 10^{-4}$	1
Zn	647	9	1.81×10^{-5}	5.08×10^{-8}	$(2.81 \pm 0.67) \times 10^{-3}$	3.5
			Alumina II with 0.14 ± 0.2 μm pores			
He	295	9	—	—	$(9.2 \pm 0.8) \times 10^{-5}$	1
Zn	647	7	1.81×10^{-5}	3.04×10^{-8}	$(1.68 \pm 0.42) \times 10^{-3}$	18

[a]Uncertainties are standard deviations.

24% higher than that obtained in the first set of runs. But the ratio of the zinc transmission coefficient to the He transmission coefficient was found to be 3.5 instead of 1.2.

We have been unable to identify the origin of this difference. It must be associated with a change in the high temperature apparatus or in the calibration. The most probable change that could account for the difference would be either a leak that increased the flux of zinc in the second apparatus or a difference in the temperature calibrations.

The runs reported under II were carried out with the alumina of smaller cross-sectional pore area in order to settle the question of whether or not zinc surface diffusion is significant in pores of the order of 0.1–1 μm cross section. In this barrier the He transmission was reduced by almost an order of magnitude, as expected from the change in average pore dimensions. Transmission for zinc was more than 18 times that for He. From the ratio of surface to Knudsen flux in the 0.14 μm barrier, 17, and the assumption that the ratio of surface to Knudsen fluxes should vary with l/r the predicted surface flux contribution in 1.2 μm pores is found to be 2.0. It can be concluded, therefore, that surface diffusion makes an important contribution to the transmission of zinc vapor in alumina pores of 1 μm or less.

Below the melting points of LiF (Fig. 4) and NaCl (Fig. 5), the temperature dependences of the enhanced transmission through porous barriers were found to be essentially the same as the temperature dependences for effusion. Above the melting point of LiF the shapes of log intensity vs $1/T$ plots for effusion through an orifice were slightly reduced, as expected because the heat of vaporization of the liquid is less than that of the solid. In contrast, above the melting points, temperature dependence of the intensities measured through porous barriers increased and became erratic. Figure 4 shows a discontinuity at the melting point. Another run made in an alumina cell that had not yet been penetrated by the LiF simply showed an increase in slope at the melting point, but then showed an upward discontinuity at a higher temperature.

If data had been collected only below the melting point, it might have been assumed that the rate-limiting process for vapor transport through the barrier is deposition of an adsorbed film on the inner barrier surface. But if that process were rate limiting, the temperature dependence of the transmission coefficient should remain the same as the temperature dependence of vaporization above the melting point as well as below. Thus, there is clearly a change in transport mechanism at the melting point.

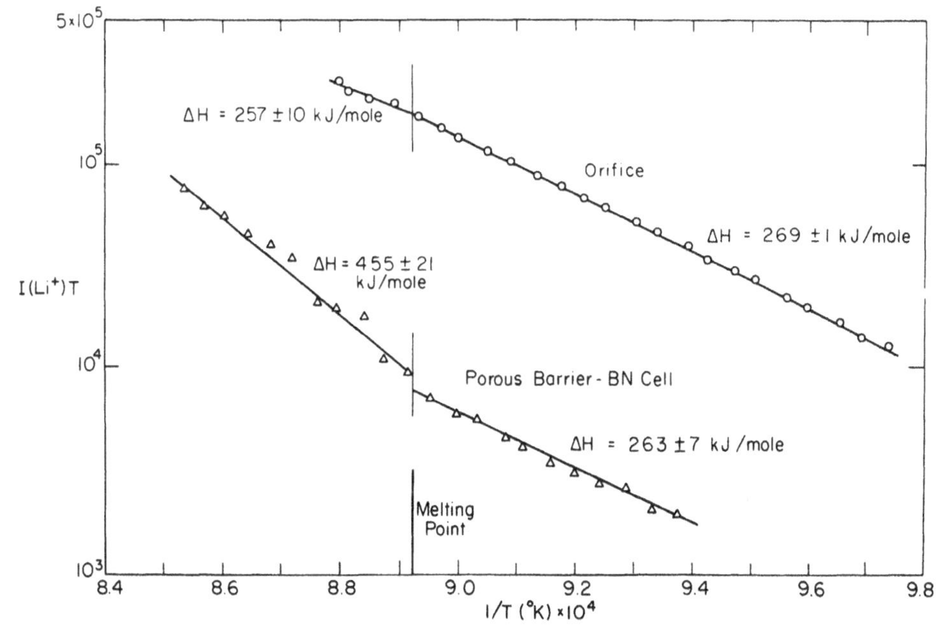

Fig. 4. Plots of ln IT vs $1/T$ for LiF through a porous barrier and an orifice.

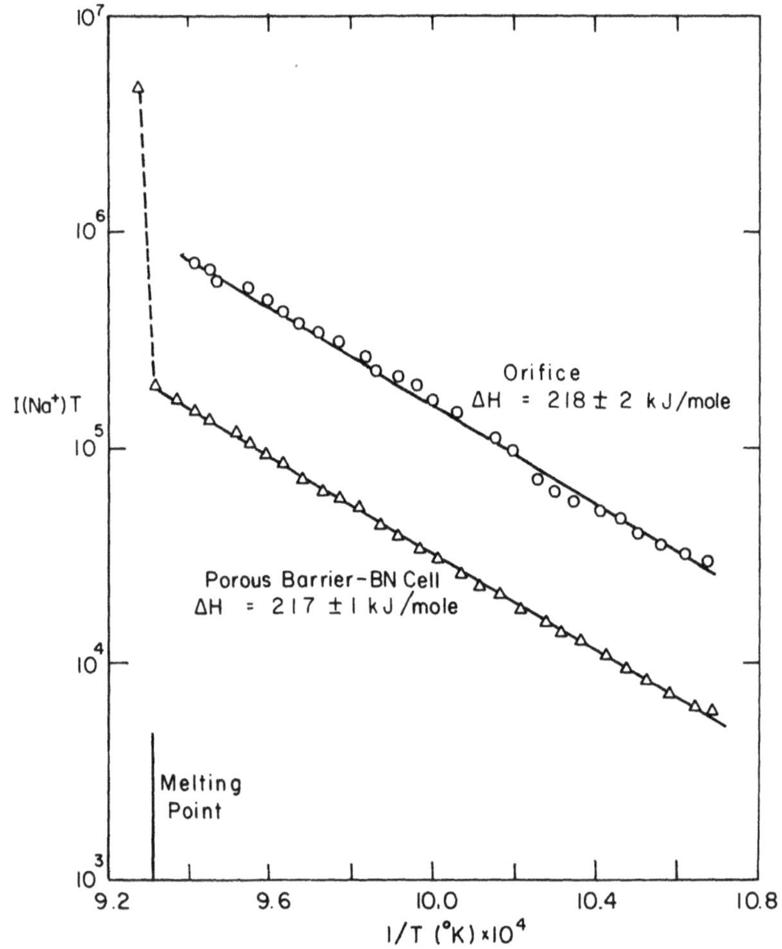

Fig. 5. Plots of ln IT vs $1/T$ for NaCl through a porous barrier and an orifice.

It seems probable that the rate-limiting process below the melting point is surface diffusion in a film of < 1 monolayer coverage and with the activated complex of the diffusional step being so loosely bound that its molecular properties approach those of the free vapor molecules. Above the melting point creep of a liquid-like film may become an additional path for partial penetration of the barrier. The distance to which the film penetrates before the alkali halide pressure gradient makes a liquid-like adsorbed film unstable would increase with temperature and would be sensitive to changes in the temperature gradient through the barrier. The rate-limiting step of surface transport thus would be diffusion in the same kind of submonolayer state that governs surface transport below the melting point, but over distances that are shortened by penetration of the liquid-like film. Further studies in which the thickness of a barrier and the temperature gradient through it are deliberately varied are planned to test this line of reasoning.

We can conclude from our data that surface diffusion would not significantly influence effusion of the four *solids* studied here through long orifices of 0.1 mm or greater diameter in alumina or similar oxides (though grain boundary penetration is serious for LiF), but we do not know the effect of channel length. In the interpretation of porous barrier studies, surface diffusion and Knudsen flow have usually been interpreted as parallel and essentially independent processes. Such a model makes the relative importance of surface and Knudsen diffusion independent of channel length. Winterbottom and Hirth's model treat surface diffusion as directly dependent upon the local vapor flux. This model makes contribution of surface diffusion important only for very short channels. We suspect that the model that will reproduce the behavior of most vapors in effusion cells will be one that modifies the independent flux model of the porous barrier literature to account for the kinds of end effects that Winterbottom and Hirth evaluated.

ACKNOWLEDGMENTS

Special thanks are due to Dr. David Meschi for his frequent advice in this study. Thanks are also due to the Lawrence Berkeley Laboratory ceramics shop for help in fabricating the porous barriers. This work was supported by the Director, Office of Energy Research, Office of Basic Energy Sciences, Materials Sciences Division of the US Department of Energy under Contract No. DE-AC03-76SF00098.

REFERENCES

1. Paule, R. C., and Margrave, J. L., in *The Characterization of High Temperature Vapors*, Margrave, J. L., ed., Wiley, New York, 1967, Chap. 6.
2. Cater, E. D., The Effusion Method at Age 61: Current State of the Art, NBS Special Publication 561, 1979.
3. Ward, J. W., and Fraser, M. V., *J. Chem. Phys.* **50,** 1877 (1969).
4. Huang, M., Meschi, D. J., and Searcy, A. W., unpublished work.
5. Winterbottom, W. L., and Hirth, J. P., *J. Chem. Phys.* **37,** 784 (1962).
6. Winterbottom, W. L., *J. Chem. Phys.* **47,** 3546 (1967).
7. Dunham, D. L., and Hirth, J. P., *J. Chem. Phys.* **49,** 4650 (1968).
8. Ward, J. W., Bivins, R. L., and Fraser, M. V., *J. Vac. Sci. Technol.* **7,** 206 (1970).
9. Ward, J. W., Fraser, M. V., and Bivins, R. L., *J. Vac. Sci. Technol.* **9,** 1056 (1972).
10. Boyer, A. J., and Meadowcroft, T. R., *Trans. Met. Soc. AIME* **233,** 388 (1965).
11. Grimley, R. T., Wagner, L. C., and Castle, P. M., *J. Appl. Phys.* **46,** 4090 (1975).
12. Ash, R., Barrer, R. M., Clint, J. H., Dolphin, R. J., and Murray, C. L., *Roy. Soc. (London) Phil. Trans.* **275,** 255 (1975).
13. Gilliland, E. R., Baddour, R. F., and Russell, J. L., *AIChE J.* **4,** 90 (1955).
14. Mohazzabi, P., and Searcy, A. W., *J. Chem. Phys.* **65,** 5037 (1977).

15. Satterfield, C. N., *Mass Transfer in Heterogeneous Catalysis,* MIT Press, Cambridge, Mass., (1970), p. 41.
16. Nicholson, D., and Petropoulous, J. H., Capillary Models for Porous Media with Special Reference to Gas Flow, in *Pore Structure and Properties of Porous Materials,* Proceedings of Internal Symposium RILEM/IUPAC, Prelim. Rpt. Part I, 1973, p. A-92.
17. Dushman, S., *Scientific Foundations of Vacuum Technique,* Wiley, New York, 1955, p. 98.
18. *Op. cit.,* Satterfield, pp. 35–41.
19. Searcy, A. W. The Kinetics of Evaporation and Condensation Reactions, in *Chemical and Mechanical Behavior of Inorganic Materials,* Searcy, A. W., Ragone, D. V., and Colombo, U., eds., Wiley Interscience, New York, 1970, Chap. 6.
20. Grimby, R. T., in *The Characterization of High Temperature Vapors,* Margrave, J. L., ed., Wiley, New York, 1967, Chap. 8.

0018–1536/84/1700–0065$04.00

Electron Impact Spectroscopy of High Temperature Species

S. TRAJMAR

Jet Propulsion Laboratory, California Institute of Technology, Pasadena, CA 91109 and Physics Department, University of Southern California, Los Angeles CA 90089

Received September 23, 1983; Accepted December 31, 1983

ABSTRACT

The application of the methods of electron impact spectroscopy to the study of high temperature species is discussed. Electron scattering by metal vapors (Ba, Bi, Cu, Mn, Tl, Zn), alkali halide molecules, and KOH are described as demonstrative examples.

Index Entries: Metal vapors, electron impact spectroscopy of; cross-sections, of high temperature species by electron impact spectroscopy; electron impact spectroscopy, of high temperature species; barium, electron impact spectroscopy of; bismuth, electron impact spectroscopy of; copper, electron impact spectroscopy of; manganese, electron impact spectroscopy of; thallium, electron impact spectroscopy of; zinc, electron impact spectroscopy of; alkali halides, electron impact spectroscopy of; potassium hydroxide, electron impact spectroscopy of.

I. INTRODUCTION

Information about the energy levels of atoms and molecules have been traditionally obtained by the methods of optical spectroscopy. The wavelength dependence of photabsorption or emission yields the desired data. It has been shown, however, in more recent years that electron impact spectroscopy can yield the same and, under certain conditions, additional information about atomic and molecular energy states. An additional benefit of electron impact studies is that optical (dipole) selection rules do not apply to electron collision processes and, therefore, optically forbidden excitations can readily take place. When the electron scattering techniques are carried out in a quantitative manner, the scattering intensities can be transformed into cross-sections that represent the time-independent probabilities for the appropriate electron impact process under the specific experimental conditions. Cross-sections for excitation by electron impact may be equivalent to photoabsorption cross-sections (high-energy, low-angle scattering), or may be very different (low-energy, high-angle scattering).

Since electron scattering experiments most frequently utilize electron beam–atomic (molecular) beam configurations, the techniques of electron impact spectroscopy are applicable to any species that can be formed into a beam. In the study of high temperature species some complications are introduced in producing the target beam, but otherwise the same techniques apply as in the case of room-temperature gases.

In this paper we give a brief background on the methods of electron impact spectroscopy, point out the differences and similarities of the optical and electron impact techniques, and demonstrate the application of electron impact spectroscopy to high temperature species by discussing electron scattering by metal vapors and alkali halide molecules.

II. METHODS OF ELECTRON IMPACT SPECTROSCOPY

In electron impact spectroscopy the excitation of atoms or molecules is caused by electron collisions and the information concerning the excitation process is obtained predominantly from the scattered electrons. The excitation takes place at the expense of the kinetic energy of the electrons. The electron scattering process is schematically depicted in Fig. 1. An energy-selected (nearly monoenergetic) electron beam is focused on a target with the desired impact energy (E_o). Electrons scattered by an an-

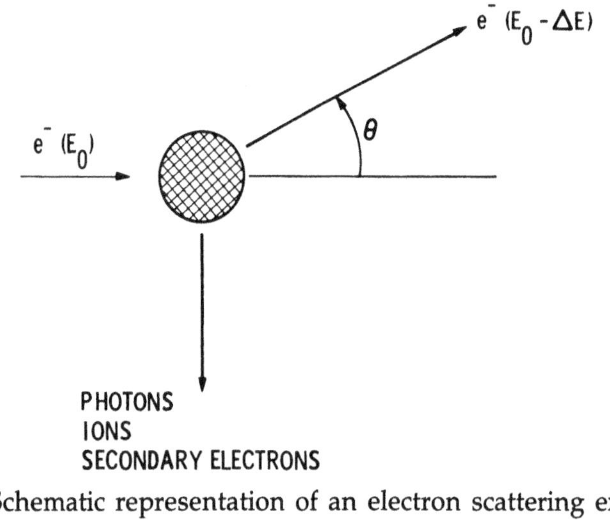

Fig. 1. Schematic representation of an electron scattering experiment.

gle (Θ) with respect to the incoming electron beam direction are detected over a small solid angle (~10^{-3} sr) and are energy analyzed. The energy and angular distribution of the scattered electrons represent information concerning the various scattering events. Electron impact may lead to a large variety of processes; elastic scattering, rotational, vibrational, and electronic excitation, superelastic scattering, ionization, dissociation, electron attachment, and so on. As a result of the electron collision, secondary particles (photons, electrons, ions and neutral fragments) may be produced. These secondary species can be detected individually or in various combinations of coincidence schemes to yield additional information about the scattering process.

A schematic diagram of an electron impact spectrometer is shown in Fig. 2. The basic components are the electron gun, electron detector, and target beam source. Electrons are extracted from a thermoionic emitter, collimated, and focused into an energy selector. A narrow section (~10 meV) of the (≈ 0.5 eV wide) electron energy distribution is selected out from the original electron beam by the energy selector. This nearly monoenergetic electron beam is then focused into the target region. The scattered electrons are collected over a small solid angle by the detector. The scattered signal is transported into the energy analyzer and then to the charged particle detector. Here each electron generates a pulse that in turn is fed into a multichannel scaler. The signal (pulse counts) as a function of some parameter (energy loss, impact energy, scattering angle) is recorded by the multichannel scaler, usually by repetitive scanning.

For the beam transport and electron energy selection and analysis, electrostatic lenses and energy selectors are generally used. The compo-

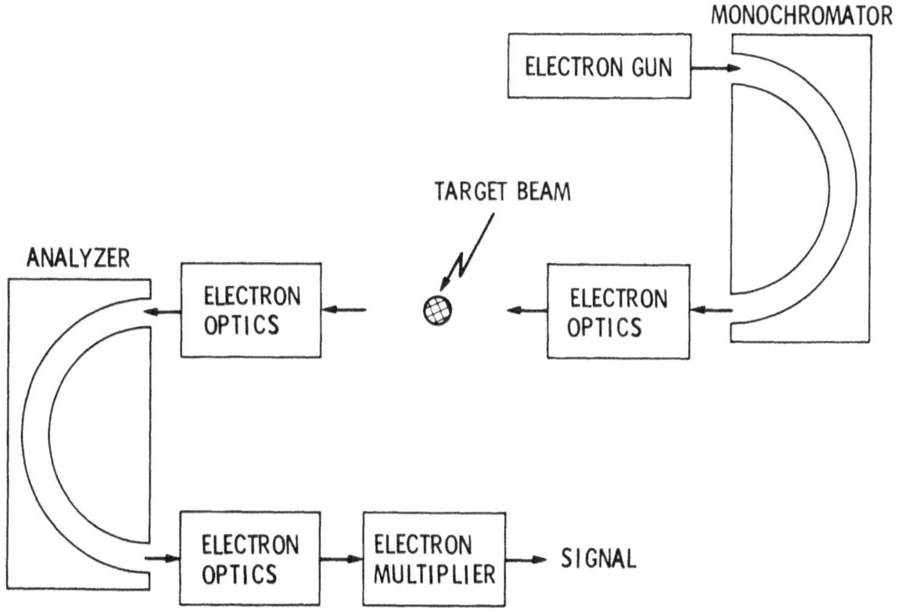

Fig. 2. Schematic diagram of an electron impact spectrometer.

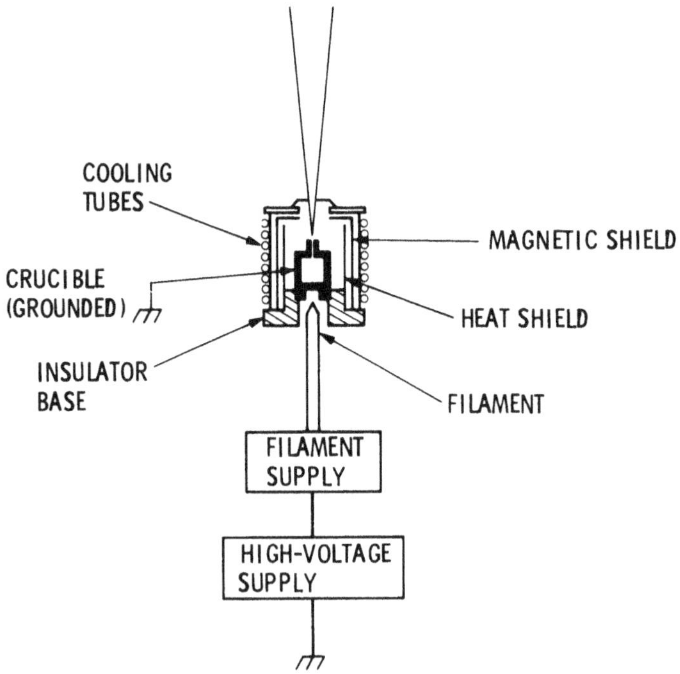

Fig. 3. A high-temperature vapor beam source.

nents of the spectrometer are enclosed in a vacuum chamber and the electron path is carefully shielded from electric and magnetic fields.

The target source is most frequently a beam generated by various effusion methods. A typical high-temperature source is shown in Fig. 3.

A more detailed discussion on the electron impact techniques can be found in refs. (*1*) and (*2*).

The spectrometer can be operated in a number of ways. In the most commonly used mode, the "energy-loss" mode, the electron impact energy and the scattering angle (position of the detector) are fixed and the scattering signal is recorded as a function of energy lost by the electron (ΔE). One obtains this way an energy-loss spectrum. Figure 4 shows typical spectra for He obtained at 20 eV impact energy and scattering angles of 30° and 135°(*3*). The energy-loss region corresponds to the excitation of the four lowest electronic states of He. The elastic peak appears far to the left at zero energy loss, while features corresponding to excitation of higher energy states and to ionization appear on the right (none of these are shown here). The location of the peaks represent the energy level scheme of the target and the scattering intensities are proportional to the corresponding cross-sections.

Other frequently used methods of operation are the "impact energy" mode and the "constant residual energy" mode. In the former case the scattering angle and the energy loss (excitation channel) are kept constant and the impact energy dependence of the excitation is studied. This mode is used to measure the energy dependence of cross-sections and to study resonances. In the constant residual energy mode, the scattering angle and the residual energy of the detected electrons are kept constant. The impact energy is varied, but the analyzer is continuously adjusted to detect only electrons with the same fixed residual energy. This mode of operation is desirable at impact energies very near threshold, where the scattered electrons leave with low kinetic energy and the detector sensitivity is a strong function to this energy.

III. COMPARISON WITH OPTICAL SPECTROSCOPY

An energy-loss spectrum is similar to an optical absorption spectrum in general appearance. Closer examination of the spectrum, however, reveals significant differences. The most important difference is that in the optical absorption spectra only electric dipole-allowed transitions appear with significant probabilities. This is not the case for the energy-loss spectra. In Fig. 4, e.g., besides the optically allowed 2^1P excitation peak,

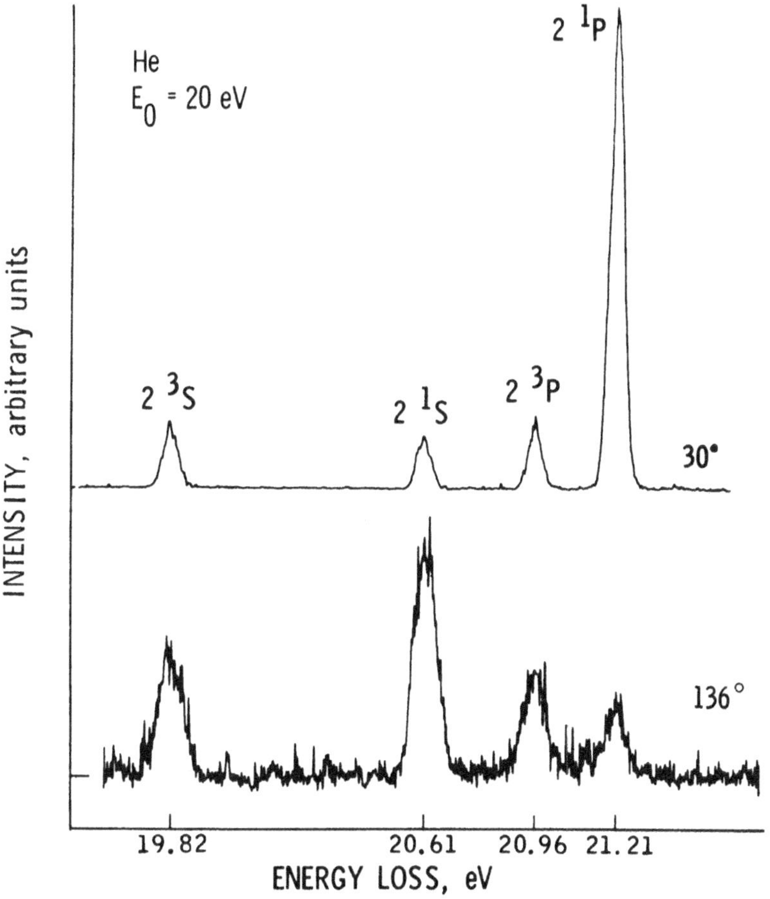

Fig. 4. Electron impact energy-loss spectra of He at 20 eV impact energy and 30° and 136° scattering angles.

the spin and/or symmetry forbidden 2^3S, 2^1S, and 2^3P excitation peaks also appear with comparable intensity. (In optical spectra, the 2^3P excitation probability is about seven orders of magnitude smaller than that of the 2^1P state.) The situation can be briefly summarized as follows. At high electron impact energies and small scattering angles (small momentum transfer), electrons excite just as do photons, and it can be shown that in the zero momentum transfer limit, the two spectra become equivalent (4). As the impact energy is lowered and the scattering angle is increased (larger momentum transfer), the energy-loss spectrum begins to deviate drastically from the optical spectrum and the spin and symmetry-forbidden transitions appear with probabilities comparable to those that are optically allowed. (See 136° spectrum in Fig. 4.) Selection rules based on group theoretical arguments can be derived for electron impact excitation, but they are applicable only under certain conditions (e.g., at certain scattering angles for arbitrary orientation of the molecule or for any scattering angle, but for specific orientation of the molecule) (5). The case in which electrons can excite optically forbidden levels has been utilized to identify molecular energy states that could not be observed by optical methods and has also been exploited to generate metastable atomic and molecular species.

Additional differences between the two spectroscopic methods are the resolution and the spectral range of coverage. Optical techniques are capable of much higher resolution in the infrared and visible regions, but in the far UV and X-ray regions, electron impact techniques yield better resolution. This occurs because the scattered electrons are slowed down and always energy-analyzed at the same low energy in the electrostatic analyzer. As far as the spectral range of coverage is concerned, optical spectroscopy requires very different techniques in various spectral regions, while in electron impact spectroscopy the whole spectrum from the IR to the X-ray region can be obtained in one single scan with the same apparatus. It should also be pointed out that in optical absorption the photon disappears, whereas in electron collision the scattered electron survives and yields additional information about the excitation process through the energy and angular dependence of the excitation cross section.

The parameter that quantitatively characterizes the electron impact process is the cross-section, which is equivalent to the photoabsorption (or photoionization) cross-section (or equivalently to the optical *f*-value, absorption coefficient, etc.) in optical spectroscopy. The electron impact cross-section, as mentioned above, is energy- and angle-dependent. The differential (with respect to angle) excitation cross-section for a particular electron impact excitation process "*n*" at a given impact energy is denoted by

$$DCS_n(E_o, \theta) \equiv [d\sigma/d\Omega]_{E_o}$$

and has the dimension of area per solid angle. Integration over all solid angles yield the integral cross-section for process *n*. Differential cross-sections with respect to energy loss for continuum transitions, and differential cross-sections with respect to the energy and angle of secondary particles in coincidence experiments, can also be defined.

IV. ELECTRON SCATTERING BY METAL ATOMS

Metal atoms represent a large variety of atomic structures in terms of the arrangements and coupling of electrons in various shells. Electron scattering studies can shed light on these structures and on the nature of electron–electron interactions both within the target atom and between the atom and scattering electron. With the exception of Hg, the electron impact study of metal atoms require elevated temperatures. Some electron scattering measurements were performed in the past on more volatile elements (alkalis, alkali earths, Hg, Zn, Cd), but very little or no information is available on the less volatile metals. Here we demonstrate through several examples the application of electron impact techniques to metals and discuss some of the pertinent points that can be deduced from such investigations.

The energy-loss spectrum of Mn was studied by Williams et al. (*6*). A typical spectrum obtained at 10 eV impact energy and 70° scattering angle is shown in Fig. 5. The ground-state electron configuration of Mn is $1s^2\ 2s^2\ 2p^6\ 3s^2\ 3p^6\ 3d^5\ 4s^2$ and it corresponds to a $^6S_{5/2}$ state. Excitation to other sextet, as well as to spin and symmetry forbidden quartet and octet states, occurs with large cross-sections. Many of these excitations occur with higher probability than elastic scattering, which is not the case with atomic and molecular gaseous species in general. This behavior can most likely be attributed to the partially unfilled subshell and to the complex electronic structure and is characteristic of heavy metal elements. In Fig. 6 the angular dependence of elastic scattering and of excitation cross-sections for the $y\ ^6P$, $a\ ^4P$, and $Z\ ^8P$ states is shown. The general pattern (well-established in the case of singlet-to-singlet and singlet-to-triplet excitation processes) holds also for the more complicated spin states of Mn. Excitations that can proceed by direct process are strong, and the DCS is strongly forward-peaked. Excitations that require spin exchange show nearly isotropic or slightly forward-peaking angular behavior. This can be understood on the basis of the electron–atom interaction potential responsible for the excitation process. For spin exchange the short-range interaction potential is important; the scattering electron has to come close to and has to intermingle with the atomic electrons. The electron that leaves the collision complex can leave with a spin different from that of the incoming electron and with nearly equal probability to any direction. (Change in spin multiplicity can also occur through spin–orbit inter-

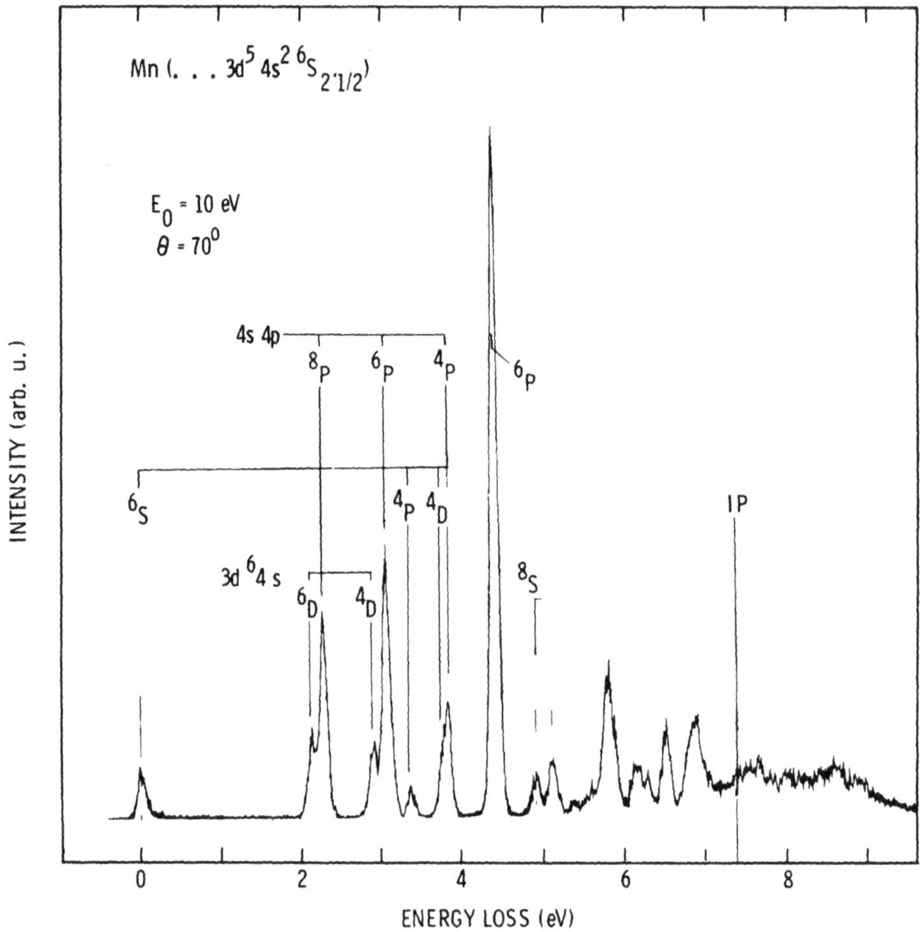

Fig. 5. Energy-loss spectrum of Mn at 10 eV impact energy and 70° scattering angle. The energy level designations are those of ref. (*21*).

action, but if this interaction is strong, the spin and orbital angular momentum quantum numbers are no good quantum numbers.) Direct excitation processes, on the other hand, are associated mainly with long-range collisions that, on the basis of the classical scattering picture, correspond to large impact parameters and to small scattering angles.

An extensive study of Cu was reported by Trajmar et al. (*7*) and an energy-loss spectrum is shown in Fig. 7. The particular interst in Cu arose because of its promising role as a laser medium for high-efficiency, high-power electric discharge or electron beam-excited lasers. In a three-level laser scheme, the upper laser levels (the ^2P states) are excited by electron impact and are stimulated to decay to the intermediate ^2D states. A large fraction of the invested excitation energy is regained in the lasing and this is the basis of high-efficiency. Another crucial question is the efficiency by which electron impact can bring about the population inversion (large cross-section for the upper and small cross section for

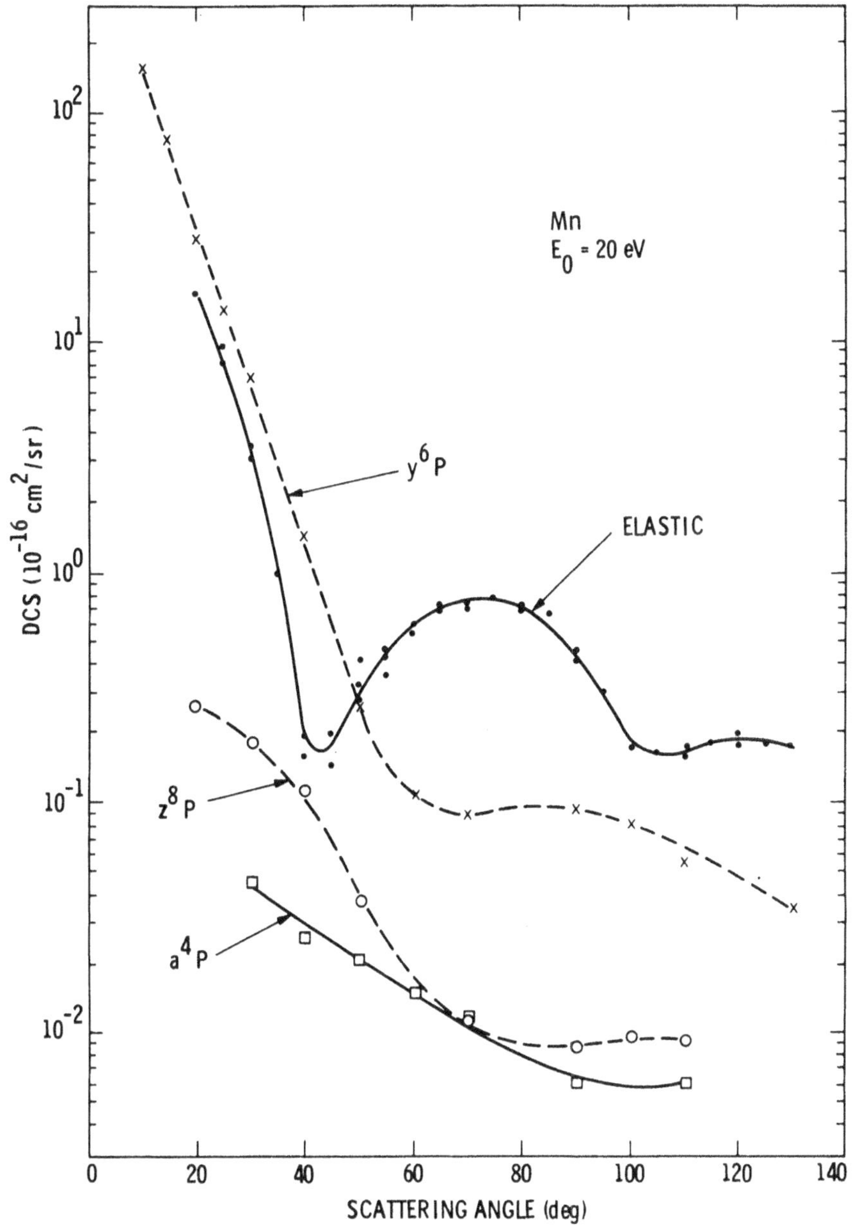

Fig. 6. Differential cross-sections for elastic scattering and for excitation of the levels indicated for Mn at 20 eV impact energy. The energy level designations are those of ref. (*21*).

the lower laser levels). As Fig. 7 shows, this requirement is well-satisfied for Cu and indeed it turned out to be an excellent laser.

An electron impact spectrum of Bi is given in Fig. 8 (*8*). One can immediately recognize that, from a lasing point of view, Bi is not nearly as suitable as Cu and no efficient lasing was found in this case. It is interesting to note that, besides the sharp atomic lines, broad features representing the excitation of Bi_2 molecule are also present in the spectrum.

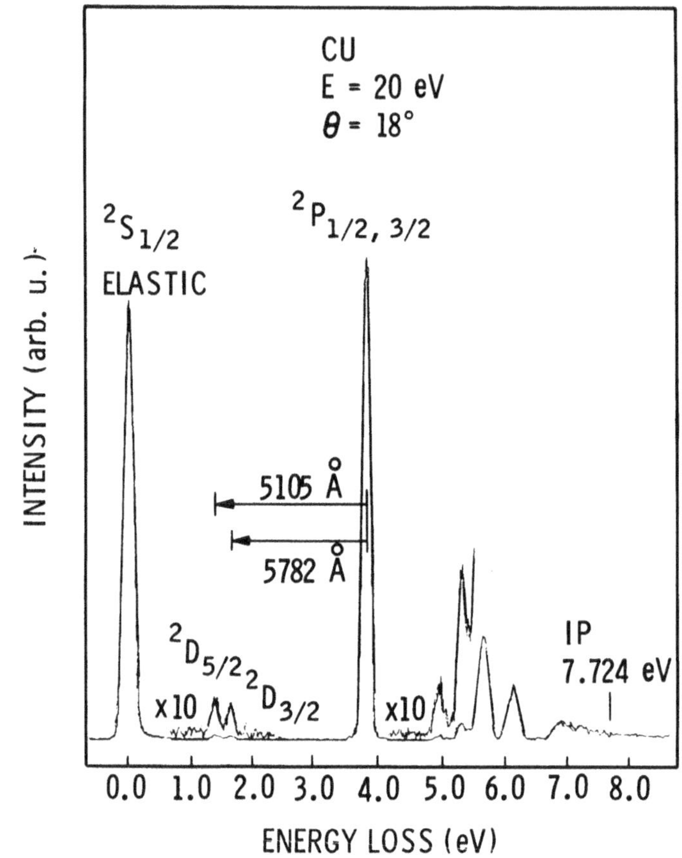

Fig. 7. Energy-loss spectrum for Cu at 20 eV impact energy and 18° scattering angle. The wavelengths for the laser transitions are indicated. The energy level designations are those of ref. (*21*).

The case of Ba is a good example of a large variety of electron impact processes. The electronic shell structure of Ba is given as . . . . $4p^6\ 4d^{10}\ 5s^2\ 5p^6\ 6s^2$, corresponding to a ^1S ground state. In Fig. 9 an energy loss spectrum is shown (*9*). The features below 5.12 eV (the first ionization threshold) correspond to various excitations of one of the 6*s* electrons. At 5.2 eV this electron is excited to the continuum and Ba^+ (. . . $5p^6 6s$; ^2S) is produced. Beyond this point up to 15.2 eV (the threshold for ionization of Ba^+), the smooth background represents the first ionization continuum of Ba and the discrete features that may appear in this region correspond to superexcited, autoionization states for which both 6*s* electrons are simultaneously excited. At 15.2 eV both 6*s* electrons can be simultaneously excited into the continuum and Ba^{2+} is produced. Between 15.2 and 21.2 eV the spectral features result from the excitation of the 5*p* subshell. Above 21.2 eV two electron excitations involving 5*p* or 5*s* electrons are found and at around 110 eV the 4*d* shell excitation shows up. Excitation and the subsequent autoionization of states that are above the first ionization limit and, therefore, overlap with one or more ionization

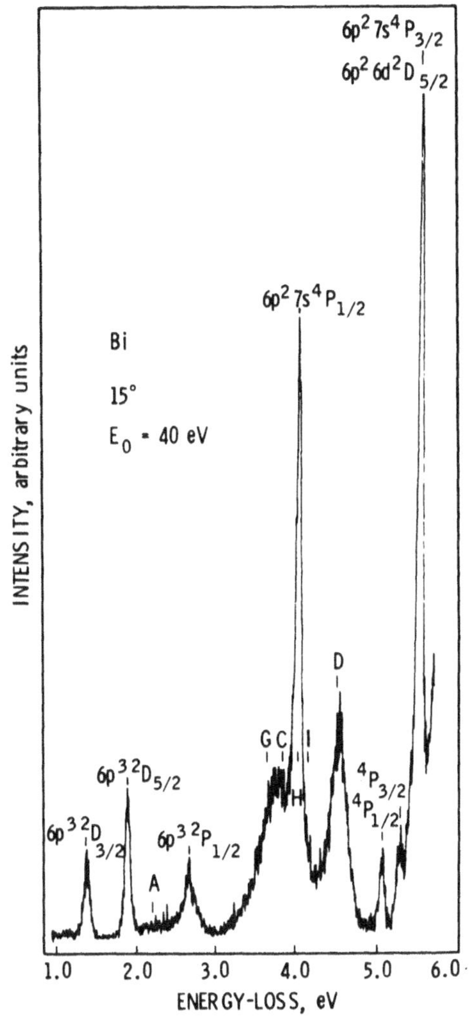

Fig. 8. Energy-loss spectrum of Bi at 40 eV impact energy and 15° scattering angle. The designation of atomic levels are those of ref. (*21*). The letters are designations for molecular Bi_2 energy levels [see ref. (22)].

continua, lead to an interference in the energy loss spectrum. (The superexcited states may also decay by photon emission, especially in the heavier elements.) The features that result from this interference generally have a Beutler-Fano resonance profile. In case of heavy elements the ionization continuum is weak compared to the autoionization features and the resonance shape looks very nearly like a regular energy-loss peak. The energy-loss spectrum clearly shows that the major contribution to the overall ionization comes from autoionization and not from direct ionization in the case of heavy metals. It is interesting to note that the very strong intra- and inter-shell couplings in these heavy elements represent a great challenge to theory. The one-electron excitation picture breaks down and no satisfactory theory has been developed yet to represent these complex interactions and the energy level structures.

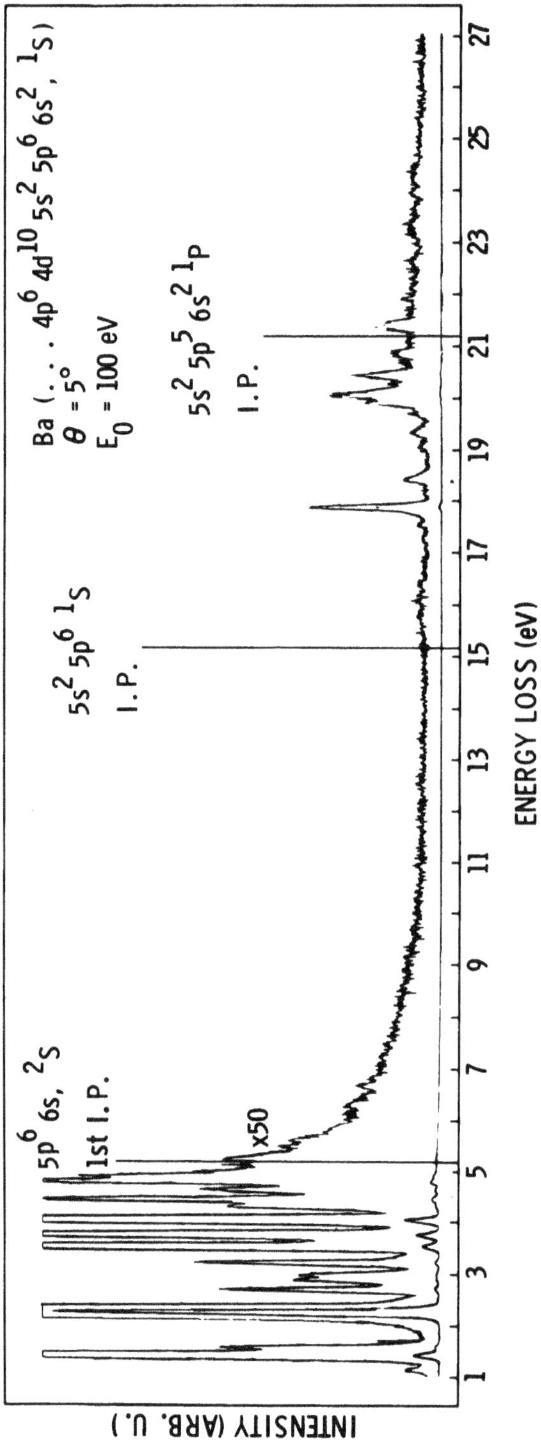

Fig. 9. Energy-loss spectrum for Ba at 100 eV impact energy and 5° scattering angle.

The energy-loss spectrum of Tl near and above the first ionization limit (*10*) is shown in Fig., 10. Again the strong autoionization features are present. The purpose of showing this figure is, however, to demonstrate the presence of both the primary and secondary electrons that appear in the conventional energy-loss spectra. At 7.8 and 8.3 eV energy-losses the excitation of the ^2D and ^2P superexcited states occurs. These features correspond to the scattered primary electrons with $E_o - 7.8$ and $E_o - 8.3$ eV residual kinetic energy. The superexcited states decay to the Tl^+ ^2S state (which is the ground state of Tl^+) by electron ejection. The ejected electrons have 1.8 and 2.3 eV kinetic energies and appear at $E_o -$ 1.8 and $E_o - 2.3$ eV energy losses, respectively. By slightly changing the impact energy of the electrons, the energy loss corresponding to the excitation of the superexcited states remains unchanged, and so do the corresponding features in the energy loss spectrum, but the features corresponding to the ejected electrons shift. Thus the secondary electrons can

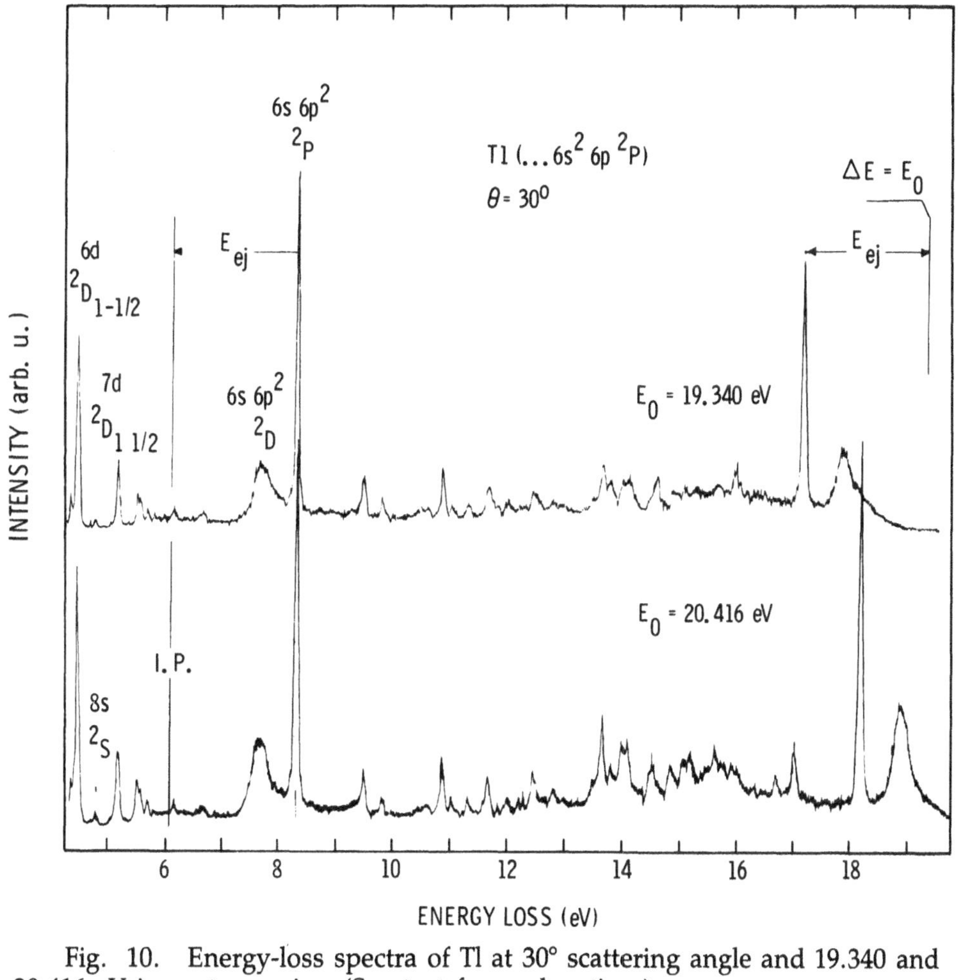

Fig. 10. Energy-loss spectra of Tl at 30° scattering angle and 19.340 and 20.416 eV impact energies. (See text for explanation.)

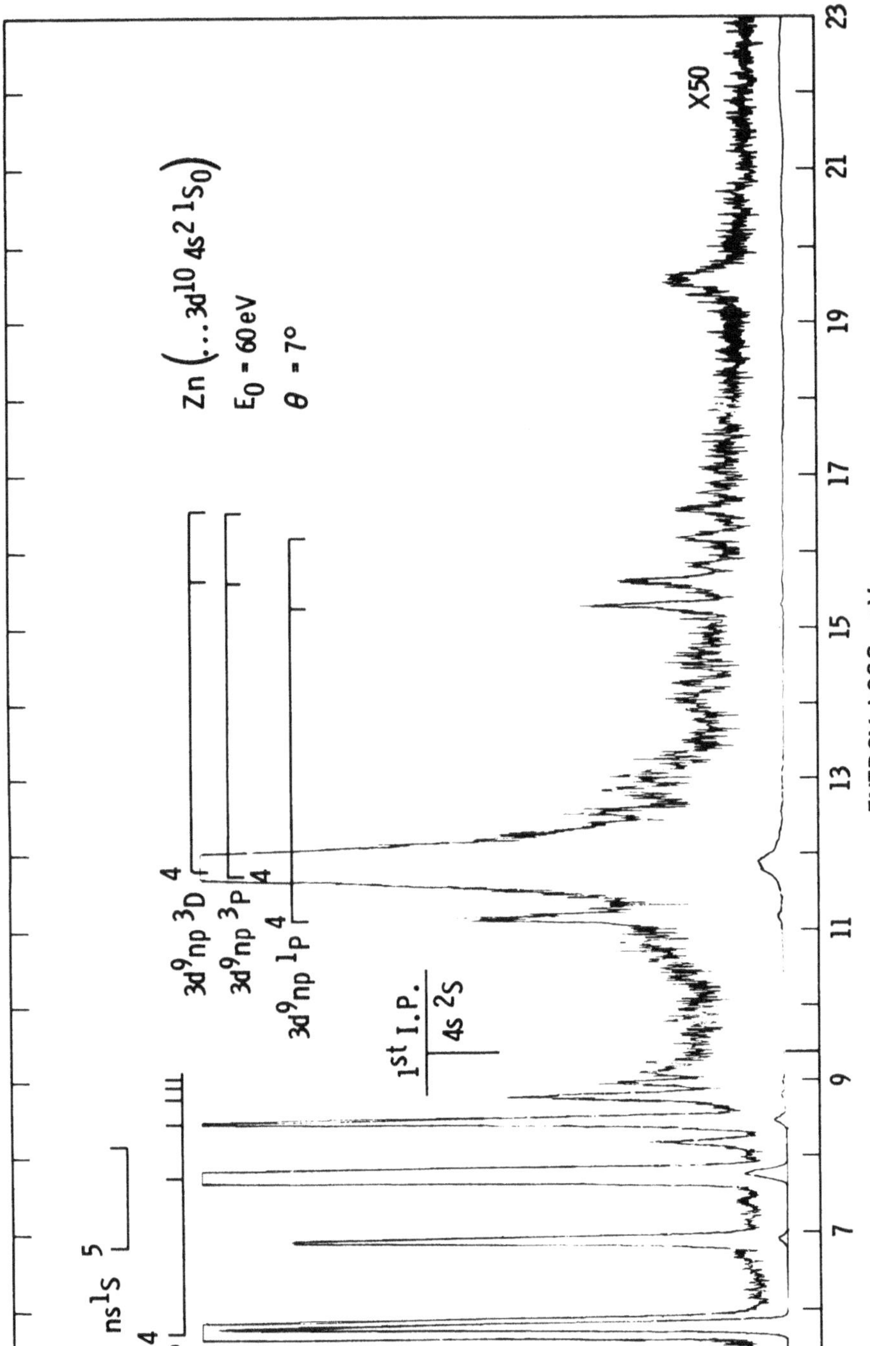

Fig. 11. Energy-loss spectrum for Zn at 60 eV impact energy and 7° scattering angle. (See text for explanation.)

be recognized. For more detailed studies concerning energy and angular correlations among primary and ejected electrons, electron–electron ($e^-/2e^-$) coincidence experiments are required.

Zinc was studied by Willims and Trajmar (*11*). The energy-loss spectrum (Fig. 11) shows the $np\ ^1P$ and $ns\ ^1S$ Rydberg excitation features and the great importance of autoionization (associated with the $3d^9\ np$ inner shell excitations) in the overall ionization process.

A discussion of the electron impact spectroscopy of metal atoms would not be complete without a few words on resonances. Resonances in electron scattering are observed when the electron is temporarily captured by the target atom for a time that is comparable to or longer than the collision time. In such cases the direct and the resonant scattering processes interfere (because of the wavelike nature of electrons) and this interference shows up in the energy dependence of cross-sections. This phenomena has been well known in nuclear physics for many years, but was discovered in low-energy electron scattering only in 1963 by Schulz (*12*). Figure 12 shows resonances appearing in the $\Delta E = 0.00$ (elastic) and $\Delta E = 4.03$ eV ($4p\ ^3P$) channels at 20° scattering angle for Zn (*11*). The resonance energies for the $4p\ ^3P$ channel and the threshold for the $4p\ ^1P$ excitation are indicated.

In recent years with the introduction of lasers into electron collision studies, a new class of experiments became possible. Here we briefly mention only the application of lasers to prepare short-lived excited spe-

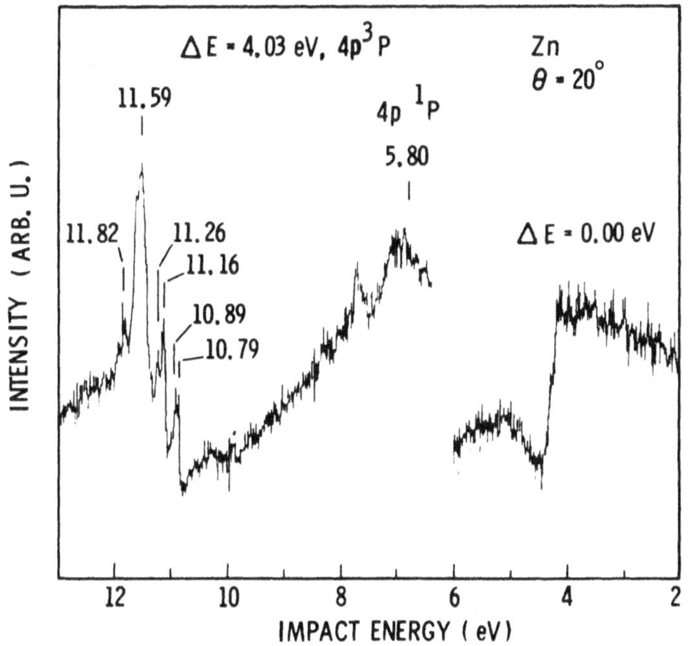

Fig. 12. Resonances for Zn at 20° scattering angle in the elastic ($\Delta E = 0.00$ eV) and $4p\ ^3P$ inelastic ($\Delta E = 4.03$ eV) channels.

cies that can be used as targets for electron collision measurements. By this method cross-sections for electron impact processes for excited species can be determined or with a slightly different view, combined photon and electron excitation processes can be studied. An example is shown in Fig. 13. The energy-loss spectrum with and without laser pumping of the $6s6p$ ^1P state of Ba is shown (*13*). When the laser excitation is present, a large number of additional features appear. To the left of the elastic peak, features are caused by superelastic scattering. Although the laser excited only the ^1P state, cascade radiation and/or electron impact de-excitation build up some population in the lower lying ^3D, ^1D, and ^3P states. All combinations of electron impact de-excitations among these and the ground state are observable. The new features ap-

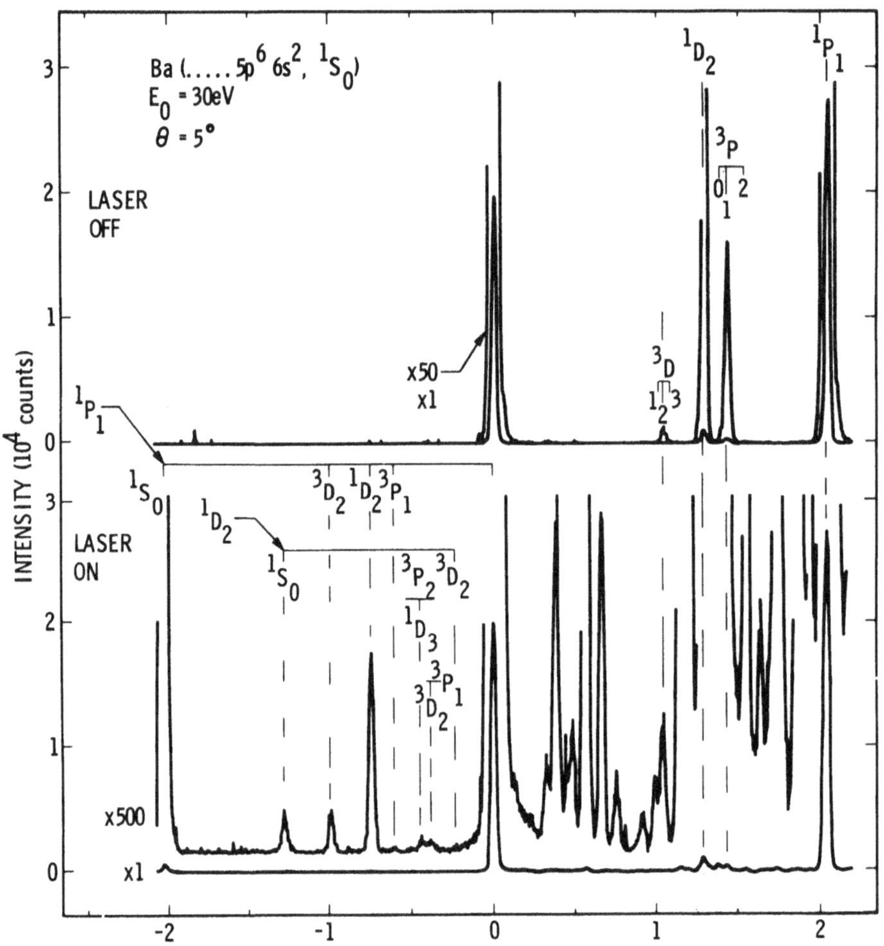

Fig. 13. The energy-loss spectra of Ba at 30 eV impact energy and 5° scattering angle. Upper spectrum (laser OFF) represents scattering by ground (^1S) state Ba, while lower spectrum (laser ON) represents scattering by a mixture of ^1S, ^3D, ^1D, ^3P and ^1P Ba. (See text for explanation.)

pearing on the right-hand side of the elastic peak correspond to excited state-to-excited state electron impact excitation (or combined photon/electron excitation) processes.

Very recently work was initiated on Na (*14*) and Ba (*15*) utilizing single-mode tunable dye lasers to populate the hyperfine level of a specific isotope. With the selection of geometry and laser polarization, the degenerate magnetic sublevel populations can also be controlled. A study of the superelastic scattering then effectively achieves an energy resolution equivalent to that of the laser linewidth (combined with natural, Doppler, and power broadening). More importantly, from these measurements magnetic sublevel excitation amplitudes (both magnitude and phase), as well as the orientation and alignment properties of the excited state prepared in the inverse electron impact inelastic process, can be determined. (The superelastic scattering experiment corresponds to the time-inverse electron–photon coincidence experiment.) These studies lead to a deeper understanding of the scattering process than conventional scattering experiments, which yield only cross-section information. Experiments along these lines have been carried out only with spin-unselected electron beams. Plans are now being made to carry out some of these scattering studies with electron spin selection both in the incoming and scattered beams. This then will represent the "perfect" scattering experiment.

V. ELECTRON SCATTERING BY ALKALI HALIDES

Alkali halides (and other highly polar molecules) constitute a special domain for electron scattering experiments. The interaction is dominated by the charge-dipole term, which is very effective in causing $\Delta J = \pm 1$ (and to lesser extent other rotational excitation processes). The cross-sections are very large (10^{-14} cm^2) at low impact energies and they are very much forward-peaked because of the long-range nature of the dominant interaction. The first substantiation of these large cross-sections was obtained from molecular beam–electron beam recoil measurements (*16*). More recently, direct electron scattering experiments yielded relative differential rotational excitation cross-sections for LiF (*17*), KI (*18*), and CsCl (*19*).

The electron scattering experiments also yielded the first information on the electronic energy levels of some of these alkali halide molecules (*18,20*).

VI. ELECTRON IMPACT INVESTIGATION OF KOH

Vušković and Trajmar (*20*) carried out an electron impact study of KOH. Although no scattering was observed by KOH molecules, the study gave an interesting demonstration of the application of electron

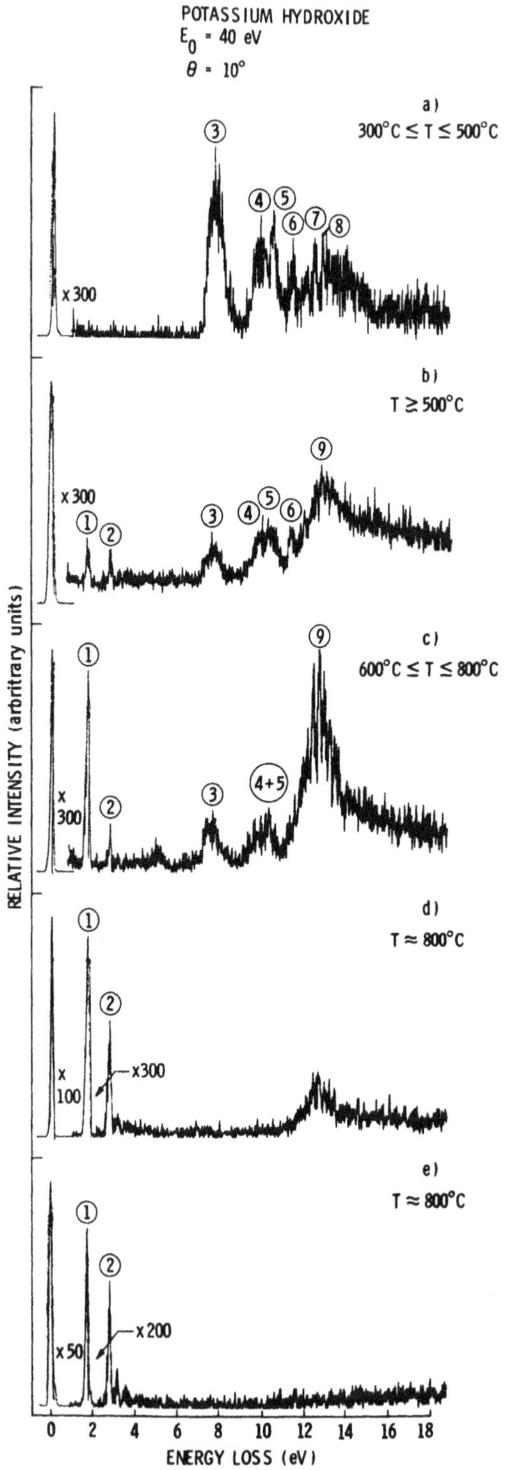

Fig. 14. Energy-loss spectra of "potassium hydroxide" at different temperatures. (See text for explanation.)

scattering as an analytical, monitoring technique. Figure 14 shows a series of energy-loss spectra obtained at increasing temperatures of the KOH oven. The spectra also represent a time sequence.

The KOH oven was degassed for several hours at 200°C. Then the temperature was raised gradually to 500°C. The energy-loss spectrum indicated as "a" in Fig. 14 was taken during this interval. In this spectrum all inelastic features (3–8) correspond to H_2O. KOH and $(KOH)_2$ may be present in the target beam, but the corresponding features are overshadowed by the strong H_2O features. At 500°C (spectrum b), dissociation processes begin as clearly indicated by the appearance of K (1 and 2) and H_2 (9) features. In the 600–800°C range (spectra c and d) dissociation dominates and the H_2O features gradually disappear. At 800°C as time progresses, H_2 also disappears (spectrum e). The K lines persisted until all sample was evacuated from the crucible.

ACKNOWLEDGMENTS

This is an appropriate time and place to express to Leo Brewer my appreciation for directing my interest into the field of electron impact spectroscopy. One of the important questions that he brought up during my graduate years with him was the location of the elusive low lying triplet states in many high temperature species. This single question attracted my interest and led me to undertake the electron collision studies that have occupied the past twenty years of my life.

REFERENCES

1. Trajmar, S., and Register, D. F., in *Electron Molecule Collisions,* Takayanagi, K., and Shimamura, I., eds., Plenum, New York, 1984, pp. 427–493.
2. Trajmar, S., Williams, W., and Srivastava, S. K., *J. Phys. B: Atom. Mol. Phys.* **10,** 3323 (1977).
3. Trajmar, S., and Chutjian, A. (unpublished).
4. Newell, W. R., Optical Oscillator Strengths by Electron Impact Spectroscopy, in *Progress in Atomic Spectroscopy,* Hal, W., and Kleinpoppen, H., eds., Plenum, New York, 1979.
5. Cartwright, D. C., Trajmar, S., Williams, W., and Huestis, D. L., *Phys. Rev. L.* **11,** 329 (1971); Goddard III, W. A., Huestis, D. L., Cartwright, D. C., and Trajmar, S., *Chem. Phys. Lett.* **11,** 329 (1971).
6. Williams, W., Cheeseborough III, J. C., and Trajmar, S., *J. Phys B: Atom. Mol. Phys.* **11,** 2031 (1978).
7. Trajmar, S., Williams, W., and Srivastava, S. K., *J. Phys B: Atom. Mol. Phys.* **10,** 3323 (1977).
8. Williams, W., Trajmar, S., and Bozinis, D. G., *J. Phys. B: Atom. Mol. Phys.* **9** L96 (1975).
9. Trajmar, S., and Williams, W. (unpublished results).
10. Trajmar, S., and Williams, W. (unpublished results).
11. Williams, W., and Trajmar, S. (unpublished results).
12. Schulz, G., *Phys. Rev. Lett.* **10,** 104 (1963).

13. Register, D. F., Trajmar, S., Jensen, S. W., and Poe, R. T., *Phys. Rev. Lett.* **41,** 749 (1978).
14. Hertel, I. V., and Stoll, W., *Adv. At. Mol. Phys.* **13,** 113 (1977).
15. Register, D. F., Trajmar, S., Csanak, G., Jensen, S. W., Fineman, M. A., and Poe, R. T., *Phys. Rev. A.* **28,** 151 (1983).
16. Slater, R. C., Fickes, M. G., and Stern, R. C., *Phys. Rev. Lett.* **29,** 333 (1972); Slater, R. C., Fickes, M. G., Becker, W. G., and Stern, R. C., *J. Chem. Phys.* **60,** 4697 (1974); Becker, W. G., Fickes, M. G., Slater, R. C., and Stern, R. C., *J. Chem. Phys.* **61,** 2283 (1974); Slater, R. C., Fickes, M. G., Becker, W. G., and Stern, R. C., *J. Chem. Phys.* **61,** 2290 (1974).
17. Vušković, L., Srivastava, S. K., and Trajmar, S., *J. Phys. B: Atom. Mol. Phys.* **11,** 1643 (1978).
18. Rudge, M. R. H., Trajmar, S. and Williams, W., *Phys. Rev. A* **13,** 2074 (1976).
19. Vušković, L., and Srivastava, S. K., *J. Phys. B: Atom. Mol. Phys.* **14,** 2677 (1981).
20. Vušković, L., and Trajmar, S., *J. Chem. Phys.* **71,** 3857 (1979).
21. Moore, C. E., National Bureau of Standards Circular No. 467, (US GPO, Washington, DC 1952.
22. Herzberg, G., *Molecular Spectra and Molecular Structure. I. Spectra of Diatomic Molecules,* Van Nostrand, New York, 1950.

0018–1536/84/1700–0085$04.60

Energetics of Silicon Oxidation Reactions

An Independent Determination of the SiO Bond Dissociation Energy

JOERG PFEIFER, GARY GREEN, AND JAMES L. GOLE*

High Temperature Laboratory, Center for Atomic and Molecular Science and School of Physics, Georgia Institute of Technology, Atlanta, Georgia 30332

Received February 7, 1984; Accepted February 8, 1984

ABSTRACT

Silicon atoms react under single collision conditions with N_2O and NO_2 to yield chemiluminescent emission corresponding to the SiO $a^3\Sigma^+$–$X^1\Sigma^+$ and $b^3\Pi$–$X^1\Sigma^+$ intercombination systems and the $A^1\Pi$–$X^1\Sigma^+$ band system. From the Si–NO_2 reaction, it has been possible to obtain a lower bound to the SiO bond energy, the first such determination independent of mass spectrometry. The result obtained, $D_0^\circ(\text{SiO}) \geqslant 186.7 \pm 1.7$ kcal/mol, is in good agreement with mass spectrometry ($D_0^\circ = 188.2$ kcal/mol). To facilitate this determination, the temperature dependence of the $A^1\Pi$–$X^1\Sigma^+$ spectrum has been analyzed to deduce the nature of the metal reactant beam and to determine the activation energy for formation of SiO* $A^1\Pi$ ($E_A = 7.4 \pm 1.2$ kcal/mol). A most striking feature of the Si–N_2O reaction is the energy balance associated with the formation of SiO product molecules in the $A^1\Pi$ and $b^3\Pi$ states. A significant energy discrepancy ($\approx$10,000 cm^{-1} = 1.24 eV) is found between the available energy to populate the highest energetically accessible excited state quantum levels and the highest quantum level from which emission is observed. It is suggested that this discrepancy may result from the formation of vibrationally excited N_2 in a concerted fast Si–N_2O reactive encounter.

*Author to whom all correspondence and reprint requests should be addressed.

Index Entries: Energetics, of SiO reactions; silicon oxide, energetics of; dissociation energy, of the SiO bond.

INTRODUCTION

It is a pleasure to contribute this paper to the Brewer Symposium Volume of High Temperature Science. Leo Brewer has had a strong influence on the manner in which our research group approaches what might be called high temperature quantum chemistry. It has always been a pleasure to discuss our experimental results with Leo and, as all of us have found, he usually contributes an added important insight and suggestion or two that places the work in even clearer perspective. This paper discusses the SiO molecule and specifically focuses on a topic over which John Chipman, Leo Brewer, John Elliott, and I have had several discussions. Although the SiO molecule is well studied and its bond dissociation energy represents an important standard in mass spectrometry on which several other measurements are based, a confirmation of this important quantity, independent of mass spectrometry, has not yet been presented. As we all know, these confirmations have been of particular interest to Leo Brewer.

For several years, the nature of the structure (*1*), energetics (*2*), and formation (*3,4*) of the SiO molecule has been the subject of a number of experimental and theoretical studies. The low-lying states of SiO have recently been discussed in detail by Field et al. (*5*). An important contribution to our knowledge of the low-lying states of SiO has come from recent studies of the chemiluminescent reaction

$$\mathrm{Si}(^3P) + \mathrm{N_2O} \rightarrow \mathrm{SiO}^*(A^3\Sigma^+, b^3\Pi\ \ldots) + \mathrm{N_2} \qquad (1)$$

by Hager et al. (*3*) and Linton and Capelle (*4*). Much of the impetus for this effort has been the desire to exploit "spin conservation" in order to produce the low-lying "$a^3\Sigma^+$" and "$b^3\Pi$" states of SiO. Although Reaction (1) is believed to strongly favor spin conservation, Linton and Capelle (*4*) have also observed some considerable emission corresponding to the $A^1\Pi$–$X^1\Sigma^+$ band system. These studies have offered the opportunity to characterize the "$a^3\Sigma^+$"–$X^1\Sigma^+$ and "$b^3\Pi$"–$X^1\Sigma^+$ intercombination band systems of SiO and investigate the possibility of using spin conservation to advantage in order to create a population inversion on the "$a^3\Sigma^+$"–$X^1\Sigma^+$ or "$b^3\Pi$"–$X^1\Sigma^+$ transitions and hence develop a medium for a visible chemical laser.

With the emphasis outlined above, previous studies have focused on the evaluation of the chemiluminescent emission from Reaction (1) primarily as a means of determining spectroscopic constants for the $a^3\Sigma^+$ and $b^3\Pi$ states (*3,4*), the relative population distributions in these states (*3*), and the total quantum yield for fluorescence from those SiO excited states populated in reaction.

It has been the purpose of our current effort, focusing both on reaction dynamics and energetics, to augment the information garnered in previous studies by characterizing the chemiluminescent emission from both the Si–N_2O and Si–NO_2 reactions at considerably lower pressures (10^{-5}–10^{-1} torr) than those associated with the studies of Hager et al. (2–4 torr) and Linton and Capelle (0.5–20 torr). As a result of our study of the "single collision" ($P(_{Tot} \leq 10^{-4}$ torr) Si–N_2O and Si–NO_2 metatheses, it appears possible to provide explanations for the rather diverse $b^3\Pi$ excited-state vibrational excitation that characterizes the spectral emissions observed by Hager et al. ($v' \leq 2$) and Linton and Capelle ($v' \leq 8$) in their "multiple collision" flame experiments. There are also subtleties in the evaluation of excited state population distributions, branching ratios for excited state formation, and quantum yields for these systems (*6*). However, in the present manuscript, we limit ourselves to the discussion of the energetics of these processes. We will use an analysis of the chemiluminescent emission from the Si–NO_2 metathesis to deduce a lower bound for the SiO bond energy.

Previous researchers have focused primarily on SiO spectroscopy and excited-state product distributions, leaving apparently unnoticed an intriguing question concerning the dynamics of the Si–N_2O reaction. One finds that there is a significant discrepancy between the available energy to populate SiO excited electronic states [Reaction (1) exoergicity $\approx$ 53,000 cm^{-1}] and the observed excitation that characterizes the Si–N_2O system. We consider this differential and offer an explanation that appears to be consistent with observations of excited-state metal oxide formation in several M + N_2O reactions.

EXPERIMENTAL

Silicon atoms distributed thermally among the ground state $^3P_{0,1,2}$ components were reacted with NO_2 and N_2O in a highly exothermic process to produce an intense blue emission. The metal beam was formed from two sources: (1) Silicon was vaporized from a cylindrical graphite crucible (c.s. grade-Micromechanisms). This crucible was heated by a graphite radiator substituted for the tantalum radiators (graphite inner sheath) previously used in our study of the Si–OCS reaction (*7,8*). The graphite radiator was surrounded by two tantalum heat shields. Maximum temperatures ranging from 1900 to 2100°C were achieved with this construction, all temperatures being measured with a Leeds and Northrup optical pyrometer to ±10°C. The graphite crucible/heat shield assembly was surrounded by a water-cooled copper jacket attached to a newly designed "quick-load" bus bar/flange assembly (*9*). The effusive silicon atom beam intersected a tenuous atmosphere (10^{-5}–10^{-3} torr) of N_2O (Matheson, >98%) and the resulting diffuse chemiluminescence resulting from the Si–N_2O reaction was observed at right angles to the metal flow through a Suprasil quartz viewing port. Light baffles were placed in the reaction chamber to minimize the interference from blackbody radiation emanating from the oven chamber (2). Using an oven configuration similar to that employed previously in multiple collision studies (*10*), the Si–NO_2 reaction was studied over the pressure range 10^{-5}–10^{-3} torr. Silicon was vaporized over the temperature range 1600–2050°C from a c.s.-grade graphite crucible. The crucible was machined to fit inside a commercial tungsten basket heater (R. D. Mathis, Long Beach, CA) that was wrapped with several layers of zirconia ZrO_2 cloth (Zircar Products, Florida, NY). The silicon vapor was reacted directly with NO_2.

Spectra were taken with a 1-m Czerny-Turner scanning spectrometer operated in first-order with a Bausch and Lomb 1200 groove/mm grating blazed at 5000 Å. For the study of the wavelength regions below 2500 Å, the spectrometer was purged with dry N_2. Both RCA 4840 and 1P28 photomultiplier tubes were used in these experiments. In order to insure high sensitivity at wavelengths less than 2500 Å, the 1P28 photomultiplier tube was coated with sodium salicylate. The sensitivity of the optical system, quartz viewing port, quartz lens system, and purged spectrometer, to short wavelength radiation was tested by monitoring the 1850 Å mercury resonance line. This emission was readily detected demonstrating our ability to easily monitor chemiluminescent emission below 2000Å. The photomultiplier signal was detected with a Keithley 417 fast picoammeter whose output signal (partially damped) drove a Leeds and Northrup stripchart recorder. All spectra were wavelength calibrated using a low pressure mercury pen lamp.

CHEMILUMINESCENT SPECTRA

Both the Si–N_2O and Si–NO_2 reactions are characterized by an intense blue chemiluminescence. The reactions of interest are

$$Si(^3P_{0,1,2}) + N_2O(^1\Sigma^+) \rightarrow SiO(a^3\Sigma^+, b^3\Pi, A^1\Pi) + N_2(^1\Sigma^+) \qquad (1a)$$

and

$$Si(^3P_{0,1,2}) + NO_2(^2B_1) \rightarrow SiO(a^3\Sigma^+, b^3\Pi, A^1\Pi) + NO(^2\Pi) \qquad (2)$$

The emission spectra, which extend from ~2250 to 4400 Å (Figs. 1–6), consist of contributions from the $A^1\Pi$–$X^1\Sigma^+$, $b^3\Pi$–$X^1\Sigma^+$, and $a^3\Sigma^+$–$X^1\Sigma^+$ band systems of SiO and are dominated by the $b^3\Pi$–$X^1\Sigma^+$ band system onsetting at 2800 Å.

$a^3\Sigma^+$–$X^1\Sigma^+$

Readily resolvable features for the $a^3\Sigma^+$–$X^1\Sigma^+$ band system extend from 3500–4400 Å (Si–N_2O, Fig. 2; Si–NO_2, Fig. 5). As has been observed by previous workers (3,4) under multiple collision conditions, the spec-

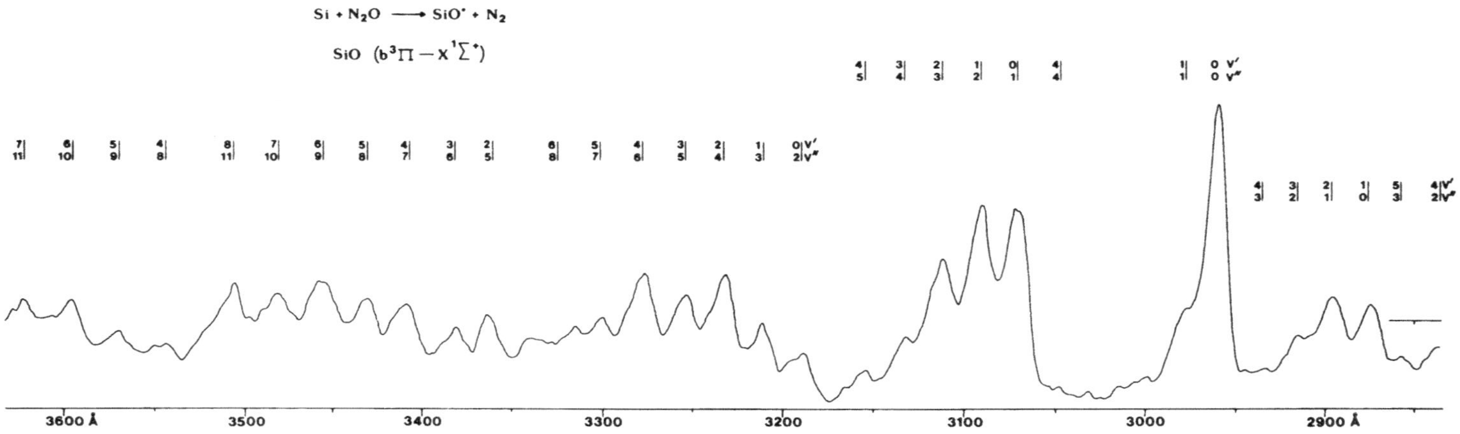

Fig. 1. (a) Chemiluminescent spectrum in the region 2850–3600 Å resulting from the reaction Si + $N_2O \rightarrow$ SiO^* + N_2 recorded under single collision conditions. The spectrum in this region corresponds to the SiO "$b^3\Pi$"–$X^1\Sigma^+$ band system. The bandheads are designated (v',v''). Spectral resolution is 10Å.

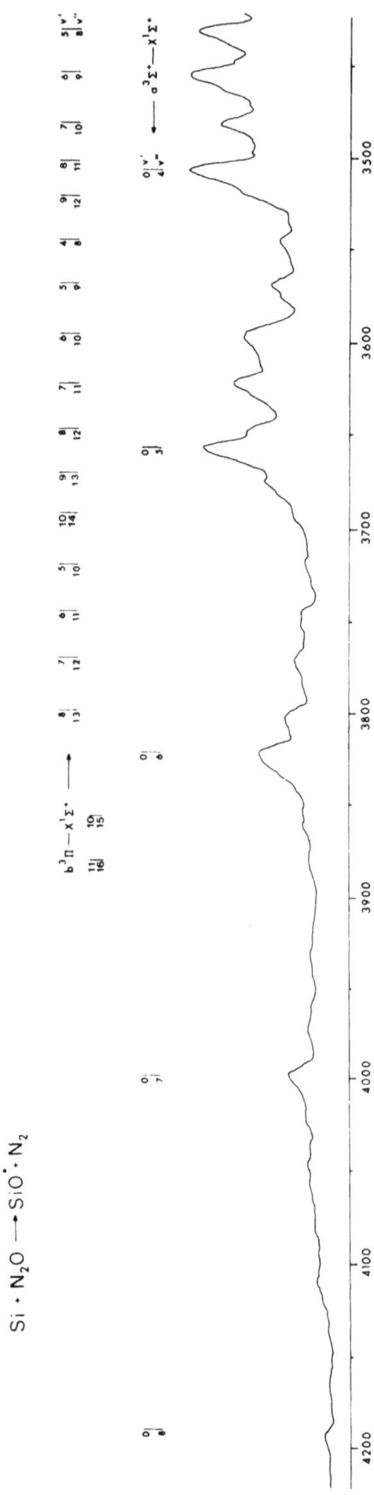

Fig. 2. Chemiluminescent spectrum in the region 3550–4200 Å resulting from the reaction $Si + N_2O \longrightarrow SiO^* + N_2$ recorded under single collision conditions. Emission from high vibrational levels of SiO "$b^3\Pi$" and the $v' = 0$ level of SiO "$a^3\Sigma^+$" is monitored. The bandheads are designated (v', v''). Spectral resolution is 10 Å. See text for discussion.

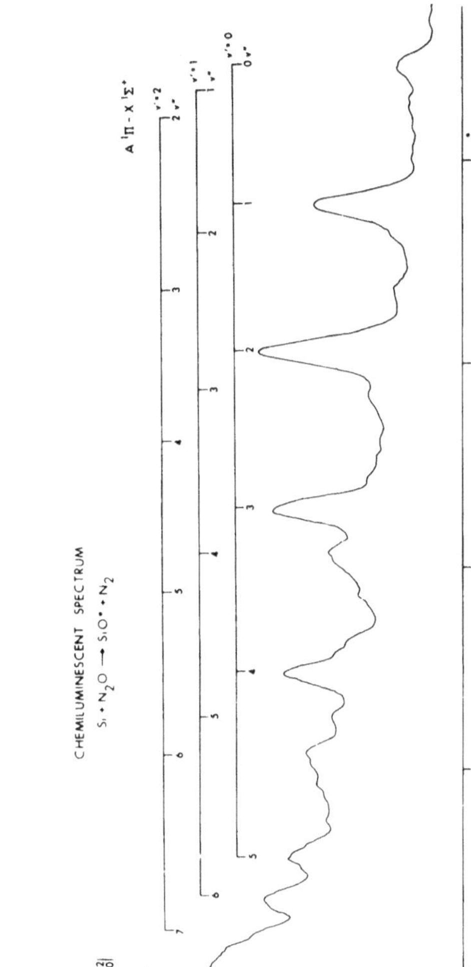

Fig. 3. Chemiluminescent spectrum in the region 2350–2950 Å resulting from the reaction Si + N_2 → SiO* + N_2 recorded under single collision conditions. The high energy end of the SiO "$b^3\Pi'$–$X^1\Sigma^+$ band system extends into the SiO $A^1\Pi$–$X^1\Sigma^+$ emission system. The bandheads in both systems are designated (v',v''). Spectral resolution is 15 Å. See text for discussion.

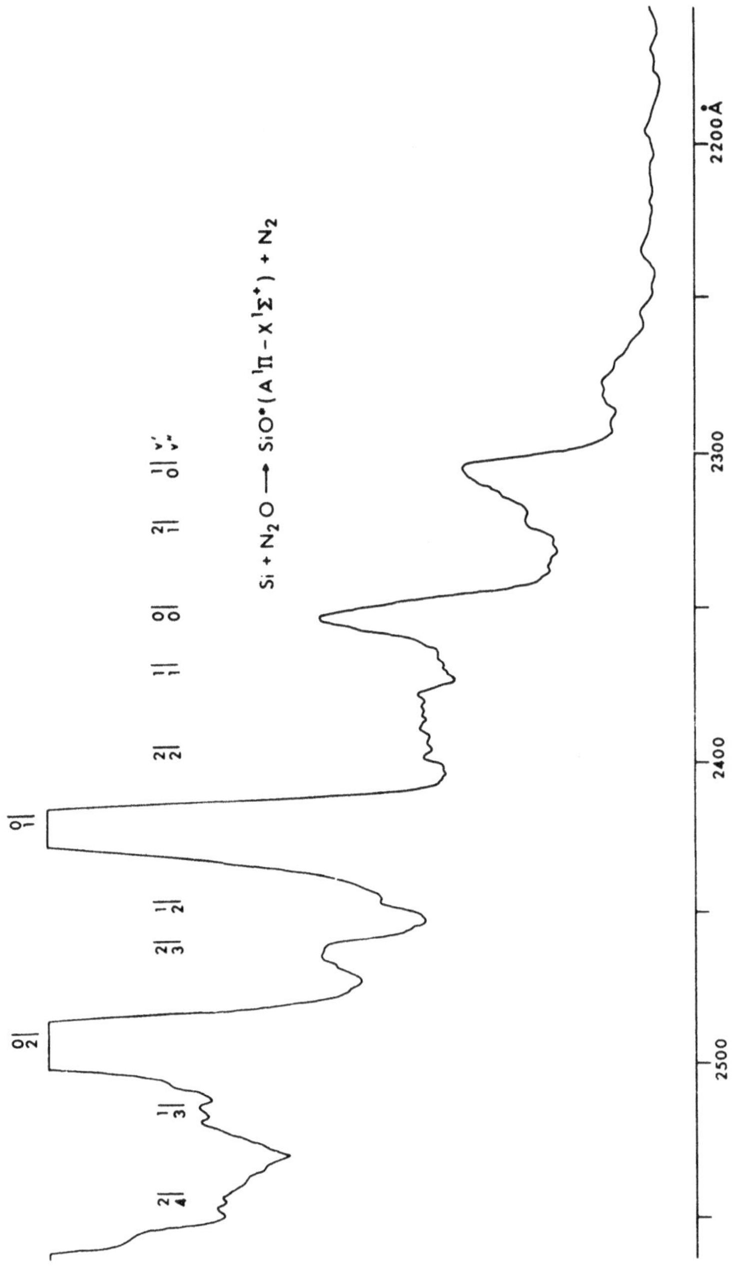

Fig. 4. Chemiluminescent spectrum in the ~2150–2550 Å resulting from the reaction Si + N_2O → SiO^* + N_2 recorded under single collision conditions. The observed spectrum corresponds to the SiO $A^1\Pi$–$X^1\Sigma^+$ band system. Bandheads are designated (v',v''). Spectral resolution is 20 Å. See text for discussion.

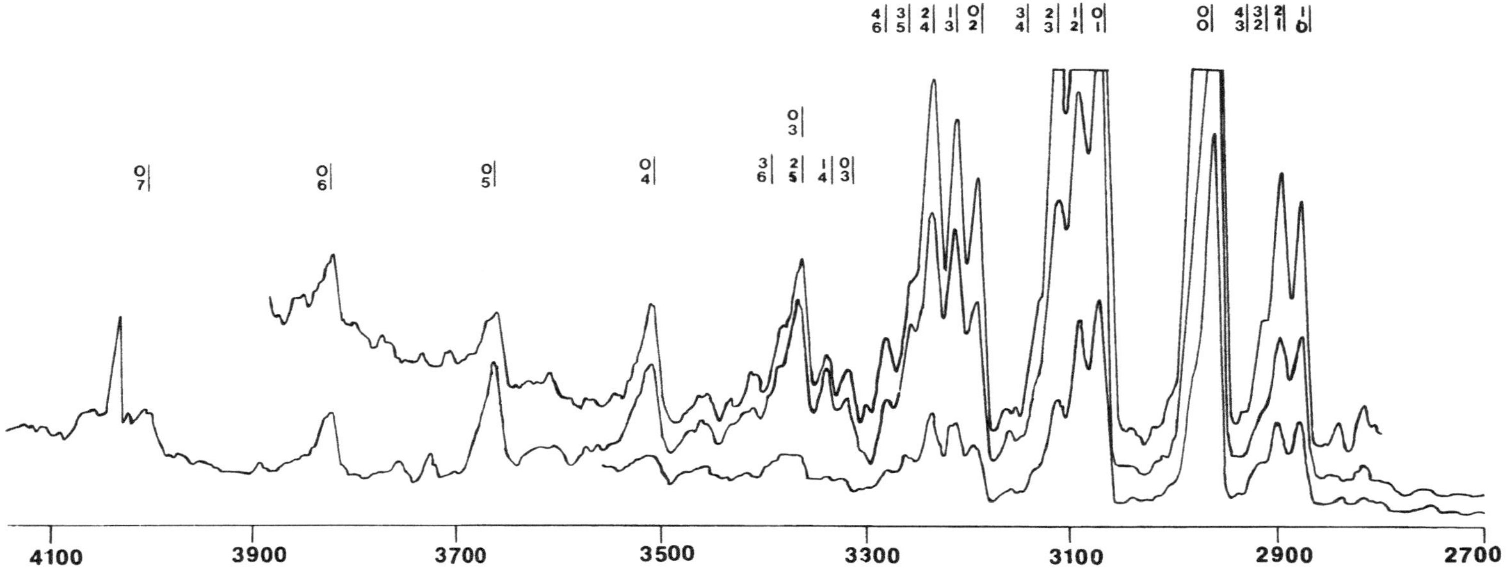

Fig. 5. Chemiluminescent spectrum in the region 2700–4100 Å resulting from the reaction Si + $NO_2 \rightarrow SiO^* + NO$ recorded under single collision conditions as a function of increasing silicon beam temperature and metal flux. The spectrum depicts both the "$b^3\Pi$"–$X^1\Sigma^+$ and "$a^3\Sigma^+$"–$X^1\Sigma$ band systems. Many fewer levels of the "$b^3\Pi$" state are populated vs the observed spectrum for the Si + N_2O reaction. Bandheads are designated (v',v''). Spectral resolution is 10 Å.

Fig. 6. Chemiluminescent spectrum in the region 2300–2750 Å resulting from the reaction Si + $NO_2 \rightarrow SiO^* + NO$ recorded under single collision conditions as a function of increasing silicon beam temperature and metal flux. The observed spectrum corresponds primarily to the SiO $A^1\Pi$–$X^1\Sigma^+$ band system. Spectral resolution is 20 Å. See text for discussion.

trum is strongly dominated by a single progression arising from the $v' = 0$ level of the $a^3\Sigma^+$ state. If emission features originating in $v' = 1$ are present in the Si–N_2O system they are weak and hidden under the $b^3\Pi$–$X^1\Sigma^+$ features. In contrast, the silicon–NO_2 reaction, possessed of both a much larger cross-section and a smaller reaction exothermicity does yield some information on $v' = 1,2$ $a^3\Sigma^+$ (6), by virtue of considerably more intense and less overlapped spectra.

$b^3\Pi$–$X^1\Sigma^+$

Emission features corresponding to the $b^3\Pi$–$X^1\Sigma^+$ band system extend from 2800 to 3950 Å (Figs. 1–3) for the Si–N_2O reaction and from 2750 to 3450 Å (Fig. 5) for the Si–NO_2 metathesis. Under single collision or near single collision conditions, the emission from the Si–N_2O metathesis is characterized by several new additional bandheads corresponding to vibrational levels $9 \leqslant v' \leqslant 11$ in the *b*–*X* system. This observation should be contrasted to the multiple collision studies of Hager et al. (3), where only emission from $v' \leqslant 2$ is observed and to the notably different vibrational excitation, $v' \leqslant 8$, found by Linton and Capelle (4). Vibrational deactivation (6) has begun to onset in the Linton and Capelle study and is quite pronounced in the experiments of Hager et al. Observed bandheads and relative intensities corrected for phototube response are given in Table 1. Some twenty new bandheads have been observed.

TABLE 1
Measured SiO Bandhead Positions and Vibrational Assignments
$b^3\Pi$–$X^1\Sigma^+$

Bandhead assignments[a] (v', v'')	Bandhead origin λ_{air}, Å	ν_{vacuum}, cm^{-1}	Relative intensity[g]
(2,0)	2791.2	35816	—
(3,1)	2811.6	35556	45.9
(4,2)	2832.6	35293	54.0
(5,3)	2853.7	35032	35.6
(1,0)[b,c]	2869.2	34843	55.2
(2,1)[b,c]	2889.8	34594	44.6
(3,2)[b]	2911.2	34340	25.0
(4,3)	2932.4	34092	12.6
(0,0)[b,c]	2953.7	33846	100
(1,1)	2973.8	33617	33.6
(2,2)	2995.2	33377	—
[(4,4)][d]	3038.9	32897	—
(0,1)[b,c]	3065.2	32615	90.1
(1,2)[b,c]	3085.8	32397	82.0
(2,3)[b,c]	3107.3	32172	59.8

(continued)

Bandhead assignments[a] (v', v'')	Bandhead origin λ_{air}, Å	ν_{vacuum}, cm^{-1}	Relative intensity[g]
(3,4)[b]	3129.2	31948	27.2
(4,5)	3151.3	31724	12.1
(0,2)[b,c]	3184.0	31398	18.8
(1,3)[b,c]	3205.2	31190	24.9
(2,4)[b,c]	3226.9	30981	34.4
(3,5)[b]	3249.2	30768	22.5
(4,6)[b]	3272.0	30554	25.1
(5,7)[b]	3295.8	30333	17.4
(0,3)[b]	3310.9	30195	12.6
(6,8)[b]	3320.2	30110	14.0
(1,4)[b,c]	3332.5	29999	11.5
(2,5)[b]	3354.8	29799	17.6
(3,6)[b]	3377.6	29598	11.4
(4,7)[b]	3401.0	29395	14.1
(5,8)[b]	3425.3	29186	16.3
(6.9)[b]	3450.5	28973	18.3
(7,10)[b]	3475.9	28761	13.4
(9,12)[e]	3530.6	28316	(6.4)
(4,8)[b]	3538.9	28249	6.4
(5,9)[b]	3564.1	28050	6.8
(6,10)[b]	3589.6	27850	9.5
(7,11)[b]	3616.1	27646	11.0
(8,12)[b]	3643.6	27438	(5.6)
(9,13)	3671.9	27226	5.2
(10,14)	3699.4	27024	2.9
(5,10)	3712.6	26928	2.9
(6,11)	3739.5	26734	3.4
(7,12)	3765.9	26547	4.1
(8,13)[f]	3792.2	26362	4.6
(10,15)	3849.7	25969	3.3
[(11,16)][d]	3883.7	25741	2.0
[(7,13)][d]	3926.8	25459	2.2
[(8,14)][d]	3952.8	25291	2.7

[a]Bandhead corresponding to the $^3\Pi_1$ component.
[b]Previously observed by Linton and Capelle, ref. (4).
[c]Previously observed by Hager et al., ref. (3).
[d]Weak feature, wavelength approximate.
[e](8,11) band under (0,4) band of $a^3\Sigma^+ - \Sigma^{1+}$ band system.
[f](9,14) band under (0,6) band of $a^3\Sigma^+ - X^1\Sigma^+$ band system.
[g]Corrected for phototube response.

$A^1\Pi$–$X^1\Sigma^+$

The $A^1\Pi$–$X^1\Sigma^+$ emission features observed in this study are depicted in Figs. 3, 4, and 6. As Fig. 3 demonstrates, the A–X emission characterizing the Si–N_2O reaction, is much weaker than that from the *b*–*X* system and is dominated by a progression emanating from $v' = 0$. The much less exothermic Si–NO_2 reaction [D_0^o(ON–O) − D_0^o(N_2–O) ≈ 1.5eV) is characterized by an even weaker *A*–*X* emission feature (relative to the *b*–*X* and *a*–*X* emission features). No emission from $A^1\Pi$ vibrational quantum levels greater than $v' = 2$ (Figs. 3 and 4) results from the Si–N_2O metathesis. If we consider the energy available to populate excited electronic states, we might expect to observe emission from $A^1\Pi$ vibrational quantum levels $v' \leqslant 15$.

REACTION KINETICS

Pressure Dependencies

We find that the observed chemiluminescent emission from the $A^1\Pi$ and $b^3\Pi$ states of SiO increases linearly with oxidant pressure over the range 3×10^{-5} to 6×10^{-4} torr, indicating that these states are formed in a process first order in N_2O and NO_2. The *a*–*X* emission features (v' = 0 emission), while dominated by direct formation in a process first-order in silicon and first-order in oxidant at the lowest pressures do display some nonlinear increase for pressures in excess of 1×10^{-4} torr. Although detailed temperature dependence studies to determine the activation energy for the Si–N_2O reaction were not performed, the chemiluminescent intensity as a function of the Si beam source temperature was followed with sufficient accuracy so as to determine that reaction with ground state Si ^3P atoms had occurred.

Temperature Dependencies: Si–NO_2 → SiO*($A^1\Pi$) + NO

The results of temperature dependence studies for the Si–NO_2 reaction are shown in figs. 7 and 8. The $A^1\Pi$ emission was monitored at 2650 Å, corresponding predominantly to an overlap of the (0,4) and 1,5) bands of the $A^1\Pi$–$X^1\Sigma^+$ band system. In Fig. 7, we plot ln *I* vs $1/T_B$ for a selection of the data obtained. Here, *I* is the chemiluminescent intensity and T_B is the beam (crucible) temperature. Data gathered for two runs are presented in Table 2. The slope and linearity of the plot in Fig. 7 shows that the Si–NO_2 reaction is first-order in metal flux.

Recently, a relation between the temperature dependence of the chemiluminescence intensity and the parameters of a beam–gas experiment has been formulated (*11*). It is

$$-R\, d(\ln I)/d(1/T_B) = \Delta H_{vap} - R_{T_B} + E_A[d(1/T_{eff})/d(1/T_B)] \quad (3)$$

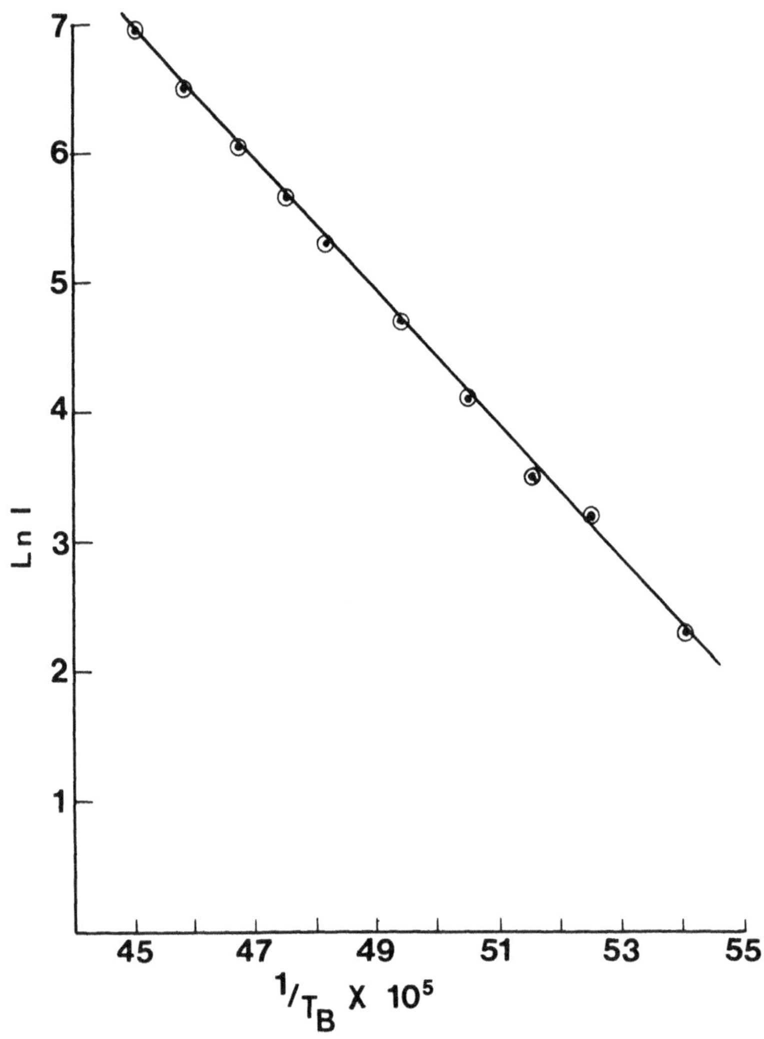

Fig. 7. Natural logarithm of relative chemiluminescent intensity versus the reciprocal of the silicon metal beam temperature at 2650 Å, corresponding to the SiO $A^1\Pi$–$X^1\Sigma^+$ band system. See text for discussion.

where E_A is the Arrhenius activation energy for the formation of the product whose chemiluminescence is monitored, ΔH_{vap} is the enthalpy of vaporization for the metal reacting to produce the observed chemiluminescence and

$$T_{eff} = (m_B T_G + m_G T_B)/(m_G + m_B) \qquad (4)$$

where T_G is the temperature of the gas (nitrogen dioxide) and m_B and m_G are the masses of the beam atoms and gas molecules, respectively.

Equation (3) may take into account the effect of low-lying electronic states (*12*), but generally considers the reaction of only ground-state metal atoms. It may also take into account possible reactions of dimeric

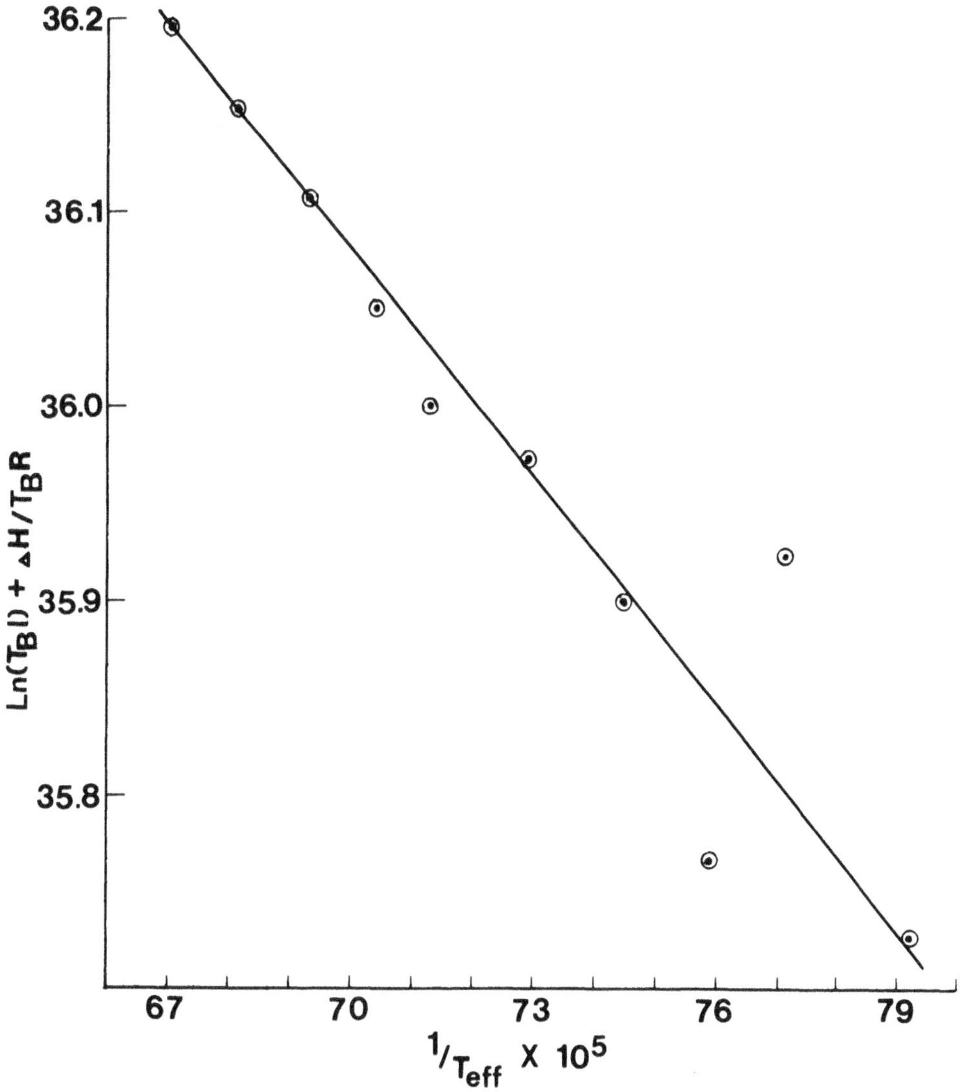

Fig. 8. Plot of [ln (IT_B) + $\Delta H_{vap}(T_B^{mean})/RT_B$] vs $1/T_{eff}$ for the Si + NO_2 reaction at 2650 Å. The slope of the plot yields the activation energy for formation of SiO* $A^1\Pi$ product molecules. See text for discussion.

and trimeric species which must be considered in the present study. The nature of the analysis (*12*) depends upon which states of the metal react to produce the observed chemiluminescence.

At 2250 K (the upper limit of the temperature dependence studies) in excess of 99% of the reacting atoms are in the ground ^3P state of silicon and the single collision Reaction (2) gives rise to the $A^1\Pi$ chemiluminescence. The overwhelming probability for ^3P atom reaction as the source of the chemiluminescence can be demonstrated directly from the plot in Fig. 7. In general, we expect E_A to be positive for reactions yielding chemiluminescence. Thus from Eq. (3)

TABLE 2
Latent Heat of Vaporization for Silicon (^3P) and Activation Energy for Formation of SiO*($A^1\Pi$)[a]

Reaction studied	$\Delta H_{vaporization}$, present study[b]	Temperature range, K	Literature[c]	E_A^d
$Si + NO_2 \rightarrow SiO^*$ ($A^1\Pi, v' = 0, 1$)	105.2 ± 1.2[e]	1850–2220	95.1	7.4 ± 1.2
$Si + NO_2 \rightarrow SiO^*$ ($A^1\Pi, v' = 0, 1$)	106.8 ± 3[e]	1880–2190	95.1	8.6 ± 3

[a]Energy in kcal/mol.
[b]Values in this column obtained from the slope of ln I vs $1/T_B$, where I is the chemiluminescent intensity and T_B is the beam temperature. The slope obtained through linear least-squares analysis is added to the value of RT_B over the temperature range of the experiment to yield a rigorous upper bound to $\Delta H_{vaporization}$; see text for discussion.
[c]ΔH_{vap} at the average temperature of the experimental runs.
[d]Values in this column obtained from the slope of $[\ln(IT_B) + \Delta H_{vap}(T_B^{mean})/RT_B]$ vs $1/T_{effective}$; see text for discussion.
[e]Spectral resolution was 30 Å.

$$-Rd \ln I/d(1/T_B) > \Delta H_{vap} - RT_B \quad (5)$$

Figure 7 denotes one of the least squares fits to obtained data points, the slope of the plot being −50,877 ± 591 K, leading to the value 101.1 ± 1.2 kcal/mol for $-Rd(\ln I)/d(1/T_B)$. Summing the slope energy and the value of RT_B over the temperature range of the experiment, we obtain a rigorous upper bound to ΔH_{vap}. These values are given in column 2 of Table 2.

The average T_B over the temperature range of the experiment was 2035 K. At these temperatures, the heat of vaporization for the equilibrium atomic vapor is 95.1 kcal/mol, only a few kcal/mol less than the observed value of $-Rd(\ln I)/d(1/T_B)$. The equilibrium vapor consists of silicon atoms in the ground ^3P and excited ^1D states (*13*) so that ΔH_{vap} for a vapor composed solely of ground state atoms will be slightly smaller. $\Delta H_{vaporization}$ for forming ^1D atoms will be considerably higher (≈18 kcal/mol) (*13*). The average value of RT_B over the range of the temperature dependence studies is ≈4.1 kcal/mol; thus the inequality given in Eq. (5) holds for reaction of ground state atoms, but it does not hold for reaction of excited ^1D atoms.

Similar arguments can be invoked when considering possible silicon dimer and trimer reactions with heats of vaporization that are 112.4 (*13*) and 104.5 (*13*) kcal/mol, respectively. The inequality of Eq. (5) does not hold if the metathesis producing chemiluminescence results from dimer or trimer reaction. The violation of the Eq. (5) inequality is even more pronounced if we consider the possible reaction of excited-state dimers and trimers. The observed chemiluminescence results overwhelmingly from the reaction of ground-state silicon atoms.

"Activation Energy" for Formation of SiO $A^1\Pi$*

In order to evaluate the activation energy for formation of SiO* $A^1\Pi$, we rearranged Eq. (3) in a manner previously described (*11*) to obtain

$$E_A \simeq -d[R \ln (IT_B) + \Delta H_{vap}(T_B^{mean})/T_B] \,/d(1/T_{eff})$$

where $\Delta H_{vap}(T_B^{mean})$ is the enthalpy of vaporization of the metal at the mean temperature over the range of the experiment. A selection of the date for evaluating the activation energy is plotted in Fig. 8 and final results are given in Table 2.

Energy Balance for the Reaction: Si + $N_2O \rightarrow$ SiO + N_2*

In this section, we focus on the large discrepancy between the *available* energy to populate SiO excited states, arising primarily from the exoergicity of the Si + N_2O reaction, and the energy for which we can account through determination of the highest excited-state quantum levels populated in reaction.

From the single collision Si–N_2O reaction it is potentially possible to set a lower bound on the dissociation energy of silicon monoxide. Using Reaction (1) and a previously discussed energy balance (14–16) we arrive at the inequality

$$D_0^o(\mathrm{SiO}) \geqslant D_0^o(\mathrm{N_2O}) + E_{int}(\mathrm{SiO}) - E_{int}(\mathrm{Si}) - E_{int}(\mathrm{N_2O}) - E_T^i - E_A(\mathrm{Si{-}N_2O}) \quad (6)$$

where the silicon atom is in the ground ^3P state and E_{int}(Si), $E_{int}(N_2O)$ and E_{int}(SiO) are the internal energies of the species noted in parenthesis.* E_T^i is the relative translational energy of the reactants and E_A(Si–N_2O) is the measured activation energy for the metathesis considered. In considering the products of the Si–N_2O reaction, SiO and N_2, neither $E_{int}(N_2)$ nor E_T^f, the final relative translational energy of product separation can be readily determined. Therefore these quantities are not included in Eq. (6) and the inequality applies.

If we neglect for the moment the activation energy for the Si–N_2O reaction (*17*) and focus on the available energy to populate SiO excited states

$$E_{Available} = D_0^o(\mathrm{SiO}) - D_0^o(\mathrm{N_2O}) + E_{int}(\mathrm{N_2O}) + E_{int}(\mathrm{Si}) + E_T^i \quad (7)$$

The available energy is partitioned to

$$E_{int}(\mathrm{SiO}) + E_{int}(\mathrm{N_2}) + E_T^f \quad (8)$$

As we have noted, the SiO molecule is among the most widely studied high temperature molecules. Its bond energy represents a standard

*For the silicon reaction, we select as our reference (zero) energy the reactants and products in Eq. (6) at 0°, that is Si(^3P_o), $N_2O(^1\Sigma^+, v = J = 0)$, SiO($^1\Sigma^+$, v = J = 0) and $N_2(^1\Sigma^+, v = J = 0)$.

for mass spectrometry. We adopt the recent determination by Hildenbrand et al. (*18*), $D_0^\circ(\mathrm{SiO}) = 8.16$ eV = 65,827 cm^{-1}. The N_2–O bond energy (*19*) is 13,400 cm^{-1}.

To calculate $E_{int}(N_2O)$, we adopt a procedure similar to that described previously (*20*) for the determination of E_{int}(OCS) and find $E(_{int}(N_2O) = 298$ cm^{-1}. To calculate E_{int}(Si) we need consider only the internal energy of the ground ^3P state since it is this state that reacts to produce the observed chemiluminescence. Using standard techniques (*21*), we find that $E_{int}(\mathrm{Si}) = 145$ cm^{-1} for a beam temperature of 2170 K.

Dagdigian et al. (*22*) have shown that, to good approximation

$$E_T^i = 3/2kT_{\mathrm{eff}} \quad T_{\mathrm{eff}} = (m_G T_G + m_B T_G)/(m_G + m_B) \tag{9}$$

where T_G is the temperature of the oxidant gas (300 K), T_B is the silicon beam temperature (2170 K), and m_B and m_G are the masses of the beam atoms and gas molecules, respectively. For the Si–N_2O reaction, we find $E_T^i = 1501$ cm^{-1}.

Summing the energies considered above, we find

$$E_{\mathrm{Available}} = 54371 \text{ cm}^{-1} \tag{8a}$$

In order to evaluate E_{int}(SiO), we seek the highest quantum level populated by the Si–N_2O reaction. Here, we exclude the use of the $a^3\Sigma^+$ state since, as we have noted (*6*), (1) the emission from this state is much weaker than and obscured by that from the $b^3\Pi$ state and (2) for levels $v' > 0$ possible strong $a^3\Sigma^+$–$b^3\Pi$ coupling may also play a role in obscuring the $a^3\Sigma^+$ emission features.

The highest vibrational quantum level populated in the SiO $A^1\Pi$ state is $v = 2$. Choosing the (2,7) band in the $A^1\Pi$–$X^1\Sigma^+$ system, we obtain

$$\begin{aligned} E_{int}(\mathrm{SiO}) &= \nu(2,7) + G''(7) - G''(0) \\ &= 35{,}951 \pm 30 \text{ cm}^{-1} + G''(7) - G''(0) \\ &= 44{,}314 \pm 35 \text{ cm}^{-1} \end{aligned} \tag{10}$$

The highest vibrational quantum level populated in the SiO "$b^3\Pi$" state is $v' = 11$. Choosing the (*11,16*) band in the "$b^3\Pi$"–$X^1\Sigma^+$ system, we obtain

$$E_{int}(\mathrm{SiO}) = 43{,}992 \pm 40 \text{ cm}^{-1} \tag{11}$$

The energy for which we can account using the SiO excited state emissions is some 10,000 cm^{-1} less than the potential available energy to the system! This implies that the SiO bond energy is over an electron volt less than the reliable value given in the literature, a highly unlikely possibility.

In general, single collision chemiluminescence studies yield dissociation energies that are in close agreement with careful mass spectrometric measurements. With few exceptions (*23*), any discrepancies are explained if one takes into account the temperature dependence of the metathesis yielding chemiluminescence and the nature of the metal atom reactant. In all cases, discrepancies have resulted in the prediction of slightly higher bond energies. Therefore the result obtained here represents a significant deviation from expectation. In the present study, we have found that ground state ^3P silicon atoms react to yield the observed chemiluminescence. If, in order to provide a correction for the possible reaction of metal atoms in the high energy tail of their translational energy distribution, we determine the activation energy for the Si–N_2O reaction, this energy increment must be added to the available energy (*24*), Eq. (8). Therefore the energy discrepancy is even larger than 10,000 cm^{-1}.

It is apparent that when the products SiO and N_2 are formed, more than one electron volt of energy must be accounted for by either the product N_2 internal energy or the relative translational energy of product separation.

Energy Balance for the Si–$NO_2 \rightarrow$ SiO ($A^1\Pi$) + NO Reaction*

In Fig. 6 we depict the SiO $A^1\Pi$–$X^1\Sigma^+$ emission that characterizes the Si–NO_2 reaction. We have found that this reaction populates the $v' =$ 0, 1, 2 levels of the SiO $A^1\Pi$ state in a process whose activation energy is 7.4 ± 1.2 kcal/mol. Using the relationship

$$\begin{aligned} D_0^\circ(\mathrm{SiO}) \geqslant\ & \underset{72.9}{D_0^\circ(\mathrm{ON{-}O})} + \underset{126.7}{E_{\mathrm{int}}(\mathrm{SiO})} - \underset{0.4}{E_{\mathrm{int}}(\mathrm{Si^*})} - \underset{.85}{E_{\mathrm{int}}(\mathrm{NO_2})} \\ & \underset{4.3}{-E_T^{i*}} - \underset{7.4 \pm 1.2}{E_A(\mathrm{Si{-}NO_2})} = 186.65 \pm 1.7\ \mathrm{kcal/mol} \end{aligned} \tag{12}$$

where E_{int}(Si), $E_{int}(NO_2)$, and E_{int}(SiO) are the internal energies of the species in parentheses, E_T^i is the relative translational energy of the reactants and $E_A(Si–NO_2)$ is the measured activation energy of the metathesis considered, we find D_0°(SiO) ⩾ 186.65 ± 1.7 kcal/mol. All energies are given in kcal/mol below the respective quantities evaluated, as outlined in the previous section. The lower bound dissociation energy is in excellent agreement with the mass spectrometric determination of Hildenbrand et al. (*18*) (D_0° = 188.2 kcal/mol). This evaluation of the SiO bond dissociation energy is the first such determination independent of mass spectrometry.

It should be emphasized that the silicon–NO_2 reaction is by no means sufficiently exothermic [D_0°(SiO)–D_0°(ON– O)] to populate levels of the SiO $A^1\Pi$ state. The necessary energy increment comes primarily from the combination of the reactant translational energy increment, E_T^i, and the fact that we must surmount a rather significant activation barrier in order that silicon and NO_2 interact. When taking into account the determined activation energy for excited-state formation, we provide a correction for the possible reaction of metal atoms in the high energy tail of their translational energy distribution. In this particular system, the SiO† $A^1\Pi$ molecules formed must emanate from the reaction of silicon atoms in this high energy tail. In effect, this reactive channel is closed to silicon atoms that do not possess sufficient energy. Once the barrier is surmounted and product formation ensues, the required energy is returned and can be used for product internal excitation.

The importance of these considered energy increments is vividly displayed if we extend the single collision studies described here to the multiple collision pressure regime. Here, the silicon beam is thermalized by a room temperature carrier gas to temperatures ranging between 300 and 600 K. This thermalization considerably lowers E_T^i and substantially decreases the concentration of sufficiently energic silicon atoms. Under multiple collision conditions, no $A^1\Pi$–$X^1\Sigma^+$ emission is observed (*6*).

DISCUSSION

It is satisfying that one can use chemiluminescent studies to confirm, independently, one of the benchmark bond energies upon which mass spectrometry relies. At the same time, it is quite intriguing that the silicon–N_2O reaction leads to the population of far fewer excited-state levels than are energetically accessible. In spite of repeated efforts, we have been unable to use the Si–N_2O reaction to populate vibrational levels $v' \geq 3$ in the $A^1\Pi$ state of SiO. In addition, we have not observed fluorescence from the $E\ ^1\Sigma_0^+$ state although it should be possible to populate two vibrational levels in this state. The Si–N_2O reaction is of sufficient exothermicity to populate 15 vibrational quantum levels in the $A^1\Pi$ state. Emission from these levels can readily be observed and recently we have employed the spin conserving Si–O_3 reaction

$$Si(^3P) + O_3(^1A_1) \rightarrow SiO(A^1\Pi) + O_2(^2\Sigma_g^-) \qquad (13)$$

to populate a minimum of 12 vibrational quanta in the $A^1\Pi$ state. In short, there appears to be little reason to suspect that the lack of $A^1\Pi$ vibrational excitation that characterizes the Si–N_2O reaction can be attributed to factors associated with the intimate molecular electronic structure of the $A^1\Pi$ state. Rather it appears that the observed effects are to be associated with the dynamics of the Si–N_2O reaction.

One can envision two possibilities: (1) a fast reactive silicon–N_2O encounter leading to the formation of SiO* $A^1\Pi$ or SiO* $b^3\Pi$ also produces ground state N_2 in a concerted process with a minimum of five vibrational quanta of excitation ($N_2\ ^1\Sigma_g^+$, $v'' \geq 5$) or (2) the Si + $N_2O \rightarrow$ SiO + N_2 surface is characterized by an extremely large translational energy barrier in the product exit channel. We favor the former possibility based

†Evaluated for a silicon beam temperature of 2170 K.

on (1) preliminary quantum chemical calculations of the Si–N_2O surface that demonstrate that the N—N bond in N_2O lengthens upon approach of a silicon atom and (2) a comparison of the N—N bond length in N—NO (1.128Å) (*25*) and that for ground state N_2, v = O (1.0977Å) (*26*). Although, in the final analysis, we must be concerned with the bond lengths corresponding to the turning points of the potentials describing the N—N bond in N_2O and N_2, it is intriguing that the vibrational excitation that must be associated with the N_2 molecule in order that the corresponding bond length be 1.128Å is $v'' = 5$. In other words, the average value of the internuclear distance for the fifth vibrational quantum level in the ground state of N_2 corresponds to the N—N bond distance in N_2O.

The results obtained in the present study cause us to focus on the nature of the energy balance in several M + N_2O reactions, where we find similar behavior. In some cases larger energy discrepancies are found to characterize several M + N_2O reactions involving the Group IIIB (Sc,Y,La) and Group IIA metals. The data are summarized in Table 3.

TABLE 3
Energy Balance Discrepancies in M + N_2O Reactions

Reaction	$E_{\text{Available}}$, cm^{-1}	Measured MO excitation	E_{int} (MO), cm^{-1}	Δ,E^a, cm^{-1}
$Sc(^2D) + N_2O$[b]	45,856[c]	Short wavelength limit ($C^2\Pi$–$X^2\Sigma^+$) (3000 Å)	33,324	12,532
$Y(^2D) + N_2O$[b]	48,154[c]	Short wavelength limit ($C^2\Pi$–$X^2\Sigma^+$) (3000 Å)	33,324	14,830
$La(^2D) + N_2O$[d]	55,862[c]	LaO*, $D^2\Sigma^+$, v' = 17[e]	39,400	16,462
$Ba(^1S) + N_2O$[f]	33,652	Short wavelength limit (4000 Å)	24,993	8659
$Ca(^3P) + N_2O$[g,h]	34,792	Short wavelength limit (3850 Å)	25,967	9095
$Sr(^3P) + N_2O$[h]	37,553	Short wavelength limit (3700 Å)	27,019	10,534
$Ti(^3F) + N_2O$	44,004	D–$X^3\Delta$ (2,1) band	34,840	9060

[a] $E_{\text{available}} - E_{\text{int}}$ (MO).
[b] Gole, J. L., and Chalek, C. L., *J. Chem. Phys.* **65**, 4384 (1976).
[c] D_0^0 (MO) + $E_{\text{int}}(N_2O) + E_T^i - D$(N–NO). See text for definitions.
[d] Chalek, C. L., and Gole, J. L., *Chem. Phys.* **19**, 59 (1977).
[e] Gole, J. L., unpublished data.
[f] Data from Jonah, C. D., Zare, R. N., and Ottinger, C., *J. Chem. Phys.* **56**, 263 (1972).
[g] Data from Irwin, J. A., and Dagdigian, P. J., *J. Chem. Phys.* **74**, 6178 (1981).
[h] Data from Wilcomb, B. E., and Dagdigian, P. J., *J. Chem. Phys.* **69**, 1779 (1978). See also Dagdigian, P. J., *Chem. Phys. Lett.* *55*, 239 (1978).

In many of these cases further characterization of the product metal oxide chemiluminescence will be required in order to better assess the actual magnitude of the energy discrepancy.

The significant fact that emerges is that the reactions outlined display an energy balance that is consistent with the formation of a vibrationally excited N_2 product. Although one might argue that the translational energy of product separation can account for a portion of these observations, it would seem somewhat surprising that a final relative translational energy of separation, E_T^f [Eq. (8)], can account for the rather sharp energy cutoffs, especially those associated with the silicon and titanium reactions.

There are several experiments that might be pursued in order to obtain direct evidence for the formation of vibrationally excited N_2 produced in M = Si, or Ti + N_2O reaction. The major problem confronting the experimenter is the detection of very low effective product N_2 concentrations ($\sim 10^{-7}$ torr). Promising detection techniques might involve the use of either electron bombardment excitation and ionization or multiphoton ionization. In the former studies one might attempt to use an electron gun to excite the Second Positive system of N_2 monitoring directly the N_2 $C^3\Pi_u$–$B^3\Pi_g$ fluorescence and back-extrapolating to the ground state N_2 population distribution (*27*). Alternatively, we are attempting to produce N_2^+ $B^2\Sigma_u^+$, monitoring the $B^2\Sigma_u^+$–$X^2\Sigma_g^+$ emission and again back-extrapolating to the N_2 population distribution (*27*). Thus far, these studies have met with only marginal success (N_2 concentrations $\leqslant 10^{-5}$ torr have been detected); however, the signal-to-noise ratio in the system can be considerably improved if the reaction zone and electron gun are magnetically confined (*28*). An alternate experiment might involve the elegant pulsed-laser multiphoton technique that Dehmer (*29*), Kay (*30*), and coworkers have used to detect CO^+. The direct assessment of the vibrational excitation characterizing the product N_2 formed in highly exothermic M + N_2O reactions will contribute important new perspectives on the nature of reactive encounters. We encourage other workers to pursue the detection of vibrationally excited N_2 in low concentration.

REFERENCES

1. Bredohl, H., Cornet, R., Dubois, I., and Remy, F., *J. Phys. B.* **7,** L66 (1974); Heil, T. G., and Schaefer III, H. F., *J. Chem. Phys.* **56,** 958 (1972); see also ref. (*5*).
2. Drowart, J., Pattoret, A., and Smoes, S., *Proc. Brit. Ceram. Soc.* **8,** 67 (1967); Hildenbrand, D. L., *High Temp. Sci.* **4,** 244 (1972); Zmbov, K. F., Ames, L. L., and Margrave, J. L., *High Temp. Sci.* **5,** 235 (1973); Kvande, H., and Wahlbeck, P. G., *High Temp.-High Press.* **8,** 45 (1976); Hildenbrand, D. L., and Murad. E., *J. Chem. Phys.* **51,** 807 (1969).
3. Hager, G., Harris, R., and Hadley, S. G., *J. Chem. Phys.* **63,** 2810 (1975); Hager, G., Wilson, L. E., and Hadley, S. G., *Chem. Phys. Lett.* **27,** 439 (1974).

4. Linton, C., and Capelle, G. A., *J. Mol. Spectr.* **66,** 62 (1977).
5. Field, R. W., Lagerqvist, A., and Renhorn, I., *Phys. Scripta* **14,** 298 (1976).
6. (a) "On the Nature of the Energy Balance and Branching Ratios Associated with the Chemiluminescent Reaction $Si(^3P) + N_2O(^1\Sigma) \rightarrow SiO^*(a^3\Sigma^+, b^3\Pi, A^1\Pi) + N_2(v'' \geq 5)$—Possible Formation of Vibrationally Excited N_2 in M + N_2O Reactions," with G. J. Green, K. D. Jordan and J. Yates, work in progress. (b) Energy Balance in the Si–NO_2 Reaction: On the Formation of SiO $a^3\Sigma^+$, $v' \geq O$," and Ultrafast $a^3\Sigma$–$b^3\Pi$ E–E Energy Transfer.
7. Green, G. J., and Gole, J. L., *Chem. Phys.* **46,** 67 (1980).
8. Gole, J. L., and Green, G. J., *Chem. Phys.* **69,** 357 (1982).
9. Pace, S. A., and Gole, J. L., unpublished.
10. See, for example: Lindsay, D. M., and Gole, J. L., *J. Chem. Phys.* **66,** 3886 (1977); Sayers, M. J., and Gole, J. L., *J. Chem. Phys.* **67,** 5442 (1977); Gole, J. L., and Pace, S. A., *J. Chem. Phys.* **73,** 836 (1980); Hanner, A., and Gole, J. L., *J. Chem. Phys.* **73,** 5025 (1980).
11. Preuss, D. R., and Gole, J. L., *J. Chem. Phys.* **66,** 880, 2994 (1977); Gole, J. L., and Preuss, D. R., *J. Chem. Phys.* **66,** 3000 (1977).
12. Dubois, L. H., and Gole, J. L., *J. Chem. Phys.* **66,** 779 (1977).
13. Hultgren, R., Desai, P. D., Hawkins, D. T., Gleiser, M., Kelley, K. K., and Wagman, D. D., *Selected Values of the Thermodynamic Properties of the Elements,* American Society for Metals, Metals Park, Ohio, 1973. Also C. E. Moore.
14. Gole, J. L., and Zare, R. N., *J. Chem. Phys.* **57,** 5331 (1972); Zare, R. N., *Ber. Bunsenges Physik. Chem.* **78,** 153 (1974).
15. Gole, J. L., and Chalek, C. L., *J. Chem. Phys.* **65,** 4384 (1976).
16. Chalek, C. L., and Gole, J. L., *Chem. Phys.* **19,** 59 (1977).
17. Preuss, D. R., and Gole, J. L., *J. Chem. Phys.* **66,** 880 (1977).
18. Hildenbrand, D. L., *High Temp. Sci.* **4,** 244 (1972).
19. Stull, D. B., and Prophet, H., *JANAF Thermochemical Tables,* 2nd ed., Natl. Stand. Ref. Data Series, Natl. Bur. Stand. 37, 1971.
20. Jones, R., and Gole, J. L., *Chem. Phys.* **20,** 311 (1977).
21. We determine E_{int}(Si) from a statistical average of the $^3P_0$(0 cm^{-1}), $^3P_1$ (77.15 cm^{-1}), and $^3P_2$ (233.31 cm^{-1}) components of the ground ^3P state.
22. Dagdigian, P. J., Cruse, H. W., and Zare, R. N., *J. Chem. Phys.* **62,** 1824 (1973).
23. Dickson, C. R., and Zare, R. N., *Chem. Phys.* **7,** 361 (1975).
24. See discussions in refs. (*7*) and (*17*).
25. Huber, K. P., and Herzberg, G., *Constants of Diatomic Molecules,* Van Nostrand-Reinhold Company, New York, 1979.
26. Pliva, J., *J. Mol. Spectr.* **12,** 360 (1964).
27. The N_2 or N_2^+ emission spectrum obtained under reaction conditions is compared to that for room temperature N_2 at low concentrations.
28. The authors are indebted to Stewart Novick for helpful discussions on the subject of increasing the signal-to-noise from electron gun sources.
29. Dehmer, J., private communication.
30. Kay, B., private communication.

0018–1536/84/1700–0109$05.00

Kinetics of Vaporization of Molten Selenium

JOHN YUN-KUANG HUANG AND PAUL W. GILLES*
Department of Chemistry, University of Kansas, Lawrence, KS 66045

AND

J. EDWARD BENNETT
Arkansas State University, PO Box 419, State University, AR 72467

Received March 1, 1984; Accepted March 3, 1984

ABSTRACT

Vaporization of molten selenium was studied under equilibrium and non-equilibrium conditions. A quadrupole mass spectrometer was used to measure the vapor composition, and a vacuum microbalance was used to measure the total rate-of-weight-loss. Both measurements were made as a function of temperature near 500 K. Experimental results indicate that the vaporization of molten selenium is a retarded reaction having vaporization coefficients of about 0.1 for all species $Se_n(g)$ (n = 2–8). Results also show that the activation free energy of vaporization for all species is approximately equal to the free energy change of depolymerization reactions of liquid molecules, as obtained by other authors. These molecules consist of long chains in thermodynamic equilibrium with octacyclic rings. Mechanisms for the vaporization of molten selenium are postulated that involve the formation of short chains or small, nearly closed loops on liquid surface as the rate-limiting step. These activated complexes have the same number of atoms as the gaseous ring molecules to be formed. A critical test for this model is proposed which, if proven correct, may lead to a novel technique for growing pure single crystals of trigonal selenium.

Index Entries: Kinetics and thermodynamics of vaporization, for Se; equilibrium and non-equilibrium vaporization, of molten Se; re-

*Author to whom all correspondence and reprint requests should be addressed.

tarded vaporization, of Se; vaporization mechanism, of Se; ring-chain equilibrium, structure for Se; polymerization reactions of liquid Se; quadrupole mass spectrometric ion intensities, of Se; enthalpy, of Se at equilibrium; entropy, of Se at equilibrium; activation enthalpy and activation entropy, of Se; free energy of Se activation and vaporization; vaporization coefficients, of Se; vapor composition, over Se; partial pressures, of Se; activated complexes, of Se; rate-limiting step, in vaporization reactions of Se.

INTRODUCTION

Many studies have been conducted on the kinetics of vaporization, and the subject has been reviewed by several authors (*1–11*). It has been well established that the surfaces of solids from which material vaporizes consist of heterogeneities such as kinks, ledges, terraces, and adatoms, and that vaporization can be considered a multistep reaction. The sequence of reaction steps becomes more complicated if the surface atom or ions must undergo rearrangements such as association, dissociation, or charge transfer before desorption from the vaporizing surface. To identify the rate-limiting step is usually difficult and requires suitable complementary experiments (*7*).

Brewer and Kane (*12*) noted that, for solids that vaporize to form polymeric vapor species, low vaporization coefficients can be expected whenever the main vaporizing species are not present as such in the condensed phase and where the atoms or molecules are bound rigidly enough to prevent easy rearrangement. Among many others, this is demonstrated in systems such as As_2O_3 (claudetite), P (red) and As where the measured vaporization coefficients are about 10^{-6} or lower.

In the case of liquids, the vaporization mechanism is usually less complex. It is usually assumed that liquids cannot sustain surface heterogeneities and, therefore, that all surface sites are equivalent. Mortensen and Eyring (*5,13*) developed a theory that predicts, for simple liquids consisting of atomic structural units or spherically symmetric molecules, that the vaporization coefficients are unity. For liquids with complex bonding, the theory predicts that the vaporization coefficient may be less than unity. McFeely and Somorjai (*14*) studied the vaporization kinetics of some hydrogen-bonded organic liquids and observed low vaporization coefficients.

Vaporization of selenium has been studied by several authors employing various techniques (*15–33*). Mass spectrometric studies have identified the existence of Se_n^+ (n = 1–10) and indicated that the intensities of Se^+, Se_3^+, Se_4^+, Se_9^+, and Se_{10}^+ are relatively low. It is generally agreed that Se_n (n = 5–8) exist as neutral molecules; however, Keller and coworkers (*23*) and Hoareau and coworkers (*24*) concluded that Se_3 and Se_4 also exist as neutral molecules. Most authors seem to agree on the composition of selenium vapor. Berkowitz and Chupka (*29,30*) showed that, at 500 K, the vapor consists of about 42% Se_6, 36% Se_5, 14% Se_7, 5.4% Se_2, 2.3% Se_8, and 0.15% Se_3 in mol%; and that, as the temperature increases, the average number of atoms in the vapor molecules decreases. They also calculated the atomization energy per bond for each of the gaseous molecules, assuming each species has ring structure, and found that the atomization energy per bond increases with the number of atoms in the molecules and that the highest value occurs at Se_8, although Se_6 is much more abundant in the vapor. They also noted that this finding is contradictory to what occurs in the sulfur system, in which S_8 is the most abundant species and also has the highest atomization energy per bond. Meschi and Searcy (*33*) investigated the magnetic moments of Se_2, Se_5, and Se_6 gaseous molecules by the Stern-Gerlach method and concluded that Se_5 and Se_6 are not diradicals. All authors favor ring structures for gaseous molecules Se_n ($n > 2$) even though ring structures have been confirmed only for Se_5 and Se_6.

The structure of solid as well as liquid selenium has been reviewed by Cooper and Westbury (34). At temperatures below its melting point, 217°C, trigonal selenium (space group $P3_1/2$) is the thermodynamically stable form. Its structural unit consists of helical chains (*35*) with a bond length of 2.373 Å, a bond angle of 103.1°, and a dihedral angle of 102° (*36*). The metastable α-monoclinic (space group $P2_1/n$) and β-monoclinic (space group $P2_1/a$) selenium have Se_8 rings as their structural units (*37*) with a bond length of 2.34 Å, a bond angle of 105.5°, and a dihedral angle of 101° (*36*). At temperatures above the melting point, long helical chains are preserved and exist in thermodynamic equilibrium with octacyclic rings. The bond length and the bond angle of the liquid chains do not differ significantly from those of the solid chains (*36*). Briegleb (*38*) studied the ring–chain equilibrium in liquid selenium and found that at the melting point the weight fraction of Se_8 rings is about 40%. Since then, many studies have covered the polymerization reactions of liquid selenium (*36,39–52*). Eisenberg and Tobolsky (*39*) presented a theory for the ring-chain equilibrium in liquid selenium. Using Briegleb's data, they obtained $\Delta H_1^\circ = 2.27$ kcal/mol and $\Delta S_1^\circ = 5.47$ cal/mol K/mol Se_8 rings for the chain-propagation reaction

$$Se_8^r\ (l) + Se_n^c\ (l) = Se_{n+8}^c\ (l) \qquad (1)$$

where the superscript r stands for rings and c for diradical chains, and the subscripts indicate the number of atoms in each species. By comparison to the values for the sulfur system, Eisenberg and Tobolsky also obtained $\Delta H_2^\circ = 25$ kcal/mol and $\Delta S_2^\circ = 23.0$ cal/mol K/mol Se_8 rings for the chain-initiation reaction

$$Se_8^r\ (l) = Se_8^c\ (l) \qquad (2)$$

Koningsberger and coworkers (*40–42*) performed ESR measurements on pure selenium liquids at temperatures between 240 and 420°C. They obtained $\Delta H_3^\circ = 29 \pm 3$ kcal/mol for the depolymerization reaction

$$Se_{n+m}^c\ (l) = Se_n^c\ (l) + Se_m^c\ (l) \qquad (3)$$

Keezer and coworkers (*44–46*) measured the viscosity of liquid selenium as functions of temperature and concentration of various monovalent dopants such as thallium, chlorine, and bromine. They found that the viscosity decreases as the temperature increases and as the dopant concentration increases. From these results, they obtained $\Delta H_2^\circ = 23$ kcal/mol and $\Delta S_2^\circ = 5.5$ cal/mol K for Reaction (*2*). They were able to grow large single crystals of trigonal selenium from these impurity-doped melts of lower viscosity.

Lucovsky (*47*) studied infrared and Raman spectra of α-monoclinic, trigonal, and amorphous selenium and confirmed the existence of Se_8 rings in liquid and amorphous selenium. This finding has been further substantiated by Waseda and coworkers' X-ray diffraction (*48*) and Tourand's neutron diffraction (*49*) studies. All studies on the structure of liquid selenium seem to have established the number-average chain length to be about 10^4 atoms at the melting point.

We chose the molten state of elemental selenium as a model system for studying the kinetics of vaporization because: (1) a liquid can be presumed to provide an ideal surface for a vaporization reaction and, hence, enable simplified mechanistic interpretations of the vaporization process; and (2) molten selenium has a complicated molecular structure consisting of long polymeric chains in equilibrium with Se_8 rings. This structure is seemingly very different from that of vapor molecules, some of which are believed to be rings.

The purposes of this study were fourfold: (1) to discover whether the vaporization of molten selenium is a retarded process; (2) to investigate whether the vapor composition is different under non-equilibrium and equilibrium conditions; (3) to learn why selenium vapor consists of only 2% of Se_8 despite the fact that about 40 wt% of Se_8 rings are in equilibrium with long polymeric chains at ~500 K, and also that Se_8 is the most stable vapor species with respect to dissociation into atoms; and (4) to postulate a mechanism for the vaporization of molten selenium.

These purposes have been accomplished by measuring as a function of temperature both the rate of vaporization (with a vacuum microbalance) and the vapor composition (with a quadrupole mass spectrometer) of molten selenium samples. Knudsen (equilibrium) and Langmuir (non-equilibrium) crucibles were used in each study.

EXPERIMENTAL

Starting Materials

The selenium sample was described as 99.9999% and was purchased from Atomergic Chemicals Company. The sample was in pellet form, and its X-ray powder diffraction pattern showed its structure to be monoclinic. Part of the sample was converted into trigonal form, laboratory designation SE-2, by annealing at 124°C for 22 h, and was ground into a fine powder before being used in vaporization experiments.

Crucibles

Crucibles used in this study were all made of Pyrex glass. For vacuum-balance experiments, the size of the crucibles was limited by the maximum load, 1 g, of the balance. The Knudsen and the Langmuir crucibles were made from 4.8-mm id and 3.0-mm id tubes, respectively; their heights were about 11 mm. The crucibles were all provided with small extensions on the outside wall so that they could be placed on a wire stirrup. The average mass of the Knudsen crucibles was about 600 mg, and that of the Langmuir crucibles was about 400 mg. Samples were loaded into the Knudsen crucibles through the orifice by a capillary tube used as a funnel. For mass spectrometric experiments, the Knudsen crucibles were made from 9-mm id tubes and were 16 mm in height; the Langmuir crucibles were made from 5.6-mm id tubes and were 7 mm in height. The Knudsen crucible had an indentation at the bottom to serve as a thermocouple well; the bottom of the Langmuir crucible was flat.

The area and the length of the orifice of Knudsen crucibles and the cross-sectional area of the Langmuir crucibles were measured with the Abbé drawing apparatus (*53*) and a Leitz Dial-pol polarizing microscope.

Vacuum-Balance Rate-of-Weight-Loss Experiments

A Cahn RG 2000 Automatic Electrobalance was used in rate-of-weight-loss experiments. It is also equipped with a Cahn Time Derivative Computer (TDC) whose output is proportional to the rate of weight change. The balance and the sample were kept in a vacuum system that was usually evacuated to pressures below 1×10^{-5} Torr before experiments were started. A heating tape wrapped around the sample hangdown tube was used to heat the sample. During the experiments, the sample temperature was usually varied randomly in a stepwise manner. Temperatures were measured with a chromel–alumel thermocouple fixed at the same vertical position as the crucible, but about 1 cm away. A Leeds and Northrup 8690 potentiometer was used to measure the thermocouple emf. The mass of the sample was continuously recorded on a Sargent stripchart recorder. The output of the TDC was recorded on another separate Sargent stripchart recorder.

Quadrupole Mass Spectrometric Experiments

An EA1 QUAD 200 quadrupole mass spectrometer (QMS) was used in these measurements. This instrument and its auxiliary apparatus have been described elsewhere (*54*). The mass range and the resolution of the mass spectrometer were calibrated using perfluorotributyl amine

(FC-43). A reference spectrum of this calibrating compound was obtained with a Varian-MAT CH-5 magnetic deflection mass spectrometer.

Samples were heated in a furnace assembly similar to that described elsewhere (*54*) but with slight modifications made to accommodate the Pyrex glass crucibles and the relatively lower temperature operation. The sample temperature was measured with a $Pt–Pt_{90}Rh_{10}$ thermocouple. The thermocouple emf was measured with a Leeds and Northrup potentiometer and was also continuously monitored with a Moseley stripchart recorder.

The ion intensities were measured with an HP 2100 series minicomputer-based data acquisition system. After the peak positions had been established, they were brought into focus at a predetermined sequence and the intensity of each peak measured for a preselected period of time, normally about 2 s. During the experiments, the intensity data were displayed on real-time basis and stored in a temporary disc file. After the experiments, the temporary file was copied to a permanent file and subsequently retrieved.

The procedures in all experiments were nearly the same. After the sample was loaded and the vacuum system sealed, the sample was evacuated for several hours until the pressure in the sample region was about 10^{-7} torr or below, and that in the ion source region was about 10^{-8} torr or below. The ion source was then outgassed for at least 2 h until the intensity of the only background selenium ion, Se_2^+, decreased to a constant low level. The sample was then heated slowly from room temperature to about 150°C until the major ion Se_6^+ was observed. This heating required about 3 h. During this period, the following operations were usually performed: the background spectrum was recorded; the data acquisition equipment and programs were tested; the shutter operation was examined; the alignment of the molecular beam relative to the shutter slit and the position of the QMS ionizer entrance slit was inspected and optimized.

The sample temperature was then raised to above the melting point and allowed to become constant, at which time readings were taken. In all experiments, the data were taken with the mass spectrometer operated at 12 eV ionizing electron energy. During the experiments, the sample temperature was usually varied randomly in a stepwise manner.

RESULTS

The results from the vacuum-balance experiments in effect were used to calibrate the mass spectrometric experiments. The vapor composition above molten selenium and the partial pressures, vaporization coefficients, enthalpies, and entropies of vaporization of individual species were obtained. The errors in these quantities were also estimated.

Crucible Dimensions

Typical dimensions for the Knudsen crucibles (e.g., crucible No. BK-18) used in the vacuum-balance experiments are the following: orifice area = 0.356 ± 0.001 mm^2; orifice length = 2.34 ± 0.03 mm; and Clausing factor = 0.255 ± 0.002. The Langmuir crucibles have a typical cross-sectional area of about 14 mm^2. The dimensions of crucibles used in mass spectrometric experiments were not measured because only the relative ion intensities were of interest.

Quadrupole Mass Spectrometric Experiments

Raw data from a mass spectrometric experiment consist of one plot for each ion intensity as a function of time. The time interval between two successive points was about 19 s. Each plot consists of segments of rising and falling intensities that mark when the sample temperature increases and decreases, respectively. In addition, there are horizontal segments that occur when the sample temperature reaches constant values. These horizontal segments have an average number of 30 intensity readings. These readings were averaged ($\bar{I}$) and the standard deviation of the mean ($\sigma_{\bar{I}}$) was calculated. The unshutterable background intensity for each species obtained from the nearby shutter check was subtracted from the mean intensity of the respective species.

To determine the equilibrium enthalpy (ΔH_v°) of vaporization from the Knudsen experiments and the activation enthalpy (ΔH_v^*) of vaporization from the Langmuir experiments for each species n, the log ($\bar{I}_n T$) vs $1/T$ data were fitted to a linear equation,

$$\log (\bar{I}_n T) = -A_n/T + B_n \tag{4}$$

by a weighted least-squares treatment. This weighting was done because, for each species, the number of intensity readings at various temperatures was not the same and the intensity was lower at lower temperatures. The data and the weighted least-squares lines are shown in Figs. 1 and 2 for Knudsen and Langmuir experiments, respectively. These plots indicate that the temperature coefficient under the non-equilibrium condition is lower than that under the equilibrium condition for each species.

Vacuum-Balance Rate-of-Weight-Loss Experiments

The data from these experiments were total rate-of weight-loss as a function of temperature. These data were used to calculate the effective total pressure P_E assuming that Se_6 was the only vapor species and also were combined with the respective mass spectrometric ion intensities versus temperature data to obtain the partial pressure for each species.

The effective pressure was calculated using the equation

$$P_E = (1/44.33w)\,(m/at)\,(T/M)^{1/2} \tag{5}$$

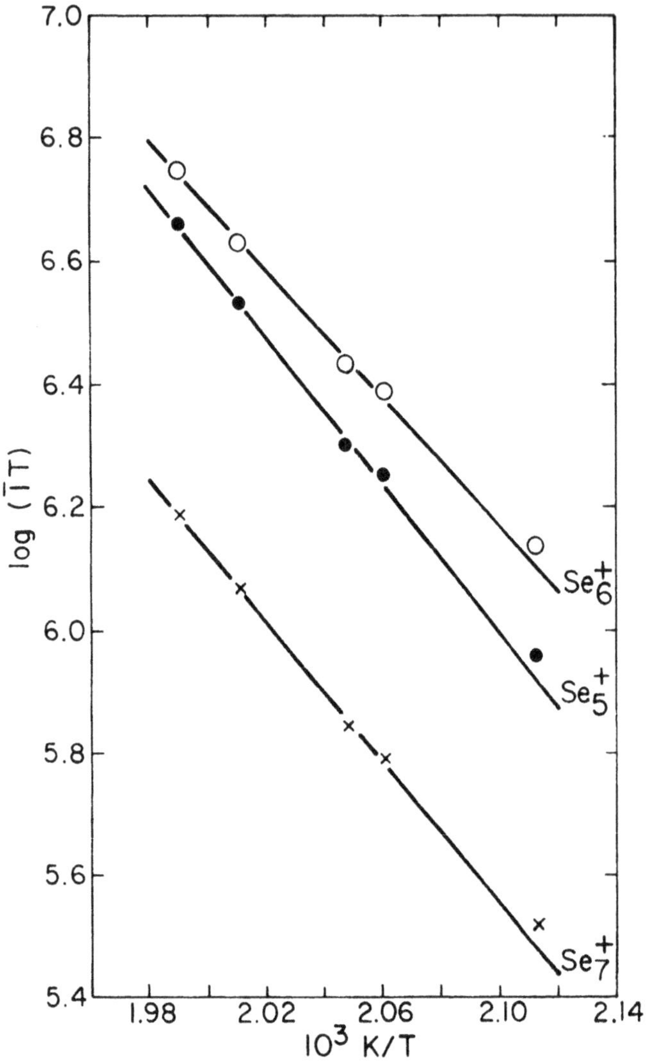

(a)

Fig. 1. Plot of log ($\bar{I}T$) vs $1/T$ for species (a) Se_5^+, Se_6^+, and Se_7^+; and (b) Se_2^+, Se_3^+, Se_4^+, and Se_8^+ obtained from Knudsen experiment KUK-9.

where P_E is in atmospheres; m/at is the rate-of-weight-loss in g/cm^2 s; w is the Clausing factor (55), which was assumed to be unity for Langmuir crucibles; and M is the molar mass of Se_6. The calculated P_E at each temperature was plotted and fitted to a linear equation

$$\log P_E = -A/T + B \tag{6}$$

by the least-squares method.

The results are shown in Fig. 3 along with the least-squares lines, with K and L standing for Knudsen and Langmuir experiments, respectively. The three points below line K and the five points below line L were not included in the data treatment. Among these points, those de-

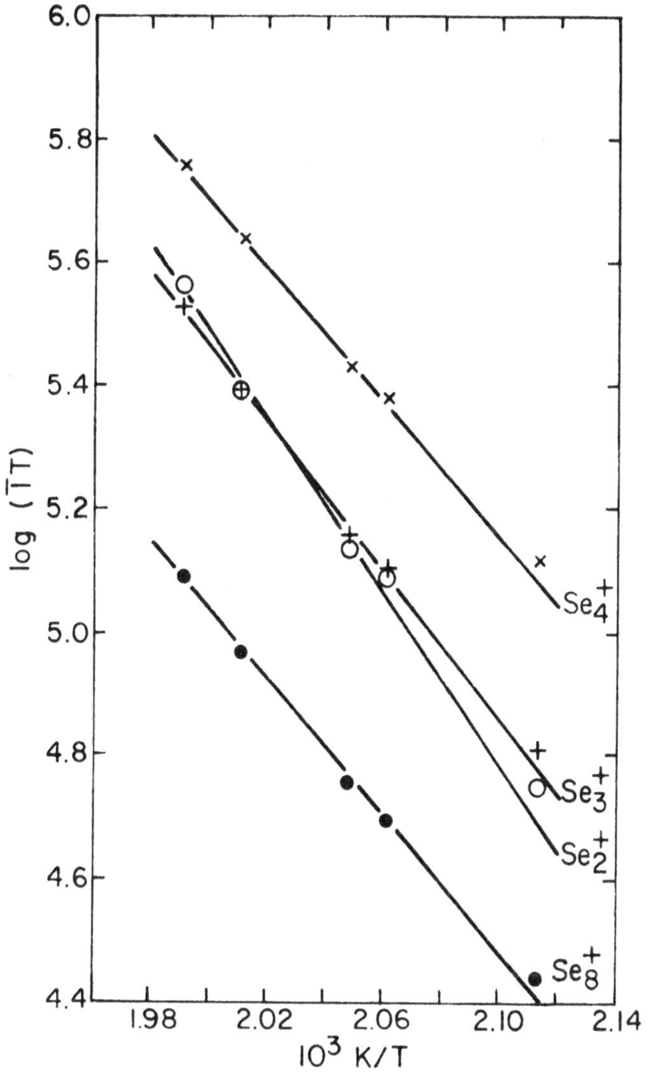

(b)

Fig. 1b.

picted as solid circles and triangles were taken while the sample was still below its melting point; the remaining one point each below lines K and L were obtained perhaps before the sample had reached its equilibrium and steady-state rate-of-weight-loss, respectively. This figure shows that the effective Langmuir pressure is one order of magnitude lower than the effective Knudsen pressure.

The rate-of-weight-loss data were also used to determine the partial pressure of each vapor species in the following manner. The partial pressure P_n of vapor species n is related to its rate-of-weight-loss, temperature, and molar mass M_n by

$$P_n = (1/44.33w)\ (m_n/at)\ (T/M_n)^{1/2} \tag{7}$$

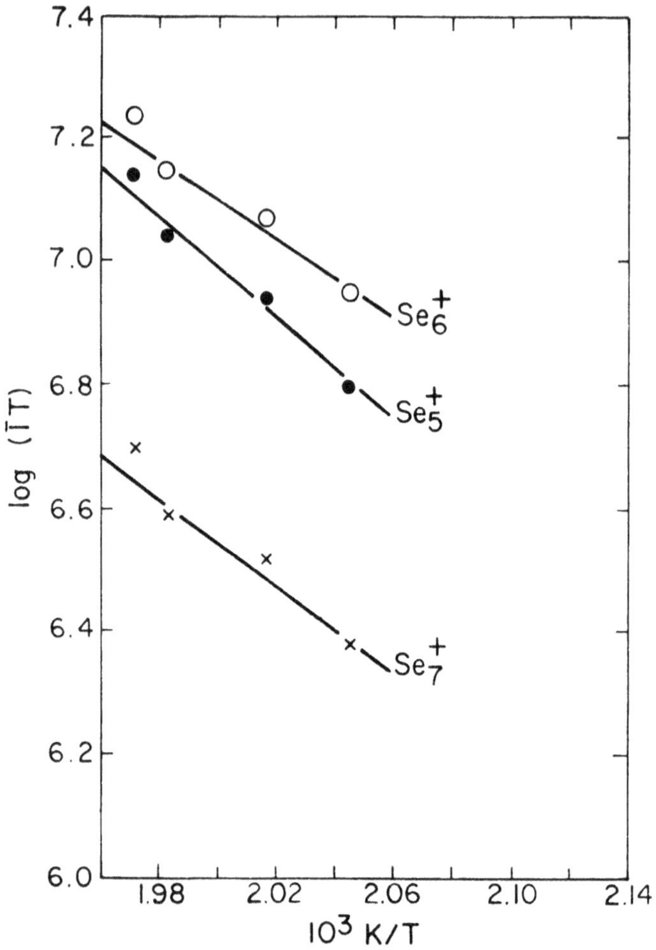

(a)

Fig. 2. Plot of log ($\bar{I}T$) vs $1/T$ for species (a) Se_5^+, Se_6^+, and Se_7^+; and (b) Se_2^+, Se_3^+, Se_4^+, and Se_8^+ obtained from Langmuir experiment KUL-16.

and is also related to the mass spectrometrically measured ion intensity $\bar{I}_n$ and temperature by

$$P_n = \bar{I}_n T/\sigma_n \gamma_n f_n \tag{8}$$

where σ_n is the ionization cross-section, γ_n is the electron multiplier efficiency, and f_n is the relative isotopic abundance.

If one assumes that (1) all the observed ions were parent ions; (2) the measured maximum ion intensity for the peak consisting of a group of isotopic masses was proportional to the relative abundance f_n of the most abundant isotope of the group (25); (3) σ_n was proportional to M_n; and (4) γ_n was inversely proportional to the square root of M_n, then

$$\sigma_n = C_1 M_n \text{ and } \gamma_n = C_2 M_n^{-1/2} \tag{9}$$

Thus, upon combining Eqs. (7)–(9), one obtains

$$m_n(T)^{1/2}/at = C\ \bar{I}'_n\ T \tag{10}$$

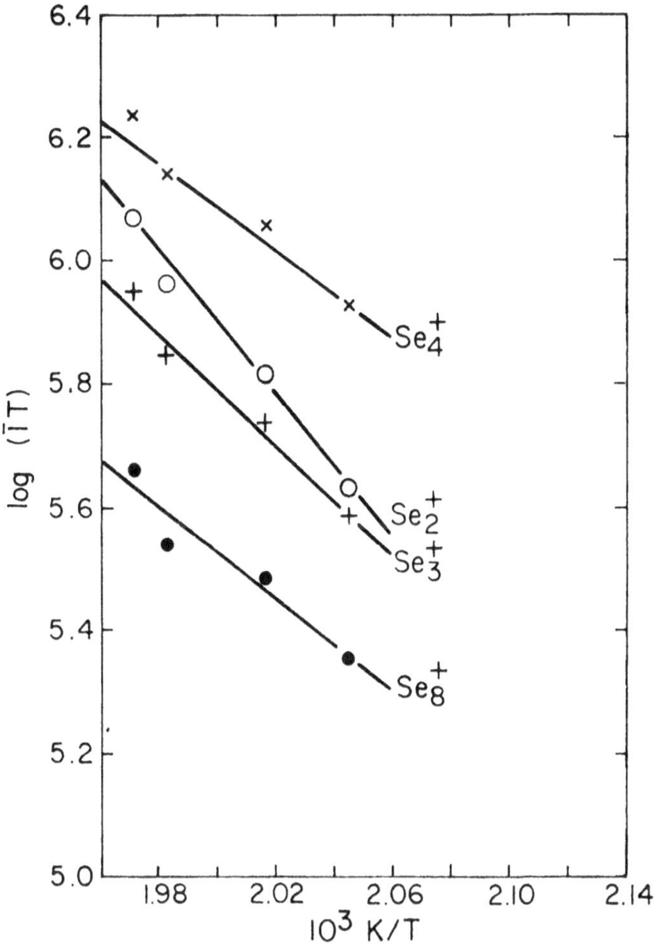

(b)

Fig. 2b.

where $C = 44.33\ w/C_1C_2$ and $\bar{I}'_n = \bar{I}_n/f_n$. By summing over all observed species, one obtains

$$C = [m(T)^{1/2}/at]/T \sum_{n=2}^{8} \bar{I}'_n \tag{11}$$

Since all the mass spectrometric experiments were performed at lower temperatures and overlapped only a narrow temperature range with those of the vacuum-balance experiments, the value of $[m(T)^{1/2}/at]$ used in Eq. (11) was either an experimental value chosen at the temperature at which mass spectrometric measurement was made or a value calculated from the equation that was fitted to the vacuum-balance measured $[m(T)^{1/2}/at]$ vs $1/T$ data.

Thus, the partial pressures are related to $(\bar{I}_nT)$ by the following equation:

$$\log P_n = -A_n/T + B_n - \log (44.33\ wM_n^{1/2}f_n/C) \tag{12}$$

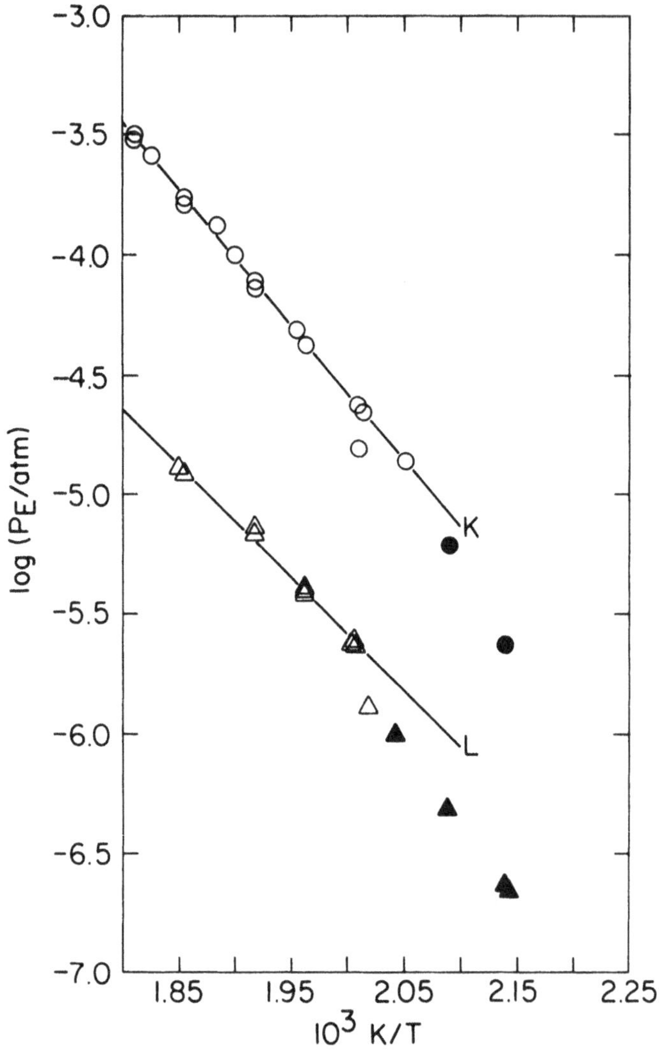

Fig. 3. Plot of log P_E(atm) vs $1/T$ obtained from Knudsen experiment ASK-1 (line K) and Langmuir experiment ASL-1 (line L).

The partial pressures at various temperatures under both Knudsen and Langmuir conditions and, consequently, the vaporization coefficient α_v° are calculated using Eq. (12).

Table 1 summarizes the calculated partial pressures and the vapor compositions at 500 K. The first column identifies the vapor species; the second column lists the Knudsen partial pressures; the third column lists the Langmuir partial pressures; the fourth column lists the equilibrium mole percent; the fifth column lists the non-equilibrium mole percent; and the sixth column lists the equilibrium mole percent obtained by Berkowitz and Chupka (*30*) for comparison. This table shows, as do the plots, that the non-equilibrium pressures are lower than the equilibrium pressures by a factor of ten. But the vapor compositions under equilib-

TABLE 1
Partial Pressures (P) of Gaseous Selenium Species and Vapor Compositions (X) under Equilibrium (K) and Non-equilibrium (L) Conditions at 500 K

Vapor species	P_K, 10^{-7} atm	P_L, 10^{-7} atm	X_K, mol %	X_L, mol %	X_K^a, mol %
Se_2	5.7 ± 0.2	0.58 ± 0.02	2.1	2.2	5.4
Se_3	6.0 ± 0.2	0.49 ± 0.02	2.2	1.8	0.15
Se_4	11.0 ± 0.4	1.04 ± 0.04	4.0	3.9	—
Se_5	94 ± 4	8.8 ± 0.4	34.2	33.0	36
Se_6	120 ± 5	11.9 ± 0.5	43.7	44.6	42
Se_7	35 ± 2	3.5 ± 0.1	12.7	13.1	14
Se_8	2.9 ± 0.1	0.35 ± 0.01	1.1	1.3	2.3

[a]Berkowitz and Chupka (*30*).

rium and non-equilibrium conditions are not significantly different. The results also indicate that the observed major species in the present study are Se_6, Se_5, and Se_7, which is in agreement with results obtained by others.

The enthalpies and the entropies of vaporization are also obtained from Eq. (12) in the usual way. The results are summarized in Table 2, where the first column identifies the vapor species; the second and the third columns list the equilibrium and the activation enthalpies of vaporization, respectively, near 500 K; the fourth and the fifth columns list the equilibrium and the activation entropies of vaporization, respectively; the sixth and the seventh columns list the calculated equilibrium and activation free energies, respectively, at 500 K; and the eighth column lists the calculated vaporization coefficients at 500 K. As shown in this table, both the activation enthalpy and the activation entropy are lower than their corresponding equilibrium quantities. Nevertheless, the activation free energy is greater than the equilibrium free energy of vaporization. Thus, there is a free energy barrier to the vaporization reaction for each species.

Error Analysis

We estimate that the relative standard error in the quantity $[m(T)^{1/2}/at]$ obtained in vacuum-balance experiments is about 3%. The relative standard error in the measured ion intensities is between 1 and 2%; this does not take into account the greater uncertainties in ionization cross-sections, electron multiplier efficiencies, and fragmentations. The estimated relative standard error in the derived quantity C is about 3%. The relative standard error in the Clausing factor w is assumed to be the same as that in the orifice length-to-diameter ratio, which is about 2%. From Eq. (12) one can estimate the relative standard error in partial pressures, which is about 4%. This gives a relative standard error in vaporization coefficients of about 6%.

TABLE 2
Equilibrium and Activation Enthalpies, Entropies, Free Energies of Vaporization, and Vaporization Coefficients for Various Species above Molten Selenium

Vapor species	ΔH°_v, kcal/mol	ΔH^*_v, kcal/mol	ΔS°_v, cal/mol K	ΔS^*_v, cal/mol K	ΔG°_v, kcal/mol	ΔG^*_v, kcal/mol	α°_v
Se_2	32.0 ± 1.7	26.6 ± 1.9	35.5 ± 3.4	20.1 ± 3.9	14.3 ± 2.4	16.6 ± 2.8	0.10 ± 0.01
Se_3	27.6 ± 1.0	20.5 ± 1.6	26.8 ± 2.0	7.6 ± 3.2	14.2 ± 1.4	16.7 ± 2.3	0.08 ± 0.01
Se_4	24.9 ± 0.9	16.2 ± 1.6	22.6 ± 1.9	0.4 ± 3.2	13.6 ± 1.3	16.0 ± 2.3	0.09 ± 0.01
Se_5	27.6 ± 1.0	18.6 ± 2.3	32.3 ± 2.1	9.4 ± 4.6	11.5 ± 1.5	13.9 ± 3.3	0.09 ± 0.01
Se_6	23.9 ± 0.8	14.5 ± 1.4	25.4 ± 1.7	1.8 ± 2.9	11.2 ± 1.2	13.6 ± 2.0	0.10 ± 0.01
Se_7	26.2 ± 0.9	16.2 ± 2.9	27.4 ± 1.9	2.9 ± 5.8	12.5 ± 1.3	14.8 ± 4.1	0.10 ± 0.01
Se_8	25.3 ± 0.8	17.3 ± 2.9	20.7 ± 1.7	0.4 ± 5.7	15.0 ± 1.2	17.1 ± 4.1	0.12 ± 0.01

DISCUSSION

The results obtained in this study are first compared to those obtained by others and then evaluated. Finally, a mechanism for the vaporization of molten selenium is proposed.

Comparison of Results

The equilibrium vapor composition at 500 K obtained in this study using 12 eV ionizing electron energy agrees well with that obtained by Berkowitz and Chupka using 70 eV ionizing electron energy, as shown in Table 1.

The equilibrium enthalpies of vaporization obtained in this study at 500 K are higher than those obtained by others at 625 K, as shown in Table 3. The reason for this difference is not clear, but several features can be cited. (1) Our purpose was to compare Knudsen and Langmuir pressures and enthalpy changes, not to obtain highly reliable values of enthalpies of vaporization. (2) We wished to keep the temperature only slightly above the melting point; hence, our temperature interval is smaller than the others. (3) Also, our temperature is lower, but heat capacity effects would not be expected to account for the difference.

The effective equilibrium total pressure P_E calculated from the present study's vacuum-balance rate-of-weight-loss data, assuming Se_6 is the only vapor species, agrees well with results obtained previously by others, which were taken from Mills' compilation (*56*) and are shown in Table 4.

Reasonableness of the Results

Before attributing the results of this study to any particular molecular process leading to vaporization behavior, it is necessary to analyze the possible consequences of other factors. These factors are discussed in this section.

TABLE 3
Comparison of Equilibrium Enthalpies of Vaporization
ΔH°_v, kcal mol^{-1}

Vapor species	Present study 500 K	Berkowitz and Chupka (*29*), 625 K	Yamdagni and Porter (*26*), 625 K
Se_2	32.0 ± 1.7	28.3	22.8
Se_3	27.6 ± 1.0	—	—
Se_4	24.9 ± 0.9	—	—
Se_5	27.6 ± 1.0	25.8	23.1
Se_6	23.9 ± 0.8	19.9	20.0
Se_7	26.2 ± 0.9	20.9	20.2
Se_8	25.3 ± 0.8	21.0	21.6

TABLE 4
Comparison of Equilibrium Total Pressure of Liquid Selenium

Workers	Method	$\log P(\text{atm}) = A/T + B$		P (at 500 K),
		$-A$	B	10^{-5} atm
Brooks (1952)	Quartz bourdon gauge	4989.5	5.208	1.69
Neumann and Lichtenberg (1939)	Torsion-effusion	5391	5.750	0.93
Niwa and Shibata (1940)	Torsion-effusion	5681	6.357	0.99
Kudryavtsev and Ustyugov (1962)	Boiling point	5011	5.23	1.61
Baker (1968)	Boiling point	5043	5.265	1.51
Yamdagni and Porter (1968)	Torsion-effusion	6066	7.556	2.65
Bonilla and Shulman (1964)	Manometric	5468	5.722	0.61
Present study	Vacuum-balance effusion	5622 ± 66	6.67 ± 0.13	2.67

1. Effect of Error in Clausing Factor on the Vaporization Coefficient

To calculate vapor pressures from rate-of-weight-loss data, one needs to correct for the reduced transmission probability of molecules from a non-ideal orifice or crucible opening. This is done by assigning an appropriate value to the Clausing factor, which is always less than or equal to unity. Since the pressure is inversely proportional to the Clausing factor, too low a value would cause the calculated pressure to be too high. This relationship means that the vaporization coefficient (which is the ratio of Langmuir pressures to Knudsen pressures) would be too low if the Clausing factor used in calculating the Langmuir pressures was too high, or if the Clausing factor used in calculating the Knudsen pressures was too low.

In this study, the Knudsen pressures are calculated using Clausing factors that are determined from the ratios of orifice length to orifice radius. The random errors in the Clausing factors caused by the errors in the ratios of orifice length to orifice radius are about 2%, whereas no systematic errors can be easily assessed. On the other hand, the Langmuir pressures are calculated assuming that the Clausing factors are unity because the samples vaporize from open crucibles rather than from orifices having finite length. One could suspect that this assumption would be invalidated as the sample surface receded and, consequently, the rate-of-weight-loss or the ion intensity decreased. Assuming that, at a constant temperature, the sample surface receded from a level about 0.4 cm from the top to the bottom of a crucible whose height is about 0.8 cm and radius is about 0.2 cm, one could estimate that the Clausing factors would change from a value of 0.8–0.5, which would cause the Langmuir pressures calculated using unity Clausing factors to be too low by a factor of about 1.3 at the beginning and about 2.0 near the end, which could not account for the observed vaporization coefficients of about 0.1. The values of ΔG_v^* would be affected, but not enough to change the conclusions drawn.

2. Effect of Systematic Errors in the Ionization Cross-Section and the Electron Multiplier Efficiency on Partial Pressures and Vaporization Coefficient

The ionization cross-section for each vapor species is assumed to be proportional to its molecular weight, and the electron multiplier efficiency inversely proportional to the square root of its molecular weight. These assumptions could, and usually do, cause significant systematic errors in the absolute values of vapor pressures. But comparisons of the results obtained in this study with those obtained by others indicate that the absolute values of the partial pressures of major vapor species, and also the total pressures calculated by combining the vacuum-balance rate-of-weight-loss data with the mass spectrometric total ion intensity data, do not deviate from those obtained by others by a factor of 10. Furthermore, the vaporization coefficient was calculated as the ratio of

Langmuir pressure to Knudsen pressure; therefore, the ionization cross-section and the electron multiplier efficiency and their associated errors cannot play a role important enough to give a vaporization coefficient of about 0.1.

3. Effect of Surface Cooling on the Vaporization Coefficient

Vaporization reactions are endothermic processes. The temperature at the liquid surface in the Langmuir experiments might be lower than that in the liquid bulk and/or that in the surroundings, which would lower the measured vapor pressure and thereby reduce the vaporization coefficient.

In the mass spectrometric measurements of this study, the thermocouples were placed against the bottom of the cell; and in the vacuum-balance measurements, beside the suspended cell. These arrangements do give a lower sample bulk temperature and a still lower sample surface temperature than the indicated temperature. Since the difference between the activation enthalpy and the equilibrium enthalpy of vaporization ($\Delta H_v^* - \Delta H_v^\circ$) for each species is about 10 kcal/mol, however, the sample surface temperature would have to be lower than the measured temperature, by about 45° at 500 K, to cause the vaporization coefficient to deviate from unity by 0.9. This is not likely to occur. Therefore, it can be concluded that the observed significant deviation of vaporization coefficient from unity cannot be attributed to any large extent to the surface cooling effect.

4. Effect of Fragmentation

The vapor compositions in this study are determined based on the assumption that all observed ionic species Se_n^+ (n = 2–8) in the mass spectrometer are formed from their respective neutral gaseous molecules. This assumption is made because all the mass spectrometric data were taken at a rather low ionizing electron energy of 12 eV.

It is difficult to assess quantitatively the extent of fragmentation caused by the complexity of the vapor phase; as a consequence of this uncertainty, the true vapor compositions are uncertain. But since the three most abundant ionic species, Se_6^+, Se_5^+, and Se_7^+, comprise about 90% of the total ion intensity, and since the existence of Se_3 and Se_4 neutral molecules have indeed been reported, one would expect only minor corrections, maybe a total of 10% at the most, to the vapor compositions reported in this study. These corrections would probably not be significant enough to change the reported enthalpy and entropy values by an amount greater than their associated errors. The effect of these corrections on the vaporization coefficients would also seem to be insignificant.

C. Summary of the Vaporization Behavior of Liquid Selenium

Based on the results obtained from this study, the vaporization behavior of liquid selenium can be summarized as follows:

1. The vapor of liquid selenium consists of about 44% Se_6, 34% Se_5, 13% Se_7, 4% Se_4, 2% Se_3, 2% Se_2, and 1% Se_8 at 500 K.
2. The vapor compositions under equilibrium and non-equilibrium conditions are about the same.
3. The vaporization coefficient α_v° for molten selenium is about 0.1 for each species at 500 K.
4. The activation enthalpy of vaporization ΔH_v^* for each species is lower than the equilibrium enthalpy of vaporization ΔH_v°; the activation entropy of vaporization ΔS_v° for each species is also lower than the equilibrium entropy of vaporization ΔS_v°. Nevertheless, the activation free energy of vaporization ΔG_v^* for each species is higher than the equilibrium free energy of vaporization ΔG_v° by about 2 kcal/mol at 500 K.

Vaporization Mechanism of Molten Selenium

A mechanism for the vaporization of molten selenium based on the transition-state theory is proposed. This mechanism is consistent with several experimental observations.

According to Searcy (*9,57*), when the activation enthalpy of vaporization ΔH_v^* is less than the equilibrium enthalpy of vaporization ΔH_v°, as is observed in this study for all species, the rate-limiting step must be a surface reaction rather than desorption. Furthermore, since the surface of a liquid cannot sustain any heterogeneities, there is no need to consider where the preferred occurrence of surface reaction is. These observations simplify the problem of proposing vaporization mechanisms to identifying the activated complexes.

We, therefore, propose that the vaporization of molten selenium proceed by the following mechanism,

$$Se_n^s \text{ (on surface, at ends of long chain)} = Se_n^{c,*} \rightarrow Se_n^{r,g} \quad (13)$$

where superscript *s* stands for liquid surface, *c* for chains, * for activated complexes, *r* for rings, and *g* for gas phase; the subscript *n* indicates the size of the molecules. Equation (13) indicates that: (a) equilibrium exists between the chain ends of the liquid molecules and the activated complexes, which is required by the transition-state theory; and (b) the formation of gaseous ring molecules from the activated complex is a fast reaction. The activated complexes are short molecules lying on the liquid surface, but still weakly bonded to long-chain liquid molecules. These short molecules contain the same number of atoms as the gaseous ring molecules they form. They are, of course, not straight and have a configuration nearly that of the rings. The formation of such short molecules would require about the same activation free energy regardless of the length. We find that ΔG_v^* for all species is about the same. This model suggests that the activation free energy of vaporization ΔG_v^* would be approximately equal to the free energy change of the depolymerization reaction. The results of previous studies by others on the depolymerization reaction show that the free energy change is about 17 ± 3 kcal/mol at 500 K. The results obtained from this vaporization study show that ΔG^*_vs vary from 13.6 kcal/mol for Se^g_5 to 17.1 kcal/mol for Se^g_8 with uncertainties of about 3 kcal/mol. The good agreement between the results of these two entirely different studies seems to substantiate this mechanism.

This mechanism is further supported by the following observations. The results of this study show that the activation enthalpies ΔH^*_v of all species are lower than the equilibrium enthalpies ΔH^o_v. The activation enthalpies of vaporization are also lower than the enthalpy of depolymerization reaction. These facts indicate that the activated complexes are still attached to the long-chain liquid molecules on the surface, but with the bond stretched to such an extent that it becomes very weak. The activation entropies of vaporization ΔS^*_v determined in the present study are much lower than the equilibrium entropies ΔS^o_v. This fact indicates that the activated complexes are more reactant-like (chains) than product-like (rings), which also supports the view that the activated complexes are still attached to the long-chain liquid molecules. Moreover, they are probably lying on the surface.

This mechanism offers an explanation to the observation that in all vacuum-balance experiments, the material collected on cold walls of vacuum systems was red and homogeneous, which appeared to be amorphous, rather than grey and crystalline. If condensation and vaporization can be assumed to occur via the same activated complexes (*9*), this mechanism suggests that the condensation process must also be a retarded reaction, because the gaseous ring molecules upon sticking to the liquid surface have to undergo ring-opening reactions to form the same-sized activated complexes before they can condense to form the thermodynamically stable phase. This retarded condensation requires a nonzero activation free energy of condensation ΔG^*_c, which will be approximately the energy required to break one Se—Se bond in the gaseous ring molecules; consequently, its rate will decrease as the surface temperature decreases. On the other hand, as the temperature of the solid surface decreases, the probability of the gaseous ring molecules sticking on the surface tends to increase (*11*). This increase happens because the weak intermolecular and gas–solid attractive forces become more important as temperature decreases. Thus, a certain temperature exists below which the gaseous molecules, upon striking the surface, will certainly stick. They preserve their ring structures and cannot be incorporated into the thermodynamically stable condensed phase, grey trigonal selenium. Previous workers (*34,58,59*) report that the films sticking on the cold surface are red and amorphous. According to the present study, red amorphous material sticking on the cold surfaces should consist of about 44% Se_6, 33% Se_5, 13% Se_7, and minute amounts of other species. The complexity of such a mixture might account for the ambiguity in the previous reports on the structure of amorphous films of selenium. And, in particular, a high abundance of Se_8 rings is very unlikely because of the low abundance of Se_8 in the gas phase.

As to the question of why the concentration of Se_8 is so low in the selenium vapor even though the liquid consists of 40 wt% Se_8 rings at 500 K and Se_8 is the most stable gaseous species with respect to dissociation into atoms, this mechanism provides an answer. The Se_8 rings in the liquid do not vaporize to form Se_8 gaseous ring molecules. They are in equilibrium with long-chain liquid molecules, which in turn have to form the weakly bonded activated complex $Se_8^{c,*}$ on the liquid surface before vaporizing.

One may ask why the vaporization coefficients are as high as 0.1 when the structures of the liquid molecules and the gaseous molecules are seemingly so different. This model suggests that to form the activated complexes, the liquid molecules do not have to undergo significant rearrangements and that, from the point of view of the long-chain liquid molecules, the structures of the gaseous molecules are not very different.

Since the activated complexes are, according to this model, molecules on the liquid surface, they must have one unpaired electron at the free end. Hence, they are paramagnetic, similar to the diradical long-chain liquid molecules that have been subjected to ESR and magnetic susceptibility studies. This model predicts, therefore, that if an electric field were applied parallel to the liquid surface, the radicals should align in the direction of the applied field. Consequently, the ring-closure step would be hindered, which in turn would cause the vaporization coefficient to decrease. On the other hand, if an electric field were applied perpendicular to the liquid surface and if the field strength were sufficient to conquer the surface intermolecular forces, these short chains would stand perpendicular to the surface and vaporize to form diradical gaseous molecules rather than rings. Further, if these short-chain diradical molecules strike a cold surface, they should condense to form grey trigonal selenium instead of red amorphous thin films. Thus, in addition to providing a critical test of the model, this idea suggests an alternative route to the preparation of high purity single crystals of trigonal selenium, which are notoriously difficult to prepare from liquid because of its high viscosity.

An equally appropriate alternative model for the activated complexes is small, nearly closed loops, interior to the chains, which have the same size as gaseous ring molecules to be formed, lying on the liquid surface that are still weakly bonded twice to long-chain liquid molecules. The remaining segments of the long-chain molecule bond together afterwards to form slightly shorter chains. This model implies higher probabilities for forming the activated complexes because more atoms are along the chain than at the ends, and for the following ring-closure step, because the end atoms are closer. This alternate model, however, faces some difficulties when considering the condensation reactions.

SUMMARY

Molten selenium samples were vaporized under both equilibrium and non-equilibrium conditions. The vapor composition was measured with a quadrupole mass spectrometer, and the total rate-of-weight-loss was measured with a vacuum microbalance. The experimental results showed that: (1) the vapor of molten selenium consists of about 44% Se_6, 34% Se_5, 13% Se_7, 4% Se_4, 2% Se_3, 2% Se_2, and 1% Se_8 at 500 K; (2) the vapor compositions under equilibrium and non-equilibrium conditions are about the same; (3) the vaporization of molten selenium is a retarded reaction, and the vaporization coefficient α_n° is about 0.1 for all species at 500 K; (4) the activation enthalpy ΔH_v^* and the activation entropy ΔS_v^* of vaporization are lower than the equilibrium enthalpy ΔH_v° and the equilibrium entropy ΔS_v° of vaporization, respectively, yet the activation free energy of vaporization ΔG_v° is higher than the equilibrium free energy of vaporization ΔG_v° for each species at 500 K; (5) the activation free energy of vaporization ΔG_v^* is approximately equal to the free energy change of depolymerization reactions in the liquid phase at 500 K.

Two alternative mechanisms are postulated for the vaporization of molten selenium that are consistent with our observations. One mechanism is that the long polymeric chains form molecules lying on the surface but still weakly bonded to liquid molecules; these molecules contain the same number of atoms as the gaseous ring molecules they form and are at the ends of long-chain liquid molecules. The rate-limiting step is the formation of these activated complexes on the liquid surface. The other mechanism involves the formation of activated complexes that are small, nearly closed loops, interior to the chains, having the same size as the gaseous ring molecules to be formed. These activated complexes are lying on the liquid surface and are still weakly bonded twice to long-chain liquid molecules. The remaining segments of the long-chain molecule bond together afterwards to form slightly shorter chains.

A critical test has also been proposed for this model that, if proven correct, might lead to a novel technique for preparing pure single crystals of trigonal selenium.

ACKNOWLEDGMENTS

The authors are pleased to acknowledge the support of the United States Department of Energy and its predecessors under Contract No. E(11-1)-1140 and of the General Research Fund of the University of Kansas. One author (JYKH) also wishes to express his appreciation to Midwest Research Institute for its support during the preparation of this manuscript.

Abstracted from the PhD Dissertation of John Yun-Kuang Huang, University of Kansas, 1981.

REFERENCES

1. Gilles, P. W., *J. Am. Ceramic Soc.* **58,** 279 (1975).
2. Stranski, I. N., and Wolff, G., *Research* **4-1,** 15 (1951).
3. Knacke, O., and Stranski, I. N., *Prog. Met. Phys.* **6,** 181 (1951).
4. Hirschwald, W., and Stranski, I. N., in *Condensation and Evaporation of Solids,* Rutner, E., Goldfinger, P., and Hirth, T. P., eds., Gordon and Breach, New York, 1964, p. 59.
5. Hirth, J. P., and Pound, G. M., in *Condensation and Evaporation,* Pergamon Press, New York, 1963.
6. Somorjai, G. A., in *Proceedings of the Third International Symposium on High Temperature Technology,* Butterworths, London, 1969, p. 73.
7. Somorjai, G. A., *Science* **162,** 755 (1968).
8. Somorjai, G. A., and Lester, J. E., *Prog. Solid State Chem.* **4,** 1 (1967).
9. Searcy, A. W., in *Chemical and Mechanical Behavior of Inorganic Materials,* Searcy, A. W., Ragone, D. W., and Colombo, U., eds., Wiley-Interscience, New York, 1970, P. 107.
10. Winterbottom, W. L., in *Techniques of Metals Research,* Bunshah, R. F., ed., Interscience, New York, Vol. 4, Part 1, 1970, p. 95.
11. Rosenblatt, G. M., in *Treatise on Solid State Chemistry,* Hannay, N. B., ed., Plenum Press, New York, Vol. 6A, 1976, p. 165.
12. Brewer, L., and Kane, J. S., *J. Phys. Chem.* **59,** 105 (1955).
13. Mortensen, E. M., and Eyring, H., *J. Phys. Chem.* **64,** 846 (1960).
14. McFeely, F. R., and Somorjai, G. A., *J. Phys. Chem.* **76,** 914 (1972).
15. Niwa, K., and Shibata, Z., *J. Faculty Sci.,* Hokkaido University, Japan, Ser. III, **3,** 53 (1940).
16. Neuman, K., and Lichtenberg, E. E., *Z. Physik. Chem.* (Leipsig) **184,** 89 (1939).
17. Metzger, F., *Helv. Phys. Acta* **16,** 323 (1943).
18. Illarionov, V. V., and Lapina, L. M., *Dolk. Akad. Nauk. SSSR* **114,** 1021 (1957).
19. Goldfinger, P., and Jeunehomme, M., *Adv. Mass Spectrom.* **1,** 534 (1959).
20. Brebrick, R. F., *J. Chem. Phys.* **43,** 3031 (1965).
21. Brebrick, R. F., *J. Chem. Phys.* **43,** 3846 (1965).
22. Brebrick, R. F., *J. Chem. Phys.* **48,** 5741 (1968).
23. Keller, H., Rickert, H., Detry, D., Drowart, J., and Goldfinger, P., *Z. Physik. Chem. Neue Folge* **75,** 273 (1971).
24. Hoareau, A., Reymond, J. M., Cabaud, B., and Uzan, R., *J. Phys.* **36,** 737 (1975).
25. Fujisaki, H., Westmore, J. B., and Tickner, A. W., *Can. J. Chem.* **44,** 3063 (1966).
26. Yamdagni, R., and Porter, R. F., *J. Electrochem. Soc.* **115,** 601 (1968).
27. Knox, B. E., *Mat. Res. Bull.* **3,** 329 (1968).
28. Saure, H., and Block, J., *Int. J. Mass. Spectrom. Ion Phys.* **7,** 145 (1971).
29. Berkowitz, J., and Chupka, W. A., *J. Chem. Phys.* **45,** 4289 (1966).
30. Berkowitz, J., and Chupka, W. A., *J. Chem. Phys.* **48,** 5743 (1968).
31. Sigai, A. G., *J. Vac. Sci. Technol.* **12,** 958 (1975).
32. Rau, H., *J. Chem. Therm.* **6,** 525 (1974).
33. Meschi, D. J., and Searcy, A. W., *J. Chem. Phys.* **51,** 5134 (1969).
34. Cooper, W. C., and Westbury, R. A., in *Selenium,* Zingaro, R. A., and Cooper, W. C., eds., Van Nostrand Reinhold, New York, 1974, p. 87.

35. Swanson, H. E., Gilfrich, N. T., and Ugrinic, G. M., *NBS Circular 539,* 4 (1955).
36. Misawa, M., and Suzuki, K., *J. Phys. Soc.* Japan **44,** 1612 (1978).
37. Unger, P., and Cherin, P., in *The Physics of Selenium and Tellurium,* Cooper, W. C., ed., Pergamon Press, New York, 1969, p. 223.
38. Briegleb, S., *Z. Physik. Chem.* **A144,** 321 (1929).
39. Eisenberg, A., and Tobolsky, A. V., *J. Polymer Sci.* **46,** 19 (1960).
40. Koningsberger, D. C., and De Neef, T., *Chem. Phys. Lett.* **4,** 615 (1970).
41. Koningsberger, D. C., Van Wolput, J. H. M. C., and Rieter, P. C. U., *Chem. Phys. Lett.* **8,** 145 (1971).
42. Koningsberger, D. C., and De Neef, T., *Chem. Phys. Lett.* **14,** 453 (1972).
43. Axmann, A., Gissler, W., Kollmar, A., and Springer, T., *Disc. Faraday Soc.* **50,** 74 (1971).
44. Keezer, R. C., and Bailey, M. W., *Mat. Res. Bull.* **2,** 185 (1967).
45. Keezer, R. C., Wood, C., and Moody, J. W., *J. Phys. Chem. Solids Suppl.* **1,** 119 (1967).
46. Keezer, R. C., in *The Physics of Selenium and Tellurium,* Cooper, W. C., ed., Pergamon Press, New York, 1969, p. 103.
47. Lucovsky, G., in *The Physics of Selenium and Tellurium,* Cooper, W. C., ed., Pergamon Press, New York, 1969, p. 255.
48. Waseda, Y., Yokoyama, K., and Suzuki, K., *Phys. Cond. Matter* **18,** 293 (1974).
49. Tourand, G., *J. Phys.* **34,** 937 (1973).
50. Kawarada, M., and Nishina, Y., *Jap. J. Appl. Phys.* **14,** 1519 (1975).
51. Popov, A. I., *J. Phys. C: Solid State Phys.* **9,** L675 (1976).
52. Deprez, J., Rialland, J. F., and Perron, J. C., *J. Phys. Lett.* **39,** L142 (1978).
53. *The Encyclopedia of Microscopy and Microtechnique,* Gray, P., ed., Van Nostrand Reinhold, New York, 1973.
54. Nordine, P. C., PhD Dissertation, University of Kansas, 1970.
55. Dushman, S., *Scientific Foundations of Vacuum Technique,* 2nd ed., Lafferty, J. M., ed., Wiley, New York, 1962.
56. Mills, K. C., *Thermodynamic Data for Inorganic Sulphides, Selenides, and Tellurides,* Butterworths, London, 1974.
57. Searcy, A. W., and Beruto, D., *J. Phys. Chem.* **78,** 1298 (1974).
58. Richter, H., and Herre, F., Z. *Naturforschg.* **7a,** 511 (1952).
59. Kaplow, R., Rowe, T. A., and Averbach, B. L., *Phys. Rev.* **168,** 1068 (1968).
60. Grimley, R. T., et al., *J. Phys. Chem.* **86,** 976 (1982).

NOTE ADDED IN PROOF

Our attention has been called to a paper by Grimley, Grindstaff, DeMercurio, and Forsman (*60*), who studied the vapor composition arising from solid hexagonal selenium by using angular distribution mass spectrometry. From the angular distribution of the ionic species as a function of ionizing electron energy, they concluded that Se_n^+ (n = 5–8) are entirely parent ions, that Se^+ and Se_3^+ are fragment ions, and that Se_2^+ and Se_4^+ are partly parent and partly fragment ions. Their gaseous mole percents at 210°C above solid Se were for (n = 2–8): 3, zero, <2.5, 47, 40, 6, and 2. These values are in fair agreement with those in Table 1 except for Se_3, and except for Se_5, which they reported to be the most abundant gas, instead of Se_6 as observed in this work and in all other studies. They reported apparent values of enthalpies of sublimation at 210°C for various species. Using the value for the enthalpy of fusion 1.4 ± 0.1 kcal gm-at^{-1} from Mills (*56*), we obtain values for the enthalpies of vaporization from the liquid for various species. Comparison of these values with those in Table 3 indicates agreement within 2 kcal, except for Se_2 where the difference is 3 kcal.

For the heavier species (n = 5–8), their work and ours are consistent, and our discussion stands. We chose an electron energy of 12 eV to eliminate the effect of fragmentation, but if the lighter and far less abundant ions (n = 2–4) are fragments under our conditions, our proposed mechanism applies to the neutral parents.

Alexsandrov and Shklyar (*61*) have studied by optical and electron microscopy the growth of hexagonal crystals from amorphous Se films 100 nm thick in an electrostatic field of 1.2×10^6 V/m. They reported preferential growth pointed toward the positively charged electrode. They found enhanced nucleation rate, growth rate and overall crystallization rate in the presence of the field, but the activation energies were larger. They proposed that the field increases the mobility of chains containing "charged unsaturated bonds." Our proposed mechanism relates not to mobility within a thin film, but rather to the orientation of neutral radicals on the surface.

REFERENCES

60. Grimley, R. T., Grindstaff, O. G., DeMercurio, T. A., and Forsman, J. A., *J. Phys. Chem.* **86,** 976 (1982).
61. Aleksandrov, V. D., and Shklyar, V. S., *Russ. J. Phys. Chem.* **55,** 1187 (1981).

0018–1536/84/1700–0135$03.80

Electron–Ion Collisions in High Temperature Plasmas

A. CHUTJIAN

Jet Propulsion Laboratory, California Institute of Technology, Pasadena, CA 91109

Received September 23, 1983; Accepted December 31, 1983

ABSTRACT

Several newly observed electron–ion collision phenomena that occur in high temperature plasmas containing electrons and ions of energies 10–10^4 eV, or temperature of 10^5–10^8 K, are reviewed. These are the phenomena of dielectronic recombination, excitation–autoionization, and electron energy-loss scattering as measured in crossed (90°) and merged electron–ion beam experiments.

Index Entries: Electron–ion collisions, in high temperature plasmas; ion–electron collisions, in high temperature plasmas; high temperature plasmas, electron–ion collisions in; plasmas, electron–ion collisions in.

INTRODUCTION

The need to understand phenomena and measure cross-sections in high-temperature electron plasmas, in which the average electron or ion temperature is greater than 10 eV (or 10^5 K), has been highlighted by several exciting discoveries and advances in the realm of astronomical observations, and in laboratory collision physics.

High temperature plasmas exist in the form of stellar objects such as coronae, chromospheres, supernovae, and quasars; shock-heated portions of the interstellar medium, and the Io–Jupiter torus. These plasmas have only recently been studied in their rich detail by earth-orbiting satellites such as SKYLAB (*1*), the International Ultraviolet Explorer (IUE)

(2), and by the Voyager 1 and 2 Spacecraft (3). The development of fusion reactors, such as the Joint European Torus (4) and the Princeton TFTR (5), requires electron temperatures of the order 10^8 K (10 keV) for ignition to occur. At these stellar-like temperatures, most ions exist in highly stripped charge states, and collisions are almost entirely between electrons and ions. Intriguing physical processes occur such as excitation, dielectronic recombination, ionization and excitation–autoionization, and bremmstrahlung (5–8). Many of the experimental aspects of excitation and ionization have been discussed in earlier reviews (9–12).

In order to characterize the temperature and electron or ion densities of astronomical objects, and to optimize the performance of fusion devices, it becomes necessary to duplicate phenomena that are observed in these plasmas and to measure their cross-sections in well-defined laboratory experiments. In such experiments one would preferentially use an electron beam–ion beam configuration in which a beam of the target ion of known velocity and charge state can be produced. The electron and ion beams are usually crossed at a 90° angle. Because of the low ion densities available (less than 10^8 cm^{-3}), one may also utilize an inclined electron–ion beam geometry. For an inclination angle of 20° one gains a factor of three in interaction length (hence signal). Another technique involves the use of a merged-beam configuration in which, for a merged length of 20 cm, one gains a factor of about 10^2 in interaction length over the usual 90° geometry. In all cases, laboratory electron–ion experiments offer a considerable variety of experimental approaches, and considerable challenges to the experimentalist as well.

In Section II we describe several new phenomena which have been observed in laboratory electron–ion collision studies. These are dielectronic recombination (Sec. IIA), excitation–autoionization (Sec. IIB), and electron energy-loss scattering (Sec. IIC). We present, in particular in Section IIC, experimental details and recent results of measurements in our own laboratory at JPL of energy-loss scattering in Zn^+ and Cd^+.

II. RECENTLY OBSERVED PHENOMENA IN ELECTRON–ION COLLISION EXPERIMENTS

A. Dielectronic Recombination

The process of dielectronic recombination (DR) in ions is often the dominant radiative cooling path for hot plasmas (13). Although the phenomenon has been the subject of considerable theoretical study (14–17), it has only recently been observed in a series of exciting, beam–beam laboratory measurements (18–21). Dielectronic recombination can be most easily demonstrated with the aid of Fig. 1 from Belić et at. (19) Using the Mg^+ ion as an example, DR involves at first (a) the approach of an elec-

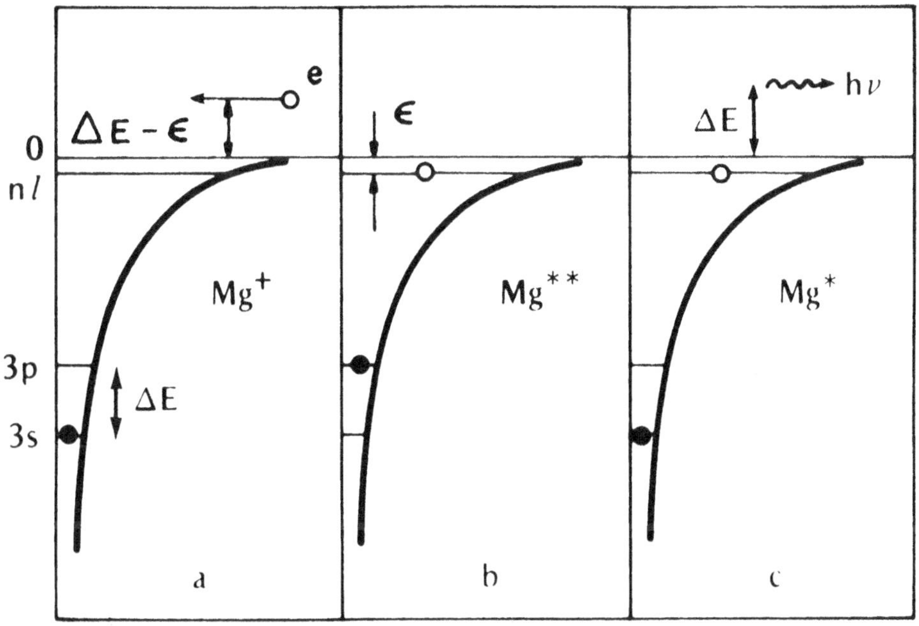

Fig. 1. Sequence of events for dielectronic recombination in the *e*–Mg^+ system (*19*) showing (a) incident electron on the $Mg^+(3s)$ ion, (b) electron capture and excitation to $Mg^+(3p)$, and (c) radiation to $Mg^*(3s)$ with high-Rydberg excitation.

tron (e) of energy $\Delta E - \epsilon$ which is slightly less than that needed to excite the Mg^+ ($3s \rightarrow 3p$) resonance transition at an excitation energy of ΔE = 4.43 eV. Because of the attractive *e*–ion Coulomb force, (b) the electron gains energy, and at distances of several Bohr radii has acquired sufficient energy to excite the 3*s*–3*p* transition (Fig. 1b). However, having acquired the deficient energy from the Coulomb interaction, the electron is obliged to remain bound, in a high-Rydberg orbital, by the energy ϵ. This combination of excited 3*p* state and high-Rydberg excitation is sufficiently long-lived (lifetime of the order 10^{-9} s) to (c) allow the $3p \rightarrow 3s$ radiative transition to occur, leaving the original ground-state Mg^+ ion as a neutral Mg^* atom in a high-Rydberg state. The sequence of events is:

$$e + Mg^+\ (3s) \rightarrow Mg\ (3p,\ nl)$$
$$Mg(3p,\ nl) \rightarrow Mg\ (3s,\ nl) + \hbar\omega$$

Measurements of Belić et al. (*19*) on Mg^+ and Williams (*20*) on Ca^+ were carried out by detecting coincidences between the emitted photon and the fast, high-Rydberg neutral atom. Measurements of Mitchell et al. (*18*) on C^+ involved detection of fast C (2*p*, *nl*) atoms using a modulated electron beam and gated counters to eliminate effects of neutralization by background charge-exchange collisions. More generally, where the target ion A is in charge-state *m*, the process is $e + A^{+m} \rightarrow A^{+(m-1)**} \rightarrow A^{+(m-1)*} + \hbar\omega$. In this case, one would detect enhanced production of charge-separated $A^{+(m-1)*}$ ions as the relative electron–ion energy is swept through energies near $\hbar\omega$. This was in fact the detection method used by Dittner et al. (21) in observing DR in the B^{2+} and C^{3+} ions. A merged beam configuration was used, and the ion beam energy varied through resonance.

In all cases, measured count rates can be converted to absolute DR cross-sections through a standard expression which, for the 90° crossed-beam case, is just (19, 20).

$$\sigma_{DR} = (\mathcal{R}/I_i I_e)\,[e^2 v_i v_e/(v_i^2 + v_e^2)^{1/2}]\,(\mathcal{F}/\zeta_\lambda \zeta_n) \qquad (1)$$

For merged beam geometries the fundamental relation between the cross-section and measured experimental quantities is (18, 21)

$$\sigma_{DR} = (\mathcal{R}/I_i I_e L)\, e^2 |(v_i v_e/\vec{v}_i - \vec{v}_e)| (\mathcal{F}/\zeta_n) \qquad (2)$$

The count rate $\mathcal{R}$ in Eqs. (1) and (2) will be the coincident count rate (19, 20), net neutral count rate (18) or net $A^{+(m-1)}$ ion count rate (21), with corresponding neutral (or ion) detection efficiencies ζ_n. All other quantities in Eq. (1) have their usual meaning: I_i, I_e are ion and electron currents, v_i, v_e their respective velocities, and e the electron charge. The quantity $\mathcal{F}$ is the so-called form factor (22) that corrects for spatial inhomogeneities in the densities of both beams, and ζ_λ the photon detection efficiency. In addition, for the merged-beam geometry one has in Eq. (2) the length L of the merged interaction region.

Recent calculations of McLaughlin and Hahn (23) on DR in C^{3+} and B^{2+} have been carried out in a Hartree-Fock distorted wave approximation, assuming isolated resonances. There is generally good agreement between these calculations and measurements, as shown here in Fig. 2 from Belić et al. (19) for DR in Mg^+, and in Fig. 3 from Dittner et al. (21) for DR in B^{2+}. A similar comparison for C^{3+} is given in Fig. 4 of Ref. 21.

Several unresolved experimental and theoretical effects remain to be studied in detail. For example, the presence of a magnetic field $\vec{B}$ in the collision center (used to confine the electron beam) will appear in the ion's rest frame as an electric field $\vec{E}'$ given by $\vec{E}' = \gamma\, \vec{v}/c \times \vec{B}$ [where $\gamma = (1 - v^2/c^2)^{-1/2}$, c the speed of light, and $\vec{v}$ the ion velocity]. This field is sufficient to cause a mixing of nearly degenerate l sublevels (24, 25). This leads to high l-state contributions to DR which, due to larger statistical weights, enhance the DR cross-section.

Also, the presence of electric or magnetic fields after the collision center (used, for example, to charge-separate the parent ion A^{+m} from $A^{+(m-1)*}$ or neutral A^*) can field-ionize the high-Rydberg nl levels populated in the DR. Thus, some of the DR cross-section is "lost" because of subsequent re-ionization to the parent charge state.

B. Ionization and Excitation–Autoionization

Electron-impact ionization of ions is a phenomenon whose study is fundamental to the understanding of laboratory and astrophysical hot plasmas. In addition to the direct ionization process, wherein the outer-

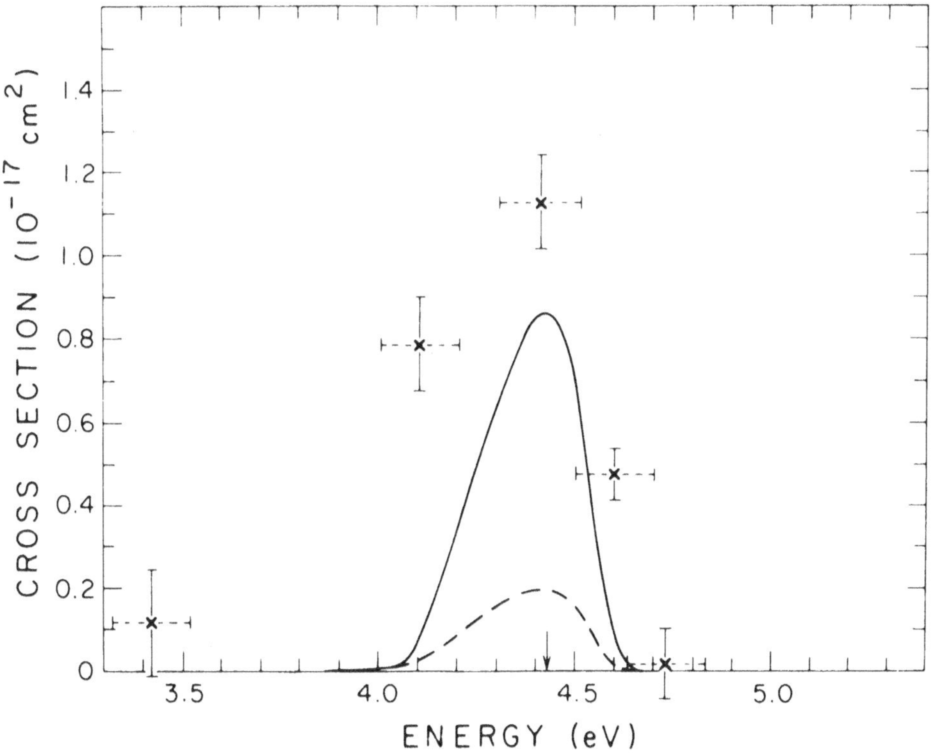

Fig. 2. Dielectronic recombination cross-section vs electron energy in Mg^+ from Belić et al. (*19*). Dashed curve is convolution of theory (*17*) for $n \leq 64$ with experimental electron energy distribution; solid curve, same as dashed curve, including all *n*. Arrow indicates 3*s*–3*p* threshold energy.

most electron is excited to the continuum by the incident electron (so called "knock-out" path), an indirect and in many cases even more probable path exists. This excitation–autoionization (EA) scheme involves (a) excitation of an inner shell electron to an electronic state lying just above the ionization continuum of the target, followed by (b) autoionization of the target to its next higher charge state. This effect was noted experimentally by Peart et al. (*26*) and by Feeney et al. (*27*) for Ca^+, Sr^+, and Ba^+ ionization, and has been recently studied in greater detail by groups at the Joint Institute for Laboratory Astrophysics (JILA) and the Oak Ridge National Laboratory.

As noted by Falk et al. (*28*) excitation–autoionization is most probable for ions in which the initial excitation to states above the continuum occurs with large oscillator strength ($f > 1$), and gives rise to autoionizing levels having energies just above the ionization threshold. These conditions are met by $p \rightarrow d$ transitions of the type $np^6nd^m \rightarrow np^5nd^{m+1}$ followed by ionization to terms of np^6nd^{m-1}. In particular the ions Ti^{3+}, Zr^{3+}, and Hf^{3+} (n = 3, 4, and 5, respectively) have been studied because of their amenability to both experimental and theoretical techniques (*29*). The energy-level structure for these ions, as well as for the series Sc^{2+}, Y^{2+}, La^{2+} and V^{4+}, Nb^{4+}, Ta^{4+} are shown in Fig. 4.

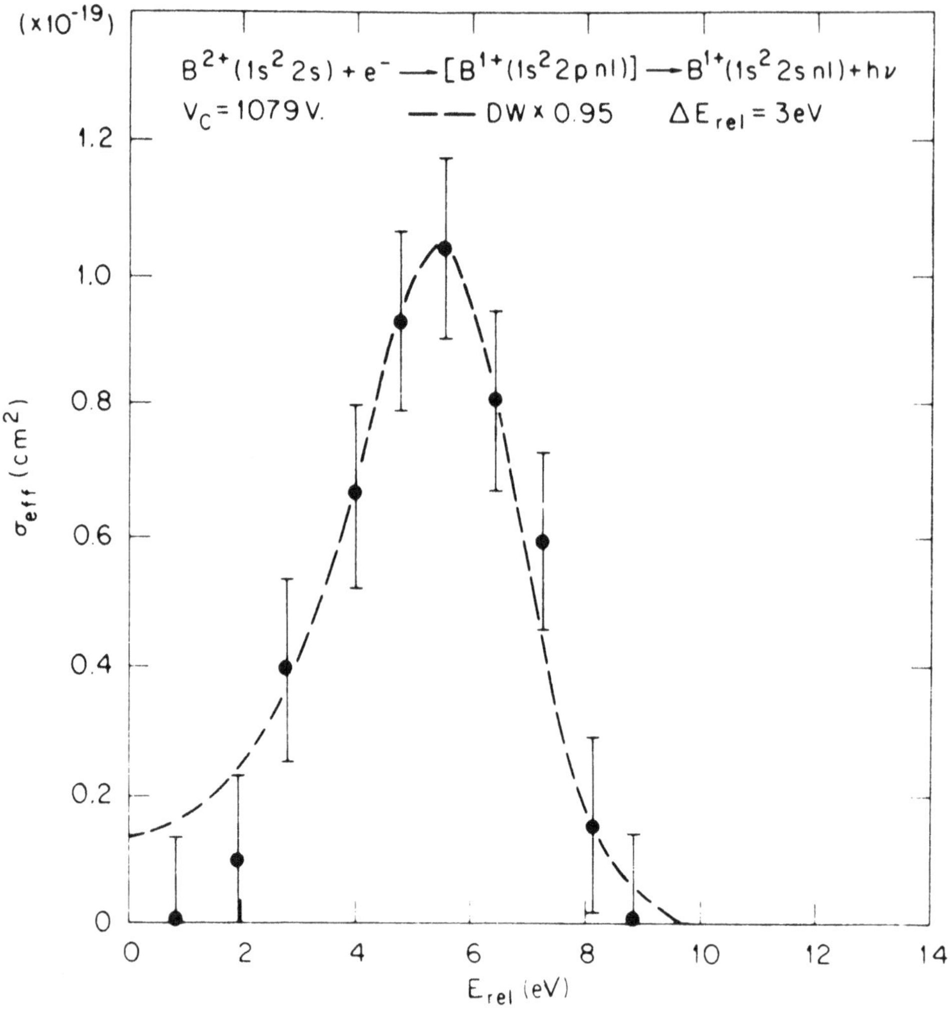

Fig. 3. Dielectronic recombination cross-section with relative electron–ion energy in B^{2+} from Dittner et al. (*21*). Dashed line is from normalized calculation of ref. *23*.

With increasing charge state, one finds that the ionization limit increases more rapidly than the energy of the levels of np^5nd^2 configuration, until these levels slip below the continuum (become bound) and cannot autoionize at all in the case of V^{4+} and Nb^{4+}.

A detailed comparison between measured ionization cross-sections (the sum of direct and EA) and the expected direct rate alone as given by reasonable theoretical calculations or by the Lotz semiempirical formula (*30*) is discussed by Falk et al. (*29*) and Gregory et al. (*31*). Experimental results from Falk et al. (*28*) are shown here in Fig. 5 where one clearly sees that measured cross-sections are factors of 10–20 times the Lotz-calculated direct cross-sections alone! Thus in many cases the EA path will be *the* major source of ionization of ions in hot plasmas, and this predominance can continue up to energies of many times the first ionization potential (*32, 33*).

C. Energy-Loss Methods

A large body of data exists for the study of electron-impact excitation of ions by detection of the emitted photon between two electronic states of the target ion. Results have been reported by the JILA group on N_2^+, Ba^+, Ca^+, Hg^+ (and Hg^{2+}), C^{3+}, Li^+, Be^+, and Zn^+ (see refs. *34–41*, respectively). Using similar techniques, Kohl and Lafyatis (*42*) have measured excitation cross sections in C^+. The principal advantages of the electron impact–photon detection approach are that one (a) obtains measurements in the threshold region where electron–ion cross-sections

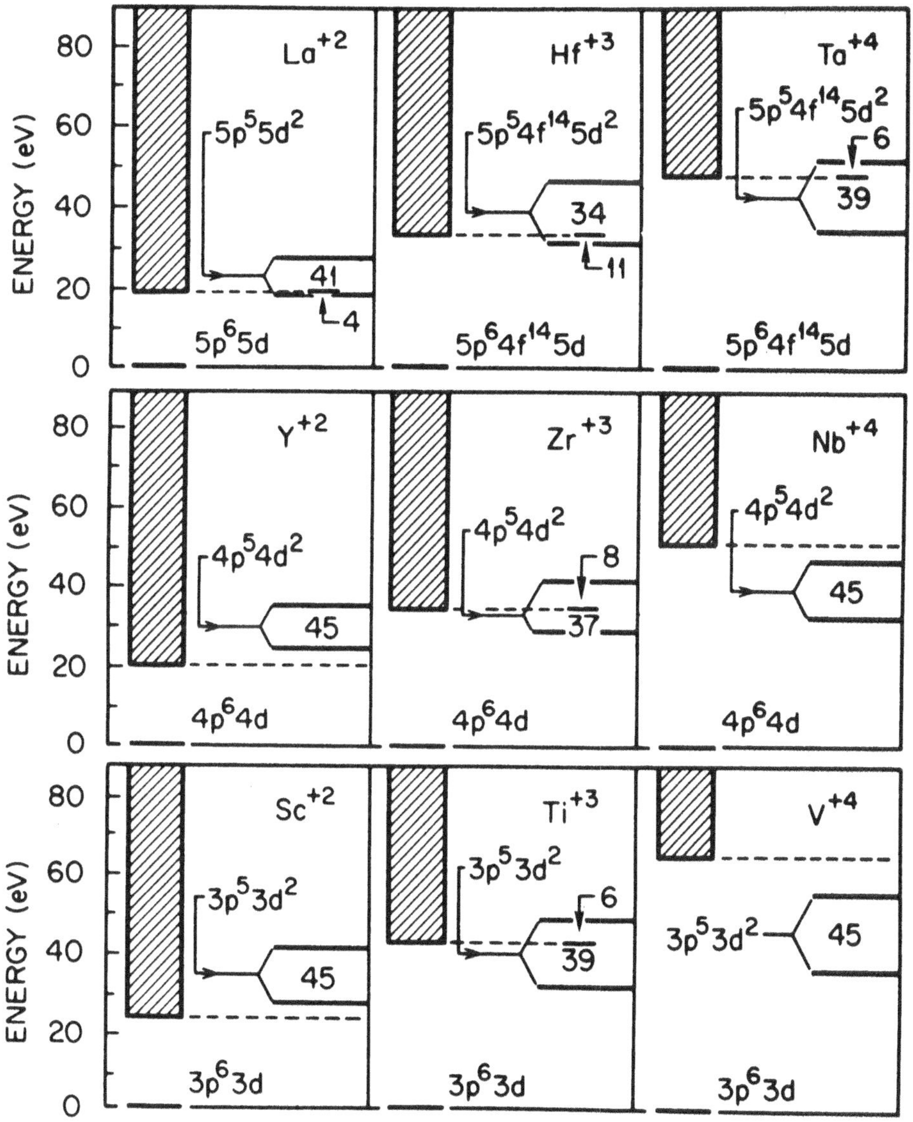

Fig. 4. Energy level structure of np^5nd^2 relative to the ionization limit for several transition metals (*28*). The number of levels above and below the respective ionization threshold is indicated for each case. Hatched region indicates the ionization continuum for the next higher ionization state.

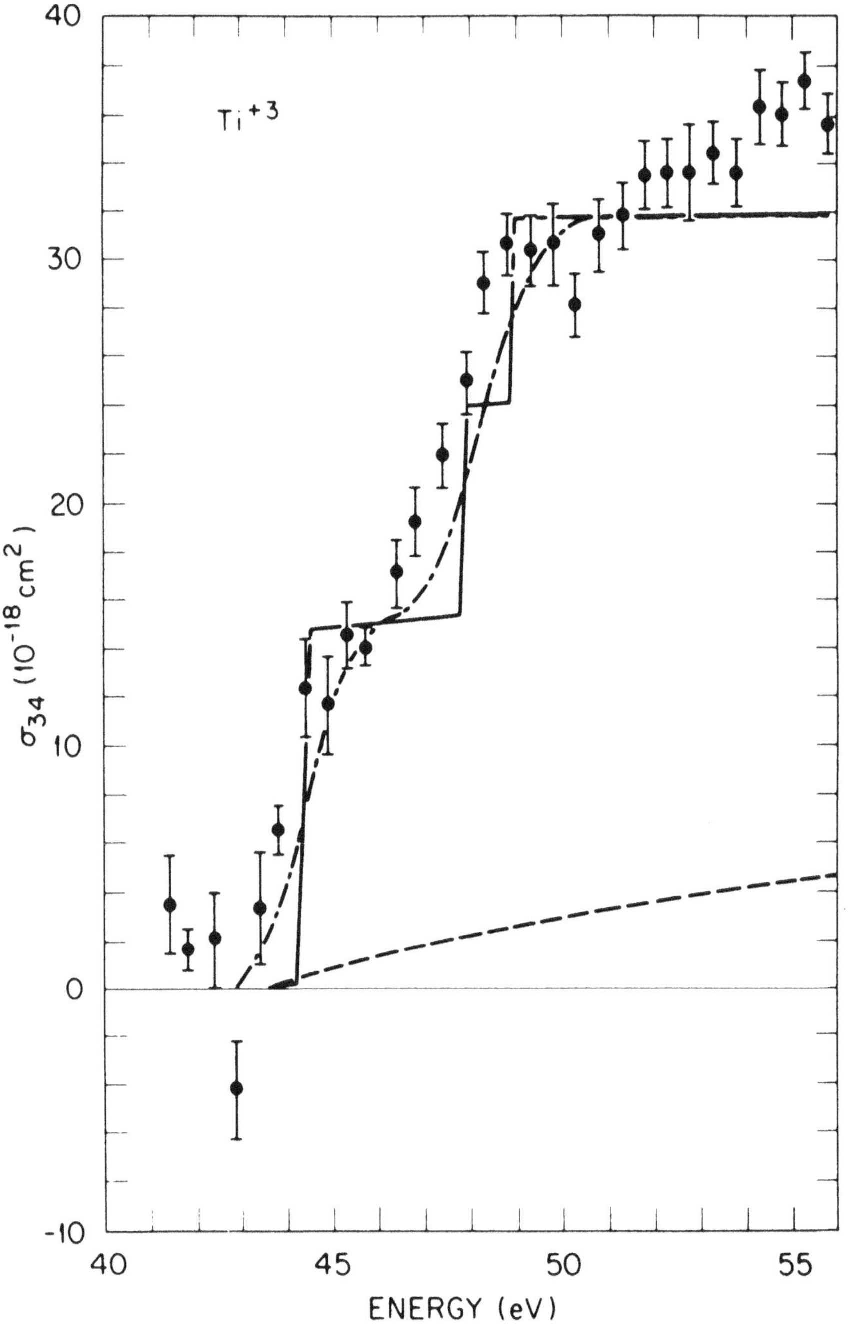

Fig. 5. Ionization cross-section σ_{34} for Ti^{3+} from Falk et al. (*28*). Solid line is the theoretical calculation divided by 2.5, and dot-dashed line (-·-) is the solid line convoluted with experimental electron energy distribution; dashed line (---) is from the semiempirical formula of Lotz (*30*).

have their maxima, and (b) measures integral cross sections directly. The principal drawbacks are that (c) cascading from higher states often contributes to the measurement, (d) with one exception (*38*), cross-sections have only been measured for dipole-allowed transitions to the ground electronic state, and (e) painstaking optical calibrations are required.

Electron–ion collision experiments reported in the literature thus far have involved detection of, as the collision product, a photon (resulting from line excitation or DR), an ion (from ionization/EA, *e*–ion recombination, or DR), or a neutral particle (from recombination or DR). Only recently, as a result of efforts at JPL, has detection of the scattered electron been added to this repertoire of techniques (*43–46*). The use of energy-loss methods in electron–ion collisions adds a level of versatility not possible before. For example, advantages relative to photon emission methods are that (a) intersystem (spin and/or symmetry-forbidden) transitions can be excited, (b) cross-sections are free of cascading contributions, (c) angular distributions of inelastically scattered electrons can be measured, and (d) resonances (i.e., temporary electron + ion states that autoionize in times $\tau \leq 10^{-12}$ s) can be accessed. The present difficulties of this technique are that (e) the threshold region—where the scattered electron must leave with a low residual energy—is more difficult to approach than in the line-excitation technique, (f) integrating an angular differential cross-section to obtain total cross-sections is tedious and involves low detection efficiencies, and (g) the interaction volume in a 90° crossed-beam geometry is small. It is hoped that as experimental techniques advance, some of the shortcomings in (e)–(g) can be overcome, and that detection of scattered electrons can be used to its full versatility.

In the energy-loss approach, one is interested in detecting the scattered electron in the process

$$e(E_o, \theta = 0°) + A^{+m}(nl) \rightarrow A^{+m}(n'l') + e(E_o - \Delta E, \theta) \qquad (3)$$

Here, the electron $e(E_o, \theta = 0°)$ has initial energy E_o and is incident along the $\theta = 0°$ direction. The target ion A^{+m} is in its ground state nl (effects of some metastable population from the ion source are neglected). Excitation to the state $n'l'$ leads to a change in energy ΔE(the so-called energy loss) and angle θ of the electron. One detects the intensity of electrons $e(E_o - \Delta E, \theta)$ as a function of their scattered energy and angle by suitable electrostatic analyzer and lens systems that can be rotated relative to the incident $\theta = 0°$ direction.

A schematic diagram of the apparatus used in the JPL measurements is shown in Fig. 6. Singly ionized ions [$m = 1$ in Eq. (3)] are generated in a discharge ion source, extracted through a small anode hole by lenses L1, and velocity analyzed in a Wien filter. Velocity-selected ions are focused by lenses L2 used as an einzel triplet. Ions after L2 are magnetically deflected and focused at the collision center by lenses L3. After the focus the ions are intercepted by L4, electrostatically deflected (ED) and focused into a deep Faraday cup.

A beam of electrons is focused onto the ion beam, and scattered electrons are detected by a hemispherical electrostatic analyzer. The hemispherical analyzer and associated lens system (*47*) are mounted on a rota-

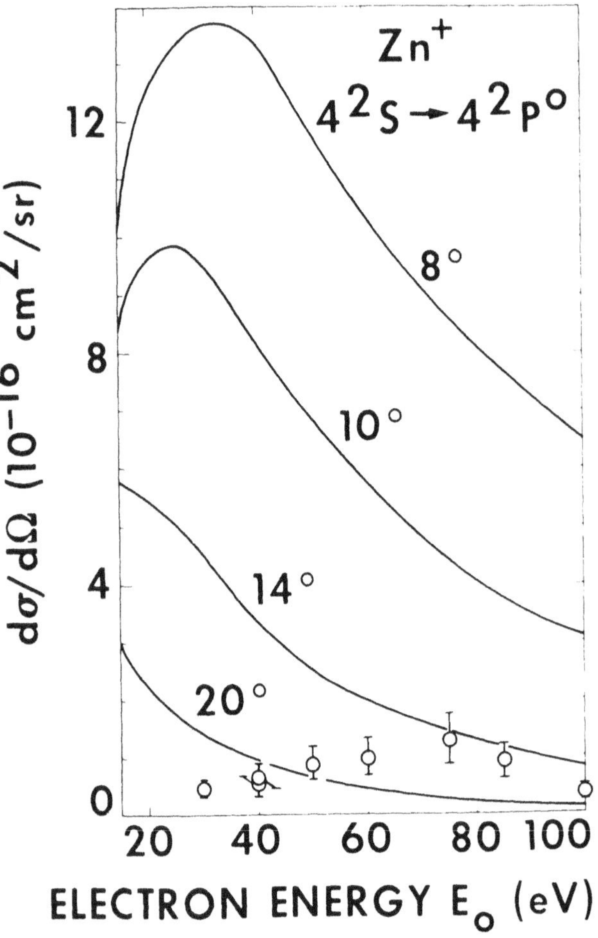

Fig. 8. Differential cross sections $d\sigma/d\Omega$ for excitation of the $4^2S \rightarrow 4^2P^o$ transition in Zn^+ at 75 eV electron energy (45). (a) Detailed comparisons of experimental results normalized to theory at $\theta = 14°$ (crosses) and theoretical calculations (solid lines) (50) for the angular range $4° \leq \theta \leq 16°$. (b) Calculated results over the range $0° \leq \theta \leq 180°$ and experimental measurements in the smaller angular range.

This reasoning was borne out by sophisticated quantum-mechanical calculations of Msezane and Henry (*49, 50*) in which a 5-state close-coupling theory with an accurate Zn^+ target wavefunction was used to calculate the differential cross section $d\sigma/d\Omega$. These results are also shown in Fig. 8. Here, experimental data have been normalized to the theoretical value of 1.57×10^{-16} cm²/sr at $\theta = 14°$. The experimental cross-section at 14° was 1.30×10^{-16} cm²/sr, as calculated from the expression

$$d\sigma/d\Omega = (\mathcal{R}/I_i I_e)\,[e^2 v_i v_e/(v_i^2 + v_e^2)^{1/2}]\,(\mathcal{F}/\Phi\Delta\Omega) \qquad (4)$$

All quantities on the right-hand side have the same meanings as in Eq. (1). In addition Φ is the detection efficiency of the electron analyzer system and channel multiplier, and $\Delta\Omega$ the solid angle subtended by the analyzer lens system at the scattering center. The quantities $\mathcal{F}$, Φ, and $\Delta\Omega$ were not measured directly in the current experiments, but were estimated from ray-tracing calculations (for $\mathcal{F}$, $\Delta\Omega$), and from the work of other groups (for Φ). The good agreement between theory and experiment in absolute magnitude, in addition to the excellent agreement in shape, is gratifying.

However, discrepancies do exist, and in Fig. 9 are shown cross-section results at electron energies E_o = 30, 40, 50, 60, 75, and 100 eV for a fixed 14° scattering angle (*44*). Comparison with theory (*49, 50*) indicates an experimental cross-section that is falling too rapidly at energies below 60 eV. The most likely cause for the discrepancy is that the experimental quantities $\mathcal{F}$, Φ, and $\Delta\Omega$, which vary with incident electron energy, were estimated for a single energy, rather than measured for each. Certainly, comparisons to a cross section in which the individual parame-

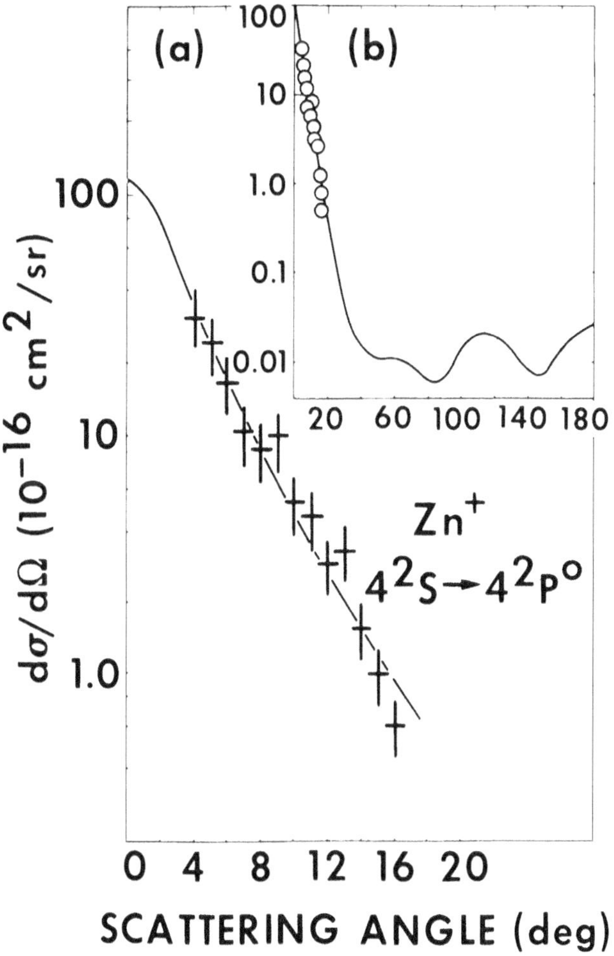

Fig. 9. Differential cross sections $d\sigma/d\Omega$ for the $4^2S \rightarrow 4^2P^o$ excitation in Zn^+ in the electron energy range $15 \leq E_o \leq 100$ eV and at the indicated scattering angles. Calculated results are indicated by solid curves (*45, 50*) and experimental data at θ = 14° are shown as open circles (*44*).

ters are measured directly is more meaningful. Such work is currently being planned at JPL.

Measurements have also been carried out on the e–Cd^+ system, in particular on the $5s \to 5p$ resonance transitions at $\Delta E = 5.47$ eV ($j = 1/2$) and 5.78 eV ($j = 3/2$) (*46*). This system is interesting since excitation involves the $n = 5$ principal quantum number, whereas Zn^+ involved $n = 4$. Thus measurement of trends between these ions can offer clues as to important effects that influence the magnitude and shape of inelastic electron–ion cross-sections. Once theoretical programs can confidently account for these trends, they can be used to calculate cross-sections in multiply charged ions that have significant density in astronomical and fusion plasmas, and for which measurements are even more difficult.

There has also been interest in the low-lying $^2P^o$ and ^2D states in Cd^+ for their role in the He–Cd^+ hollow-cathode laser. Surprisingly, these states appear to be populated by electrons rather than the usual Penning ionization process (*51*). One thus expects cross-section measurements in Cd^+ and other singly-ionized ions to have bearing on the performance and modeling of many hollow-cathode laser systems.

In Fig. 10 we present first measurements of energy-loss spectra in Cd^+. Spectra of the $5^2S \to 5^2p^o$ transition are shown for a 75 eV incident energy at scattering angles of 4°, 6°, 8°, and 12°. As in Zn^+, the electron energy resolution (0.55 eV FWHM) was insufficient to resolve the two fine-structure components [although in the next ($n = 6$) member of the series, $6^2S \to 6^2P^o$ in the Hg^+ ion, these levels could be resolved!].

One integrates the relative scattering intensity in spectra such as shown in Fig. 10 and obtains a differential cross section from a combination of measured and calculated quantities in Eq. (4). The quantities $\mathcal{R}$, I_i, I_e, v_i, and v_e ($v_e >> v_i$) are measured or nominally known, whereas for present purposes the ratio $\mathcal{F}/\Phi\Delta\Omega$ was evaluated by calibrating to a calculated cross-section (*49, 50*) for the $^2S \to {}^2P$ transition in Zn^+.

Differential cross-sections in the angular range $4° \leq \theta \leq 16°$ are shown in Fig. 11. As in the Zn^+ case, the cross-sections are strongly forward-peaking for this dipole-allowed transition. For comparison, the calculated $^2S \to {}^2P$ cross-section in Zn^+ is shown in the same threshold units. One sees a striking difference in both magnitude and shape of the two cross-sections: the Cd^+ data are larger in magnitude and more steeply rising than for Zn^+. Thus one's expectation that results in these two ions should be comparable is certainly not met!

A similarly large cross-section was noted by Hane et al.[51] in their optical emission measurements in Cd^+. One can integrate the differential results of Fig. 11 and obtain an integral cross section σ_I for the two fine-structure components. The results at 75 eV are shown in Fig. 12 for the sum σ_I ($^2P_{3/2}$) + σ_I ($^2P_{1/2}$), and for σ_I ($^2P_{3/2}$) alone. The latter quantity was derived from the measured sum by assuming LS coupling and a ratio of cross sections of 2 : 1 for the $j = 3/2$ and $j = 1/2$ components, i.e., just the ratio of statistical weights. The quantity measured by Hane et al.

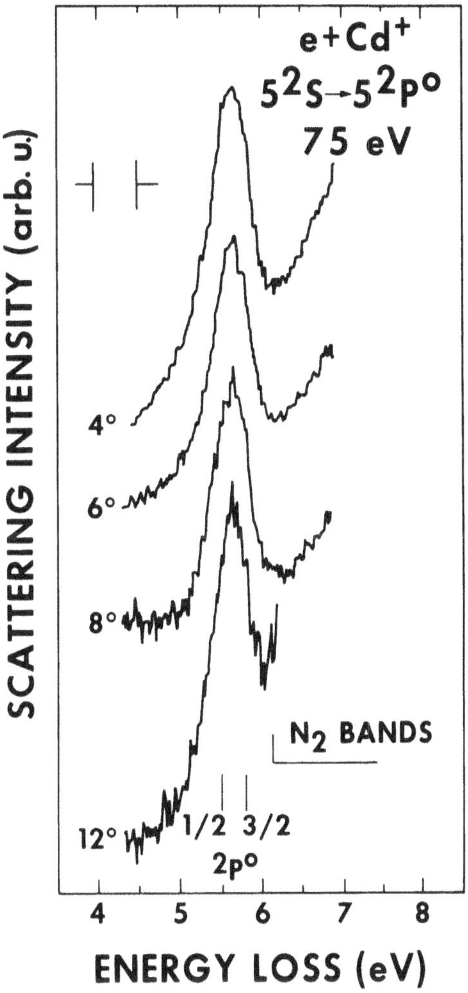

Fig. 10. Energy-loss spectra at the indicated scattering angles for the $^2S \rightarrow$ $^2P^o$ resonance transition in Cd^+ at an incident electron energy of 75 eV (*46*). The overall resolution in the spectra is shown.

(and shown in Fig. 12) is an integral cross-section for the $j = 3/2$ component alone and includes a small cascade contribution. The present measurements are seen to continue the trend, as given by the Hane et al. results, the Wigner threshold law (*52*), and Bethe (*53*) theory, of decreasing cross-section with increasing electron energy.

Although no theoretical calculations are available in the e–Cd^+ system, it is interesting to compare our results to the semiempirical Gaunt theory (*54–56*), widely used in astrophysics, and to Gaunt-factor expressions given by Mewe (*57*). In this theory, the integral cross section σ_I can be written as

$$\sigma_I = (8\pi^2 a_0^2/\sqrt{3})(E_H/\Delta E)^2 f_{ij}\, g(U)\, U^{-1} \tag{5}$$

where a_o is the Bohr radius, E_H the Rydberg energy (13.61 eV), ΔE the transition energy between states i and j, and f_{ij} the corresponding oscillator strength. The quantity U is the incident electron energy in threshold units ($U = E_o/\Delta E$), and $g(U)$ the so-called Gaunt factor that gives the probability of the incident electron going from energy E_o to final energy $E_o - \Delta E$. It may be evaluated either by direct computation through spherical harmonics, or by semiempirical comparisons with accurate theoretical data and experimental measurements. Mewe has taken the latter approach, and we may compare the Cd^+ 5*s*–5*p* transition cross-section to an approximate expression for $g(U)$ suggested by Mewe. This expression is just

$$g(U) = A + B + C + D \ln U \tag{6}$$

where the semiempirical constants recommended (*57*) are $A = 0.6$, $B = C = 0$, and $D = 0.28$ for an allowed transition with no change in principal quantum number. Results are shown in Fig. 12. The calculated cross-section attains a peak at threshold of 2.73×10^{-15} cm^2, which is a factor

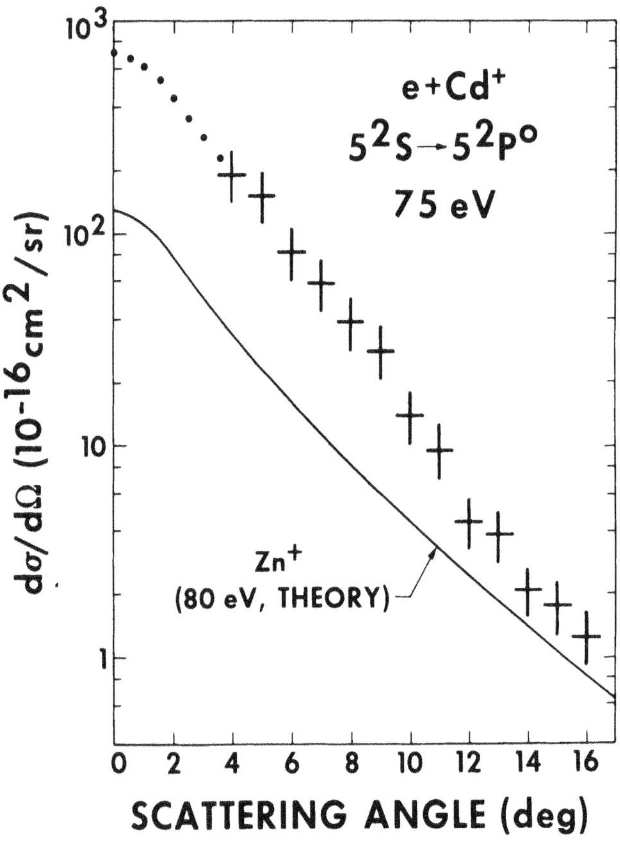

Fig. 11. Experimental differential cross-sections in Cd^+ for the $5^2S \rightarrow 5^2P^o$ transition (crosses) (*46*). Comparison is made with calculated cross sections for the $4^2S \rightarrow 4^2P^o$ transition in Zn^+ at an electron energy (80 eV) corresponding to a comparable fractional energy loss (*45, 50*). Dots represent extrapolation to 0° to obtain the integral cross section.

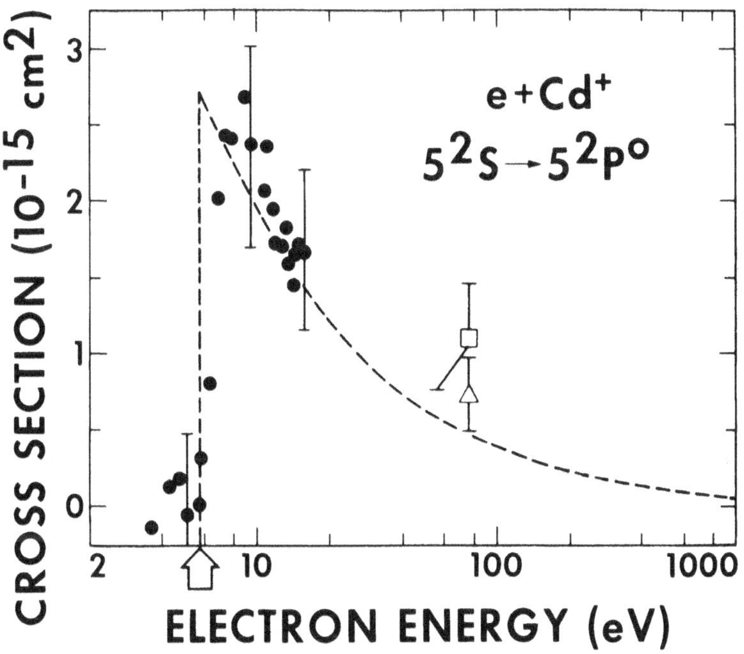

Fig. 12. Integral cross sections (*44*) for the $^2S \rightarrow {}^2P^o$ transition in Cd^+. Filled circles (●) are emission results of Hane et al. (*51*) for the $^2S_{1/2} \rightarrow {}^2P^o_{3/2}$ fine-structure component, including cascade. Open square (□) is present result for the $^2P^o_{3/2}$ and $^2P^o_{1/2}$ components, while open triangle (△) is present result for the $^2P^o_{3/2}$ component alone, assuming LS coupling. Dashed line (---) is the Gaunt-factor result (*57*) for the $^2P^o_{3/2}$ and $^2P^o_{1/2}$ components. The threshold energy is indicated by the vertical arrow.

of 0.7 that of the peak of approximately $(1 + 1/2)\ (2.5 \times 10^{-15}) = 3.75 \times 10^{-15}$ cm^2 from the work of Hane et al. At higher energies, calculation underestimates the cross-section by a factor of 2.5. Both bounds are within the advertised accuracy of approximation in Eqs. (5) and (6), but it would be valuable to compare the Gaunt-factor calculated results with a more *ab initio* theory in order to assess the form of $g(U)$ for comparison to Cd^+. Although such calculations for the many-electron e–Cd^+ system are difficult, they do appear to lie within the realm of possibility (*58*). Considering the wide usage of the Gaunt theory in high temperature plasmas, such comparisons with cross-sections from energy-loss measurements would be quite valuable.

ACKNOWLEDGMENTS

This research was carried out by the Jet Propulsion Laboratory, California Institute of Technology, under contract with the National Aeronautics and Space Administration.

REFERENCES

1. Fawcett, B. C., *Physica Scripta* **24,** 663 (1981).
2. Boksenberg A., et al., *Nature* **275,** 404 (1978).
3. Shemansky, D. E., *Astrophys. J.* **236,** 1043 (1980).
4. Pease, R. S., *Contemp. Phys.* **18,** 113 (1977).
5. Cecchi, J. L., *J. Nucl. Mater.* **93–94**(A), 28 (1980); Post, D. E., "The Role of Atomic Collisions in Fusion" in *Physics of Ion–Ion and Electron–Ion Collisions* Brouillard, F., and McGowan, J. W., eds., Plenum, New York, 1983, p. 37.
6. Hinnov, E., *Phys. Rev. A* **14,** 1533 (1976).
7. Drawin, H. W., *Phys. Scripta* **24,** 622 (1981).
8. Crandall, D. H., "Electron-Ion Collisions of Fusion Interest", Nucl. Instr. Meth. in press.
9. See, for example, Dolder, K. T., *Case Studies in Atomic Collision Physics,* McDaniel, E. W., and McDowell, M. R. C., eds., North-Holland, Amsterdam, 1969, Ch. 5; Dolder, K. T., and Peart, B., *Rep. Prog. Phys.* **39,** 693 (1976); Dunn, G. H. in *The Physics of Ionized Gases,* Matic, M., ed., Boris Kidrić Inst. Nucl. Sciences, Belgrade, 1980, p. 49.
10. Dunn, G. H., *IEEE Trans. Nucl. Sci.* **NS-23,** 929 (1976).
11. Dolder, K. T., *Contemp. Phys.* **17,** 237 (1976).
12. Crandall, D. H., *Phys. Scripta* **23,** 153 (1981); "Electron Impact Excitation of Ions" in *Physics of Ion–Ion and Electron–Ion Collisions,* Brouillard, F., and McGowan, J. W., eds., Plenum, New York, 1983, p. 201.
13. Dubau, J., and Volonte, S., *Rep. Prog. Phys.* **43,** 199 (1980).
14. Seaton, M. J., and Storey, P. J., in *Atomic Processes and Applications* Burke, P. G., and Moiseiwitsch, B. L. eds., North-Holland, Amsterdam, 1976, Ch. 6.
15. Bely-Dubau, F., Gabriel, A. H., and Volonte, S., *Mon. Not. Royal Astronom. Soc.* **189,** 801 (1979).
16. LaGattuta, K., and Hahn, Y., *Phys. Rev. A* **26,** 1125 (1982).
17. LaGattuta, K., and Hahn, Y., *J. Phys. B: Atom. Molec. Phys.* **15,** 2101 (1982).
18. Mitchell, J. B. A., Ng, C. T., Forand, J. E., Levac, D. P., Mitchell, R. E., Sen, A., Miko, D. B., and McGowan, J. W., *Phys. Rev. Lett.* **50,** 335 (1983).
19. Belić, D. S., Dunn, G. H., Morgan, T. J., Mueller, D. W., and Timmer, C., *Phys. Rev. Lett.* **50,** 339 (1983).
20. Williams, J. F., "Dielectronic Recombination Measured with a Crossed Electron and Calcium Ion Beams Coincidence Technique," in *XIIIth Int. Conf. Phys. Electron. Atom. Collisions: Book of Abstracts,* Eichler, J., Hertel, I. V., and Stolterfoht, N. eds., North-Holland 1983, p. 209.
21. Dittner, P. F., Datz, S., Miler, P. D., Moak, C. D., Stelson, P. H., Bottcher, C., Dress, W. B., Alton, G. D. Nesković, N., and Fou, C. M., *Phys. Rev. Lett.* **51,** 31 (1983).
22. Harrison, M. F. A., *Brit. J. Appl. Phys.* **17,** 371 (1966).
23. McLaughlin, D. J., and Hahn, Y., *Phys. Rev. A* **27,** 1389 (1983) (C^{3+}); *Phys. Rev. A* **28,** 493 (1983) (B^{2+}).
24. Burgess, A., and Summers, H. P., *Astrophys. J.* **157,** 1007 (1969).
25. Jacobs, V. L., Davis, J., and Kepple, P. C., *Phys. Rev. Lett.* **37,** 1390 (1976); LaGattuta, K., and Hahn, Y., *Phys. Rev. Lett.* **51,** 558 (1983).
26. Peart, B., and Dolder, K. T., *J. Phys. B: Atom. Mol. Phys.* **1,** 872 (1968); **8,** 56 (1975); Peart, B., Stevenson, J. G., and Dolder, K. T., ibid. **6,** 146 (1973).
27. Feeney, R. K., Hooper, J. W., and Elford, M. T., *Phys. Rev. A* **6,** 1469 (1972).
28. Falk, R. A., Dunn, G. H., Griffin, D. C., Bottcher, C., Gregory, D. C., Crandall, D. H., and Pindzola, M. S., *Phys. Rev. Lett.* **47,** 494 (1981).

29. Falk, R. A., Dunn, G. H., Gregory, D. C., and Crandall, D. H., *Phys. Rev. A* **27,** 762 (1983).
30. Lotz, W., *Astrophys. J. Suppl.* **14,** 207 (1967); *Z. Phys.* **206,** 205 (1967); **216,** 241 (1968); **220,** 466 (1969).
31. Gregory, D. C., Dittner, P. F., and Crandall, D. H., *Phys. Rev. A* **27,** 724 (1983).
32. Griffin, D. C., Bottcher, C., and Pindzola, M. S., *Phys. Rev. A* **25,** 154 (1982).
33. Bottcher, C., Griffin, D. C., and Pindzola, M. S., *J. Phys. B: Atom. Molec. Phys.* **16,** L65 (1983).
34. Crandall, D. H., Kauppila, W. E., Phaneuf, R. A., Taylor, P. O., and Dunn, G. H., *Phys. Rev. A* **9,** 2545 (1974).
35. Crandall, D. H., Taylor, P. O., and Dunn, G. H., *Phys. Rev. A* **10,** 141 (1974).
36. Taylor, P. O., and Dunn, G. H., *Phys. Rev. A* **8,** 2304 (1973).
37. Crandall, D. H., Phaneuf, R. A., and Dunn, G. H., *Phys. Rev. A* **11,** 1223 (1975); Phaneuf, R. A., Taylor, P. O., and Dunn, G. H., *Phys. Rev. A* **14,** 2021 (1976).
38. Taylor, P. O., Gregory, D., Dunn, G. H., Phaneuf, R. A., and Crandall, D. H., *Phys. Rev. Lett.* **39,** 1256 (1977).
39. Rogers, W. T., Østgaard-Olsen, J., and Dunn, G. H., *Phys. Rev. A* **18,** 1353 (1978).
40. Taylor, P. O., Phaneuf, R. A., and Dunn, G. H., *Phys. Rev. A* **22,** 435 (1980).
41. Rogers, W. T., Dunn, G. H., Østgaard-Olsen, J., Reading, M., and Stefani, G., *Phys. Rev. A* **25,** 681 (1982).
42. Kohl, J. L., and Lafyatis, G. P., *XIIIth Int. Conf. Phys. Electron. Atom. Collisions: Book of Abstracts,* Eichler, J., Hertel, I. V., and Stolterfoht, N., eds., North-Holland, 1983, p. 195.
43. Chutjian, A., "Angular Distributions for Electron Impact Excitation of Ions from Energy-Loss Spectra Measurements," *XIIIth Int. Conf. Phys. Electron. Atom Collisions: Book of Invited Papers,* Eichler, J., Hertel, I. V., and Stolterfoht, N., eds., North-Holland, 1984, in press.
44. Chutjian, A., and Newell, W. R., *Phys. Rev. A* **26,** 2271 (1982).
45. Chutjian, A., Msezane, A. Z., and Henry, R. J. W., *Phys. Rev. Lett.* **50,** 1357 (1983).
46. Chutjian, A., *Phys. Rev. A.* **29,** 64 (1984).
47. Chutjian, A., *Rev. Sci. Instr.* **50,** 347 (1979).
48. Chutjian, A., and Srivastava, S. K., *J. Phys. B: Atom. Mol. Phys.* **8,** 2360 (1975).
49. Msezane, A. Z., and Henry, R. J. W., *Phys. Rev. A* **25,** 692 (1982).
50. Msezane, A. Z., and Henry, R. J. W., private communication; *Bull. Am. Phys. Soc.* **28,** 798 (1983); *XIIIth Int. Conf. Phys. Electron. Atom. Collisions: Book of Abstracts,* Eichler, J., Hertel, I. V., and Stolterfoht, N., eds., North-Holland, 1983, p. 193.
51. Hane, K., Goto, T., and Hattori, S., *Phys. Rev. A* **27,** 124 (1983).
52. Wigner, E. P., *Phys. Rev.* **73,** 1002 (1948).
53. Bethe, H. A., *Ann. Phys.* **5,** 325 (1930).
54. Seaton, M. J., in *Atomic and Molecular Processes,* Bates, D. R., ed., Academic Press, New York, 1962, Ch. 11.
55. van Regemorter, H., *Astrophys. J.* **136,** 906 (1962).
56. Younger, S. M., and Wiese, W. L., *J. Quant. Spectrosc. Radiat. Transfer* **22,** 161 (1979).
57. Mewe, R., *Astron. Astrophys.* **20,** 215 (1972).
58. Msezane, A. Z., and Henry, R. J. W., *Bull. Am. Phys. Soc.* **29,** 825 (1984).

0018–1536/84/1700–0155$02.00

Absorption Spectra of Diphenylacetylene and 1,4-Diphenylbutadiyne Cations in Solid Argon

BENUEL J. KELSALL, ROBERT T. ARLINGHAUS, AND L. ANDREWS*

Chemistry Department, University of Virginia, Charlottesville, Virginia 22901

Received September 23, 1983; Accepted December 31, 1983

ABSTRACT

One and two-photon matrix photoionization techniques have been used to prepare diphenylacetylene and 1,4-diphenylbutadiyne cations. Identification of these cations was confirmed by photoelectron spectra and radiolysis studies. Near-infrared absorption bands for each cation correlate with photoelectron spectra and suggest structural relaxation about the molecular axis on ionization, and strong blue absorptions that do not correlate with photoelectron spectra are probably caused by $\pi \rightarrow \pi^*$ transitions.

Index Entries: Matrix photoionization; diphenylacetylene cation, absorption spectra of; 1,4-diphenylbutadiyne cation, absorption spectra of; absorption spectra, of diphenylacetylene and 1,4-diphenylbutadiyne cations.

INTRODUCTION

High temperature chemistry is first concerned with evaporative and dissociative processes; however, at extremely high temperatures ionization becomes an important energy reservoir. Over the last decade, we have developed methods for producing and trapping molecular cations in solid argon for absorption spectroscopic studies aimed at a better un-

*Author to whom all correspondence and reprint requests should be addressed.

sorption appeared at 26.4 (A = 0.56, fwhm = 400 cm^{-1}) with weak satellites at 27.0, 27.4, and 28.1. In addition to these absorptions in the near UV region, trace (b) shows a broad band near 12.7, and a sharper superposed band at 16.0. In a similar experiment with diphenylacetylene, the sharp precursor absorption origin at 34.0 decreased from A = absorbance = 0.20 to 0.08 while the sharp product absorption appeared at 24.0 (A = 0.25) with the broad counterpart at 12.7.

Argon resonance photoionization experiments with diphenylacetylene gave all of the absorptions shown in Fig. 1(b) with consistently lower intensities than in experiments employing the two photon ionization technique. In addition the argon resonance lamp produced a series of bands beginning at 26.6 that overlap with the 26.4 band and its satellites. The 26.6 series of absorptions has been identified as a hydrogen addition radical formed by the reaction of a photodetached hydrogen atom and a neutral diphenylacetylene molecule (7). In the argon resonance photoionization experiments, the 23.9 shoulder was relatively more intense than in the matrix two-photon ionization experiments.

The 1,4-diphenylbutadiyne precursor was used in only matrix two-photon ionization experiments. Spectra for one experiment are shown in Fig. 2 where trace (a) refers to the spectrum obtained immediately after condensation and trace (b) represents the spectrum after 30 min of 10–45 ($\times 10^3$ cm^{-1}) photolysis. As with the diphenylacetylene, the freshly prepared sample exhibited no absorption until the onset of the precursor bands. After irradiation, though, new absorptions were produced in the visible and near IR regions. The principal product absorption near 21.1 (A = 0.32, fwhm = 300 cm^{-1}) has a shoulder at 21.2 and a series of much weaker bands at 21.5, 22.0, 22.3, 22.7, and 23.3. In addition to these absorptions, the visible region shows a moderately intense peak at 23.8 (A = 0.1, fwhm = 430 cm^{-1}) with very weak structure at 24.7, and 25.4, and a weak 26.1 band. The infrared region shows an absorption centered at 10.9 (A = 0.15, fwhm = 700 cm^{-1}) with unresolved shoulders at 10.5 and 11.1 and a broad weak band at 13.0.

DISCUSSION

The product absorptions will be identified, assignments will be given, and comparisons to other related molecules will be made for structural considerations.

Assignments

All of the absorptions produced in this investigation by intense near-ultraviolet photolysis are assigned to parent cations derived from the precursor molecules. The matrix two photon ionization technique has been used to produce cations in high yield from similar precursors by a mechanism that involves resonant excitation of the neutral molecule followed

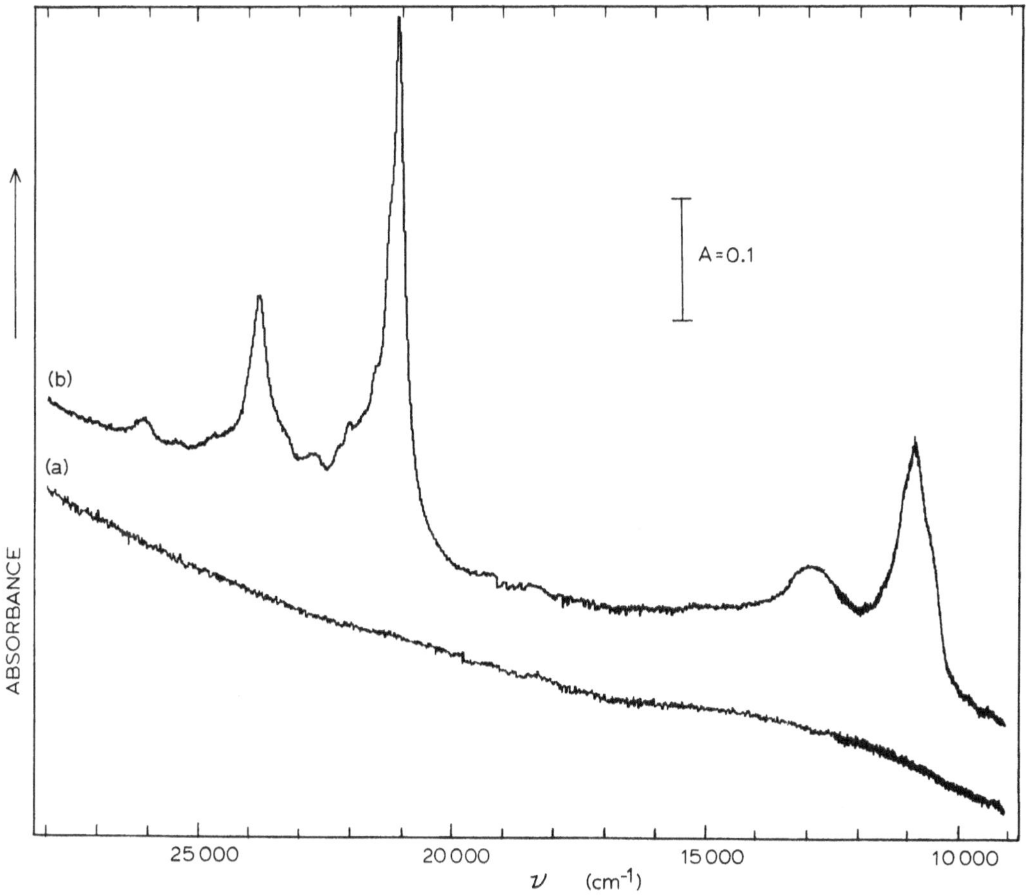

Fig. 2. Absorption spectra in the 9000–28,000 cm^{-1} region for Ar/ CH_2Cl_2/1,4-diphenylbutadiyne sample. (a) After deposition at 20 K for 3 h, (b) after 10,000–45,000 cm^{-1} high-pressure mercury arc irradiation for 30 min.

by relaxation to a lower excited state where absorption of a second resonant photon excites the molecule into an autoionizing state. The role of the halocarbon electron trap is to scavenge detached electrons and thus prevent the neutralization of other cations (*4*). For both diphenyl acetylene and 1,4-diphenylbutadiyne, other photochemical processes such as hydrogen detachment or photofragmentation are unfavorable because of the strong internal bonding. Therefore, as with biphenyl, the parent cation is the most reasonable product to be formed with these precursors.

In addition, the major cation bands detected in the study are in excellent agreement with those reported for the subject cations produced by radiolysis in organic glasses (*8*). The diphenylacetylene cation exhibited a broad 11.9 band and a structured 22.4 absorption system and 1,4-diphenylbutadiyne cation gave two broad bands at 10.7 and 12.2 and three sharper absorptions at 20.7, 23.5, and 25.8. These glassy matrix spectra are similar in appearance, but red-shifted from the present solid argon observations owing to a stronger interaction with the organic matrix.

The photoelectron spectrum (pes) of diphenylacetylene has been obtained by two groups (*9,10*). The first band is relatively broad with adiabatic and vertical ionizations of 7.90 and 8.06 eV and higher vertical π ionizations have been reported at 9.30, 9.55, 10.87, and 11.7 eV before the onset of σ ionization. The first pes band has been assigned to ionization from the highest occupied b_{3u} molecular orbital, which gives a $^2B_{3u}$ ground electronic state for the cation, assuming planar D_{2h} symmetry (z is molecular axis, y is in molecular plane) as found for neutral molecules in the crystalline phase (*11*). The remaining pes bands predict possible hole promotion ($\pi \rightarrow \pi$) transitions near 11.3, 13.3, 24.0, and 30.7 ($\times 10^3$ cm^{-1}). The 11.3 transition is for promotion of an electron from the nearby b_{1g} or a_u levels to the b_{3u} orbital; only the former transition is allowed by symmetry. Although there is a substantial blue shift, the 12.7 absorption is assigned to this $^2B_{3u} \rightarrow {}^2B_{1g}$ transition. A 13.3 transition would therefore be caused by the forbidden $^2A_u \rightarrow {}^2B_{1g}$ transition, which was not observed.

The weak 16.0 band, which appears as a shoulder on the broad 12.7 band, is somewhat sharper and probably caused by a different electronic transition from the 12.7 band. The 16.0 band is accordingly assigned to the next allowed transition, $^2B_{3u} \rightarrow {}^2B_{2g}$, in reasonable agreement with the 16.2 prediction from a weak unassigned pes band at 10.0 eV (*10*).

The strong, sharp 24.0 optical absorption is not expected from the pes for phenylacetylene. Although a transition is predicted near 24.0, this transition is not dipole allowed, based on HMO orbitals and calculations (*12*), and it cannot explain the strong 24.0 absorption, which must be caused by an allowed transition. Consequently, the 24.0 absorption is assigned to the $^2B_{3u} \rightarrow {}^2B_{2g}$ transition in which the electron in the half-filled highest occupied orbital of the cation is excited to the lowest unoccupied molecular orbital. This type of intense $\pi \rightarrow \pi^*$ transition has been observed for a number of substituted aromatic and conjugated olefinic cations, including biphenyl and hexatriene cations. The 26.4 peak is broader than the vibrational structure associated with the 24.0 band system and the interval is too large for a vibrational component. Accordingly, this weaker band system is probably caused by another $\pi \rightarrow \pi^*$ transition.

The photoelectron spectrum for 1,4-diphenylbutadiyne consists of bands reported at 8.23, 8.55, 8.87, 9.48, 9.72, 10.48, and 11.64 eV (*13*). The first band, assigned to a b_{2g} level, again assuming a planar cation represents the binding energy of an electron in the highest occupied molecular orbital and consequently determines the symmetry for the ground state cation. The differences between the first band and the others represent the energies required to excite the ground-state cation into some higher electronic configurations that may be detected in the optical spectrum. The first two transitions predicted from the photoelectron spectrum are not considered here, since they require a lower energy than the spectral region examined in this study. The 10.1 ($\times 10^3$ cm^{-1}) difference between the first and fourth pes is in reasonable agreement with the 10.9 optical absorption; this difference probably represents a lower limit for the transition energy because the first adiabatic ionization energy is not available for use in the predictions. The fourth pes band is probably due to ionization from an orbital with a_u symmetry although a specific assignment has not been given. Figure 2(b) shows that the 10.9 band is broad with some evidence of poorly resolved vibronic structure; the 10.9 band is accordingly assigned to the allowed $^2B_{2g} \rightarrow {}^2A_u$ transition. The other photoelectron bands are of no help in assigning the absorption spectrum.

As with diphenylacetylene, the photoelectron spectrum does not predict the most intense absorption in the optical spectrum. Because of its intensity and position, the 21.1 peak is assigned to the $^2B_{2g} \rightarrow {}^2B_{3u}$ ($\pi \rightarrow \pi^*$) transition. This characteristic transition exhibits a systematic red shift from biphenyl to diphenylacetylene and 1,4-diphenylbutadiyne cations. The 23.8 band is probably caused by another related $\pi \rightarrow \pi^*$ transition; the separation between the stronger and weaker $\pi \rightarrow \pi^*$ transitions increases in the series from biphenyl (*5*) to diphenylbutadiyne cations.

Although the photoelectron spectra for diphenylacetylene and 1,4-diphenylbutadiyne show general agreement with the observed optical spectrum for the cations, the agreement is not as good as with molecules like naphthalene, phenylacetylene, and 1-phenylpropyne (*14,15*). Discrepancies of this type have been attributed to structural relaxation from the neutral molecule geometry following ionization. Structurally hindered molecules such as biphenyl and α−methylstyrene show red optical bands that are blue-shifted from the positions predicted from photoelectron spectra (*5,6,16*). These shifts, which have been shown to be structure sensitive, demonstrate that the cations are more planar than their neutral counterparts. Consequently, the blue shifts for the red optical bands of the diphenylacetylenes indicate that the cations are more planar than the neutrals and that stabilization provided by the delocalization of the cationic charge over both phenyl rings inhibits rotation about the alkyne axis. The relative widths of the lower energy optical bands indicate that there is some steric hindrance to coplanarity in the cations, but that the steric effect is reduced as the physical separation between the two phenyl rings is increased.

Vibrational Assignments

Table 1 lists frequencies for electronic transitions and the vibronic intervals associated with them for diphenylacetylene and 1,4-diphenylbutadiyne cations. The table shows that only the near UV transitions have well-defined vibronic structure. Although definitive assignments may not be made because of the lack of isotopic data, some

TABLE 1
Absorption Band Positions (cm^{-1}) and Spacings in Near-Ultraviolet Two-Photon Ionization Experiments with Diphenylacetylene and 1,4-Diphenylbutadiyne in Solid Argon at 20 K

Diphenylacetylene		1,4-Diphenylbutadiyne	
Absorption	Spacings	Absorption	Spacings
12,700		10,900	
16,000		13,000	
24,000	0	21,070	0
200	200	240	170
590	590	520	450
680	680	22,030	960
25,180	1180	250	1170
620	1620	720	1650
		23,300	2230
26,430	0		
27,070	640	23,830	0
430	1000	24,750	920
28,050	1620	25,450	1620
		26,100	2270

correlations can be drawn. Both cations have a low frequency mode for the $\pi \rightarrow \pi^*$ transition. The 200 cm^{-1} mode for diphenylacetylene could correspond to the 257, 152, or 155 cm^{-1} vibrations for the lowest frequency modes of the a_g, b_{1g}, and b_{2g} symmetry groups, respectively (*17*). Even though all three modes could couple with the electronic transition, the observed band is probably due to the a_g mode. This mode, which represents a stretching between the two phenyl groups, will show a substantial decrease in frequency as an electron is removed from a predominantly bonding orbital and as the distance between the two phenyl groups increases. The 590 cm^{-1} band of the diphenylacetylene cation shows a small shift from the 625 cm^{-1} b_{1g} mode of the molecule. The 1,4-diphenylbutadiyne cation counterpart is deplaced a considerable energy, which suggests that the mode is caused by a bending motion that is strongly dependent upon the separation of the phenyl groups. The 680 cm^{-1} peak can be assigned to a mode analogous to the Jahn-Teller active ν_{18} (e_{2g}) mode of benzene (*18*). This vibration has been observed in the UV spectra of alkyl substituted phenylacetylene cations at 652 cm^{-1} (*15*). The 1180 cm^{-1} feature is in agreement with the 1180 cm^{-1} a_g vibration of the neutral, and the associated 1170 cm^{-1} diphenylbutadiyne band indicate that the mode is unaffected by the addition of another acetylene, thus demonstrating that the mode involves the phenyl rings; similar vibronic intervals were observed in the spectra of phenyl alkyne cations

(*15*). The 1620 and 1650 cm^{-1} bands for the two cations are assigned to ring stretching modes derived from the benzene ν_{16} (e_{2g}) mode at 1585 cm^{-1} (*18*).

Although the 26.4 and 23.8 absorptions appear as though they could be part of the vibronic structure associated with the intense 24.0 and 21.1 origins, their displacements of 2430 and 2760 cm^{-1}, respectively, are unreasonably large to be caused by an alkyne stretching mode. The C≡C stretch for neutral diphenylacetylene is 2220 cm^{-1}, and vibronic structure in the 34.0 band spaced 2095 cm^{-1} is probably caused by the C≡C stretching mode. Since the C≡C stretching frequency for a variety of halo- and alkyl-substituted acetylenes and butadiynes exhibit a significant shift to lower energy upon ionization (*19*), it is unlikely that the diphenyl analogs would show such a large shift to higher energy. The 2230 cm^{-1} band seen in the 1,4-diphenylbutadiyne experiments, however, can be assigned to C≡C alkyne stretching mode. The vibronic bands associated with the 26.4 and 23.8 origins are reasonable for aromatic ring skeletal vibrations, whereas the final interval of 2270 cm^{-1} in the 23.8 band is likely because of the C≡C stretching mode in this excited cation state.

CONCLUSIONS

Diphenylacetylene and 1,4-diphenylbutadiyne cations were prepared by matrix one and two-photon ionization techniques, and their optical absorption spectra were recorded between 9000 and 30,000 cm^{-1}. Identification of these cations was confirmed by photoelectron spectra and radiolysis studies. The present study shows that stabilization as a result of delocalization of the cation charge between the two phenyl rings inhibits rotation about the alkyne axis. In addition, this study demonstrates the usefulness of matrix absorption spectra of cations in interpreting photoelectron spectra.

ACKNOWLEDGMENTS

The authors gratefully acknowledge National Science Foundation support for this work.

REFERENCES

1. Andrews, L., Grzybowski, J. M., and Allen, R. O., *J. Phys. Chem.* **79,** 904 (1975).
2. Prochaska, F. T., and Andrews, L., *J. Chem. Phys.* **67,** 1091 (1977).
3. Prochaska, F. T., and Andrews, L., *J. Am. Chem. Soc.* **100,** 2102 (1978).
4. Kelsall, B. J., and Andrews, L., *J. Chem. Phys.* **76,** 5005 (1982).
5. Puiu, A. C., Andrews, L., Chupka, W. A., and Colson, S. D., *J. Chem. Phys.* **76,** 3854 (1982).

6. Andrews, L., Arlinghaus, R. T., and Payne, C. K., *J. Chem. Soc. Faraday Trans. 2* **79,** 885 (1983).
7. Kelsall, B. J., and Andrews, L., to be published.
8. Shida, T., and Hamill, W. H. *J. Chem. Phys.* **44,** 4372 (1966).
9. Eland, J. H. D. *Int. J. Mass. Spect. Ion Phys.* **9,** 214 (1972).
10. Maier, J. P., and Turner, D. W., *J. Chem. Soc. Faraday Trans. 2,* 196 (1973).
11. Mavridis, A., and Moustakali-Mavridis, I. *Acta Cryst.* **B33,** 3612 (1977).
12. Cotton, F. A., *Chemical Applications of Group Theory,* Wiley-Interscience, New York, 1971, 2nd ed., Chapter 7.
13. Andreocci, M. V., Biceo, P., Cauletti, C., and Piancastelli, M. N., *Gazz. Chim. Ital.* **110,** 31 (1980).
14. Andrews, L., Kelsall, B. J., and Blankenship, T. A., *J. Phys. Chem.* **86,** 2916 (1982).
15. Andrews, L., Kelsall, B. J., and Harvey, J. A., *J. Chem. Phys.* **77,** 2235 (1982).
16. Andrews, L., Harvey, J. A., Kelsall, B. J., and Duffey, D. C., *J. Am. Chem. Soc.* **103,** 6415 (1981).
17. Kellerer, B., and Hacker, H. H., *J. Mol. Struct.* **13,** 79 (1972).
18. Herzberg, G., *Infrared and Raman Spectra,* Van Nostrand, Princeton, 1945, p. 364.
19. Maier, J. P., *Accts. Chem. Res.* **15,** 18 (1982).

0018–1536/84/1700–0165$02.00

Study of the Disproportionation of Sodium Thiosulfate by X-Ray Photoelectron Spectroscopy

X. B. XU† AND S. G. CHANG*

Lawrence Berkeley Laboratory, University of California, Berkeley, CA 94720

Received September 23, 1983; Accepted December 31, 1983

ABSTRACT

The disproportionation of sodium thiosulfate to form sodium sulfite and elemental sulfur has been studied in the temperature range 495–548°K by means of X-ray photoelectron spectroscopy (XPS). The disproportionation started with an induction period, followed by a steady-state disproportionation rate regime, and ended up with a diffusion rate-controlling process. The duration of the induction period was inversely proportional to the temperature of heating. The disproportionation in the steady-state regime followed a first-order reaction rate law with the enthalpy and entropy of activation $H^* = 26.0$ kcal/mol and $S^* = -14.9$ eu.

Index Entries: Disproportionation, of $Na_2S_2O_3$; sodium thiosulfate, disproportionation of; X-ray photoelectron spectroscopy, of disproportionating $Na_2S_2O_3$; photoelectron spectroscopy, of disproportionating $Na_2S_2O_3$.

INTRODUCTION

Thiosulfates are thermally unstable and decompose to form sulfite and elemental sulfur with heat treatment. The kinetics of disproportionation of thiosulfates has not been investigated, however. This knowl-

*Author to whom all correspondence and reprint requests should be addressed.
†On leave from the Institute of Environmental Chemistry, Chinese Academy of Sciences, Beijing, People's Republic of China.

edge is needed in the power plant flue-gas cleanup industry, where thiosulfates can be generated or are employed as an additive in some flue gas desulfurization processes (*1*). In this paper, we report this kinetic information obtained by means of X-ray photoelectron spectroscopy.

EXPERIMENTAL METHODS

Sodium thiosulfate samples were prepared by spraying an aqueous solution of 1% $Na_2S_2O_3$ from a glass nebulizer onto clean aluminum foils (~6×12 mm) and then dehydrated by heat treatment in an oven at 338 K for at least 24 h. The average thickness of the sample material on the aluminum foil was calculated from the total sulfur content determined by X-ray fluorescence analysis and the morphology of samples depicted by an ISI Model DS-130 scanning electron microscope (SEM).

The kinetics of disproportionation of sodium thiosulfate was studied by means of an AEI ES200 photoelectron spectrometer utilizing Al K X-ray (1486.6 eV) with a sample chamber pressure of less than 10^{-8} torr during measurements. Samples were heated by a resistance heating element in the sample holder under vacuum. The XPS spectrum of the S ($2p_{1/2,3/2}$) region of $Na_2S_2O_3$ samples was taken periodically after samples had reached a desired temperature. The temperature range studied was 495–548 K. The kinetic data were obtained from the change of XPS spectrum intensity of reduced sulfur (S^{2-}) in $Na_2S_2O_3$ as a function of heating time at a given temperature.

RESULTS AND DISCUSSION

SEM results showed that sodium thiosulfate was in clusters of aggregates scattered on the aluminum foil. An average thickness of about 20 nm was calculated for these aggregates from the area projected for the aggregates and the determinations of their sulfur contents by X-ray fluorescence technique. The density of sodium thiosulfate was assumed to be 1.667 (*2*).

Figure 1 illustrates the use of XPS in monitoring the changes in concentrations of oxidized and reduced sulfur species during the course of the sodium thiosulfate disproportionation. The binding energies of the Sulfur $2p_{1/2,3/2}$ electrons of oxidized (S^{6+}) and reduced (S^{2-}) sulfur in sodium thiosulfate are 168.4 and 162.3 eV, respectively. Both types of sulfur species show approximately the same intensity (Fig. 1a). XPS intensity of the reduced sulfur decreased along with the heating, as shown in Fig. 1B–f. Whereas the intensity (measured by integrating the peak area) of the oxidized sulfur remains constant, there is a small shift in the binding energy toward the low energy side. We attribute this shift of binding energy to the chemical conversion of $Na_2S_2O_3$ to Na_2SO_3. The binding energy of the sulfur $2p_{1/2,3/2}$ electrons in Na_2SO_3 is 166.7 eV

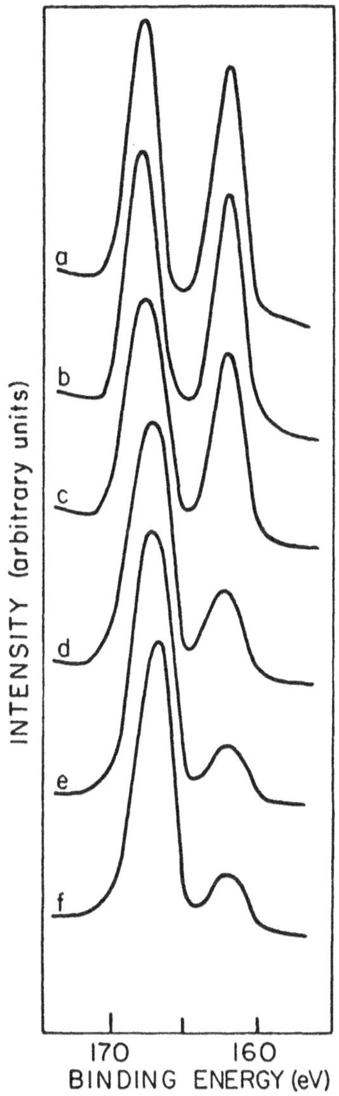

Fig. 1. Sulfur (2*p*) photoelectron spectra of $Na_2S_2O_3$ at 523 K showing changes in intensity of the reduced sulfur and shift in binding energy of oxidized sulfur as a function of time of heating: a, b, c, d, e and f correspond to 0, 3, 6, 11, 15, and 20 min heating, respectively.

which is too close to that of oxidized sulfur in $Na_2S_2O_3$ to be resolved with the XPS spectrometer employed for this experiment.

The production of elemental sulfur could not be detected by XPS techniques because of the vaporization under the high vacuum and temperature conditions employed. The major sulfur vapor species at temperatures employed in this experiment (495–548 K) is S_2 (*3*). The disproportionation of thiosulfate can be represented as

$$Na_2S_2O_3 \rightleftarrows Na_2SO_3 + \tfrac{1}{2}S_2 \uparrow$$

The rate equation of a first-order reaction can be expressed as

$$\ln C_o / C = k(t - t_o)$$

where C_o and C is the concentration of reactant at time t_o and t, respectively, k is the rate constant. Figure 2 is a typical curve of the plot of ln C_o / C vs t for the disproportionation of sodium thiosulfate. The concentration ratio of reactant C_o / C was taken as the intensity ratio of the undecomposed reduced sulfur at time t_o to t (normalized to the intensity of the oxidized sulfur) as observed by XPS. This figure suggests that the disproportionation might begin with an induction period (section I), followed by a steady-state regime (section II), and end with a period when the diffusion rate of species from the bulk to surface seems to be the controlling factor (section III). The duration of the induction period was shorter and the steady-state disproportionation rate was faster with the increase in temperature.

The effect of temperature changes on the steady-state disproprotionation decomposition rate constants is shown in Fig. 3. When the transition-state theory (4) is applied, the rate constant for the disproportionation of $Na_2S_2O_3$ can be expressed as

$$k \geqslant (k'T/h)\, e^{S^*/R}\, e^{-H^*/RT}$$

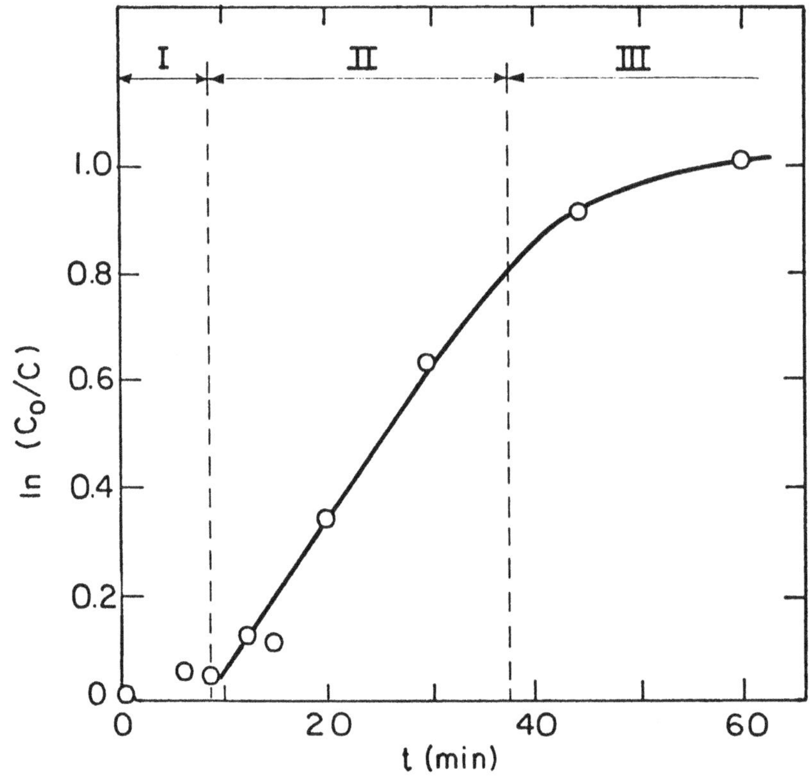

Fig. 2. A plot of ln C_o/C as a function of time of heating of $Na_2S_2O_3$ at 498 K.

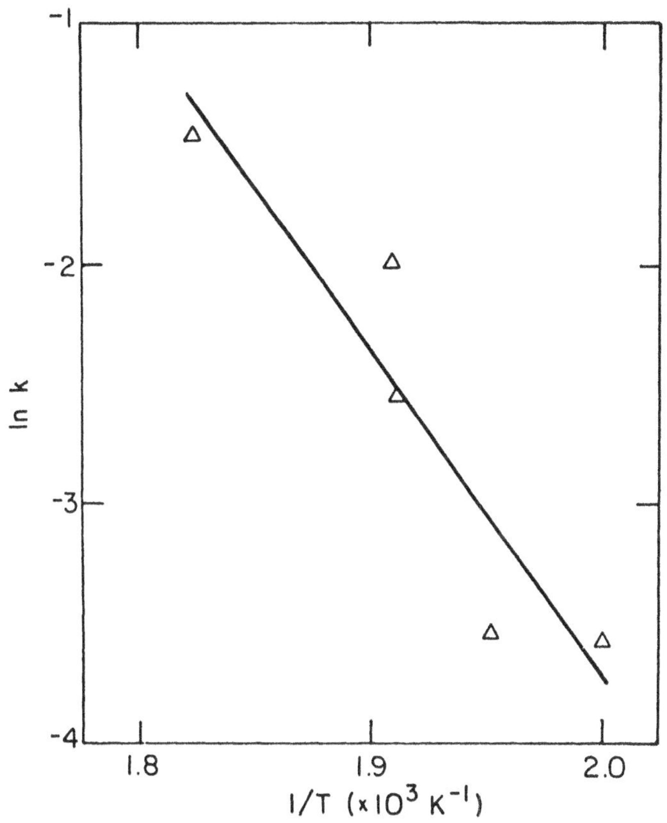

Fig. 3. Temperature dependence of disproportionation rate constant of $Na_2S_2O_3$.

where k' and h are the Boltzmann and Planck constants, respectively. The enthalpy and entropy of activation for the disproportionation can be calculated as $H^* = 26.0$ kcal/mol and $S^* = -14.9$ eu. The lower limit for the rate constant was used because in several much thicker sample specimen (average thickness were not determined), we obtained slower disproportionation rates. We attributed this result to the slow heating rate of the samples.

In conclusion, aside from providing the kinetic information for thiosulfate disproportionation in this paper, we have also demonstrated the application of X-ray photoelectron spectroscopy for the study of kinetics of solid-phase reactions at high temperatures.

ACKNOWLEDGMENTS

We thank Drs. T. Novakov and R. Dod for helpful discussions and our use of the X-ray photoelectron spectrometer; and Mr. R. Giauque for analyzing the X-ray fluorescence data. This work was supported by the Assistant Secretary for Fossil Energy, Office of Coal Research, Advanced

Environment Control Division of the US Department of Energy under Contract Number DE-AC03-76SF00098 through the Pittsburgh Energy Technology Center, Pittsburgh, Pennsylvania.

REFERENCES

1. Martin, A. E., ed., *Emission Control Technology for Industrial Boilers,* Noyes Data Corporation, Park Ridge, NJ, 1981.
2. Handbook of Chemistry and Physics, 44th Edition, The Chemical Rubber Publishing Co., Cleveland, OH, 1962, p. 661.
3. Wiedemeir, H., and Csillag, F., *High Temp. Sci.* **12,** 277 (1980).
4. Benson, S., *The Foundations of Chemical Kinetics,* McGraw-Hill, NY, 1960, pp. 225–266.

0018–1536/84/1700–0171$02.00

Observations of Excited Metastable and Radiative States of He_2, Ne_2, and Ar_2 by Neutralized Ion-Beam Spectroscopy

GREGORY I. GELLENE and RICHARD F. PORTER*

Department of Chemistry, Cornell University, Ithaca, NY 14853

Received October 26, 1983; Accepted December 31, 1983

ABSTRACT

Rare gas dimers formed when a fast beam of X_2^+ (X = He, Ne, Ar) is neutralized in the reaction

$$X_2^+ + K_{(g)} \rightarrow X_2^* + K^+$$

have been studied by neutralized ion beam techniques. From measurements of the kinetic energy released in the dissociation of X_2^* and with consideration of available spectroscopic information, it is proposed that the neutral dimers are formed initially in their first excited singlet and triplet electronic states. Estimates of the radiative lifetimes of the $^3\Sigma_u^+$ states of He_2^*, Ne_2^*, Ar_2^* are >3.2, >7.2, and 2.1 μs, respectively. Radiation from the $^1\Sigma_u^+ \rightarrow X^1\Sigma_g^+$ transitions is predicted to occur at $\lambda \approx$ 83, 80, and 114 nm for He_2, Ne_2, and Ar_2, respectively. Ion beam attenuation cross-sections for He_2^+, Ne_2^+, and $Ar_2{}^+$ are 250, 180, and 80 $Å^2$, respectively, indicating that electron transfer from $K_{(g)}$ is a highly efficient process. The capability of generating relatively intense beams of metastable He_2 and Ne_2 dimers may have applications for other beam experiments.

Index Entries: Ion beam spectroscopy, of He_2, Ne_2, and Ar_2; spectroscopy, of He_2, Ne_2, and Ar_2 by ion beams; He_2, ion beam spectroscopy of; Ne_2, ion beam spectroscopy of; Ar_2, ion beam spectroscopy of; metastable excited states, of He_2, Ne_2, and Ar_2; radiative states, of He_2, Ne_2, and Ar_2.

*Author to whom all correspondence and reprint requests should be addressed.

INTRODUCTION

Neutralized ion beam spectroscopy is a technique for investigating the modes of energy disposal when a fast beam of positively charged ions is neutralized in the electron transfer reaction

$$AB^+ + M(g) \rightarrow AB^* + M^{+\ddagger} \quad (1)$$

where AB^+ is any diatomic or polyatomic ion and M(g) is a thermalized target particle. The state of AB^* produced, which is strongly dependent on the ionization potential of the target, may be characterized as unstable, metastable, or stable by observing the angular distribution of particles in the neutral beam. In general, AB^* and $M^{+\ddagger}$ may be produced in excited electronic states and the principles of energy resonance and Franck-Condon factors are important considerations in qualitatively assessing the relative importance of various product states. By energy resonance it is meant that those reactions for which $\Delta E \approx 0$ proceed with highest efficiency. The "verticalness" of the transition is related to the short time scale of the interaction (estimated to be $\approx 10^{-15}$s) and implies that the neutral is produced initially in the same geometry of the ion. Several applications of this technique to derive new information about the stability of species produced in Reaction (1) have been summarized in a recent review (*1*).

In this paper we present a further application of neutralized ion beam spectroscopy to obtain information on radiative lifetimes of excited molecules. This technique, which is particularly well-suited to species with unstable ground states, will be illustrated by results obtained for rare gas dimers produced by the reaction

$$X_2^+ + K_{(g)} \rightarrow X_2^* + K^+ \qquad (X = \text{He, Ne, Ar}) \quad (2)$$

EXPERIMENTAL

The apparatus employed in this investigation has been extensively described elsewhere (*2*) and will only briefly be mentioned here. Ions were produced by electron impact in a low temperature (−150°C), high pressure (1–2 torr) ion source, accelerated through 6 kV, mass resolved, and focused into a collision chamber containing K in the pressure range of 1–10 mtorr. The chamber could be floated at a large negative potential (typically −4 kV), thus increasing the kinetic energy of the neutral beam. Unreacted ions eggressing from the collision chamber could be removed from the beam by electostatic deflection, so that only neutral species continued toward the magnetic electron multiplier located either 51.3 or 16.9 cm from the center of the collision chamber. The detector could be translated normal to the beam axis to sample beam intensity as a function of angle giving rise to a neutral beam profile [$I(\Theta)$ vs Θ].

Dimer ions were produced by ionizing high purity samples of rare gases under high pressure, low temperature conditions. Typical dimer ion intensities were a few percent of the respective monomer signal. The number density of K in the collision chamber was calculated from available thermodynamic data (3) and the temperature of the chamber was determined by a chromel–alumel thermocouple referenced to 0°C.

RESULTS

Cross-Sections

Ion beam attenuation cross-sections (σ_a) were calculated for He_2^+, Ne_2^+, and Ar_2^+ by the relation

$$\sigma_a = [-1/(n_1 - n_2)l] \ln (I_{t1}/I_{t2}) \tag{3}$$

where l is the interactive path length (4) and I_t is the transmitted ion intensity determined at two values of n, the target number density, under conditions where $I°$, the total beam intensity, was constant. These cross-sections, which are summarized in Table 1, are all large, indicating that electron transfer from K is an efficient process for all three dimer ions investigated. However, a trend of decreasing cross-section with increasing mass of the dimer is noted.

Analysis of Profiles

Under our experimental conditions, the transit time required for a neutral to reach the detector is about 4×10^{-7} to 4×10^{-6} s. If the neutral product of Reaction (2) survives longer than about 10^{-5} s then the full width at half-maximum will essentially be the same as that of the primary ion beam. Alternatively, a neutral that dissociates or radiatively decays to a dissociative state on a time scale shorter than about 10^{-7} s will give rise to a broadened beam profile because of the components of fragment velocity normal to the beam axis. A combination of these events, or

TABLE 1
Analysis of He_2, Ne_2, and Ar_2 Neutral Beam Profiles

Species	FE, eV[a]	σ_a, Å[2b]	σ_s/σ_d	Flight Time, μs	σ_s/σ_d	Flight Time, μs	τ_{rad}, μs
He_2	−3.1	250	0.88	1.35	1.09	0.44	>3.2
Ne_2	−1.2	180	0.60	3.01	0.68	0.99	>7.2
Ar_2	−0.5	80	0.021	4.26	0.087	1.40	2.1

[a]Relative error ± 10%.
[b]Relative error ± 30%.

a single dissociation process occurring on a 10^{-6} s time scale, will give rise to a composite beam profile such as those observed for all three rare gas dimers investigated (Figs. 1 and 2). These two mechanisms resulting in a composite beam profile can be distinguished by observing the stable/dissociative neutral branching ratio as a function of the beam transit time. When a homonuclear diatomic molecule undergoes dissociation, the maximum kinetic energy released (the Fragmentation Energy) can be calculated by the expression (5)

$$FE = -M_1 (V_{cm} \sin \Theta_{max})^2 \quad (4)$$

where V_{cm} is the center of mass velocity and M_1 is the mass of the fragment observed at the maximum laboratory scattering angle, Θ_{max}.

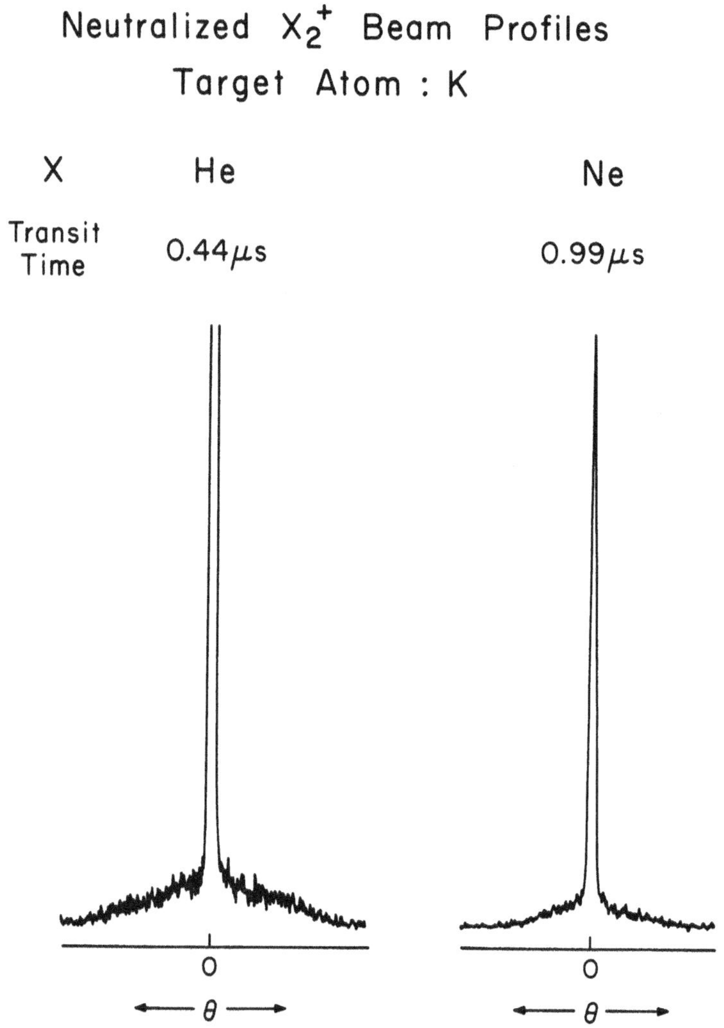

Fig. 1. Neutral beam profiles for He_2 and Ne_2 following electron transfer from $K_{(g)}$ to their precursor ions [Reaction (2)].

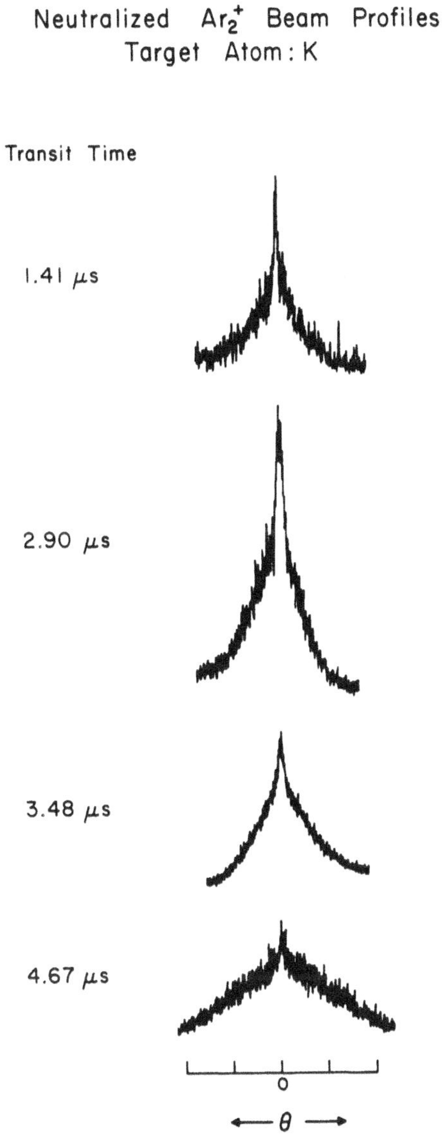

Fig. 2. Neutral beam profiles for Ar_2 following electron transfer from $K_{(g)}$ to Ar_2^+ [Reaction (2)].

Near resonant, vertical electron transfer to a rare gas dimer ion (X_2^+) from a low ionization potential target (e.g., K, IP = 4.34 eV) would be expected to produce primarily the first excited singlet and triplet state of the neutral dimer (X_2^*) directly or by radiative cascading from higher states. This can be understood by examining Fig. 3, which depicts qualitatively the relationship between the ground and excited potential curves for the general case. It can be seen that the direct near-resonant production of the $^1\Sigma_u^+$ ground state of X_2 at the intersection point with the ground state level of K would lead to dissociation with a very large kinetic energy release (approximately 17.8, 15.9, and 10.4 eV for He_2, Ne_2,

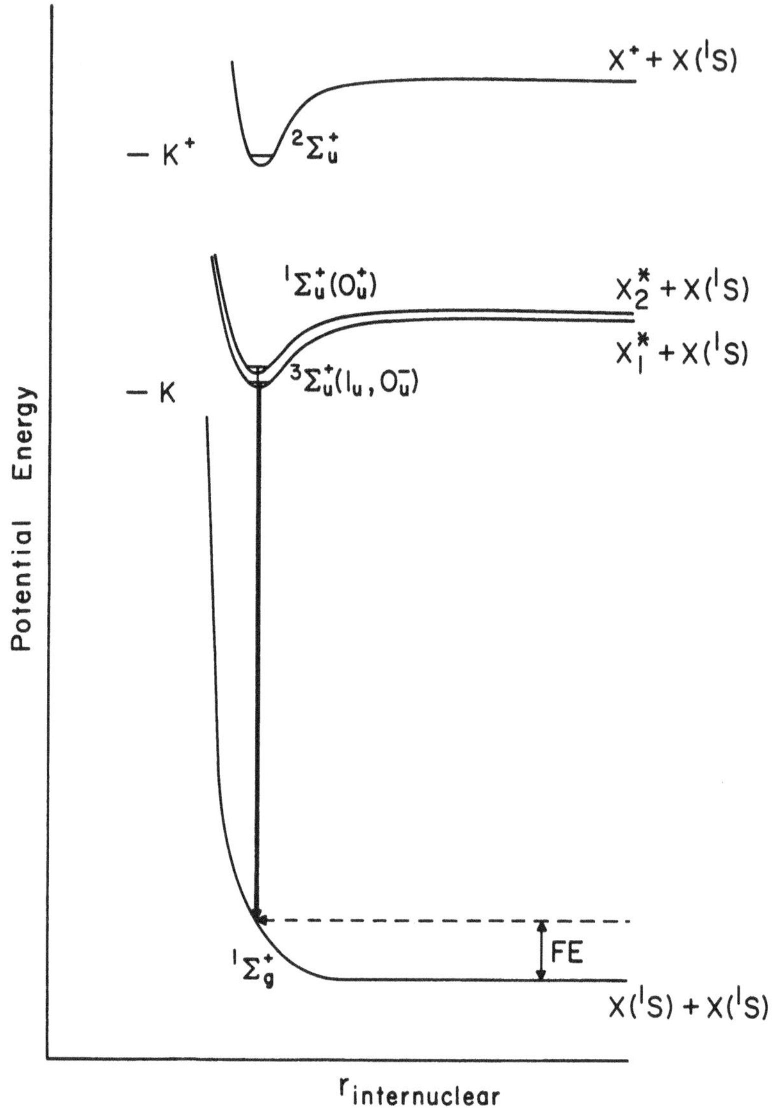

Fig. 3. Qualitative energy level diagram depicting electronic states of rare gas dimers relevant to this study. X_2^* and X_1^* are the $2s^1S$ and $2s^3S$, $3s'[1/2]°$ and $3s[3/2]°$, and $4s'[1/2]°$ and $4s[3/2]°$ atomic states for He_2, Ne_2, and Ar_2, dissociation products respectively.

and Ar_2, respectively) (*6*). Fragmentation energies calculated by Eq. (4) and the maximum scattering angles observed for He_2, Ne_2 (Fig. 1), and Ar_2 (Fig. 2) are −3.1, −1.2, and −0.50 eV, respectively (Table 1). These relatively small FE values can be understood in terms of a radiative transition to the essentially repulsive ground state, following formation of excited electronic states of the neutral dimer, as depicted in Fig. 3. It can be noted that direct vertical electron transfer to the ground state would require a substantial energy defect with the prediction of much lower cross sections than those observed. Further, since the ${}^1\Sigma_u^+ \rightarrow X^1\Sigma_g^+$ transition is strongly allowed, whereas the ${}^3\Sigma_u^+ \rightarrow X^1\Sigma_g^+$ is spin forbidden, production of the ${}^1\Sigma_u^+$ and ${}^3\Sigma_u^+$ excimer states would give rise to a composite beam profile.

He_2 and Ne_2

Beam profiles for He_2 and Ne_2 (Fig. 1) were resolved into two intensity components, I_s, the signal arising from nondissociating molecules (the central spike) and I_d, the signal from dissociated species (the broadened base portion). I_s and I_d are related to the primary ion beam intensity (I°) and the transmitted ion intensity (I_t) by the relation

$$I^\circ - I_t = \alpha I_s + I_d \tag{5}$$

where α is a correction factor for X_2^+/X_2 detection efficiency. When I_t is measured at two collision chamber temperatures, we obtain the difference relationship for a fixed I°

$$\Delta I_t = (\alpha)\Delta I_s + \Delta I_d \tag{6}$$

values of α, calculated from the empirical relationships for secondary electron yields of Beuhler and Friedman (7) are 1.16 and 1.19 for He_2 and Ne_2, respectively. Using this relationship and the measured quantities ΔI_t and ΔI_s, we obtain the values of $\Delta I_s/\Delta I_d = \sigma_s/\sigma_d$, the cross-section branching ratio, given in Table 1 for two collision chamber–detector distances. This method of analysis avoids the necessity of applying a multiplier sensitivity correction for the dissociated component of the beam. It should be noted that for a flight time variation of a factor of 3, σ_s/σ_d is essentially unchanged within ±25%. This behavior is not expected for a single first-order radiative decay process occurring over the beam flight time. The observations can be interpreted as the occurrence of a fast radiative process leading to ground state atoms and a slower radiative process from a metastable state of the dimer. For a simple first-order decay process, the lifetime of the species can be calculated from the expression

$$\tau = t/\ln\,(I_n/I_s) \tag{7}$$

where I_n is the total neutral intensity (I_s+I_d) and t is the beam flight time. Assuming that the fraction of metastables that dissociated is within the experimental error in I_s/I_d, we estimate a lower bound on τ_{rad} for the ${}^3\Sigma_u^+$ state of He_2 and Ne_2 of 3.2 and 7.0 μs, respectively.

Ar_2

Neutral beam profiles obtained for Ar_2 (Fig. 2) are composite, but for this system the branching ratio σ_s/σ_d increases as the neutral beam flight time decreases. This is evidence that the metastable state is dissociating at an observable rate during its transit. If a single radiative process is assumed, we obtain from σ_s/σ_d and Eq. (7), lifetimes of 1.10 and 0.55 μs for the long and short flight times, respectively (Table 1). This discordance can be rationalized on the basis of two competing radiative processes terminating at the dissociative ground state. For long flight times the dissociative component of the beam arises from both decay processes, but for short flight times σ_s/σ_d approaches the correct branching ratio for the two states. To evaluate the lifetime of the long-lived metastable state, we first normalize the beam intensities for different flight times and obtain the ratio $(I_s)_{t_1}/(I_s)_{t_2} = \gamma$. From the expression

$$\tau_{rad} = (t_2 - t_1)/\ln \gamma \tag{8}$$

we find $\tau_{rad} = 2.1$ μs. Experimental errors in σ_s/σ_d place the limits on this quality between 1.7 and 3.1 μs.

DISCUSSION

A comparison of the FE values in Table 1 indicates that the kinetic energy release decreases as the mass of the dimer increases. This result can be understood in terms of a corresponding increase in bond length of the dimer (r_e is approximately 1.04, 1.80, and 2.43 Å for He_2^*, Ne_2^*, and Ar_2^*, respectively) (*6*), which results in vertical radiative transitions terminating at lower points on the steep repulsive wall of the ground-state potential curve. Oka et al. (*8*) using optical spectroscopic techniques obtained a τ_{rad} of 6.6 and 3.2 μs for the $^3\Sigma_u^+$ states of Ne_2 and Ar_2, respectively. Although we are not certain, it seems most probable that we are observing the same metastable states. However, vibrational excitation may have a significant effect on the radiative lifetime as theoretically determined values τ_{rad} for Ne_2 ($^3\Sigma_u^+$) varied from about 1–12 μs depending on the vibration state (*9*). Gray and Tomlinson (*10*) observed a metastable state of He_2 following He_2^+/Li electron transfer reactions that they identified as the $^3\Sigma_u^+$ state with $\tau > 10^{-8}$s. The present measurements extend the lifetime boundary more than two orders of magnitude. Since the metastable states of He_2^* and Ne_2^* are not observed to undergo a significant decay over the transit times investigated, the σ_s/σ_d can be taken as a measure of the triplet/singlet production ratio.

The radiative energy of the $^1\Sigma_u^+ \rightarrow {}^1\Sigma_g^+$ transition ($E_{h\nu}$) occurring in these experiments can be estimated by the energy balance relationship

$$E_{h\nu} = E_{X_2}\,(^1\Sigma_u^+) + FE \tag{9}$$

The results of this calculation are compared with observed emission continua for He_2, Ne_2, and Ar_2 in Table 2. The calculated values of $E_{h\nu}$ for each dimer lies within the range of its observed emission continuum. It should be noted, however, that the emission spectra from the excimer states produced in Reaction (2) need not extend through the same wavelength region as that observed in rare gas discharges, since the mechanisms of excimer formation are different. In general we would expect a narrower continuum in the emission spectrum of X_2^* produced from Reaction (2). Finally it can be noted that the results of this study

TABLE 2
Estimation of $^1\Sigma_u^+ \rightarrow {}^1\Sigma_g^+$ Radiative Transition Energies of X_2^*, Following Electron Transfer Reaction (2)

Species	$E_{h\nu}$, eV	Reported continuum, eV[a]
He_2	≥14.9	12.4–20.6
Ne_2	≥15.4	11.8–16.6
Ar_2	≥10.9	7.5–11.6

[a]Reference (*11*).

demonstrate that relatively intense beams of He_2^* and Ne_2^*, with approximately 18.0 and 16.5 eV of electronic energy, respectively, can be produced by neutralized ion beam techniques. These beams may find applications in additional beam experiments investigating energy transfer.

ACKNOWLEDGMENTS

We thank Nancy Kleinrock Schneider for her assistance. This work was supported by NSF through the Materials Science Center (Grant GH-33637), Cornell University, and NSF Grant CHE-8215184.

REFERENCES

1. Gellene, G. I., and Porter, R. F., *Acct. Chem. Res.* **16,** 200 (1983).
2. Gellene, G. I., Cleary, D. A., and Porter, R. F. *J. Chem. Phys.* **77,** 3471 (1982).
3. Stoll, D. R., and Prophet, H., *JANAF Thermochemical Tables,* Natl. Stand. Ref. Data Ser. Natl. Bur. Stand. 37, 1971, 2nd ed.
4. In this calculation, we used an effective interaction length of 0.41 cm, which is the geometric length of the collision chamber plus a 30% increase to account for atoms effusing through the slits in the path of the beam.
5. Fluendy, M. A. D., and Lacoley, K. P., *Chemical Applications of Molecular Beam Scattering,* Chapman and Hall, London, 1973, pp. 37–43.
6. Huber, K. P., and Herzberg, G., *Constants of Diatomic Molecules,* Van Nostrand Reinhold Company, New York, 1979.
7. Beuhler, R. J., and Friedman, L., *Internat. J. Mass Spectrom. Ion Phys.* **23,** 81 (1977).
8. Oka, T., Rama Rao, K. V. S., Redpath, J. L., and Firestone, R. F., *J. Chem. Phys.* **61,** 4740 (1974).
9. Schneider, B., and Cohen, J. S., *J. Chem. Phys.* **61,** 3240 (1974).
10. Gray, J., and Tomlinson, R. H., *Chem. Phys. Lett.* **3,** 523 (1969).
11. Tanaka, Y., Jursa, A. S., and LeBlanc, F. J., *J. Opt. Soc. Am.* **48,** 304 (1958).

0018–1536/84/1700–0181$02.20

ScNi and TiCo Molecules

Ground States, Bonding, and Brewer-Engel Theory

R. J. VAN ZEE and W. WELTNER, JR.*

Chemical Physics Center, University of Florida, Gainesville, FL 32611

Received November 23, 1983; Accepted December 31, 1983

To ***Leo Brewer:*** *great scientist, courageous gentleman, and good friend*

ABSTRACT

The isoelectronic ScNi and TiCo molecules have been trapped in argon matrices near 4 K and their electronic ground states established as $^2\Sigma$ via electron spin resonance (ESR) spectroscopy. These ground states are indicative of multiple metal–metal bonding. The hyperfine splittings at the ^{47}Ti^{59}Co nuclei indicate an approximate $3d\sigma_{Ti}^{0.33}$ + $4s\sigma_{Ti}^{0.33}$ + $3d\sigma_{Co}^{0.07}$ + $4s\sigma_{Co}^{0.25}$ unpaired electron configuration. Likewise the ^{45}ScNi nuclear hyperfine interactions indicate an approximate $3d\sigma_{Sc}^{0.50}$ + $4s\sigma_{Sc}{}^{0.35}$ + $(4s\sigma + 3d\sigma)_{Ni}^{0.15}$ configuration. The spin distribution, electronic structures, and bonding in these molecules are discussed relative to the Brewer-Engel theory of metals and to calculations made on such alloy systems. It may be inferred from these results that the ScCo, TiFe, TiNi, and VCo molecules probably have $^1\Sigma$ multiply bonded ground states.

Index Entries: Electron spin resonance, of ScNi and TiCo molecules; electronic properties, of ScNi and TiCo molecules; Brewer-Engel theory, of metallic molecules of ScNi and TiCo; ScNi, ESR determination of bonding and ground states in; TiCo, ESR determination of bonding and ground states in; NiSc, ESR determination of ground states and bonding in; CoTi, ESR determination of bonding and ground states in; bonding, ESR determination for ScNi and TiCo; electronic ground states, ESR determination for ScNi and TiCo; Brewer-Engel theory, applied to TiCo and ScNi.

*Author to whom all correspondence and reprint requests should be addressed.

INTRODUCTION

The electronic properties of transition-metal molecules, beginning with the diatomics, are largely unknown, and it is of interest to correlate if possible their properties with those of the bulk metals and alloys. Brewer (*1,2*) has considered and extended the theories of Pauling (*3*) and particularly of Engel (*4*) to provide an understanding of the structures and thermodynamics of intermetallic compounds from the properties of the atoms. He has also suggested that the principles applied to the metallic state should be useful in predicting the properties of transition-metal molecules (*5,6*). Cocke and Gingerich (*6*) have measured the dissociation energy of the TiRh molecule, confirming the high stability predicted by the Brewer-Engel theory. Still, more basic understanding of the success of the theory could be obtained from knowledge of the electronic structure and bonding in such cases.

In our electron spin resonance (ESR) studies of transition metal molecules trapped in the solid rare gases at 4 K, we have successfully observed many of the homo- and heteronuclear diatomics, such as Sc_2 (*7*), Mn_2 (*8*), CrCu, CrAg, and CrAu (*9*) (referred to as I), MnAg, and CrZn (*10*) (referred to as II). Others have probably also been prepared, but were not detected because of the nature of their ESR spectra (that has been speculated upon in II). Transition-metal molecules tend to be of high-spin, reflecting the open-shell *d* orbitals of the atoms, and those that are of low-spin [Cr_2 (*11*), Mo_2 (*12*)] exhibit multiple bonding and high dissociation energies. ScNi and TiCo are shown here to lie in the latter category, as expected from Brewer's theory of the corresponding alloys (*2*).

EXPERIMENTAL

The procedure, apparatus, liquid helium cryostat, and X-band ESR spectrometer have been described in I (*9*) and in other papers (*13*). Scandium chips [1000 ppm trace metal impurities: 100% ^{45}Sc (I = 7/2)], titanium sponge [>99.9% purity: 7.4% ^{47}Ti (I = 5/2), 5.5% ^{49}Ti (I = 7/2), 87.1% other isotopes (I = 0)], cobalt powder [10 ppm trace metal impurities: 100% ^{59}Co (I = 7/2)] and nickel powder [> 99.99% purity: 1.1% ^{61}Ni (I = 3/2), 98.9% other isotopes (I = 0)] were all purchased from Spex. Titanium foil enriched in ^{47}Ti [99% purity: 80.1% ^{47}Ti (I = 5/2), 1.1% ^{49}Ti (I = 7/2), 18.8% other isotopes (I = 0)] was purchased from Oak Ridge National Laboratories, Oak Ridge, Tennessee.

Two tantalum resistance-heated cells were used for the simultaneous vaporization of two metals. The tantalum cell used for scandium had a 20-mil wall thickness, the cell for titanium had a 40-mil wall thickness, and the tantalum cells for cobalt and nickel had a 15-mil wall thickness with 60-mil wall thickness alumina liners. Typical vaporization temperatures were 1450°C (Sc), 1800°C (Ti), 1600°C (Co), and 1500°C (Ni), as measured with an optical pyrometer (uncorrected for emissivity).

Neon and argon [see I (9)] were used as matrix gases. Spectra reported are from argon only because of their greater signal intensity.

RESULTS

The ESR powder pattern spectrum of randomly oriented ScNi molecules isolated in argon at 4 K is shown in Fig. 1. The spectrum is pieced together from the observed spectra because of severe overlapping at lower fields from Sc_2 (7). The spectrum extends from ~2000 to 4500 G and is centered about 3300 G, indicative of a $^2\Sigma$ molecule. The spectrum results from the interaction of an unpaired electron with the ^{45}Sc nucleus, which has $I = 7/2$, resulting in eight perpendicular and eight parallel hyperfine (hf) lines. Observed perpendicular and parallel line positions are given in Table 1. The spectrum of ScNi isolated in neon was very weak in intensity compared to that in argon and was not analyzed in detail.

The ESR spectrum of TiCo in argon is shown in Fig. 2. It extends from roughly 1400 G to over 5000 G and is again centered near 3300 G,

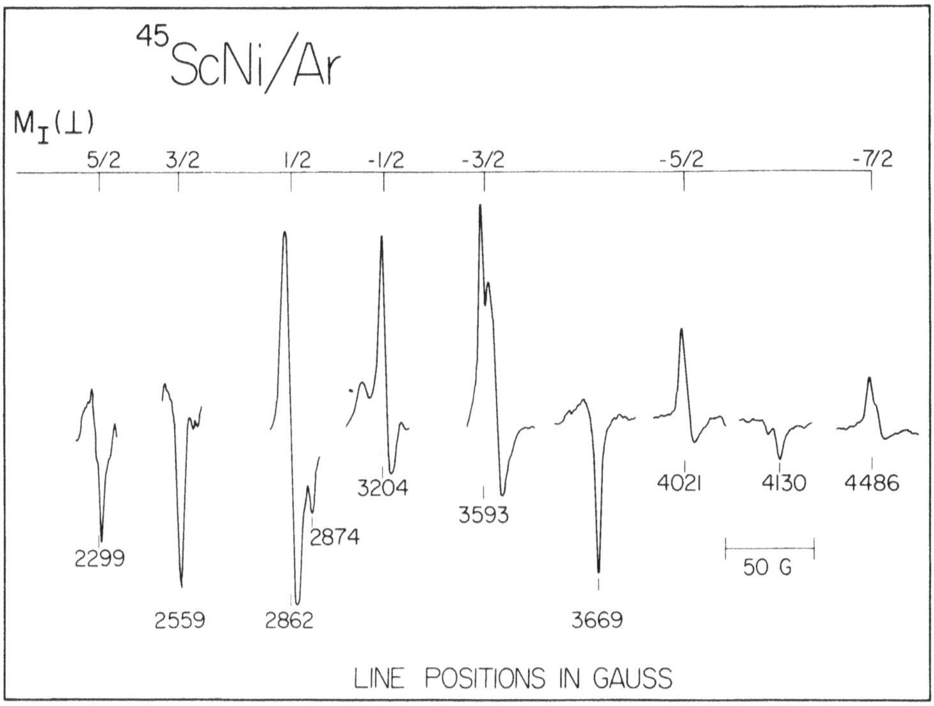

Fig. 1. ESR spectrum of $^{45}ScNi$ molecules trapped in solid argon at 4 K (ν = 9.3750 GHz).

TABLE 1
Comparison of Observed and Calculated ^{45}Sc Hyperfine Line Positions (in Gauss) for $^2\Sigma$ ^{45}ScNi Molecules Isolated in Solid Argon at 4 K (ν = 9.3750 GHz) Using the Listed Magnetic Parameters[a]

$M_I(^{45}Sc)$	⊥ Lines observed, G	⊥ Calculated, G	∥ Lines observed, G	∥ Calculated, G
7/2	—	2078	—	1955
5/2	2299(2)	2299	—	2226
3/2	2559	2559	—	2530
1/2	2862	2860	2874(2)	2872
−1/2	3204	3204	3252	3253
−3/2	3593	3592	3669	3671
−5/2	4021	4020	4130	4127
−7/2	4486	4487	—	4617

[a] $g_{\parallel}$ = 2.001(1), $g_{\perp}$ = 1.9990(5), $A_{\parallel}$ = 1062(3) MHz, $A_{\perp}$ = 960(3) MHz.

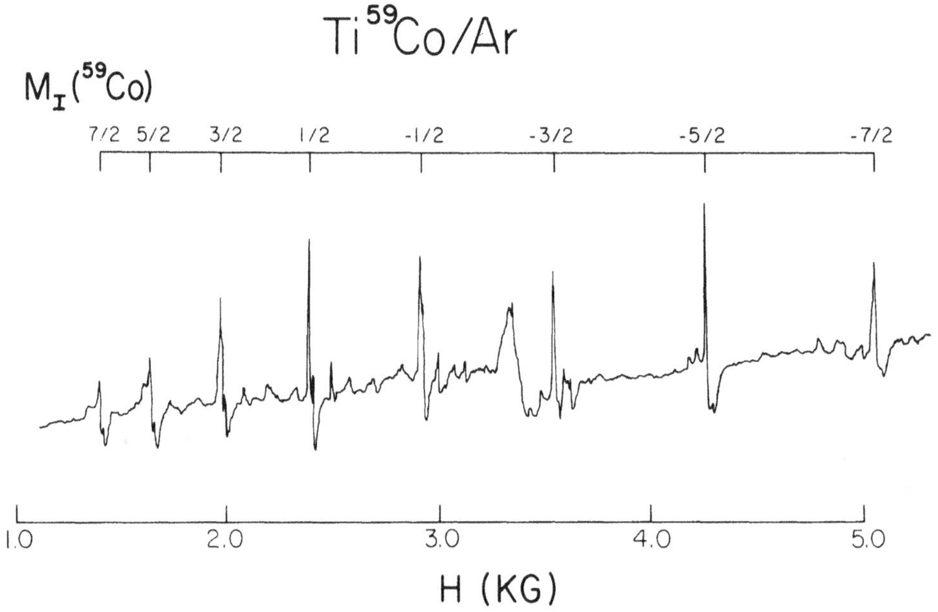

Fig. 2. ESR spectrum of Ti ^{59}Co molecules trapped in solid argon at 4 K (ν = 9.3700 GHz).

indicating another $^2\Sigma$ molecule. Hyperfine interaction with the ^{59}Co (I = 7/2) nucleus splits the ESR signal into eight lines. One major and several minor sites are detected in the argon matrix and several also in neon. The low natural abundance of any Ti isotopes with nuclear spin precludes observing additional hfs from Ti.

Upon incorporating ^{47}Ti (I = 5/2) enriched to 80.1% into TiCo, additional hfs is observed. The ^{47}Ti hfs of one ^{59}Co hf line is shown in Fig. 3.

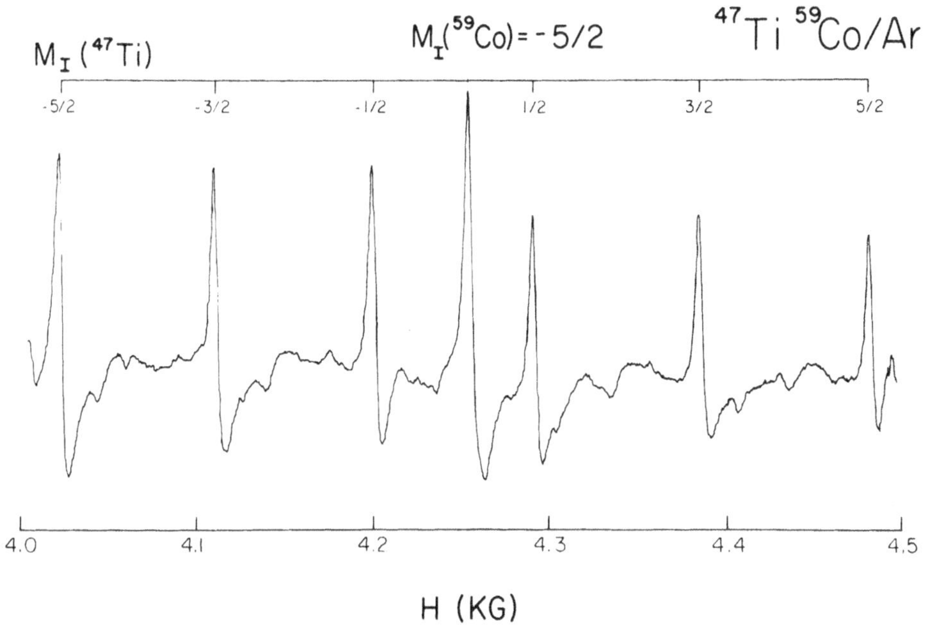

Fig. 3. Hyperfine structure produced on one of eight ^{59}Co (I = 7/2) lines when ^{47}Ti ^{59}Co molecule is formed using titanium enriched to 80% in ^{47}Ti. Central line is for Ti ^{59}Co molecule as in Fig. 2. [Argon matrix at 4 K (ν = 9.3725 GHz).]

The sextet of lines from ^{47}Ti (I = 5/2) is observed to have a splitting near 100 G for the parallel lines and about 90 G for the perpendicular. Observed ^{59}Co hf lines from TiCo are listed in Table 2, and ^{47}Ti hfs of one of those ^{59}Co lines is given in Table 3.

TABLE 2
Comparison of Observed and Calculated ^{59}Co Hyperfine Line Positions (in Gauss) for the $^2\Sigma$ Ti^{59}Co Molecule Isolated in Solid Argon at 4 K (ν = 9.3700 GHz) Using the Listed Magnetic Parameters[a]

$M_I(^{59}(Co)$	⊥ Lines observed, G	⊥ Calculated, G	‖ Lines observed, G	‖ Calculated, G
7/2	1394(3)	1394	1327(3)	1330
5/2	1642	1642	1602	1605
3/2	1964	1967	1952	1953
1/2	2387	2385	—	2390
−1/2	2909	2908	—	2928
−3/2	3536	3535	3565	3566
−5/2	4255	4256	4293	4293
−7/2	5054	5054	5094	5094

[a] $g_\parallel$ = 1.9970(5), $g_\perp$ = 1.9866(5), $A_\parallel$ = 1494(3) MHz, $A_\perp$ = 1445(3) MHz.

TABLE 3
Comparison of Observed and Calculated ^{47}Ti ($I = 5/2$) Hyperfine Line Positions (in Gauss) for $M_I = -5/2$ (assuming $A_\perp$ ^{59}Co is positive) of the $X^2\Sigma$ $^{47}Ti^{59}Co$ Molecule Isolated in Solid Argon at 4 K ($\nu = 9.3725$ GHz) Using the Following Parameters[a]

$M_I(^{47}Ti)$[b]	⊥ Lines observed, G	⊥ Calculated, G	‖ Lines observed, G	‖ Calculated, G
−5/2	4024(2)	4025	4045(2)	4045
−3/2	4111	4112	4140	4140
−1/2	4200	4201	4235	4236
+1/2	4291	4292	4333	4334
+3/2	4385	4385	4434	4434
+5/2	4481	4481	4537	4536

[a] $A_\parallel = 274(3)$ MHz, $A_\perp = 254(3)$ MHz.
[b] Assuming the hyperfine parameters of ^{47}Ti have the same sign as the nuclear moment.

ANALYSIS

The spin Hamiltonian used in the analysis of the ScNi and TiCo spectra is:

$$H = g_\parallel \beta H_z S_z + g_\perp \beta(H_x S_x + H_y S_y) + \sum_i [A_\parallel^i I_z^i S_z^i + A_\perp^i (I_x^i S_x^i + I_y^i S_y^i)]$$

Quadrupole and additional Zeeman terms were found to be unnecessary in fitting the observed lines.

Initial estimates of $g\|$, $A\|$, $g\perp$, and $A\perp$ were obtained from second-order perturbation treatments (*14*). Then the 16 × 16 (for $S = 1/2$, $I = 7/2$) spin matrix was diagonalized and the energy level differences compared to $h\nu$ at particular line positions. $A\|$, $g\|$, $A\perp$, and $g\perp$ were then varied and the matrix again diagonalized. This was repeated for each line position until all calculated line positions using one set of parameters were fit to within the observed line uncertainty. For $Ti^{59}Co$ in argon only the major site was fit. For ^{47}Ti hfs the second-order treatment was found to be satisfactory (*14*). Calculated line positions for ^{45}Sc, ^{59}Co, and ^{47}Ti hfs are given in Tables 1, 2, and 3, respectively, where they are compared to the observed line positions. Parameters obtained from the fitting procedures are shown in Table 4.

It is possible that the ESR specra are being incorrectly attributed to $^2\Sigma$ molecules and that the ground states are really $^4\Sigma$ or even $^6\Sigma$ with very small zero-field splitting parameters ($D < \sim 0.02$ cm^{-1}). The supposition would then have to be made that only the one fine-structure line with $M_S = 1/2 \longleftrightarrow -1/2$ is being observed in each case and that the weaker adjacent transitions are not detected. This is possible but unlikely; first because a small D value in transition-metal molecules having considerable spin-orbit coupling is improbable (*15*), and secondly be-

TABLE 4
Derived Magnetic Parameters for the Indicated Nuclei in the Following Bimetallic Molecules/argon at 4 K

Parameter	^{45}ScNi	^{47}Ti^{59}Co
$g_{\parallel}$	2.001(1)	1.9970(5)
$g_{\perp}$	1.9990(5)	1.9866(5)
$\lvert A_{\parallel}\mathrm{Sc}\rvert$ (MHz)	1062(3)	
$\lvert A_{\perp}\mathrm{Sc}\rvert$ (MHz)	960(3)	
$\lvert A_{iso}\mathrm{Sc}\rvert$ (MHz)	994(3)	
$\lvert A_{dip}\mathrm{Sc}\rvert$ (MHz)	34(2)	
$\lvert\psi_s(0)\rvert^2$Sc (au)	0.914(3)	
$\lvert<3\cos^2\theta - 1/r^3>\rvert$Sc (au)	0.52(3)	
% s character Sc	35.2	
$\lvert A_{\parallel}\mathrm{Ti}\rvert$ (MHz)		274(3)
$\lvert A_{\perp}\mathrm{Ti}\rvert$ (MHz)		254(3)
$\lvert A_{iso}\mathrm{Ti}\rvert$ (MHz)		261(3)
$\lvert A_{dip}\mathrm{Ti}\rvert$ (MHz)		7(2)
$\lvert\psi_s(0)\rvert^2$Ti (au)		1.03(1)
$\lvert<3\cos^2\theta - 1/r^3>\rvert$ Ti (au)		0.5(2)
% s character Ti		33.3
$\lvert A_{\parallel}\mathrm{Co}\rvert$ (MHz)		1494(3)
$\lvert A_{\perp}\mathrm{Co}\rvert$ (MHz)		1445(3)
$\lvert A_{iso}\mathrm{Co}\rvert$ (MHz)		1461(3)
$\lvert A_{dip}\mathrm{Co}\rvert$ (MHz)		16(2)
$\lvert\psi_s(0)\rvert^2$Co (au)		1.384(3)
$\lvert<3\cos^2\theta - 1/r^3>\rvert$ Co (au)		0.25(3)
% s character Co		24.6

cause the g tensor components are very reasonable for $^2\Sigma$ molecules. If the molecules were of high spin, even a small value of D would shift the observed fine-structure line and appear as a shift in $g\perp$ from g_e. However, the values of $\Delta g_{\perp} = -0.003$ to -0.016 found assuming $^2\Sigma$ ground states are essentially in accord with expectations for such molecules simply on the basis of angular momentum coupling with excited electronic states (see below).

DISCUSSION

Spin Distribution

The hyperfine splitting in the ESR spectra, when compared to those in the atoms as given by Morton and Preston (*16*), yield approximate spin density distributions in the molecules. For ScNi one find roughly 85% ($0.35\ s\sigma_{Sc} + 0.50\ d\sigma_{Sc}$) of the unpaired spin is on the scandium atom. For TiCo, where hfs for both nuclei were observed, about 66% (0.33

$s\sigma_{Ti}$ + 0.33 $d\sigma_{Ti}$) is on titanium and 25% exhibits s character on Co (see Table 4). Thus the MO occupied by the unpaired electron is in each case highly $s\sigma$–$d\sigma$ hybridized. Also, the unpaired spin is more localized on the light element in each case and is more localized on Sc in ScNi than on Ti in TiCo.

g Tensor Components

From Table 4 one finds $\Delta g_{\perp} = g_{\perp} - g_e = -0.0033(5)$ and $-0.0157(5)$ for ScNi and TiCo, respectively. Corresponding $\Delta g_{\parallel}$ values are $-0.001(1)$ and $-0.0053(5)$. Thus, the magnitudes of the Δg values for the two molecules differ by a factor of five, which is a little unexpected for two such similar molecules. The TiCo values are slightly more accurate because hfs for both nuclei were observed for it.

The second-order expression for $\Delta g_{\perp}$ is (*17*)

$$\Delta g_{\perp} = -2\sum_n <0|\xi l_x|n> <n|l_x|0> /(E_n - E_o), \qquad (1)$$

where ξ is the molecular spin-orbit coupling constant, l_x the angular momentum operator, and o and n the ground state and nth excited $^2\Pi$ state, respectively. The observed negative values of $\Delta g_{\perp}$ im ply that the dominant excited states coupling with the ground state are form ed by excitation of the unpaired spin into an unoccupied π orbital, i.e., the excited states are regular $^2\Pi_r$ states. The difference in $\Delta g_{\perp}$ for the two molecules must then be attributed mainly to differences in ξ and in $E_n - E_o$. The values of ξ might be different by a factor approaching two at the most and then only because the unpaired spin populations are largely on the light atoms. (Approximate values of the spin-orbit coupling of the atoms from Dunn's tables (*18*) are Sc 60 cm^{-1}, Ti 100 cm^{-1}, Co 500 cm^{-1}, and Ni 600 cm^{-1}. Taking ξ as 100 cm^{-1} for ScNi and 200 cm^{-1} for TiCo yields $\Delta E \cong$ 60,000 and 25,000 cm^{-1}, respectively, if the matrix elements contribute no additional factors. The latter is again an unknown since the properties of the excited $^2\Pi$ state(s) are of course not known. Then this discussion only serves to point out the unknowns and that there is a significant difference in the electronic properties of ScNi and TiCo, since differences in their spin-orbit coupling do not appear to account for the $\Delta g\perp$ values.

Spin-Rotation Constant

Curl's relationship (*19*)

$$\gamma = -2B\Delta g\perp$$

makes it possible to calculate γ, the spin rotation constants of these molecules if their rotational constants, B, can be approximated. If the bond lengths are taken as 2Å (*20*), then one finds that γ(ScNi) = +0.0011 cm^{-1} and γ(TiCo) = +0.0050 cm^{-1}. In general γ determined from $\Delta g_{\perp}$values in matrices are accurate to about 10% (*21*), but here the possible errors could be a little larger because of the necessity of approximating the *B* values.

Bonding and the Brewer-Engel Theory

The Brewer-Engel theory of transition metal alloys predicts that intermetallic compounds formed from elements on the left and right sides of the Periodic Table should be particularly stable because of the formation of multiple *d* bonds (*1,2*). Brewer has also applied this to gas-phase diatomics (*5*), and the high dissociation energy of the TiRh molecule is evidence of its validity (*6*).

Among the transition-metal diatomics there is, in general, competition between obtaining as high a spin as possible and the formation of multiple bonds utilizing *d* orbitals. Also entering is the necessity of promoting the atoms from filled *s* orbitals to their valence states, e.g., $d^{n-2}s^2$ to $d^{n-1}s$ so that an *s–s* bond can form (*22*). This promotion energy is apparently easily made up by the energy of bonding in most cases. [An example where this is not the case is Mn_2, where the high promotion energy prevents the formation of a valence bond (*8*).] The strength of atomic *d* bonding roughly decreases toward d^5 and increases after d^5 (*2*), thus Sc combining with Ni should favor stronger *d* bonding than Ti and Co. This is reflected in Miedema's Table 1 of predicted dissociation energies (*23*) for those two diatomics and in the calculated heats of formation of the alloys (*24*). One then reasons that the high promotion energy of Sc is offset by the low promotion energy of Ni and by the strong *d* bonding.

A further factor in heteronuclear diatomics is charge transfer, which has been considered and calculated in alloys of ScNi and TiCo having the CsCl structure (*25*). It has been suggested that the stability of such structures is a consequence of this ion-like contribution. Such charge transfer can also be expected to occur from the light to the heavier elements in both molecules; more so in ScNi than TiCo because the former contains elements further apart in the first row and differing more in electron ionization energy and affinity (see Table 5).

In view of the above remarks it is significant that ScNi and TiCo have low-spin $S = 1/2$ ground states. The simplest way of viewing the bond in these molecules is to place four valence electrons in $s\sigma$ and $d\sigma$ bonding orbitals (both hybridized) and one unpaired spin in an antibonding hybrid $s\sigma^*$ orbital. The remaining eight electrons are then in $d\pi$ and $d\delta$ orbitals, which must be partially bonding if the predicted high-bond energies in these molecules are correct (*23*). This picture would satisfy the observed hyperfine and *g* tensor information, but it is not in accord with the variation of predicted dissociation energies of related diatomics given by Miedema (*23*). That variation implies that the unpaired spin in ScNi and TiCo is in a **bonding** hybrid $s\sigma$–$d\sigma$ orbital since its removal (to form ScCo, TiFe) leads to a decrease in the predicted dissociation energy, and the addition of another electron to that orbital (to form TiNi) leads to an in-

TABLE 5
Atomic Electronic Parameters from the Literature

Parameter	Sc	Ti	Co	Ni
Ionization potential (eV)[a]	6.56	6.83	7.86	7.63
Electronegativity[b]	1.2	1.2	1.4	1.5
Electron affinity (eV)[c]		0.080(14)	0.622(10)	1.157(10)
Metal radii (Å)[d]	1.60	1.46	1.26	1.24
Promotion energy[e] to $d^{n-1}s$ config. (in cm^{-1})	11,520	6,560	3,480	205

[a]Leighton, R. B., *Principles of Modern Physics*, McGraw-Hill, New York, 1959.
[b]Boyd, R. J., and Markus, C. E., *J. Chem. Phys.* **75**, 5385 (1981).
[c]Feigerle, C. S., Corderman, F. R., Bobashev, S. V., and Lineberger, W. C., *J. Chem. Phys.* **74**, 1580 (1981); Corderman, R. R., Engleking, P. C., and Lineberger, W. C., *J. Chem. Phys.* **70**, 4474 (1979).
[d]Pauling, L., *The Nature of the Chemical Bond*, Cornell Univ., Ithaca, 1960.
[e]Brewer, L., and Winn, J. S., *Faraday Symp. Chem. Soc.* **14**, 126 (1980).

crease in the predicted dissociation energy. Of course, it is possible that the variation among Miedema's predictions is incorrect, although this seems unlikely.

Also, since charge transfer is expected from the lighter to the heavier element, why is the spin density largest on the lighter element in both molecules? One must reason that the charge transfer is occurring among the inner MOs and the orthogonality of the odd-electron MO to them means that to a first approximation the magnitude of the coefficients of the light element AOs will be increased. For the same reason, the more "ionic" molecule, ScNi, would have more spin density on Sc than would Ti in TiCo. Our hypothesis is then that in these molecules the spin density distribution is the opposite of the charge transfer.

ScCo, ScCu, TiFe, TiNi, VCo: $^1\Sigma$ Molecules?

The addition to or subtraction of one electron from ScNi and TiCo should lead to the formation of diatomics with $^1\Sigma$ ground states. One then predicts that the neutral molecules ScCo, ScCu, TiFe, TiNi, and VCo have this closed shell lowest state. Among these ScCu may be doubtful because of the stabilizing effect of its d^{10} shell. Although it does form multiple bonds (*26*), Cu is usually considered an alkali-like transition metal. Also, Miedema's predicted dissociation energy is lower by almost 100 kJ mol^{-1} for ScCu than ScNi, and 40 kJ mol^{-1} lower than its isoelectronic counterpart TiNi (*23*).

ACKNOWLEDGMENTS

We gratefully acknowledge the support of the National Science Foundation (Grant CHE-8210707) and some computational support provided by the Northeast Regional Data Center at the University of Florida.

REFERENCES

1. Brewer, L., in *Electronic Structure and Alloy Chemistry of the Transition Elements*, Beck, P. A., ed., Interscience, New York, 1963, p. 211; *High-Strength Materials*, Proc. Second Berkeley Intern. Mat. Conference, June, 1964, Zackay, V. F., ed., Wiley, New York, 1965, Chapter 2, p. 12; *Acta Metall.* **15,** 553 (1967).
2. Brewer, L., *Science* **161,** 115 (1968).
3. Pauling, L., *The Nature of the Chemical Bond*, Cornell University Press, Ithaca, New York, 1960.
4. Engel, N., *Kem. Maanedsbl.* **30,** 53, 75 (1949); **30,** 97, 105, 113 (1949); *Powder Metall. Bull.* **7,** 8 (1964); *Am. Soc. Metals, Trans. Quart.* **57,** 610 (1964).
5. Brewer, L., personal communication, 1968, as referenced in Gingerich, K. A., *Faraday Symp. Chem. Soc.* **14,** 109 (1980).
6. Cocke, D. L., and Gingerich, K. A., *J. Chem. Phys.* **60,** 1958 (1974).
7. Knight, L. B., Jr., Van Zee, R. J., and Weltner, W., Jr., *Chem. Phys. Lett.* **94,** 296 (1983).
8. Baumann, C. A., Van Zee, R. J., Bhat, S. V., and Weltner, W., Jr., *J. Chem. Phys.* **74,** 6977 (1981); **78,** 190 (1983).
9. Baumann, C. A., Van Zee, R. J., and Weltner, W., Jr., *J. Chem. Phys.* **79,** 5272 (1983).
10. Baumann, C. A., Van Zee, R. J., and Weltner, W., Jr., *J. Phys. Chem.* (in press).
11. See Michalopoulos, D. L., Geusic, M. E., Hansen, S. G., Powers, D. E., and Smalley, R. E., *J. Phys. Chem.* **86,** 3914 (1982); Bondybey, V. E., and English, J. H., *Chem. Phys. Lett.* **94,** 443 (1983); Delley, B., Freeman, A. J., and Ellis, D. E., *Phys. Rev. Lett.* **50,** 488 (1983).
12. See Hopkins, J. B., Langridge-Smith, P. D. R., Morse, M. D., and Smalley, R. E., *J. Chem. Phys.* **78,** 1627 (1983); Bursten, B. E., Cotton, F. A., and Hall, M. B., *J. Am. Chem. Soc.* **102,** 6348 (1980).
13. Easley, W. C., and Weltner, W., Jr., *J. Chem. Phys.* **52,** 197 (1970); Knight, L. B., Jr. and W. Weltner, Jr., *ibid.* **54,** 3875 (1971); Graham, W. R. M., Dismuke, K. I., and Weltner, W., Jr., *Astrophys. J.* **204,** 301 (1976).
14. Low, W., *Paramagnetic Resonance in Solids*, Academic Press, New York, 1960, p. 72.
15. *See* Weltner, W., Jr., *Magnetic Atoms and Molecules*, Van Nostrand Reinhold, New York, 1983, p. 198 ff.
16. Morton, J. R., and Preston, K. F., *J. Mag. Reson.* **30,** 577 (1978).
17. Stone, A. J., *Proc. Roy. Soc. (London)* **A271,** 424 (1963).
18. Dunn, T. M., *Trans. Far. Soc.* **57,** 1441 (1961).
19. Curl, R. F., *Mol. Phys.* **9,** 585 (1965).
20. For multiple metal–metal bonds, the bond lengths are known only for Cr_2 (1.69 Å, see ref. *11*) and Mo_2 (1.93 Å, see ref. *12*). The effective radius of Cr decreases from 1.25 Å in the metal to 0.85 Å in Cr_2 or by 32%. Using this ratio on ScNi, where the combined metal radii is 2.84 Å, gives 1.93 Å, justifying our use of 2.0 Å.
21. See ref. *15*, page 51.
22. Brewer, L., and Winn, J. S., *Faraday Symp. Chem. Soc.* **14,** 126 (1980).
23. Miedema, A. R., *Faraday Symp. Chem. Soc.* **14,** 136 (1980).
24. Alonso, J. A., and Girifalco, L. A., *J. Phys. Chem. Solids* **39,** 79 (1978).
25. See ref. 24 and ref. there to Beck, Yamashita, and Asano, Moruzzi, Papaconstantopoulos, Giner and their coworkers.
26. See ref. 9 and Engel, N., *Acta Metall.* **15,** 557 (1966).

0018–1536/84/1700–0193$02.00

Temperature Dependence of Pb($^3P_1^\circ$)

Spin-Orbit Relaxation by Atomic Hydrogen

M. J. BIRD AND L. F. PHILLIPS*

Chemistry Department, University of Canterbury, Christchurch, New Zealand

Received December 30, 1983; Accepted December 31, 1983

ABSTRACT

The interconversion of Pb($^3P_1^\circ$) and Pb($^3P_0^\circ$) in a series of premixed $H_2/N_2/O_2$ flames has been monitored by measuring the relative intensities of emission at 364.0 and 368.3 nm, following excitation at 283.3 nm. Improved rate constants have been obtained for the processes

$$\mathrm{Pb(^3P_1^\circ) + M \rightarrow Pb(^3P_0^\circ) + M}$$

(where M is either atomic hydrogen or a bulk flame constituent) over the temperature range 1570–2110 K. In contrast to previous results, in which the rate constant for hydrogen atoms was found to have a negative temperature dependence, we now find a small positive temperature dependence of the rate constant for both M = H and M = bulk flame gases. The implications of this finding are discussed in terms of a model that involves non-adiabatic transitions between excited states of PbH.

Index Entries: Temperature, and Pb($^3P_1^\circ$) spin-orbit relaxation; Pb($^3P_1^\circ$) spin-orbit relaxation, temperature dependence of; spin-orbit relaxation, of Pb($^3P_1^\circ$); hydrogen, Pb($^3P_1^\circ$) spin-orbit relaxation by atomic.

INTRODUCTION

Previously (*1,2*) we reported the results of experiments in which atomic lead, present as a trace additive in a series of premixed $H_2/N_2/O_2$ flames, was excited from the $^3P_2$ state to the $^3P_1^\circ$ state by 405.8 nm radia-

*Author to whom all correspondence and reprint requests should be addressed.

tion from a nitrogen-pumped dye laser. Fluorescence was observed from both the $^3P_1^\circ$ state and the nearby ($\Delta E = 327.3$ cm^{-1}) $^3P_0^\circ$ state and, using Einstein transition probabilities and cross-sections for quenching of Pb($^3P^\circ$) from the literature, the rate of interconversion of the two states was calculated from the observed ratio of the emission intensities. The interconversion rate was found to vary markedly with atomic hydrogen concentration, both from flame to flame and within a single flame, so that separate rate constants k_H and k_M could be derived for spin-orbit relaxation by collision with atomic hydrogen and with the other "bulk" flame gases (mainly H_2O, H_2, and N_2).

The values obtained for k_H (which were rather scattered) decreased with increasing temperature; the values found for k_M increased with temperature. The negative temperature-dependence of k_H could be accounted for in terms of a postulated curve-crossing between bound excited states of PbH. However, the apparent rate of decrease with increasing temperature was appreciably greater than the approximately $T^{-1/2}$ dependence that was predictable on the basis of the usual theory of non-adiabatic processes (*3–6*). The aim of the present work was to improve the precision of the measurements of k_H, to enable a more reliable comparison between the experimental and theoretical temperature coefficients. In fact the present results show that the previous measurements of k_H were in error, for reasons that will be explained shortly, and that actually both k_M and k_H increase slowly with increasing temperature (roughly as $T^{1.5}$).

Two components of the previous experimental system clearly needed improvement—one was the measurement of the $I_{368.3}/I_{364.0}$ fluorescence intensity ratio; the other was the measurement of atomic hydrogen concentration. Our first attempt to improve the precision of the measurements of fluorescence intensity involved a lead halide photodissociation laser pumped by an argon fluoride laser (*7,8*). Despite numerous trials, with both longitudinal and transverse excitation of lead iodide vapor, we did not succeed in producing a usable excitation source by this method and the attempt was abandoned. We finally settled on a flowing, microwave-powered discharge through helium containing a trace of lead iodide vapor, as described in the experimental section. Apart from stability of operation, this source had the additional advantage of exciting the $^3P_1^\circ$ state directly from the ground state by way of the 283.3 nm Pb resonance line, rather than from the metastable $^3P_2$ state, as was the case with the dye laser operating at 405.8 nm, and so was unaffected by the relatively low population of the $^3P_2$ state in cooler flames.

Our previous measurements of atomic hydrogen concentration were based on measured intensities of CuH emission at a large number of points in each flame (*9*), scaled by means of absolute measurements at a few points in each flame. The absolute measurements employed the Li/LiOH method (*10*), based on the relative emission intensities from lithium and sodium atoms when a mixed $NaNO_3$ + $LiNO_3$ solution was atomized into the flame. This method is subject to error in cool flames (below about 1700 K) because of chemiluminescent excitation of the sodium, leading to values of H-atom concentration that are too low. In the previous work it was believed that the absolute H-atom concentration measurements were made sufficiently far above the reaction zone for this source of error to be unimportant. Our present results, which are based on absolute H-atom concentrations measured by the Na/NaCl method (*11*), show that this assumption was incorrect, and that the apparent negative temperature dependence of k_H is mainly attributable to the use of erroneously low H-atom concentrations for the cooler flames.

With the demise of the negative temperature dependence of k_H, it is no longer necessary to postulate that the spin-orbit relaxation occurs by way of curve-crossing between bound excited states of PbH at some point well below the energy level of the separate Pb($^3P^\circ$) and H atoms. In fact the states involved are already so close to one another in energy that an actual crossing of potential energy curves may not be required to bring about the observed non-adiabatic process.

EXPERIMENTAL

The burner, atomizer and flow system, and range of flame composition were as described previously (2). The only significant departure from previous practice was that chlorine was added to the flame in the form of hydrogen chloride gas (Matheson technical grade, 99% minimum purity) metered directly with a Hastings flowmeter. The local concentration of hydrogen atoms in the flame was determined from the effect of added HCl on the intensity of sodium *D*-line emission via the usual expression.

$$I_0/I = 1 + K[\mathrm{HCl}]/[\mathrm{H}] \tag{1}$$

where I_0 and I are the intensities in absence and presence of HCl, the equilibrium constant K for the reaction

$$\mathrm{Na + HCl \rightleftharpoons NaCl + H} \tag{2}$$

was calculated using Gibbs free energy data from the JANAF tables (*12*), and the concentration of HCl was calculated iteratively from the measured flow rate with allowance for the occurrence of the equilibrium

$$\mathrm{H + HCl \rightleftharpoons H_2 + Cl} \tag{3}$$

This method of determining [H] is not subject to error arising from sodium chemiluminescence, since, in contrast to the situation with thallium (*13*), the chemiluminescence of sodium is not significantly enhanced when chlorine is introduced into a flame.

The fluorescence of atomic lead was excited using the flowing electrodeless discharge lamp shown in Fig. 1. The lamp was normally op-

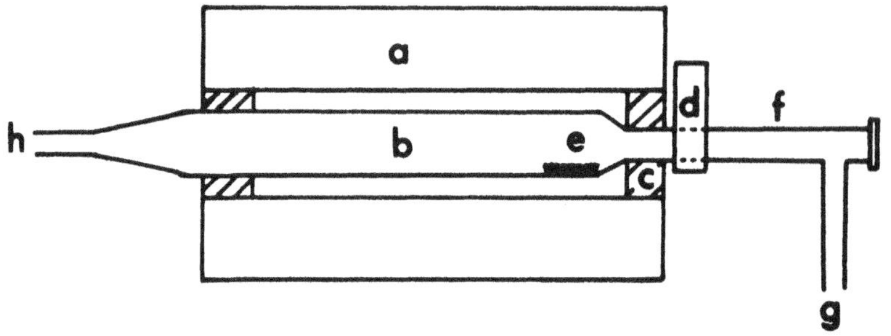

Fig. 1. Microwave-powered lead resonance lamp. Key: (a) tube furnace with temperature controller; (b) silica tube, 2.5 × 50 cm; (c) asbestos tape; (d) microwave cavity; (e) PbI_2 reservoir; (f) silica tube, 1.5 × 20 cm; (g) exit to pump; (h) helium inlet.

erated with the oven temperature controlled in the range 400–450°C, with a slow flow (<1 sccm) of helium and a microwave power of 175–200 W. The intensity and stability of this lamp were such as to make it a promising source for use in analytical applications of flame fluorescence. The lamp output was chopped at 650 Hz and the fluorescence signal, observed at right angles to the exciting beam with a Spex "Minimate" monochromator and EMI type 6256SA photomultiplier, was measured with a PAR model HR-8 lock-in amplifier. The 368.3 nm intensity was corrected for scattered lamp emission as follows. A Corning 7-59 filter was inserted between lamp and flame to cut out the 283.3 nm exciting light. This typically reduced the observed intensity at 368.3 nm by a factor of five. The resulting intensity, corrected for the filter transmission at 368.3 nm, was then subtracted from the 368.3 nm intensity measured with the filter absent. Separate experiments with other filters showed that scattering was negligible in comparison with fluorescence at 364.0 nm, and that the excitation of 368.3 nm fluorescence by lamp emissions at 364.0 and 405.8 nm was also negligible.

RESULTS AND DISCUSSION

Representative curves showing the decay of the intensity ratio r (= $I_{368.3}/I_{364.0}$), with increasing distance from the reaction zone, are given in Fig. 2. The curves are theoretical fits based on the detailed mechanism given in ref. (2). The scatter of the points about the theoretical curves is significantly less than in previous work. The fitting process yields values of k_H and k_M for the processes

$$\mathrm{Pb}(^3P_1^\circ) + \mathrm{H} \rightarrow \mathrm{Pb}(^3P_0^\circ) + \mathrm{H} \tag{4}$$

and

$$\mathrm{Pb}(^3P_1^\circ) + \mathrm{M} \rightarrow \mathrm{Pb}(^3P_0^\circ) + \mathrm{M} \tag{5}$$

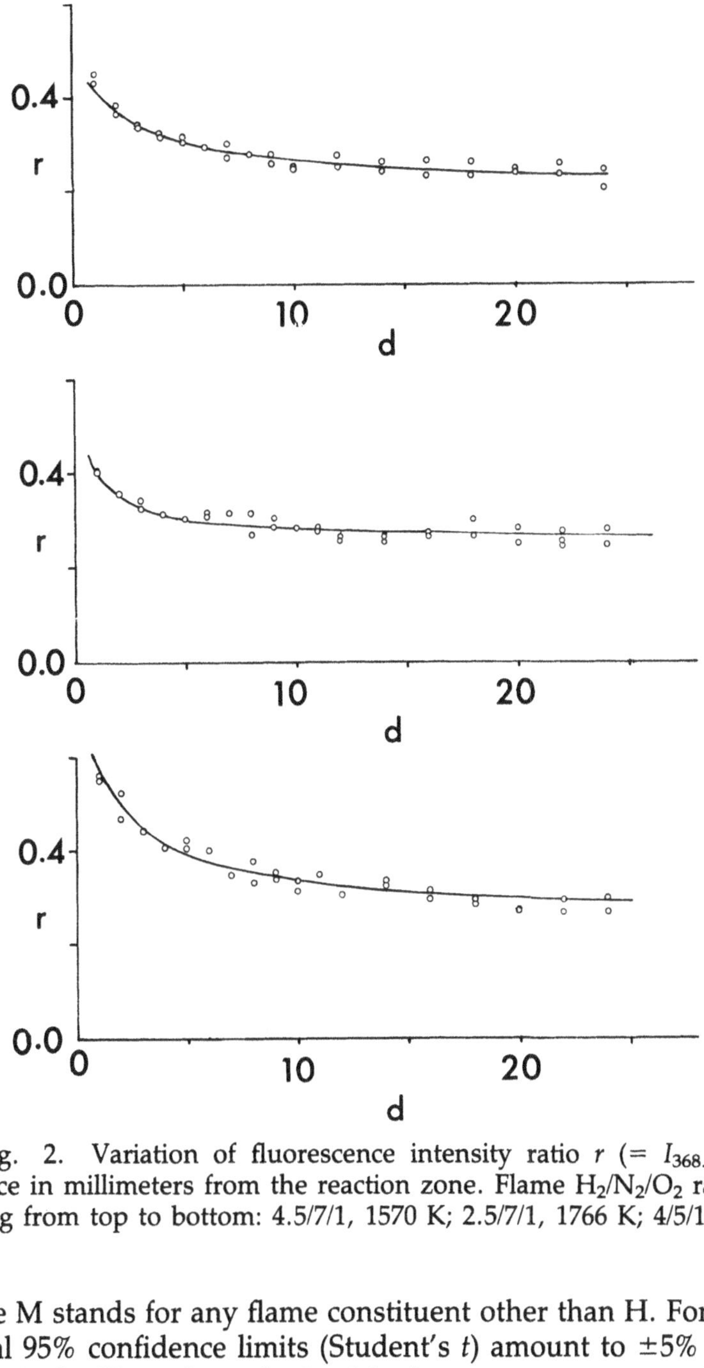

Fig. 2. Variation of fluorescence intensity ratio r (= $I_{368.3}/I_{364.0}$) with distance in millimeters from the reaction zone. Flame $H_2/N_2/O_2$ ratios, and T, reading from top to bottom: 4.5/7/1, 1570 K; 2.5/7/1, 1766 K; 4/5/1, 1870 K.

where M stands for any flame constituent other than H. For k_M and k_H, typical 95% confidence limits (Student's t) amount to ±5% and ±10%, respectively. The values obtained in the temperature range 1570–2110 K are given in Fig. 3, where both lines through the data points have been drawn to correspond to a $T^{1.5}$-dependence of the rate constant. The values found for k_M are about 20% larger overall than those obtained using laser excitation; the reason for this is obscure. The values of k_H are all considerably lower than found previously, this difference arising from the use of markedly different values of H-atom concentration. The re-

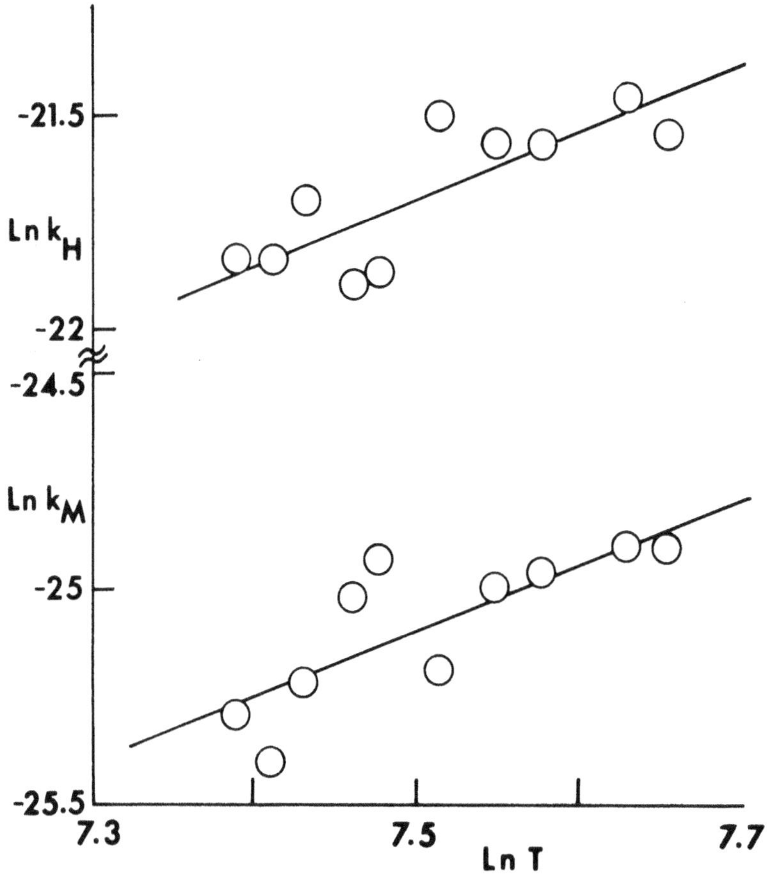

Fig. 3. Temperature dependence of k_H and k_M.

sults for k_H are about 25% lower in general because of the intentional use of a "high" value for D^0_{LiOH} in the previous work, while at the low temperature end there is the additional effect of removing the error caused by sodium chemiluminescence, which led to the apparent negative temperature dependence of k_H.

The model calculations described in ref. (2) demonstrated that, with realistic potential functions, the crossing of bound potential curves at a point well below the asymptotic energy level for the separated atoms can be expected to give rise to a curve-crossing rate constant that decreases with increasing temperature. This type of system is well-described by Landau-Zener theory, which assumes a localized crossing of diabatic potential energy curves that can be regarded as linear in the crossing region. The positive temperature dependence now found for k_H implies that in fact the crossing does not occur at negative energy, but instead probably occurs at fairly large internuclear distance in a qualitatively similar manner to the spin-orbit relaxation of alkali metals by rare gases, as discussed by Nikitin and coworkers [see, for example, ref. (4), p. 142]. The correlation diagram given previously [Fig. 5 of ref. (2)] is still applicable in this situation, but the Landau-Zener theory itself may not be applicable because of the unlocalized nature of the region of nonadiabatic interactions. Further theoretical work is required both to establish the nature of the potential curves for interaction of Pb($^3P_{1,0}^\circ$) with H($^2S_{1/2}$) at large internuclear distance and to predict the dependence of the crossing rate on temperature.

ACKNOWLEDGMENT

This work was supported by the New Zealand Universities Research Committee.

REFERENCES

1. Brady, T. J., and Phillips, L. F., *Chem. Phys. Lett.* **57,** 535 (1978).
2. Brady, T. J., and Phillips, L. F., *Proc. Roy. Soc. (London)* **A376,** 545 (1981).
3. Nikitin, E. E., *Chemische Elementarprozesse,* Hartman, H., ed., New York, Springer-Verlag, 1974.
4. Nikitin, E. E., *Theory of Elementary Atomic and Molecular Processes in Gases* (Transl. by Kearsley, M. J.), Oxford University Press, London, 1974.
5. Child, M. S., *Disc. Faraday Soc.* **53,** 18 (1972).
6. Child, M. S., *Molecular Collision Theory,* Academic Press, London, 1974.
7. Erlich, D. J., and Osgood, R. M., Jr., *J. Quantum Electron.* **16,** 257 (1980).
8. Hemmati, H., and Collins, G. J., *J. Quantum Electron.* **16,** 594 (1980).
9. Bulewicz, E. M., James, C. G., and Sugden, T. M., *Proc. Roy. Soc. (London)* **A235,** 89 (1956).
10. Bulewicz, E. M., and Sugden, T. M., *Trans. Faraday Soc.* **52,** 1475 (1956).
11. Bulewicz, E. M., Phillips, L. F., and Sugden, T. M., *Trans. Faraday Soc.* **57,** 921 (1961).
12. *JANAF Thermochemical Tables,* NSRDS-NBS37, National Bureau of Standards, Washington, DC, 1971.
13. Phillips, L. F., and Sugden, T. M., *Trans. Faraday Soc.* **57,** 2188 (1961).

species. Nonreactive ground-state molecular interactions are interesting in that there is no formal chemical bonding involved so that the weakly cohesive attractions are primarily derived from London dispersive forces. However, electronic excitation of the IIA metal atom often leads to a significant enhancement of the bonding interaction with the other moiety such that these species may be categorized as excimers or exciplexes. Interest in these complexes is piqued by their potential as active laser mediums since population inversion is facile for species that are very weakly bound and hence dissociative in the ground state. The recent rare gas–halide excimer lasers illustrate this point and have sparked spectroscopic study of similar species. For instance, there have been extensive spectroscopic studies of the Group IIA heteronuclear and homonuclear diatomic species that have recently been reviewed by Miller and Andrews (*1*). Additionally, there has been intense spectroscopic research aimed toward understanding the chemical exit channels of photosensitized decomposition reactions of small polyatomic and diatomic molecules with excited Group IIA metals (*2*). Along this line, Mandl and Hyman have carried out gain/absorption measurements on the green emission from the $(Hg_2NH_3)^*$ exciplex to determine its potential as an energy storage laser (*3*). Spectroscopic analysis of excimers and exciplexes will continue to be popular in view of the continuing search for different active laser mediums.

Reactive interactions of the Group IIA metals with small molecules have also been extensively studied along theoretical and experimental avenues. Theoretical studies have modeled chemical bonding on transition metal surfaces and clusters with the Group IIA metal–olefin interactions (*4*). This research is motivated by the hypothesis that bonding in catalytic systems primarily involves the 4*s* and 4*p* orbitals of the transition metals in question, with the *d* orbitals playing a minimum role (*5,6*). Since beryllium can be treated with a fair degree of theoretical rigor, Swope and Schaefer have studied this atom interacting with an ethylene molecule with Be in its ground ^1S state and threefold degenerate excited ^3P state. These calculations indicate that the ground-state interaction is essentially repulsive, but the excited-state interaction ($^3B_2$) is rather well bound (19 kcal/mol). Such investigations intimate the potential for photosensitized heterogeneous catalysis.

There have been a number of experimental and theoretical investigations of the Group IIA metal–water interactions. Kauffman (*7*) has performed infrared matrix isolation studies on the IIA metal–water interactions, demonstrating that initially a weakly bonded metal–water adduct could be trapped in the matrix upon cocondensation, as evidenced by a decrease in the ν_2 bending mode frequency of water. Experimental results suggested that the bonding of the metal–water complex involves an interaction through the oxygen of water. Selective photolysis of the rare gas matrix employing Corning cutoff filters induced the metal to insert oxidatively into the O—H bond of water to form the hydroxy metal hydride, HMOH, for Mg, Ca, Sr, and Ba metal atoms. In addition to HMOH, the metal monohydroxide, MOH, was generated upon irradiation of the strontium– and barium–water adducts. Continued photolysis of the matrix with ultraviolet radiation ($\lambda > 3000A$) was unsuccessful in photodissociating any of these products. Theoretical investigators have evaluated the M—O bond distances and energies of dissociation for the metal–water adduct with the metal in its ground and excited states, as well as the equilibrium conformations for the HMOH molecule. Bentley (*8*) has examined the ground-state interactions of Be and Mg with water in a C_{2v} point group and found that bonding is essentially nonexistent; however, he reports that the bonding of Be . . . OH_2 in the $^3B_2$ excited state is attractive by about 19 kcal/mol. Curtiss and Frurip (*9*) have calculated the bonding interactions of the adducts for beryllium and magnesium to be 0.62 and 2.33 kcal/mol, respectively, which can be accounted for in terms of dispersive forces. They find definite minima in the potential surfaces for these adducts that correspond to a C_{2v} configuration for Mg . . .OH_2 and a C_s configuration for Be . . .OH_2 with the Be atom above the plane defined by the water. Curtiss and Frurip (*9*) also calculated bond distances and geometries for HMgOH and HBeOH and reported that HMgOH is linear, whereas HBeOH has a BeOH bond angle of 145°. These investigators also carried out limited calculations that indicated there is a potential energy barrier to insertion that they believe is apparently overcome by the energy provided in photolysis so the reaction follows a ground-state potential energy surface.

In view of these fascinating investigations of the Group IIA metal atom chemistry as they relate to the spectroscopic characterization of exciplexes, the theoretical modeling of heterogeneous catalysis, with hints of photoactivation, and the detailed experimental and theoretical exploration of metal–water interactions and reactions, it is of considerable interest to probe the electronic structures of the Group IIA metal atom hydration reaction intermediates and products isolated in rare gas matrices. Such an investigation may be of assistance in (1) elucidating the nature of the photochemical exit channels pursued on a collision of an excited Group II metal atom with a small molecule, as well as (2) characterizing the detailed electronic structure of the exciplex. This study will also form a useful adjunct to the previous infrared matrix isolation studies since it will examine the perturbations of the electronic structure of metal atoms upon interaction with water as well as possibly characterize new reaction products not apparent in the infrared studies. Molecular orbital and electronic state correlation diagrams will be employed to help understand the photolytically induced reaction pathways and the electronic structures of the intermediates. Additionally, the electronic spectra of the Ba atom and dimer will be presented and discussed since previous studies (*10*) were quite tentative in their assignments.

EXPERIMENTAL

The matrix isolation apparatus employed in this study is described in detail elsewhere (*21*). Briefly, cryogenic temperatures are achieved utilizing an Air Products Displex Closed Cycle Helium Refrigerator, Model CS-202. Matrices are formed on aluminum-coated, second surface, UV grade sapphire mirrors in good thermal contact with an aluminum block that, in turn, is thermally bound to the cold station of the refrigerator. UV–VIS absorption spectra are taken utilizing a ¼-m Perkin-Elmer Model 210 grating monochromator (calibrated with mercury emission lines), a Hamamatsu R-928 PMT, and a pair of PAR phase sensitive amplifiers (Models 128A and 186). The optical design allows referencing of a host matrix to circumvent severe light scattering problems that normally plague the UV spectra. A quartz crystal microbalance mounted on the aluminum cold block allows measurement of the background contaminant, the host, and the guest species deposition rates, thereby permitting highly accurate calculation of the M/R ratios.

Purified grade magnesium metal was obtained from J. T. Baker. Purified calcium metal was procured from Sargent-Welch, SC1190. Both the strontium and the barium metals were acquired from Alfa Ventron at 99.99% purity. All four metals were vaporized from a stainless-steel crucible at temperatures that corresponded to a vapor pressure of about 1×10^{-4} torr for the metal investigated. Thirty mil-thick carbon cloth was wrapped around the crucible to insulate it electrically from the Ta foil furnace assembly. The furnace assembly was outgassed for 2 h prior to deposition at temperatures about 50 degrees below that during deposition. H_2O and D_2O were vacuum distilled going through at least five freeze–thaw cycles. High purity krypton was used (Matheson Research Grade, 99.995%) with a liquid nitrogen cold trap. Guest-to-host ratios for the metal and water were accurately evaluated by measuring the deposition rates for krypton, water, and the metal of interest with the quartz crystal microbalance and will be reported for each experiment. Each matrix sample preparation required a 15 min deposition. Selective photolysis studies employed the monochromator used in the spectroscopic studies having its slits widened to allow a 20 Å bandpass with a tungsten–halogen light source. Photolysis times and wavelengths will be reported for each experiment.

RESULTS

Magnesium and Water

Cocondensation of magnesium metal vapor with water vapor on a sapphire substrate maintained at 15 K in an excess of krypton generates a transparent, colorless matrix.

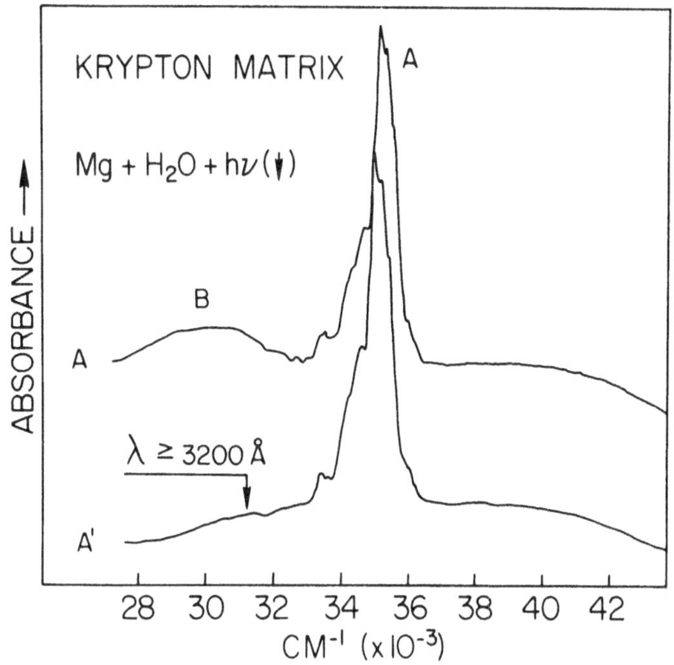

Fig. 1. Electronic absorption spectra for magnesium and water isolated in a krypton matrix: trace A, $Mg:H_2O:Kr = 1:50:1550$; trace A′, Same as trace A, but photolyzed for 2 s with light of $\lambda > 3200$ Å.

Figure 1 shows two electronic absorption spectral traces corresponding to two experiments: $Mg/H_2O/Kr$ and $Mg/H_2O/Kr + h\nu$. Table 1 lists the assignments corresponding to band maxima for each of these spectra. The assignments were made after the matrix has been annealed to 42 K over a period of 3 min. The annealing process was effective in sharpening some of the bands with corresponding minor red and blue shifts in band positions. Band "A" in trace A corresponds to a strongly allowed electronic resonance transition for the Mg metal atom since this is the only band that persists in high dilution studies of Mg in Kr without water. Comparison of this band energy with gas-phase transition energies for

TABLE 1
Assignments for the Optical Absorption Spectra of the $Mg/H_2O/Kr$ and $Mg/H_2O/Kr + h\nu$ Systems

Peak	λ(Å)	$\bar{\nu}(cm^{-1})$[a]	Species	Assignment[c]	Gas phase[b] (cm^{-1})
A	2854	35 032	Mg	$3s3p(^1P) \leftarrow 3s^2(^1S)$	35 051
B	3326	30 068	$Mg \,.\,.\, OH_2$	$^1B_2$ & $^1B_1 \leftarrow X\ (^1A_1)$	?

[a] $\pm$ 15cm^{-1}.
[b] From ref. *11*.
[c] Energies of transition are taken from band maxima unless noted otherwise.

Mg (*11*) allows the assignment of band "A" to the $3s3p(^1P) \leftarrow 3s^2(^1S)$ atomic resonance transition of Mg. The major "A" band is only slightly blue-shifted (20 cm^{-1}) relative to the gas-phase transition and is triplet in structure. This absorption spectra is in excellent agreement with those of other authors (*12,13*). Although various theories have been projected to explain the triplet structure of this $^1P \leftarrow {}^1S$ transition (e.g., multiple trapping sites, metal–metal interactions, crystal field effects, etc.), a recent magnetic circular dichroism (MCD) study (*14*) of Mg in various inert gas matrices indicated that the characteristic triplet structure arises from Mg atoms at a single site and can probably be interpreted in terms of a Jahn-Teller effect assuming an octahedral site symmetry. No other atomic transitions were observed.

Because it grew in as a consequence of increasing the water concentration, the band denoted by "B" may be assigned to the reaction product of magnesium and water which, in confirmation of previous infrared studies (*7*), may be identified with the 1:1 magnesium–water adduct. High water dilution studies failed to change the structure of band "B" thereby eliminating the possibility of a higher-order metal–water interaction causing the broad character of this band, which has a red onset at about 27,000 cm^{-1} and a blue cutoff at about 32,000 cm^{-1}. Irradiation of the matrix sample with light of $\lambda > 5000$ Å (CS 3-70 and Pyrex-water filters) for 10 min does nothing to destroy the integrity of the "B" band. A 0.5 s photolysis with light of $\lambda > 3200$ Å (CS1-64 filter) destroys about 30% of the intensity of band "B." Trace A′ in Fig. 1 shows that a 2s photolysis with the same color light totally eliminates the adduct band without subsequent generation of any new absorption bands between 210 and 800 nm. The final photolysis product observed in the infrared studies was HMgOH, which apparently has its first electronic transitions in the ultraviolet region. Irradiation of the matrix causes a slight red shift in the "A" band, which is probably attributable to a photo-induced annealing process. With the consideration that the infrared studies show that the adduct is initially formed upon cocondensation, and that the "B" band displays very similar photosensitivity to that of the adduct band in the infrared work, the assignment of the "B" band to the metal–water complex is quite plausible.

Calcium and Water

Cocondensation of calcium and water vapor on a sapphire substrate maintained at 15 K in an excess of krypton generates a transparent, red matrix.

Figure 2 shows an electronic absorption trace corresponding to one experiment: Ca/H_2O/Kr. Figure 3 shows five expanded-scale absorption traces corresponding to the Ca/H_2O/Kr matrix sample being selectively photolyzed with different wavelengths of light. Table 2 lists the assignments corresponding to the band maxima for each of these spectra. The assignments were made after the matrix had been annealed to 42 K over a

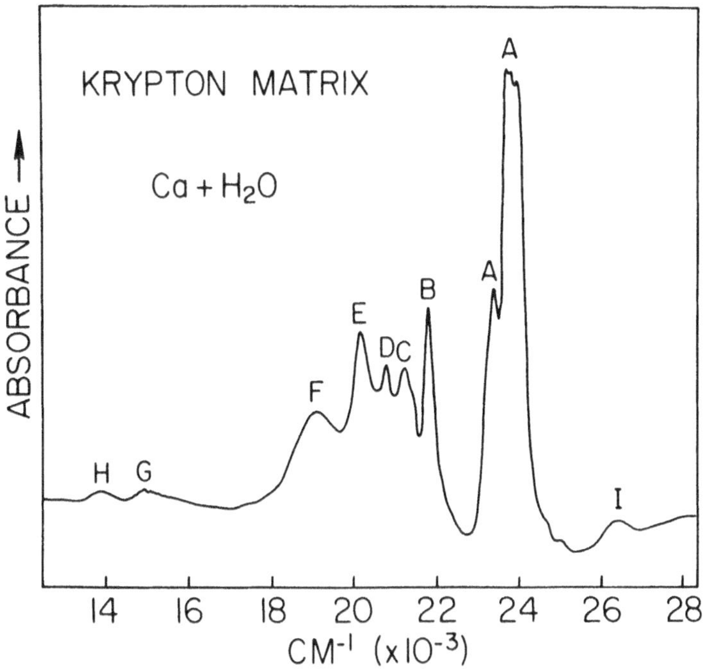

Fig. 2. Electronic absorption spectra for calcium and water isolated in a krypton matrix: $Ca:H_2O:Kr$ = 1:60:3700.

period of about 3 min. The annealing process was effective in sharpening some of the bands with corresponding minor red and blue shifts in position. The bands denoted by "A" and "B" in Fig. 2 can be attributed to atomic resonance transitions for calcium since only these bands persist during the course of high Ca dilution experiments without water. Previous absorption, emission, and MCD (*15*) studies of Ca atoms in inert matrices are in excellent agreement with these spectra and have established the nature of the observed transitions. From these studies, bands "A" and "B" are assigned to the $4s4p(^1P) \leftarrow 4s^2(^1P)$ and the $4\ s3d(^1D) \leftarrow 4s^2(^1S)$ Ca atomic resonance transitions. The MCD study has unequivocally established that the absorption doublet observed in the "A" band is caused by the Ca atom residing in two different sites in the rare gas solid. The more important role that multiple sites play in Ca than Mg precluded a MCD analysis of the characteristic triplet observed in the major "A" and of Ca. The $^1D \leftarrow {}^1S$ assignment to the "B" and has been questioned since it is a rather intense band for a parity-forbidden atomic transition. However, this assignment has been bolstered by the MCD study (*15*) that concludes it is likely that the $^1D \leftarrow {}^1S$ intensity is borrowed from the $^1P \leftarrow {}^1S$ transition via a static site mechanism where the site symmetry has no center of inversion and is sufficiently high to resolve the components of the *P* and *D* states into the same irreducible representation to permit intensity borrowing. Andrews et al. (*15*) suggest that a D_{3h} site symmetry is the most reasonable in a krypton solid. No other atomic transitions

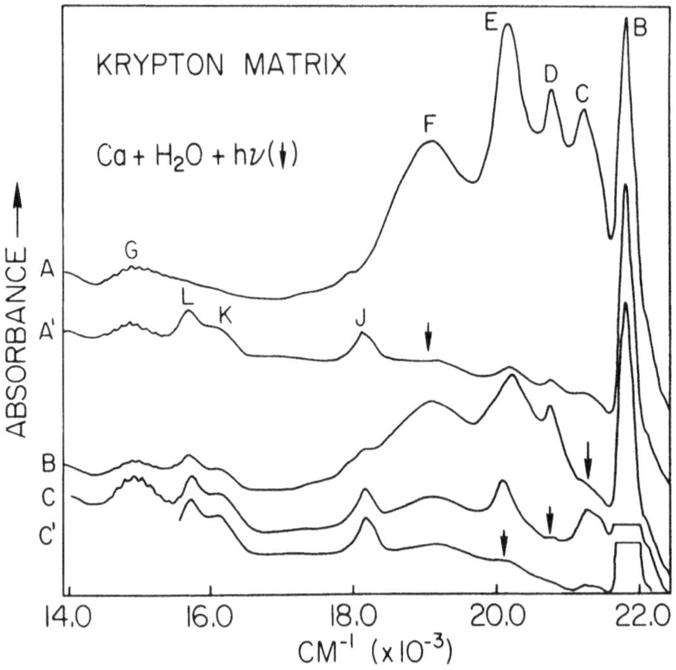

Fig. 3. Expanded-scale electronic absorption spectra for calcium and water isolated in a krypton matrix with selective photolysis. Trace A, Ca:H_2O:Kr = 1:60:3700; trace A′, same as trace A, but photolyzed for 40 min with light of λ = 5237–5257 Å; trace B, same as trace A, but photolyzed for 30 min with light of λ = 4698–4718 Å; trace C, same as trace A, but photolyzed for 30 min with light of λ = 4813–4833 Å; trace C′, same as trace C, but photolyzed for 15 min with light of λ = 4953–4973 Å.

were observed. The bands denoted by "G" and "I" are assigned to electronic transitions of the calcium dimer, since upon increasing the Ca concentration, these bands are the first to grow in after the atomic bands and display a quadratic intensity dependence on the metal deposition rate. The "G" band shows a vibronic progression with $\Delta G'_{v+1/2}$ = 119 cm $^{-1}$. The "G" and "I" bands have been quite extensively studied in the gas and solid phases (*1*) and are assigned to the $A(^1\Sigma^+{}_u) \leftarrow X(^1\Sigma^+{}_g)$ (ω_e = 118 cm^{-1} in krypton) and the $^1\Pi \leftarrow X(^1\Sigma^+{}_g)$ transitions for Ca_2, respectively.

The bands denoted by "C," "D," "E," and "F" may be assigned to a 1:1 reaction product of calcium and water since these bands are the first bands to grow in as a consequence of doping the matrix sample with water and maintain identical relative intensities with different water dilutions. In conjunction with the previous IR studies (*7*), these bands may be identified with the calcium–water adduct, Ca . . . OH_2. From Ca concentration studies, it is apparent that the "H" band also grows in as a consequence of water doping, but only if there is some calcium dimer present in the matrix; consequently, the "H" band is assigned to a 2:1

TABLE 2
Assignments for the Optical Absorption Spectra of the $Ca/H_2O/Kr$ and $Ca/H_2O/Kr + h\nu$ Systems

Peak	λ(Å)	$\bar{\nu}(cm^{-1})^a$	Species	Assignment[c]	Gas Phase[b] (cm^{-1})
A	4209	23 756	Ca	$4s4p(^1P) \leftarrow 4s^2(^1S)$	23 652
A	4277	23 378	Ca	$4s4p(^1P) \leftarrow 4s^2(^1S)$	23 652
B	4582	21 826	Ca	$4s3d(^1D) \leftarrow 4s^2(^1S)$	21 850
C	4705	21 256	$Ca \ldots OH_2$	$C(^1B_1) \leftarrow X(^1A_1)$	?
D	4809	20 794	$Ca \ldots OH_2$	$C(^1B_1) \leftarrow X(^1A_1)$	?
E	4947	20 215	$Ca \ldots OH_2$	$B(^1A_1) \leftarrow X(^1A_1)$	?
F	5231	19 116	$Ca \ldots OH_2$	$A(^1B_1) \leftarrow X(^1A_1)$	?
G	6906	14 480	Ca_2	$A(^1\Sigma_u^+) \leftarrow X(^1\Sigma_g^+)$	~15 150[d]
H	7178	13 931	$Ca \ldots OH_2$	$A \leftarrow X$	?
I	3794	26 357	Ca_2	$^1\Pi_u \leftarrow X(^1\Sigma_g^+)$	~26 700[d,e]
J	5517	18 126	CaOH	$B(^2\Sigma+) \leftarrow X(^2\Sigma+)$	?
K	6213	16 096	CaOH	$A(^2\Pi) \leftarrow X(^2\Sigma+)$	?
L	6359	15 727	CaH	$B(^2\Sigma+) \leftarrow X(^2\Sigma+)$	15 755[d]

[a] $\pm 15\ cm^{-1}$.
[b] From ref. 1, 11, and 16.
[c] Energies of transition are taken from band maxima unless noted otherwise.
[d] ν_{00}.
[e] In an argon matrix.

reaction product of calcium and water. Although higher-order calcium cluster–water were slightly discussed in the infrared study, the "H" band assignment is bolstered by the fact that it is red shifted from the Ca_2 $^1\Sigma_u^+ \leftarrow {}^1\Sigma_g^+$ transition and is easily photolyzed away with long wavelength light as was possible in the IR studies.

Isotopic substitution studies were conducted with D_2O and H_2O^{18} in an effort to characterize better the nature of the adduct bands. H_2O^{18} substitution was unsuccessful in shifting the adduct band structure; however, D_2O substitution generated relatively dramatic changes in the band structure, red shifting the "C," "D," "E" bands and blue shifting the broad "F" band. The "C," "D," and "E" bands were red shifted by 90, 42, and 75 cm^{-1}, respectively, and the "F" band was blue-shifted by 143 cm^{-1}. Attempts to associate these isotopic shifts in the band positions with vibronic progressions of an excited state in different normal modes were unsuccessful; therefore, these bands probably correspond to different electronic states of the metal–water adduct, different geometric isomers for the same 1:1 adduct, matrix site effects for the same electronic transition of the adduct, or, in all probability, a composite of all of these possibilities.

Selective photolysis of each of these adduct bands through a monochromator having a 20 Å bandpass was instrumental in establishing the interrelationship between these bands and distinguishing between their respective natures. Figure 3 displays five traces corresponding to four independent photolysis studies. Trace A′ shows that selectively photolyzing the "F" band for 40 min was effective in eliminating the "C," "D," "E," and "F" bands equally with the concomitant generation of the "K," "J," and "L" product bands, which suggests that all these bands correspond to different electronic transitions for the same adduct without site effects. However, trace B shows that a 30 min selective photolysis of the "C" band was effective in totally eliminating the "C" band with only partial decreases in the intensities of the "D," "E," and "F" bands. This suggests that "C" may not be correlated with the "D" or "E" bands. Trace C shows that selectively photolyzing the "D" band for 30 min was effective in totally eliminating the "D" band with only partial decreases in the intensities of the "C," "E," and "F" bands, which suggests that the "D" band cannot be associated with the "E" or "C" bands. However, the structure of this "E" band is much narrower than the "E" band observed in the photolysis of the "C" band, which suggests that both the "D" and "C" bands correlate with the "E" band that is actually comprised of two superimposed bands. This may explain the anomalous intensity of the "E" band in trace A. Trace C′ confirms the association between the "E" and "C" bands, since this trace shows that a 15-min selective photolysis of band "E" after band "C" has been totally eliminated (trace B) is successful in totally destroying bands "F," "E," and "D," thus supporting the association of the "E" and "D" bands. In summary, photolysis of the "F" band suggests that it is equally associated with each of the other adduct bands, whereas selective photolysis of the "C," "D," and "E" bands suggests that the "C" and "D" bands are not associated, but that each is associated with the "E" band, which is probably composed of two superimposed bands. Photolysis of each of these adduct bands was effective in generating the "J," "K," and "L" product bands. From this selective photolysis behavior it is possible to postulate that the adduct resides in two matrix sites or exists in two geometric forms and has three excited electronic states that are observed in the matrix. Two of the transitions are not significantly split by the site/isomer and correspond to bands "F" and "E" (however, band "E" is broadened in one of the states), whereas the third transition is split by the site/isomer and corresponds to bands "D" and "C".

The infrared studies only report the formation of HCaOH from the photolysis of the Ca . . . OH_2 adduct. Since the first excited states of HCaOH are probably quite high in energy, the low energy "J," "K," and "L" product bands probably originate from other photolysis products not identified in the infrared studies. Reasonable photolysis products include diatomic CaO and the CaOH and the CaH radicals. CaO has two low energy electronic transitions around 11,549 and 25,913 cm^{-1}; however, these energies do not correlate well with the product bands observed in this study. The electronic structure of the CaOH radical should be quite similar to that of the CaF diatomic, which has several low energy excited states at about 16,500 cm^{-1} ($^2\Pi$) and 18,850 cm^{-1} ($^2\Sigma^+$). These values correspond well with the assigned energies for the product bands observed in the matrix at 16,096 and 18,126 cm^{-1}. Thus, in all probability, the CaOH radical is generated upon photolysis of the calcium–water adduct, which extends a trend initially noted by Kauffman (7) in that he observed SrOH and BaOH formation upon photolysis of their respective metal–water adducts, although he did not observe CaOH formation. Since H is homolytically cleaved during photolysis from the adduct in the formation of CaOH, it is reasonable to believe that CaH would be formed by reaction of mobile H atoms and uncomplexed Ca atoms. The "L" product band at 15,727 cm^{-1} correlates well with the v_∞ = 15,755 cm^{-1} $B(^2\Sigma^+) \leftarrow X(^2\Sigma^+)$ transition for the CaH diatomic and hence is assigned accordingly. Therefore, this study reports the formation of both CaOH and CaH radicals as a consequence of selective photolysis of each of the calcium–water adduct bands.

Strontium and Water

Cocondensation of strontium and water vapor on a sapphire substrate maintained at 15 K in an excess of krypton generates a transparent, yellow-brown matrix.

Figure 4 shows an electronic absorption spectral trace corresponding to one experiment: Sr/H_2O/Kr. Figure 5 shows six expanded-scale absorption traces corresponding to the Sr/H_2O/Kr matrix being selectively

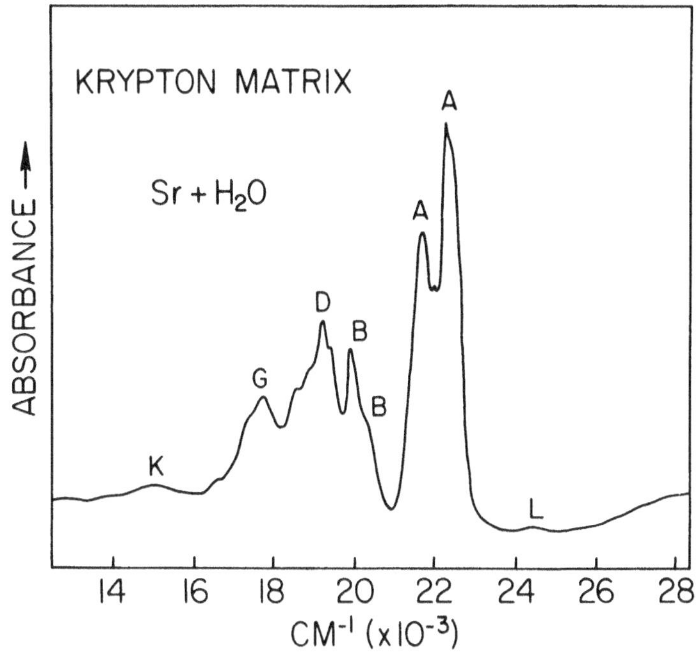

Fig. 4. Electronic absorption spectra for strontium and water isolated in a krypton matrix: Sr:H_2O:Kr = 1:98:3500.

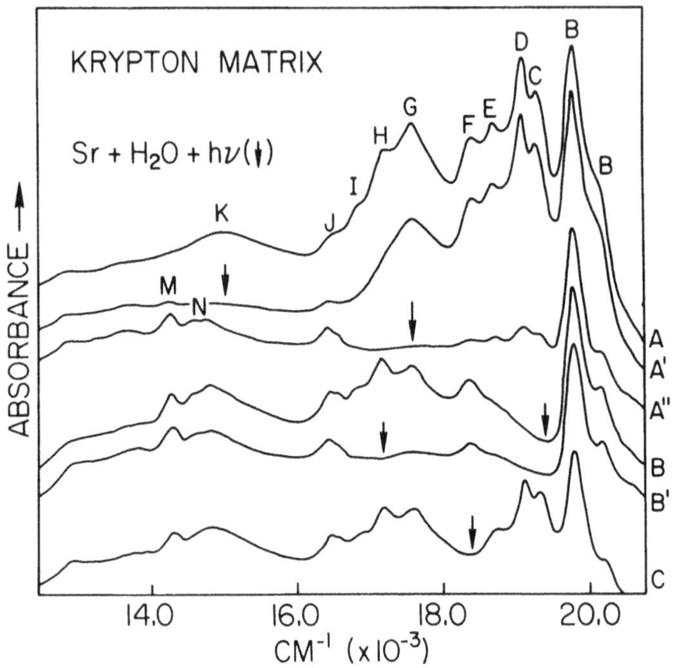

Fig. 5. Expanded-scale electronic absorption spectra for strontium and water isolated in a krypton matrix with selective photolysis: Trace A, $Sr:H_2O:Kr$ = 1:98:3500; trace A′, same as trace A, but photolyzed for 15 min with light of λ = 6643–6663 Å; trace A″, same as trace A′, but photolyzed for 30 min with light of λ = 5716–5736 Å; trace B, same as trace A, but photolyzed for 30 min with light of λ = 5146–5166 Å; trace B′, same as trace B, but photolyzed for 15 min with light of λ = 5800–5820 Å; trace C, same as trace A, but photolyzed for 15 min with light of λ = 5416–5436 Å.

photolyzed with different wavelengths of light dispersed by a monochromator. Table 3 lists the assignments corresponding to band maxima for each of these spectra. The assignments are made after the matrix has been annealed to 42 K over a period of 3 min. The annealing process was not only effective in sharpening some of the bands (especially the "E" band: cf., Figs. 4 and 5) with corresponding minor red and blue shifts in band positions, but also resulted in an overall decrease in the intensities of the "C," "D," "E," "F," and "G" bands with a concurrent increase in the intensity of the "H," "I," and "K" bands, as can be seen from a comparison of Figs. 4 and 5. High water dilution studies demonstrate that the "H," "I," and "K" bands are associated with higher-order water interactions, $(H_2O)_n$ where $n > 1$, since these bands are not apparent with water concentrations less than 150:1 for $Kr:H_2O$, whereas the "C," "D," "E," "F," and "G" bands persist with the same relative magnitudes. Hence, annealing the matrix is evidently effective in allowing water to migrate through the matrix generating higher-order water interactions.

TABLE 3
Assignments for the Optical Absorption Spectra of the $Sr/H_2O/Kr$ and $Sr/H_2O/Kr + h\nu$ Systems

Peak	λ(Å)	$\bar{\nu}(cm^{-1})$[a]	Species	Assignment[e]	Gas Phase[b] (cm^{-1})
A	4478	22 330	Sr	$5s5p(^1P) \leftarrow 5s^2(^1S)$	21 698
A	4504	22 204	Sr	$5s5p(^1P) \leftarrow 5s^2(^1S)$	21 698
A	4618	21 654	Sr	$5s5p(^1P) \leftarrow 5s^2(^1S)$	21 698
B	4965	20 140	Sr	$5s4d(^1D) \leftarrow 5s^2(^1S)$	20 150
B	5041	19 838	Sr	$5s4d(^1D) \leftarrow 5s^2(^1S)$	20 150
C	5177	19 318	$Sr\cdots OH_2$	?	?
D	5228	19 129	$Sr\cdots OH_2$	?	?
E	5332	18 755	$Sr\cdots OH_2$	?	?
F	5414	18 470	$Sr\cdots OH_2$	?	?
G	5676	17 619	$Sr\cdots OH_2$	$A(^1B_2) \leftarrow X(^1A_1)$	?
H	5797	17 249	$Sr\cdots(OH_2)_2$	?	?
I	5917	16 901	$Sr\cdots(OH_2)_2$	?	?
J	6041	16 553	SrOH	$^2\Sigma^+ \leftarrow {}^2\Sigma^+$	?
K	6627	15 090	$Sr\cdots(OH_2)_2$	?	?
L	4121	24 268	Sr_2	$^1\Pi_u \leftarrow X(^1\Sigma_g^+)$	24 600[c]
M	7003	14 279	SrH	$^2\Sigma^+ \leftarrow X(^2\Sigma^+)$	14 352[g]
N	6768	14 777	SrOH	$^2\Pi \leftarrow A(^2\Sigma^+)$	?
N	6850	14 598	SrOH	$^2\Pi \leftarrow A(^2\Sigma^+)$	?
O[f]	7310	13 680	Sr_2	$^1\Sigma_u^+ \leftarrow X(^1\Sigma_g^+)$	13 224[d]

[a] $\pm$ 15 cm^{-1}.
[b] From refs. 1, 11, and 16.
[c] Band maximum in argon matrix.
[d] ν_{00} in krypton matrix.
[e] Energies of transition are taken from band maxima unless noted otherwise.
[f] Not shown in the spectra.
[g] ν_{00}.

The bands denoted by "A" and "B" in Figs. 4 and 5 can be attributed to atomic resonance transitions for strontium since only these bands persist during the course of high strontium dilution experiments that exclude water doping. Previous absorption, emission, and MCD (*15*) studies of strontium atoms in inert matrices are in good agreement with these spectra and have helped to establish the nature of the observed transitions corresponding to the "A" and "B" bands. From these studies, the bands denoted by "A" and "B" are assigned to the $5s5p(^1P) \leftarrow 5s^2(^1S)$ and the $5s5d(^1D) \leftarrow 5s^2(^1S)$ strontium atomic resonance transitions. The "A" band has triplet structure whereas the "B" band has a doublet character that becomes a blue shoulder with water-doping. The multiplet structures for these two transitions can probably be accounted for in terms of multiple site effects since the MCD studies (*15*) demonstrate that the three distinct components observed in absorption for strontium in an argon matrix are generated by the Sr atom residing in three distinct matrix sites. Multiple sites for strontium in krypton probably reflect a poor fit for the atom in the lattice because of the atom's larger size. By analogy with Ca, the MCD study suggests that the three sites are of approximate D_{3h} symmetry, which would allow intensity borrowing from the $^1P \leftarrow {}^1S$ transition to the parity forbidden $^1D \leftarrow {}^1S$ transition. The bands denoted by "L" and "D" are correlated with electronic transitions of the strontium dimer since they are the first bands to grow in after the atomic bands, and their associated energies correlate well with electronic transitions for Sr_2 observed in previous matrix and gas-phase studies (*1,15*). These studies allow one to assign the "O" band to the $^1\Sigma_u^+ \leftarrow X(^1\Sigma_g^+)$ transition of Sr_2, and the "L" band to the $^2\Pi_u \leftarrow X(^1\Sigma_g^+)$ transition. The MCD studies (*15*) were instrumental in verifying the nature of latter transition.

The bands denoted by "C," "D," "E," "F," and "G" in Fig. 5 may be correlated with the 1:1 reaction product of strontium and water since these bands are the first to grow in, as a consequence of having doped the matrix with water, and maintain identical relative intensities with different water dilutions. The "H," "I," and "K" bands are associated with a 1:2 metal:water interaction since these are the second set of bands to grow while raising the water concentration and increase in intensity during temperature cycling. In collaboration with the previous IR studies, the first set of bands may be associated with electronic transitions of the strontium–water adduct, $Sr \ldots OH_2$. The "G" band is quite broad, which may be indicative of a strong interaction with the matrix cage. In contrast, the "C," "D," "E," and "F" bands are relatively narrow, which suggests discrete, bound states for the adduct.

Isotopic substitution studies were conducted with D_2O and H_2O^{18} in lieu of H_2O^{16} in hopes of obtaining a better characterization of the adduct bands. The H_2O^{18} substitution was uneventful in that no adduct bands shifted except for the "F" band, which was slightly shifted to the red (30 cm^{-1}); however, D_2O substitution was effective in significantly red-shifting most of the 1:1 adduct bands. The "C," "D," "E," "F," and "G" bands were red-shifted by 17, 60, 92, 80, and 63 cm^{-1}, respectively. Attempts to rationalize these isotopic shifts in band position in terms of a vibronic progression of an excited state in different normal modes were unsuccessful. Therefore, as was discussed for the calcium–water adduct, these bands probably correspond to a combination of different effects including multiple excited states, geometric isomers, and matrix site effects for the 1:1 $Sr–H_2O$ complex.

Selective irradiation of each of these adduct bands through a monochromator was invaluable in establishing the interdependence of these bands which, in turn, was of assistance in understanding their respective natures. Figure 5 shows six spectral traces corresponding to five independent photolysis experiments with light having a 20 Å bandpass. Trace A′ shows that selectively photolyzing the "K" band for 15 min was effective in totally eliminating the "K," "I," and "H" without affecting any of the bands, which substantiates the previous association of these bands obtained by water dilution and annealing studies. No product bands were generated from this photolysis, which bears witness that these bands correspond to a different molecular species than the blue set of adduct bands since their irradiation generated new product bands (*vide infra*). Trace A′′ shows that a subsequent 30-min selective irradiation of the "G" band was successful in almost totally eliminating all the adduct bands ("G," "F," "E," "D," and "C") with a simultaneous generation of the "J," "M," and "N" product bands. Superficially, these results suggest that the five adduct bands correspond to different electronic transitions for the same 1:1 adduct; however, further selective photolysis was illuminating. Trace B shows that a 30 min irradiation of the "C" band was effective in totally eliminating the "C," "D," and "E" bands with only partial destruction of the "G" and "F" bands. This suggests that the "C," "D," and "E" bands are associated. Further photolysis (15 min) of this matrix hitting the "H" and unavoidably the "G" band is shown in trace B′ and results in a significant decrease in the "G" and "F" bands, which substantiates the association of these two bands. Trace C shows that a 15-min selective photolysis of the "F" band is effective in totally eliminating the "F" band with only partial decreases in the intensities of the "G," "C," "D," and "E" bands, which reinforces the independent nature of this band. Unfortunately, the "C," "D," and "E" bands are too closely positioned to allow a meaningful selective photolysis investigation of them. In summary, selective photolysis of the "G" band suggests a direct association of all the adduct bands; however, further selective photolysis of the other adduct bands indicates that the "G," "E," "D," and "C" bands and the "G" and "F" bands are associated. Photolysis of each of the 1:1 adduct bands was effective in generating the "J," "K," and "M" product bands. The selective photolysis behavior observed for the strontium–water system was not as conclusive as the calcium–water system. In general, at least two electronic transitions are observed corresponding to the "G" and "F" bands and the "G" and "E," "D," and "C" bands. It is not possible from the information gleaned in this study to decipher the nature of the "E," "D," and "C" bands; however, it is likely that they arise from multiple sites and/or geometric isomers.

The infrared studies (*7*) observed the generation of the HSrOH molecule and the SrOH radical from irradiation of the Sr . . . OH_2 adduct with light of λ = 520–580 nm. Subsequent photolysis of the HSrOH molecule with light of λ = 200–300 nm was successful in increasing the yield of the SrOH radical. In this study, selective photolysis of the Sr . . .OH_2 complex generated the "M," "N," and "J" bands, which can probably be associated with the SrOH radical and the SrH diatomic by analogy with the observed photolysis products of the Ca–H_2O system. The electronic structure of the SrOH radical should be similar to that of the analogous SrF diatomic, which has several low energy excited states at 17,264 cm^{-1} ($^2\Sigma^+$) and 15,352–15,072 cm^{-1} ($^2\Pi$). Since SrOH was a photolysis product in the IR studies and these values correlate fairly well with the assigned energies for the product bands observed in the matrix at 16,443, 14,777, and 14,598 cm^{-1}, and "J" and "N" bands are tentatively assigned to the $^2\Sigma^+ \leftarrow X(^2\Sigma^+)$ transitions for SrOH, respectively. The 14,279 cm^{-1} "M" band correlates quite well with the 14,352 cm^{-1} $^2\Sigma^+ \leftarrow X(^2\Sigma^+)$ transition for CaH and hence is assigned accordingly. The lowest predissociative excited states of HSrOH are evidently above 30,000 cm^{-1} since this energy of light was required to photodissociate the molecule. Although this region was studied, no product bands were observed, which is consistent with the predissociative nature of these high energy states.

Barium and Water

Cocondensation of barium and water vapor on a sapphire substrate maintained at 15 K in an excess of krypton generates a transparent, colorless matrix.

Figure 6 shows an electronic absorption trace corresponding to one experiment: Ba/D_2O/Kr. Figure 7 shows two traces corresponding to expanded scale scans of two different isotopic experiments: Ba/D_2O/Kr and Ba/H_2O^{16}/Kr. Figure 8 shows five absorption traces corresponding to the Ba/D_2O/Kr matrix being selective photolyzed with different colors of light. Table 4 lists the assignments corresponding to band maxima for each of these spectra. The assignments were made after having annealed

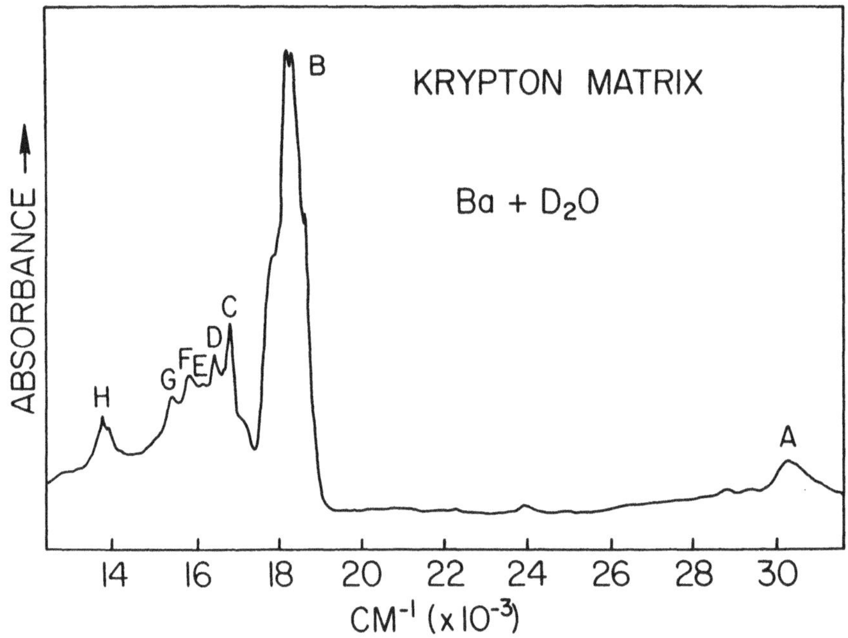

Fig. 6. Electronic absorption spectra for barium and water isolated in a krypton matrix: Ba:D_2O:Kr = 1:54:2750.

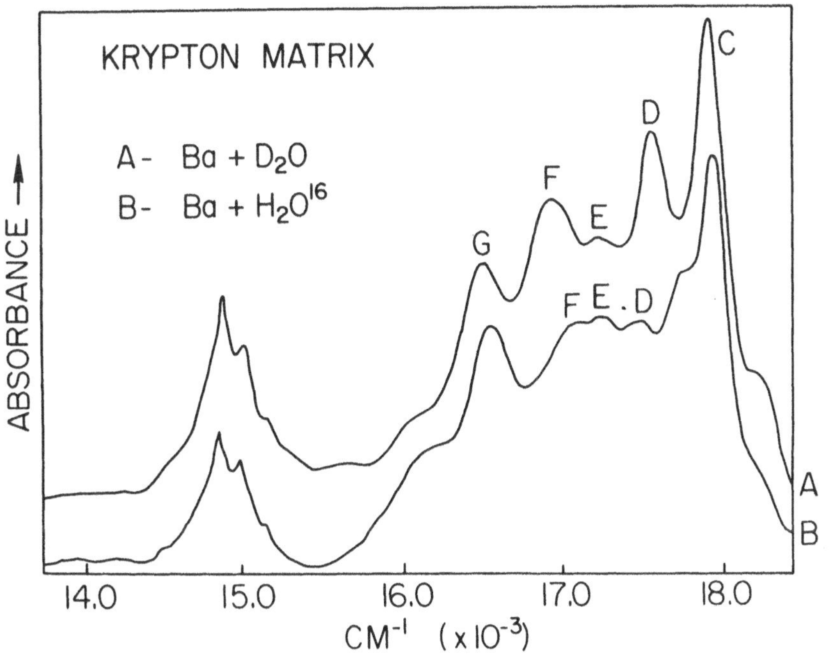

Fig. 7. Expanded-scale electronic absorption spectra for barium and isotopic water isolated in a krypton matrix: trace A, Ba:D_2O:Kr = 1:54:2750; trace B, Ba:H_2O^{16}:Kr = 1:274:14,500.

the matrix to 42 K over a 3 min period. This process was effective in decreasing the intensity of the "F" band while increasing the intensity of the "G" band. Also, the triplet structure of the "H" band and the red shoulder on the "C" band were partially washed out. The bands that were discriminated against in the temperature cycling process probably correspond to unstable matrix site effects.

The bands denoted by "A" and "B" in Fig. 6 can be attributed to atomic resonance transitions for barium since only these bands persist during the course of high Ba dilution experiments without water doping. Comparison of the energy positions of these bands with gas-phase transition energies (*11*) allows the probable assignment of the "A" and "B" bands to the $6s6d(^1D) \leftarrow 6s^2(^1S)$ and the $6s6p(^1P) \leftarrow 6s^2(^1S)$ atomic resonance transitions for barium. The "A" band is blue-shifted (253 cm^{-1}) relative to the gas-phase transition, as is typical according to relatively strong repulsive matrix cage interactions, and is quite weak and broad compared to the "B" band. The weak character of this band is consistent with the parity-forbidden nature of the transition from which it originates. Matrix site symmetry was effective in allaying the forbidden character of the low energy $^1D \leftarrow {}^1S$ transitions of calcium and strontium in argon and krypton matrices; consequently, it is quite conceivable that the same static symmetry mechanism plays an important role in breaking down the selection rules for barium's high energy $^1D \leftarrow {}^1S$ transition even though the transitions arise from different electron configuration

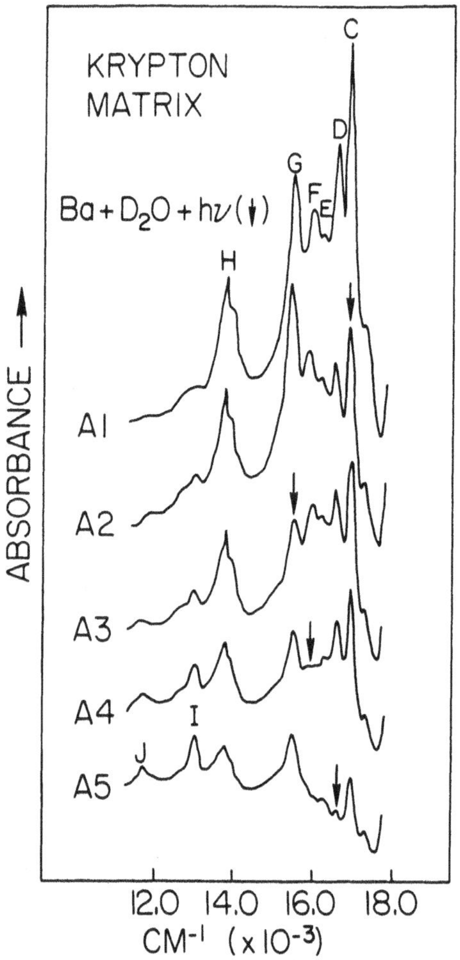

Fig. 8. Electronic absorption spectra for barium and water isolated in a krypton matrix with selective photolysis: trace Al, Ba:D_2O:Kr = 1:54:2750; trace A2, same as trace Al, but photolyzed for 1.75 min with light of λ = 5901–5921 Å; trace A3, same as trace A2, but photolyzed for 1.75 min with light of λ = 6463–6483 Å; trace A4, same as trace A3, but photolyzed for 2 min with light of λ = 6258–6278 Å; trace A5, same as trace A4, but photolyzed for 1.75 min with light of λ = 6036–6056 Å.

rearrangements. It was not possible to learn whether or not the low energy $6s5d(^1D) \leftarrow 6s^2(^1S)$ transition for Ba had also partially shed its forbidden character by virtue of matrix site symmetry, as was noted for Ca and Sr, since this near-infrared transition is just barely beyond the low energy range where the UV–VIS spectrometer can reliably probe. The major "B" band is blue-shifted by 209 cm^{-1} relative to the gas-phase transition and has triplet structure with a red shoulder. The triplet structure may possibly be interpreted in the same manner as was discussed for the band structure of magnesium. The red shoulder is probably

TABLE 4
Assignments for the Optical Absorption Spectra of the $Ba/H_2O/Kr$ and the $Ba/H_2O/Kr + h\nu$ Systems

Peak	λ(Å)	$\bar{\nu}(cm^{-1})^a$	Species	Assignment[c]	Gas Phase[b] (cm^{-1})
A	3280	30 490	Ba	$6s6d(^1D) \leftarrow 6s^2(^1S)$	30 237
B	5341	18 722	Ba	$6s6p(^1P) \leftarrow 6s^2(^1S)$	18 060
B	5426	18 428	Ba	$6s6d(^1D) \leftarrow 6s^2(^1S)$	18 606
B	5474	18 269	Ba	$6s6d(^1D) \leftarrow 6s^2(^1S)$	18 606
C	5899	16 952	$Ba\cdots OH_2$	$B(^1B_2) \leftarrow X(^1A_1)$	?
C	5961	16 775	$Ba\cdots OH_2$	$B(^1B_2) \leftarrow X(^1A_1)$	?
D	6070	16 473	$Ba\cdots OH_2$	$B(^1B_2) \leftarrow X(^1A_1)$	?
E	6152	16 255	$Ba\cdots OH_2$	$B(^1B_2) \leftarrow X(^1A_1)$	?
F	6216	16 087	$Ba\cdots OH_2$	$B(^1B_2) \leftarrow X(^1A_1)$	?
G	6431	15 550	$Ba\cdots OH_2$	$B(^1B_2) \leftarrow X(^1A_1)$	?
G	6573	15 215	$Ba\cdots OH_2$	$B(^1B_2) \leftarrow X(^1A_1)$	?
H	7083	14 118	$Ba\cdots OH_2$	$A(^1B_1) \leftarrow X(^1A_1)$	?
H	7153	13 979	$Ba\cdots OH_2$	$A(^1B_1) \leftarrow X(^1A_1)$	?
H	7229	13 833	$Ba\cdots OH_2$	$A(^1B_1) \leftarrow X(^1A_1)$	?
I	7631	13 105	BaOH	$^2\Sigma^+ \leftarrow X(^2\Sigma^+)$	?
J	8409	11 892	BaOH	$^2\Pi \leftarrow X(^2\Sigma^+)$	?
K[d]	7697	12 992	Ba_2	$^1\Sigma^+_\mu \leftarrow X(^1\Sigma^+_g)$	?

[a]$\pm$ 15 cm^{-1}.
[b]From refs. 1, 11, and 16.
[c]Energies of transition are taken from band maxima unless noted otherwise.
[d]Not shown in the spectra.

caused by the Ba atom residing in a different matrix site since its large size generates a poor fit in the rare gas lattice.

The structureless band denoted by "K" is assigned to an electronic transition of the barium van der Waals dimer since this is the first band to grow in after the atomic bands upon increasing the Ba concentration, and the growth displays a quadratic dependence on the metal deposition rate. Only one other study of barium systems has been reported (*10*), but the results are largely inconclusive because of Sr and Mg impurities and failure to isolate Ba sufficiently to be able to discriminate against higher-order metal aggregates. Four unstructured absorptions in argon to the red of atomic features at 570, 600, and 850 nm were reported. From the trend in Mg_2, Ca_2, and Sr_2 absorptions and the intermediate positions of the IIA mixed $^1\Sigma_u{}^+(^1S + {}^1P) \leftarrow {}^1\Sigma_g{}^+(^1S + {}^1S)$ absorption for Ba_2 is around 760 nm, in very good agreement with the 7697 Å "K" band for Ba_2 in a krypton matrix reported in this study. Although the previous study could not eliminate the possibility that the 735–753 nm band could be assigned to an atomic $^3P \leftarrow {}^1S$ transition, concentration studies in this work establish that the "K" band arises from Ba_2 and hence may be assigned to the $^1\Sigma_u^+ \leftarrow {}^1\Sigma_g^+$ transition. Assuming this assignment is correct, a lower limit to the dissociation energy for the excited state dissociating to ^1P and ^1S atoms may be estimated from

$$D_0'' = E(^1P \leftarrow {}^1S) - E_{00} + D_0''$$

D_0' calculates to be approximately 7400 cm^{-1} assuming $D_0'' = 1600$ cm^{-1} (*10*) and $E_{00} = 12{,}250$ from the red cutoff of the "K" band observed in absorption.

The bands denoted by "C," "D," "E," "F," "G," and "H" in Fig. 6 may be associated with the 1:1 reaction product of barium and water since these bands grow in as a consequence of having doped the matrix with water and maintain identical relative intensities with different water dilutions. In collaboration with the previous IR studies (7), these bands may be identified with the barium–water adduct, Ba . . . OH_2. The "C," "D," "E," and "G" bands are clustered together to the red of the atomic transition and have relatively narrow bandwidths, whereas the "H" band is red-shifted considerably from this cluster of bands and has a broad, triplet structure.

Isotopic substitution studies were conducted with D_2O and H_2O^{18} in lieu of H_2O^{16} in hopes of deriving a better characterization of the nature of the adduct bands. H_2O^{18} substitution was uneventful in that none of the adduct bands significantly shifted or changed in intensity; however, D_2O substitution effected a number of changes in the band structure, as is depicted in Fig. 7. First, the intensities of the "F" and "D" bands were enhanced dramatically upon D_2O substitution with associated shifts of 163 cm^{-1} to the red and 73 cm^{-1} to the blue, respectively. Second, substitution was also effective in eliminating the red shoulder on the major "C" band with a simultaneous resolution of a blue shoulder. Finally, the "C," "E," and "G" bands were also red-shifted by 34, 126, and 62 cm^{-1}, respectively. The "H" bands were not affected by isotopic substitution. Attempts to interpret these isotopic shifts in band position in terms of a vibronic progression of an excited state in different normal modes were unsuccessful; consequently, these bands probably correspond to a combination of multiple excited states, geometric isomers, and matrix site effects for the 1:1 Ba–H_2O complex.

Selective irradiation of each of these adduct bands through a monochromator was of assistance in understanding their interrelationships and their respective natures. Figure 8 shows five spectral traces corresponding to four photolysis experiments with different colored light. Trace A2 shows that a 1.75 min selective photolysis of the "C" band effected about a 50% decrease in the intensity of the "C" band without affecting the other bands except, perhaps, the "D" band that rides on the red shoulder of the "C" band and the "H" band, which decreased in intensity. This suggests that the "D" band is only associated with the "H" band. Trace A3 shows that a 105 s photolysis of the "G" band decreases the intensity of the "G" band by over 50% without affecting the intensities of the other bands, except the "H" band that again decreases slightly in intensity. This suggests an association of the "G" band with the "H" band also. Similar behavior is observed for the "F," "E," and "D" bands so that one may summarize the overall photolysis behavior by stating that the "C," "D," "E," "F," and "G" bands are all independent of each other, but all of these bands are associated with the "H" band. This statement is not rigorous with regard to the "F" and "E" bands, since it is not possible to distinguish between them photolytically. In another experiment not shown, photolysis of the "H" band was discovered to be effective in eliminating all of the adduct bands, which bolsters the correlation of the "H" band with all of the other bands. Photolysis of each of the adduct bands was proficient in generating both the "I" and "J" product bands. The photolytic behavior of the adduct bands suggests the presence of two electronic states for the adduct corresponding to the "H" band and the "C," "D," "E," "F," and "G" band cluster. This band cluster may correspond to an electronic state split by geometric isomers or matrix sites.

The infrared investigations of the barium–water chemistry (7) reported the generation of the HBaOH molecule and the BaOH radical from irradiation of the Ba . . . OH_2 complex with light of λ = 580–620 nm. Only subsequent photolysis of HBaOH with light of λ = 200–300 nm was successful in increasing the yield of the BaOH radical. In this study, selective photolysis of the Ba . . . OH_2 complex generated the "I" and "J" product bands, which are probably only associated with the BaOH and BaH products (as was observed for the calcium- and strontium–water systems) since the HBaOH product probably has electronic states beyond the range and sensitivity of the spectrometer. The analogous BaF molecule has two low-energy electronic states at 11,647–12,278 cm^{-1} ($^2\Pi$) and 14,063 cm^{-1} ($^2\Sigma^+$). The BaH molecule has electronic states at 15,055–14,605 cm^{-1} ($^2\Pi$) and 11,092 cm^{-1} ($^2\Sigma^+$). The "I" and "J" bands occur at 13,105 and 11,892 cm^{-1}, respectively, and probably cannot be correlated to the transitions for BaH since the shifts that would be required to make a correspondence are too large in magnitude to be reasonable. Since only the $^2\Sigma^+ \leftarrow {}^2\Sigma^+$ system was observed in absorption for CaH and SrH in this study, it is quite reasonable that BaH was being formed, but just could not be observed since the potentially perspicuous $^2\Sigma^+ \leftarrow {}^2\Sigma^+$ transition for BaH is probably less than 11,000 cm^{-1}, which is beyond the range of the spectrometer. Therefore, by elimination, the "I" and "J" bands are tentatively assigned to the $^2\Sigma^+ \leftarrow {}^2\Sigma^+$ and the $^2\Pi \leftarrow {}^2\Sigma^+$ transitions for the BaOH radical, respectively. Both the $^2\Pi$ and $^2\Sigma^+$ excited states were observed for the CaOH and SrOH radicals.

DISCUSSION

Nature of the Electronic States of the Group IIA Metal–Water Complex

Preliminary Considerations

Mg, Ca, Sr, and Ba metal atoms initially formed a complex with water such that the metal is coordinated to the $3a_1$ lone electron pair on the oxygen atom of the water molecule. Calculations for Mg indicate (9) that the metal is symmetrically coordinated in the plane of the water molecule according to the C_{2v} point group. Hydrate or adduct formation perturbs the electronic structure of the metal atom in such a manner that the $^1P \leftarrow {}^1S$ and the $^1D \leftarrow {}^1S$ atomic transitions are apparently split into several states by the interaction and red-shifted from their respective uncomplexed resonance transitions. Selective photolysis studies were fruitful in establishing the interrelationships of the sets of multiple bands generated upon adduct formation. Insight into the nature of the perturbation that leads to red-shifted spectra may be gleaned from an examination of Fig. 9, which depicts the metal in its ground and excited states approaching the $3a_1$ lone pair orbital of water from infinite separation. At suitable metal–water separations, the metal is unperturbed by the water interaction, and a normal atomic transition is observed that is of energy E_2 in the diagram. As the metal approaches the water molecule in closer

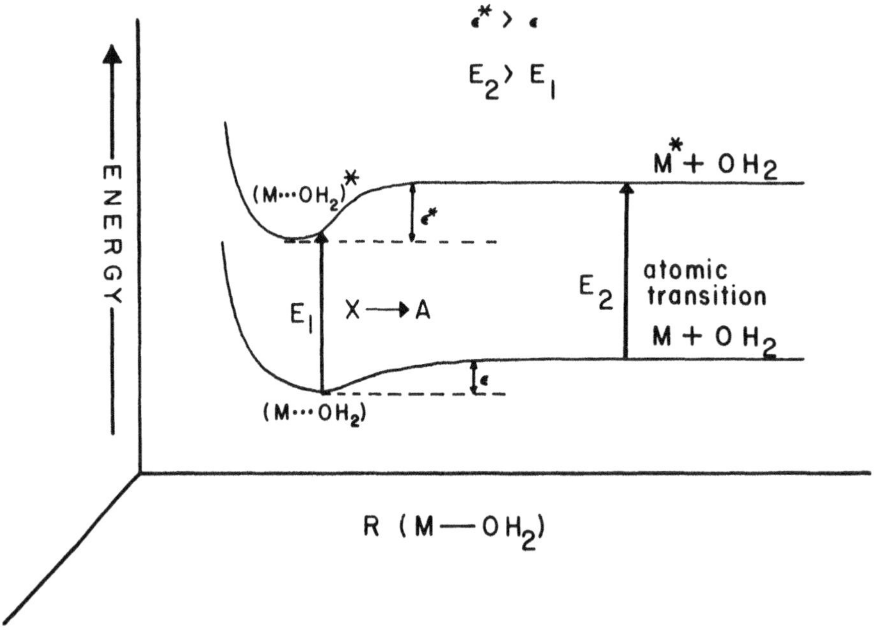

Fig. 9. Schematic diagram of the potential energy surfaces for the metal–water interaction with the metal in its ground and excited states.

proximity, an attractive interaction is encountered such that the eventual equilibrium arrangement finds the metal–water complex bound in a potential energy well. For reasons that will be developed later, the water is more highly attracted to the metal in its excited state than its ground state, so that $\epsilon^* > \epsilon$, which implies that $E_2 > E_1$. Consequently, the $A \leftarrow X$ metal–water complex transition is red-shifted relative to the atomic lines. Since the ground-state interaction with water, ϵ, is very weak for the Group IIA metals, the difference between E_1 and E_2 is a measure of the minimum dissociation energy of the exciplex in its bound excited state, ϵ^*, dissociating to M* (^1D or ^1P) + $H_2O(^1A_1)$. More precisely,

$$D'_0 = E(^1D \text{ or } ^1P \leftarrow {}^1S) - E_{00} + D''_0$$

where E_∞ may be estimated from the red cutoff of the adduct bands observed in absorption for a particular state, and D''_0 may be evaluated from previous calculations (9). The D'_0 values for the metal–water complexes in different electronic states are tabulated in Table 5.

The Group IIA metal–water complex has an unequivocal electronic ground state of $^1A_1$ since the *s* orbital is nondegenerate under any symmetry resolution. However, the metal–water complex potentially has eight low-energy excited states corresponding to the water interaction splitting the *p* orbital and *d* orbital degeneracy of the metal atom in the C_{2v} point group. Splitting of both the *p* and *d* orbitals is possible for the

TABLE 5
Dissociation Energy Estimates[a] for the Group IIA Metal-Water Complex Excited States in Solid Krypton

Molecule	State	$^1P \leftarrow {}^1S^{b,e}$	$^1D \leftarrow {}^1S^{b,e}$	O-O[b,c]	D'_0[b]	$(D'_0)^{b,d}$
$Mg\cdots OH_2$	?($^1B_2$)[f]	35 032	—	27 380	7 652 + D''_0	8 457
$Ca\cdots OH_2$	^1B_g[g]	—	21 826	18 135	3 691 + D''_0	6 211
$Ca\cdots OH_2$	$^1A_1$	—	21 826	19 810	2 016 + D''_0	4 536
$Ca\cdots OH_2^i$	$^1B_1$[g]	—	21 826	20 650	1 176 + D''_0	3 696
$Ca\cdots OH_2^j$	$^1B_1$[g]	—	21 826	21 115	711 + D''_0	3 231
$Sr\cdots OH_2$	$^1B_2$[g]	—	19 838	16 790	3 048 + D''_0	—
$Ba\cdots OH_2$	$^1B_2$	18 269	—	13 436	4 833 + D''_0	—
$Ba\cdots OH_2^k$	$^1B_1$	18 269	—	15 300	2 970 + D''_0	—
$Ba\cdots OH_2^j$	$^1B_1$	18 269	—	16 775	1 494 + D''_0	—

[a]For $(M\cdots OH_2)^*$ dissociating to M* + $H_2O(^1A_1)$.
[b]cm^{-1}.
[c]Estimated from red cutoff of absorption band.
[d]D''_0 values for $Mg\cdots OH_2$ and $Ca\cdots OH_2$ of 805 cm^{-1} and 2520 cm^{-1}, respectively.
[e]For multiple bands, energy values are taken from the major band position.
[f]$^1B_2$ is the logical choice since it is the strongest bound state for the Ca-, Sr-, and Ba-H_2O complexes.
[g]Assumes adiabatic correlation to M* (^1D) from an avoided surface intersection.
[i]Peak labeled "D".
[j]Peak labeled "C".
[k]Peak labeled "G".

heavier Group IIA metals since they have closely spaced, low-energy ^1P and ^1D atomic states that can be perturbed by interaction with water. There are three electronic states generated by perturbation of the p orbital by water; these are labeled $^1B_2$, $^1B_1$, and $^1A_1$ corresponding to the promoted electron residing in the b_2–p_y, the b_1–p_x and the a_1–p_z orbitals, respectively. Occupation of the a_1–p_z orbital by the promoted electron directs electron density toward the $3a_1$ lone electron pair on water, which suggests that the $^1A_1$ state is repulsive because of electron repulsion. Both the $^1B_2$ and the $^1B_1$ states are expected to be attractive since these states correspond to an orbital occupancy that disperses the electron density away from the nucleus so the $3a_1$ lone pair on water may achieve an enhanced Coulombic attraction with the metal's nucleus. *Ab initio* studies for the (Li . . . OH_2)* adduct (*8*) report that the $^2A_1$ state is repulsive whereas the $^2B_2$ and $^2B_1$ states are attractive. These calculations also indicate that the $^2B_2$ state is more stable than the $^2B_1$ state because occupation of the b_2-p_y orbital allows for a favorable quadrupole–quadrupole interaction with water's inplane

$$\overset{+}{\mathrm{H}}\text{—}\overset{-}{\mathrm{O}}\text{—}\overset{+}{\mathrm{H}}$$

quadrupole along the y-axis, whereas occupation of the b_1–p_x orbital of the metal generates a repulsive quadrupole–quadrupole interaction with the filled, out-of-plane, nonbonding b_1 orbital of water. For the lithium–water interaction, the calculated energy difference between the $^2B_2$ and the $^2B_1$ excited states is 4.3 kcal/mol.

There are five electronic states generated through perturbation of the metal d orbitals by water, splitting the fivefold degeneracy in the C_{2v} point group. These states are labeled $^1A_1(z^2)$, $^1A_1(x^2 - y^2)$, A_2, B_1, and B_2 and correspond to the promoted electron residing in the a_1–$d_z{}^2$, the a_1–$d_{x^2y^2}$, the a_2–d_{xy}, the b_1–d_{xz}, and the b_2–d_{yz} orbitals, respectively. However, in view of what has been learned regarding the stabilities of the $^1A_1$, $^1B_1$, and $^1B_2$ states related to p orbital occupancy, there is strong reason to believe that a number of these states that arise from d orbital occupancy are also repulsive in nature. Certainly, the $^1A_1(z^2)$ state is repulsive since it corresponds to a_1–d_{z^2} orbital occupancy that directs electron density toward the $3a_1$ lone electron pair on water as was true for the a_1–p_z orbital. The $^1B_1$ state is also predictably unstable since it corresponds to b_1–d_{xy} orbital occupancy. Although this occupancy disperses electron density away from the nucleus core of the metal, it does so by directing the charge density toward the b_1 lone pair on water, inducing electron repulsion again. The $^1B_2$ state is also expected to be unstable since the corresponding b_2–d_{yz} orbital occupancy places charge density, in-plane, indirectly toward the $3a_1$ lone pair. However, the $A_1(x^2$–$y^2)$ and the A_2 states are probably relatively stable since they correspond to a_1–$d_{x^2-y}{}^2$ and a_2–d_{xy} orbital occupancies directing charge density away on the water molecule. The energy of the $^1A_1$ state is probably degenerate with the $^1A_2$ state since they have equally favorable and repulsive quadrupole–quadrupole components with the water's in-plane and out-of-plane quadrupoles. Although the $^1A_2$ state is probably stable, it can not be observed in absorption if C_{2v} symmetry is sustained since a $^1A_2 \leftarrow {}^1A_1$ electronic transition is electric-dipole forbidden. In summary, for a Group IIA metal–water interaction, there exists an unequivocal $^1A_1$ ground state arising from the $^1S + {}^1A_1$ combination; however, there are three bound excited states that can be accessed from the $^1A_1$ ground state that are derived from $^1P + {}^1A_1$ and $^1D + {}^1A_1$ combinations, specifically: $^1B_2$, $^1B_1$, and $^1A_1$. Therefore, the multiple set of absorption bands generated upon adduct formation should, at least in part, correlate to these three electronic states that arise from symmetry splitting of p and d orbital degeneracy.

In studying the interaction of the metal–water complex, Bentley (*8*) states that the principal noncovalent attractive force between the metal and water is electrostatic in origin, with minor polarization and charge transfer components. Electrostatic attraction should increase dramatically by electronic excitation of the adduct since electron promotion involves charge density dispersal from the vicinity of the metal nucleus by vacating a a_1–s orbital. This generates a hole in the molecular orbital structure of the metal that makes it more electropositive to the $3a_1$ electron pair on water. The ground state of the Group IIA metal–water complex is calculated to be very weakly bound, i.e., primarily van der Waals attraction. Therefore, adduct electronic transitions should be accompanied by significant increases in the bonding interactions and decreases in the internuclear M—O bond distances; in turn, this implies large off-axis Franck-Condon factors for the transition. These factors may play an important role in the photo-induced reactions of the metal–water complex as will be discussed shortly.

Magnesium and Water

The magnesium–water adduct band has no evident band structure; rather, it displays a rather broad continuous absorption encompassing a 4500 cm^{-1} range. This broadness should be anticipated from the van der Waals nature of bonding between Mg and H_2O in that continuous absorption spectra are characteristic of molecular species that are very weakly bound in the ground state but have a well-defined potential energy minimum in the excited state. This is the case since absorption may take place from a wide, continuous range of internuclear separations in the weakly bound ground state, which, along the excited state potential energy surface, corresponds to a wide, continuous range of energies that is reflected in a broad, continuous absorption band. For this reason as well, it would be difficult to detect the presence of several closely spaced, bound excited states in absorption from the ground state of a molecular species with a poorly defined internuclear separation. However, it is also possible to generate continuous spectra by absorption into a dissociative excited state. The first excited-state potential energy surfaces for the

TABLE 6
Estimated Heats of Reaction[a,b,c] (kJ/mol)

	ΔH_r
$Mg + H_2O \rightarrow HMgOH$	−92
$Mg + H_2O \rightarrow MgOH + H$	+88
$HMgOH \rightarrow MgOH + H$	+180
$Ca + H_2O \rightarrow HCaOH$	−113
$Ca + H_2O \rightarrow CaOH + H$	+58
$HCaOH \rightarrow CaOH + H$	+171
$Sr + H_2O \rightarrow HSrOH$	−96
$Sr + H_2O \rightarrow SrOH + H$	+63
$HSrOH \rightarrow SrOH + H$	+159
$Ba + H_2O \rightarrow HBaOH$	−171
$Ba + H_2O \rightarrow BaOH + H$	+17
$HBaOH \rightarrow BaOH + H$	+188

[a]Heats of reaction were estimated using bond energy data from references 16, 19 and 20.
[b]Group IA metal hydride dissociation energies were used for the H-MOH bond energies.
[c]At very low temperatures, $\Delta H_r \approx \Delta G_r$, since $T\Delta S_r$ is very small.

HMgOH species are probably in the UV, which one may be able to correlate with the excited state of the adduct at about 30,000 cm^{-1} since the insertion reaction is exothermic by about 7500 cm^{-1} (see Table 6). Thus, the excited states of the adduct may be unbound with respect to rearrangement to form the insertion product in an excited state in which case a continuous spectra would be observed in absorption. Although from preliminary considerations several adduct excited states might be expected to be apparent an electronic absorption states, these excited states are probably washed out by the observed broad absorption continuum.

Calcium and Water

Band intensity behavior observed during selective photolysis studies of the Ca . . . OH_2 adduct bands suggested that the adduct has three excited electronic states, one of which is split by a site or geometric isomer effect. From preliminary considerations, it was noted that the adduct should have three stable excited electronic states, $^1B_2$, $^1B_1$, and $^1A_1$, which should be correlatable to the three matrix transitions. From previous arguments that are corraborated by quantum-chemical calculations, the $^1B_2$ state should be lower in energy relative to the $^1B_1$ state, both of which are diabatically derived from a splitting of the *p* orbital degeneracy. Adiabatically, however, the $^1B_1$ and the $^1B_2$ surfaces correlate into the M(^1D) + $H_2O(^1A_1)$ combination since these bound surfaces have an avoided intersection with the $^1B_1$ and $^1B_2$ surfaces that arise from breaking the *d* orbital degeneracy. The $^1A_1$ surface has no surface crossings and hence correlates directly into the $M(^1D)$ + $H_2O(^1A_1)$ combination. The $^1B_2$ state cannot be assigned to the "C" and "D" bands since this implies that the $^1B_2$ state is less stable than the $^1B_1$ state. The $^1B_2$ state is more appropriately assigned to the "F" band since such an association allows for a reasonable energy separation between the $^1B_2$ and $^1B_1$ states that is not achievable otherwise. From the barium–water discussion to be developed shortly, it is apparent that the $^1B_1$ state is heavily influenced by slightly different geometric conformations for the adduct that generate splittings of this state; therefore, the "C" and "D" bands are both tentatively assigned to the $^1B_1$ state split by two geometric isomers of the adduct with the "C" band corresponding to the most stable $^1B_1$ state. By elimination, the $^1A_1$ state is tentatively assigned to the "E" band. Different geometric conformations of the adduct that would split the $^1B_1$ state yet leave the $^1B_2$ state unperturbed correspond to the metal tilting out of the plane, defined by the water molecule breaking the C_{2v} symmetry and giving C_s symmetry so the "C" band may more appropriately correspond to an A' state. Quantum-chemical calculations by Curtiss and Frurip (9) indicate that this conformation is the most stable for the Be–H_2O adduct, but very little energy is required to place the metal atom back in-plane. For the ground state of the adduct, stable tilted conformations are reasonable since, with this geometry, more significant ligand–metal bonding may be achieved by a mixing of the filled b_1–a' orbital of water with an unfilled metal orbital of b_1–a' symmetry. The fact that a definite band structure is observed for the Ca . . . OH_2 adduct as opposed to the Mg . . . OH_2 adduct suggests that the ground state has a relatively well-defined potential energy minimum, the lack of which was probably partially responsible for washing out discrete band structure for the Mg . . . OH_2 adduct. In summary, the "F," "E," and "D" and "C" bands are tentatively assigned to $^1B_2 \leftarrow {}^1A_1$, $^1B_1 \leftarrow {}^1A_1$ and $^1A_1 \leftarrow {}^1A_1$ transitions for the adduct, respectively, where the $^1B_1 \leftarrow {}^1A_1$ transition is believed to be split by the presence of two geometric isomers for the adduct, the existence of which will be shortly substantiated in the discussion of the band structure for the Ba . . . OH_2 adduct.

Isotopic substitution generated various red and blue shifts in the electronic absorption band positions for the calcium–water complex that may be rationalized, in a general manner, in the context of isotopic frequency shifts for several vibrational modes of the adduct. The Y-shaped adduct (C_{2v} point group) has six nondegenerate vibrations, five of which are in the plane of the adduct molecule; however, only three of these vibrations are expected to have both a difference in their vibrational frequency for the ground and the excited state and display a deuterium isotope shift. These three vibrations correspond to three bending modes; (1) the A_1 bending mode primarily composed of the bending mode of water; (2) the B_2 bending mode, which corresponds to an in-plane wagging motion of the water molecule with respect to the calcium atom; and (3) the B_1 mode that can be identified with an out-of-plane rocking motion of the water molecule with respect to the calcium atom. There is strong reason to believe that the A_1 bending mode of water should be significantly different in the ground as opposed to the excited state of the adduct. Adduct formation in the infrared studies (*7*) was evidenced by a decrease in the bending mode frequency of water. This decrease was postulated to have been induced by a partial charge transfer from the $3a_1$ orbital of water (which is the orbital accounting for the bent configuration of water) to the metal atom caused by Coulombic attraction of the metal nucleus. As per previous discussions, electronic excitation to attractive states of the adduct disperses charge from the vicinity of the metal nucleus to enhance the Coulombic forces of attraction, thereby increasing the partial electron transfer from the $3a_1$ orbital to the metal with a concomitant further decrease in the A_1 bending mode of water. The vibrational frequencies of both the B_2 and B_1 bending modes should increase significantly in going from the ground to the excited state since the Ca—O bond strength increases dramatically for this same transition according to previous arguments. Briefly, the vibrational frequency of the A_1 mode should be lower in the excited state relative to the ground state, whereas the B_1 and B_2 frequencies should be higher; therefore, for the A_1 mode, deuterium substitution should induce blue shifts in a O—O transition, whereas, considering only the B_1 or B_2 modes, deuterium substitution should cause red shifts in a O—O transition. Of course, the actual overall blue or red shifts observed in the absorption band positions correspond to a composite of these three effects, where a blue shift indicates that the A_1 vibrational frequency change dominates over the B_2 and the B_1 frequency changes, and the red shift indicates the B_2 and the B_1 frequency changes are predominant. With this in mind, it is evident that the ${}^1B_2 \leftarrow {}^1A_1$ electronic transition corresponding to the "F" band of the adduct, which was blue-shifted upon deuterium substitution, causes a dominating change in the bending mode of water, whereas the ${}^1A_1 \leftarrow {}^1A_1$ and the ${}^1B_1 \leftarrow {}^1A_1$ transitions corresponding to the "C," "D," and "E" bands, which were red-shifted upon substitution, do not.

Table 5 lists dissociation energy estimates for several excited states of the Ca . . . OH_2 adduct in solid krypton. Assuming the dissociation energy of the ground state is 2520 cm^{-1} (*17*), the ${}^1B_2$, ${}^1A_1$, and ${}^1B_1$ states are bound by 6211, 4536, and 3696 cm^{-1} (for the most stable ${}^1B_1$ state), respectively. The dissociation energies for the ${}^1B_2$ and the ${}^1B_1$ states are calculated relative to the ${}^1D \leftarrow {}^1S$ transition since, although these states diabatically arise from the $M^*({}^1P) + H_2O({}^1A_1)$ combination, they encounter an avoided surface crossing with the ${}^1B_2$ and the ${}^1B_1$ states arising from the former combination are probably destabilized by the avoided crossing. These dissociation energies are in good qualitative agreement with a number of quantum-chemical studies that have reported dissociation energies for metal–water complexes in their excited and ground states. For example, the calculated dissociation energy of (Be . . . OH_2)* in the ${}^3B_2$ excited state is 18.6 kcal/mol (*8*), which compares well with the 17.8 kcal/mol value for the $^1B_2$ state of (Ca . . . OH_2)*. These calculations (*8,22*) also indicate there is a 1500 cm^{-1} and a 1855 cm^{-1} energy spacing between the $^2B_2$ and the $^2B_1$ states for (Li . . . OH_2)* and Al . . . OH_2, respectively. These values are in good agreement with the 1675 cm^{-1} spacing observed between the $^1B_2$ and the $^1B_1$ states of Ca . . . OH_2. *Ab initio* calculations have not considered the $^1A_1$ state arising from the ^1D + $^1A_1$ combination for the Group IIA metals.

Strontium and Water

Band intensity behavior observed during selective photolysis studies of the Sr . . .OH_2 adduct was not as informative as the behavior observed in the Ca–H_2O system, but suggested that at least two excited electronic states exist. The relative position and the photolysis behavior of the "G" band is identical to that displayed by the "F" band for the Ca . . . OH_2 adduct. Photolysis of the "G" band was successful in eliminating all of the other adduct bands, and it is the most red-shifted band from the atomic resonance transition for the 1:1 Sr . . . OH_2 adduct suggesting that it may be associated with the most stable excited state of the adduct. In view of these analogous associations and the arguments presented in the Ca–H_2O adduct discussion, the "G" band is tentatively assigned to the $^1B_2 \leftarrow {}^1A_1$ transition of the adduct. Although the $^1B_2$ state diabatically arises from a symmetry splitting of the *p* orbital degeneracy, adiabatically it correlates into the M*(^1D) + $H_2O(^1A_1)$ combination because of a surface crossing previously described. The assignments that correspond to the "C," "D," "E," and "F" bands are not unambiguous since selective photolysis studies were unsuccessful in deconvoluting their interrelationships. However, in analogy to the Ca . . . OH_2 adduct study, these bands probably can be associated with transitions to a $^1A_1$ state and perhaps several $^1B_1$ states. Multiple $^1B_1$ states probably correspond to different out-of-plane geometric isomers of the adduct. Multiple matrix sites may also be further complicating this analysis.

Barium and Water

Band intensity behavior observed during the selective photolysis studies of the Ba . . . OH_2 bands suggests that the adduct has two excited electronic states, one of which is split forming a band cluster. The analysis of the electronic structure of the Ba . . .OH_2 adduct is simplified by the absence of the low energy ^1D state in the vicinity of the ^1P state, which complicated the interpretation of the electronic structure of the Ca– and Sr–H_2O adducts. The ^1D state is too low in energy (less than 11,000 cm^{-1}) to be apparent in the UV–VIS spectra. According to preliminary considerations, the absence of the ^1D state means that the Ba . . . OH_2 adduct should have only two stable excited electronic states, $^1B_2$ and $^1B_1$, which should correlate with the two systems of matrix transitions. Since the $^1B_2$ state is believed to be the most stable, the "H" band can be tentatively assigned to a $^1B_2 \leftarrow {}^1A_1$ transition for the adduct, and by elimination, the "C," "D," "E," "F," and "G" bands can be tentatively assigned to a $^1B_1 \leftarrow {}^1A_1$ transition for the adduct. These states correlate adiabatically into the $M^*(^1P) + H_2O(^1A_1)$ combination since there is no avoided surface crossing with other $^1B_2$ or $^1B_1$ potential surfaces. Although the splitting noted for the $^1B_1 \leftarrow {}^1A_1$ transition may be partially understood in terms of matrix site effects, some of the bands are probably attributable to different geometric isomers of the Ba–H_2O adduct, as was noted for the Ca–H_2O adduct. Different geometric conformations of the adduct that would split the $^1B_1$ state without perturbing the $^1B_2$ state correspond to the metal tilting out of the plane defined by the water molecule breaking the C_{2v} symmetry for C_s symmetry. Any out-of-plane tilt would destabilize the $^1B_1$ state since the tilt would enhance the electron repulsion between the b_1 orbital of the metal and the b_1 orbital of water; therefore, the "G" band may correspond to a $^1B_1 \leftarrow {}^1A_1$ transition for a C_{2v} adduct, whereas the "C" band may correspond to a very bent geometric isomer for the adduct. Since Ba has very low energy empty d and p orbitals that have $b_1 - a'$ symmetry components able to mix with the filled $b_1 - a'$ orbital of water, it is reasonable to expect stable geometric isomers for the adduct of C_s symmetry.

Figure 7 shows that deuterium substitution not only red- and blue-shifted the adduct bands, but also significantly increased some of the intensities. As was previously discussed, the red and blue shifts can probably be interpreted in the context of differing zero point energy shifts for the A_1, B_1, and B_2 vibrational modes in the ground and the excited states; however, the lack of quantitative information regarding these vibrational frequencies in the different states precludes any further analysis in this regard. The dramatic intensity changes for the "F" and "D" bands is interesting and was not observed for the other metal–water adduct bands. It may be suggested that these bands correspond to two different out-of-plane geometric isomers that have such shallow potential energy wells that lowering the zero point energy level for the B_1 vibrational mode via deuterium substitution is effective in significantly stabilizing these particular conformations. This interpretation for the observed isotopic behavior is in keeping with the conjectured isomeric origin of the band cluster, and the calculated shallowness of the potential energy well for the out-of-plane configuration of the Be . . . OH_2 adduct.

Table 5 lists dissociation energy estimates for several excited states of the Ba . . . OH_2 adduct in solid krypton. The $^1B_2$ state and the most stable $^1B_1$ state have dissociation energies of 4833 $cm^{-1} + D''_0$ and 2970 $cm^{-1} + D''_0$, respectively. These energies cannot be meaningfully compared to the $^1B_2$ and the $^1B_1$ states for the Ca– and Sr–H_2O adducts since the latter adducts dissociate into the $M^*(^1D) + H_2O(^1A_1)$ combination;

however, the general qualitative agreement of the dissociation energies is good among these experimental systems, as well as when compared to the *ab initio* dissociation energies calculated for other metal–water complexes. The 1863 cm^{-1} energy spacing between the $^1B_2$ and $^1B_1$ states for Ba . . . OH_2 is also in good agreement with other B_2–B_1 separations evaluated experimentally (1675 cm^{-1} for Ca . . . OH_2) and through quantum-chemical techniques (1500 and 1855 cm^{-1} for (Li . . . OH_2)* and Al . . . OH_2, respectively).

Orbital and State Correlations

The reaction dynamics of the Group IIA metal–water chemistry may be interrogated in the light of orbital and state correlations that elucidate the reaction pathways that are photolytically induced. Initially, it is appropriate to examine the evolution of the molecular orbital structure of the reactants into the MO structure of the products that allows one to evaluate the configurations, and hence the electronic states, of the products that are generated by chemical reaction. A suitable molecular orbital correlation diagram for the investigation of magnesium–water interactions is shown in Fig. 10. This MO correlation diagram analyzes the metal atom approaching the water molecule along the perpendicular bisector of the O—H bond so the system remains in the C_s point group even with in-plane deviations from the perpendicular bisector. It is reasonable to believe that the C_{2v} adduct precursor must rearrange to this C_s approach before the insertion reaction may occur. The molecular orbital structure of the HMgOH species is expected to be quite similar to a bent HAB molecule (*18*) identifying the OH fragment with the *B* atom. The ground-state electron configuration for the ten electron HMgOH species is expected to be

$$(1a'')^2(6a')^2$$

which corresponds to a $^1A'$ electronic ground state with a doubly occupied highest occupied molecular orbital. The first excited electron configuration is expected to be

$$(1a'')^2(6a')(7a')$$

which corresponds to a $^1A'$ excited electronic state that is predictably quite high in energy above the ground $^1A'$ state since (1) no bands attributable to HMgOH were observed in the UV–VIS spectra, and (2) light of $\lambda < 2000$ Å was unsuccessful in dissociating the HMgOH to form MgOH + H in the IR studies as it was for the Sr and Ba insertion products. In the C_s approach, the filled 4*s*–*a'* orbital of Mg correlates into the 6*a'* orbital for HMgOH. The unfilled $4a_1$–a' antibonding valence orbital of water correlates into the 7*a'* orbital for HMgOH. According to this orbital occupancy, a $^1A'$ potential energy surface is generated that correlates the $Mg(^1S)$ + $H_2O(A_1)$ reactants into the $^1A'$ ground state of HMgOH (see

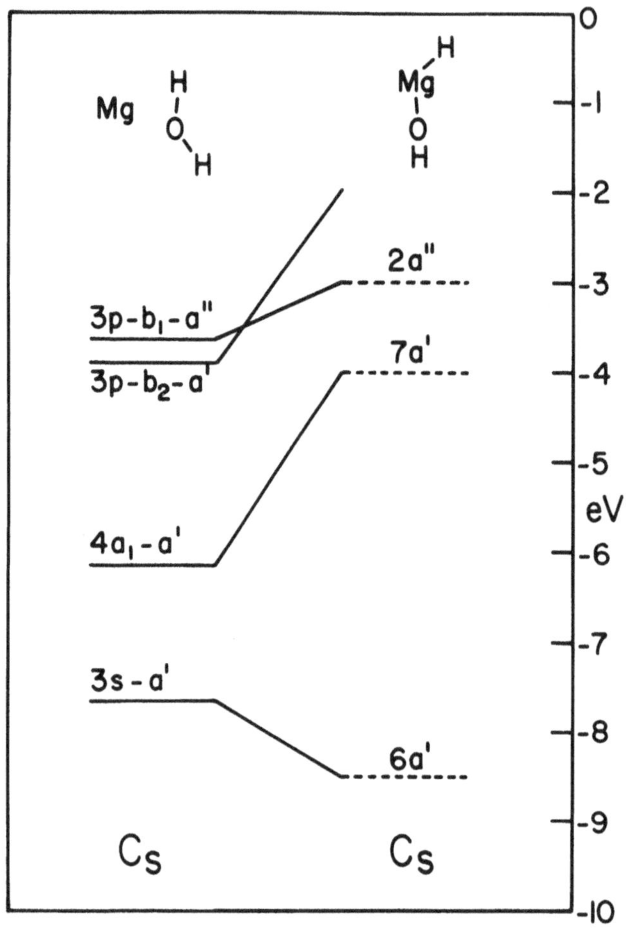

Fig. 10. Molecular orbital correlation diagram for the magnesium–water interactions.

Fig. 11). Table 6, which tabulates the heats of reaction for M + $H_2O \rightarrow$ HMOH, indicates that this correlation is an exothermic process. The correlation diagrams indicate that no potential energy barrier is encountered along this surface; however, a small barrier predictably exists since some promotional energy is required to break up the *s*-orbital electron pair and subsequently to hybridize. Curtiss and Frurip (*9*) have carried out limited calculations that also indicate that there exists a barrier to insertion of the Mg and Be metal atoms into the water molecule; therefore, isolation of the Mg . . . OH_2 adduct is expected. Figure 11 shows that the metal–water interaction splits the ^1P excited state degeneracy into a $^1A''$ excited state and two $^1A'$ states, one of which adiabatically correlates into the $^1A'$ excited state of HMgOH; however, the molecular orbital correlations indicate the diabatic correlation of the ^1P–$^1A'$ state is to a higher product $^1A'$ state, but encounters an avoided surface intersection with a $^1A'$ charge transfer surface generating a potential energy barrier to insertion along this surface. Consequently, this excited $^1A'$ potential surface is

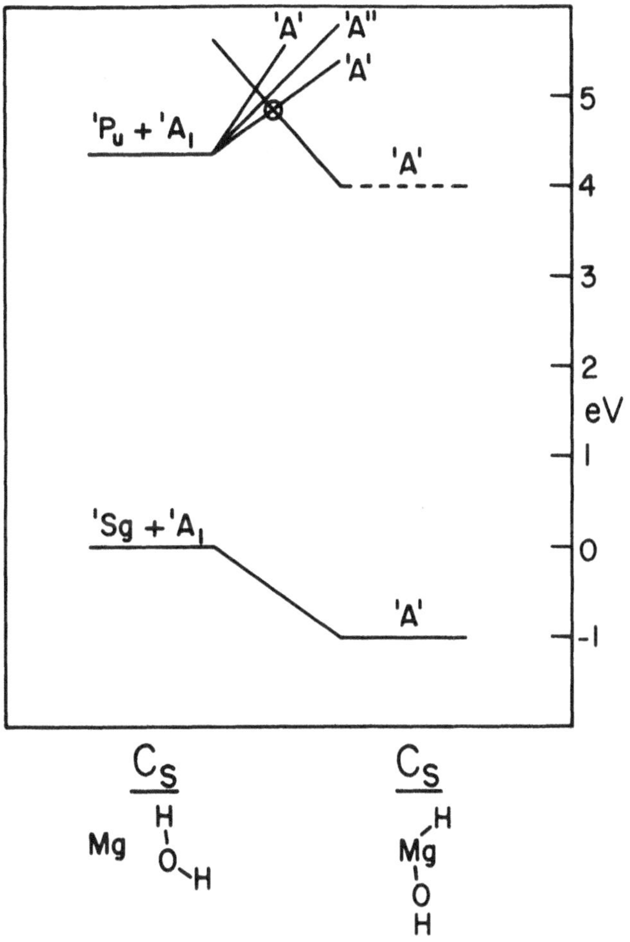

Fig. 11. Electronic state correlation diagram for the magnesium–water interaction: (--) denotes states that have not been observed; (0) denotes an avoided surface intersection.

not expected to be responsible for the insertion reaction. The insertion reaction probably then evolves along the ground $^1A'$ potential surface since the excited states of HMgOH are too high in energy to allow a exothermic correlation from the metal's excited ^1P state to the lowest energy excited states of HMgOH along $^1A'$ or $^1A''$ potential surfaces. Since bonding for the magnesium–water complex is much greater in the excited state as opposed to the ground state (800 versus 8500 cm^{-1}), large off-axis Franck-Condon factors probably exist for the $^1B_1$ or $^1B_2 \leftrightarrow {}^1A_1$ adduct transition, which in emission creates a vibrationally excited adduct. Since excited states are stable with respect to reaction, this excess vibrational energy is probably effective in overcoming any small potential barrier.

The molecular orbital and electronic state correlations for the Ca–, Sr–, and Ba–H_2O systems are significantly different from those for the Mg–H_2O system. The *ns–a'* orbital has a lower ionization energy that positions it above water's $4a_1$–a' orbital. This unfilled orbital correlates into the $6a'$ HOMO, whereas the filled ns–a' orbital correlates into the $7a'$ LUMO of HMOH. According to these orbital occupancies, two A' potential surfaces are generated: (1) an A' surface that correlates the ground state reactants, $M(^1S) + H_2O(^1A_1)$, into the . . .$(1a'')^2(6a')(7a')$ excited electron configuration of HMOH and (2) an A' surface that correlates excited reactants $M^{+2} + H_2O^{-2}$, into the . . . $(1a'')^2(6a')^2$ ground-state configuration for HMOH. A pure state correlation indicates the ground states of the reactants and products correlate, however; illuminating this with the MO correlation allows one to predict a significant potential energy barrier to reaction along the ground potential surface generated by an avoided surface intersection of the two A' surfaces. The photochemical pathways that may exist from excited states of the adduct to the insertion product are not unambiguous because of the more important role the d orbitals play in the electronic structures of these species for Ca, Sr, and Ba, where they severely complicate state and orbital correlations. However, the fact that photolysis of the metal–water adduct generated the MOH radicals for M = Ca, Sr, and Ba but not for Mg—may suggest that a different pathway is followed by the heavier Group IIA metals, perhaps along an excited potential surface. The heats of reaction for MOH formation from $M + H_2O$, shown in Table 6, are very similar for Mg, Ca, and Sr, which suggests the lack of MgOH formation probably cannot be rationalized in terms of simple thermodynamics. In general the Mg–H_2O insertion reaction is expected to proceed along the ground $^1A'$ potential surface, whereas there is strong reason to believe a potential energy barrier exists along this same ground $^1A'$ surface for the Ca, Sr, and Ba–H_2O systems that may preclude a ground-state reaction.

ACKNOWLEDGMENTS

This work was supported financially by the National Science Foundation and by the Robert A. Welch Foundation.

REFERENCES

1. Miller, J. C., and Andrews, L., *Appl. Spectrosc. Rev.* **16,** *1* (1980), and references therein.
2. Okabe, H., *Photochemistry of Small Molecules,* Wiley, New York, 1978.
3. Mandl, A., and Hyman, H. A., *J. Chem. Phys.* **74,** 3167 (1981).
4. Swope, W. C., and Schaefer III, H. F., *J. Am. Chem. Soc.* **98,** 7962 (1976).
5. Kunz, A. B., Guse, M. P., and Blint, R. J., *Chem. Phys. Lett.* **36,** 191 (1975).
6. Fassaert, D. J. M., Verbeck, H., and Van Der Avoird, A., *Surf. Sci.* **29,** 501 (1972).
7. Kauffman, J. W., PhD Thesis, Rice University, 1982.
8. Bentley, J., *J. Am. Chem. Soc.,* to be published.
9. Curtiss, L. A., and Frurip, D. J., *Chem. Phys. Lett.* **75,** 69 (1980).
10. Miller, J. C., Ault, B. S., and Andrews, L., *J. Chem. Phys.* **67,** 2478 (1977).

11. Moore, C. E., *Circ. US Nat'l Bur. Stand.* **467,** (1949).
12. Brith, M., and Schnepp, O., *J. Chem. Phys.* **39,** 2714 (1963).
13. Knight, L. B., and Ebner, M. A., *J. Mol. Spectr.* **61,** 412 (1976).
14. Mowery, R. L., Miller, J. C., Krausz, E. R., Schatz, P. N., Jacobs, S. M., and Andrews, L., *J. Chem. Phys.* **70,** 3920 (1979).
15. Miller, J. C., Mowery, R. L., Krausz, E. R., Jacobs, S. M., Kim, H. W., Schotz, P. N., and Andrews, L., *J. Chem. Phys.* **74,** 6349 (1981), and references therein.
16. Huber, K. P., and Herzberg, G., *Constants of Diatomic Molecules,* Van Nostrand, New York, 1979.
17. Curtiss, L. A., and Frurip, D. J., private communication.
18. Gimarc, B. M., *Molecular Structure and Bonding,* Academic, New York, 1979.
19. Jackson, D. D., *Thermodynamics of the Gaseous Hydroxides,* UCRL-51137, (1975) Lawrence Livermore Laboratory, Livermore, California, 94500.
20. Gaydon, A. G., *Dissociation Energies and Spectra of Diatomic Molecules,* Chapman and Hall, London, 1968.
21. Douglas, M. A., Hauge, R. H., and Margrave, J. L., *High Temp. Sci.* **16,** 35 (1983).
22. Trenary, M., Schaefer III, H. F., and Kollman, P. A., *J. Chem. Phys.* **68,** 4047 (1978).

0018–1536/84/1700–0237$02.60

Reactions of Iron Atoms and Iron Dimers with Methane, Ammonia, and Hydrogen Fluoride in Low Temperature Matrices

J. W. KAUFFMAN, R. H. HAUGE, AND J. L. MARGRAVE*

Department of Chemistry, Rice University, Houston, Texas 77251

Received February 7, 1984; Accepted February 8, 1984

ABSTRACT

Infrared matrix isolation studies of the cocondensation of iron atoms with CH_4, NH_3, and FH indicate only formation of $Fe{\cdot}NH_3$ and Fe·FH adducts. Photolysis of the matrices with visible/near UV light causes iron to insert into the HX bond, where X is CH_3, NH_2, or F, to produce the $HFeCH_3$, $HFeNH_2$, and HFeF species. Molecular vibrational frequencies and mode assignments are given in accompanying tables. Tentative assignments were also made for HFe_2F, $HFeNH_2(NH_3)$, and HFe_2NH_2.

Index Entries: Iron atoms and dimers, reactions in low temperature matrices; methane, reaction with iron and iron dimers; ammonia, reaction with iron and iron dimers; hydrogen fluoride, reaction with iron and iron dimers

INTRODUCTION

We have previously reported that metal atoms react with water molecules at 15°K to form an $M{\cdot}OH_2$ adduct (*1*). On photolysis, the adduct rearranges to form the HMOH species by oxidative insertion of the metal into the O—H bond of water. The results of the present study show that metal atoms react with other molecules, isoelectronic with water, in

*Author to whom all correspondence and reprint requests should be addressed.

a similar manner—methane, ammonia, and hydrogen fluoride. Only methane, which lacks a nonbonding electron pair, did not form an identifiable adduct with iron. In agreement with our work, Barrett found that iron atoms did not react with methane on cocondensation (*8*). Formation of an NH_3 adduct is not surprising in view of previous EPR experiments and theoretical calculations for lithium and ammonia (*2,3*). These studies showed that lithium metal atoms form a stable adduct with ammonia at 15°K. Mossbauer matrix isolation studies by Barrett (*4*) also provide evidence that iron metal atoms and ammonia react to form $Fe \cdot NH_3$ adducts. In addition, theoretical calculations by Trenary indicate that Li metal atoms form an adduct with hydrogen fluoride (*5*).

All of the molecules, including methane, underwent a metal atom insertion reaction on photolysis to form similar insertion products, HMX, where X is CH_3, NH_2, or F. A preliminary report of the insertion of Fe atoms in CH_4 to form $HFeCH_3$ has been published (*6*) and confirmed by Ozin et al. (*7*). We report here the infrared spectra and vibrational assignments of the iron metal reaction products with CH_4, NH_3, and FH. Also, the reaction pathways that lead to the insertion products are described.

EXPERIMENTAL

Methane, ammonia, and hydrogen fluoride were cocondensed with iron atoms/dimers in excess argon for 1 or 2 h on a mirrored copper surface cooled to ~15 K with an Air Products closed-cycle helium refrigerator, Model CSW-202. Isotopically substituted methane gases ($^{13}CH_4$, $^{12}CD_4$) and ammonia gases ($^{15}NH_3$, $^{14}NH_2D$, $^{14}NHD_2$, $^{14}ND_3$) were used to facilitate the assignment of vibrational bands and help identify reaction products. All of these gases were purchased from US Services, Inc., at 99.9% purity. HF gas (99.9%) was purchased from Matheson Gas Co. All gases were further purified with several freeze–pump–thaw cycles using liquid N_2 and by passing them through a cold trap before use. Matheson argon gas (99.9%) was used as the matrix material and was passed through a liquid N_2 trap before deposition. The gases effused through 0.125 in. stainless steel orifices, about 1.5 in. away from the cold surface, and the flow rates were regulated with in-line pressure thermocouple gages and needle valves.

The iron was vaporized from a 0.25 in. id alumina crucible over the range 1230–1460°C, as measured with an optical pyrometer. Relative metal concentrations were determined approximately by matrix colors, as listed in Table 1. A detailed description of the furnace, matrix isolation apparatus, and experimental procedure has been previously reported (*1*).

TABLE 1
Matrix Colors Showing Relative Fe Concentrations

Furnace temperature, °C	Matrix color	
1230	Light orange	Increasing Fe concentration ↓
1300	Orange	
1350	Reddish-orange	
1400	Red	
1440	Dark red	
1460	Black	

RESULTS

Methane

Iron metal atoms/dimers were cocondensed with methane gas in excess argon and in a pure methane matrix. High concentrations of methane were needed when using an argon matrix to produce strong product absorption bands. Although no absorption band was observed that could be attributed to an adduct, the "e" reaction product bands appeared on photolysis of the matrix with UV light. The "e" bands and frequencies are shown in Fig. 1 and are listed in Table 2.

Product bands were quite intense in the methane matrix, and five vibrational modes were assigned to the "e" species. The ~2900 cm^{-1} bands are obviously assignable to the C—H stretching mode, shifting ~738 and ~9 cm^{-1} on deuterium and carbon-13 substitution. The isotopic shifts of the other four vibrational mode frequencies are illustrated in Fig. 1. The 1653 cm^{-1} band shows a large deuterium shift and a slight carbon-13 dependence, indicating it is mostly a hydrogen motion. It lies in close proximity and shows isotopic shifts that are analogous to the Fe—H stretching mode in the HFeOH molecule (*9*), so it is similarly assigned to an Fe—H stretching mode in the "e" molecule. The 1151 cm^{-1} band shows strong carbon-13 and deuterium shifts of ~10 and ~271 cm^{-1}. It also appears in the region of the methane v_4 gaseous CH_2 bending mode at 1306 cm^{-1}, thus suggesting assignment to a CH_2 bending mode. The 549 cm^{-1} band is best assigned to an Fe—C—H rocking mode, shifting about ~4 and ~129 cm^{-1} on carbon-13 and deuterium substitution, respectively. The 521 cm^{-1} band has a ~13 cm^{-1} carbon-13 shift and 46 cm^{-1} deuterium shift, suggesting assignment to the Fe—C stretching mode. The "e" product bands are assigned to $HFeCH_3$ molecules, resulting from an oxidative insertion of iron into the methane C—H bond.

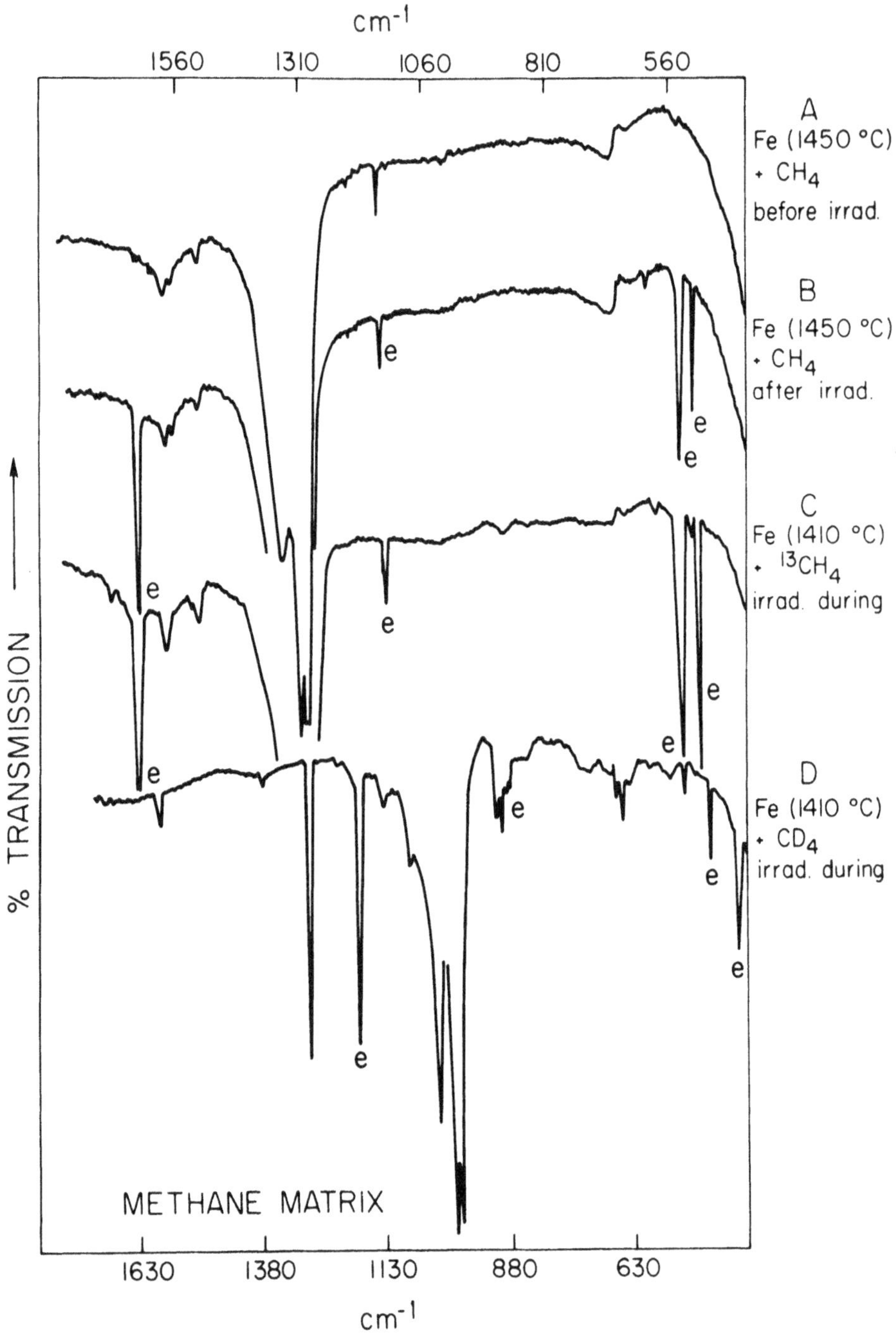

Fig. 1. Infrared spectra of iron–methane reaction product $HFeCH_3$ in methane matrix: (A) matrix with high iron metal concentration before photolysis; (B) photolysis of matrix A with UV light; (C) photolysis of $^{13}CH_4$ matrix + high iron metal concentration with UV light; (D) photolysis of $^{12}CD_4$ matrix + high iron metal concentration with UV light.

TABLE 2
Observed (obs.) and Calculated[a] (calc.) Frequencies (cm^{-1}) for Iron + Methane Reaction Products in Methane and Argon Matrices

"e" $HFeCH_3$

	C—H stretch, obs.	Fe—H stretch, obs.	H—C—H bend, obs.	Fe—C—H rock, obs.	F—C stretch obs.	F—C stretch calc.
			Methane Matrix			
$^{12}CH_4$	2929.8[b]	1653.4	1151.1	549.5	521.4	521.4
	2895.4			546.1[b]		
	2878.0					
$^{13}CH_4$	2921.2[b]	1652.8	1141.7	546.0	508.2	508.5
	2893.8			543.2[b]		
	2873.5					
$^{12}CD_4$	2191.4	1189.1	879.9	420.6	475.4	480.9
				417.6[b]		
			Argon Matrix			
$^{12}CH_4$		1684.2[b]		543.5	528.7	
		1675.7		540.2[b]	524.3[b]	
$^{13}CH_4$		1683.7[b]		540.3	512.0	511.3
		1675.5		537.7[b]		
$^{12}CD_4$		1210.8[b]			480.2	483.7
		1205.5				

[a]Calculated frequencies were obtained using the relationship for an *xyz* molecule, x = H, y = Fe, z = CH_3; i = isotopic molecule.

$$\frac{[v_1^2(i)v_3^2(i)]}{[v_1^2v_3^2]} = \frac{\left[\dfrac{m_y^i + m_y^i + m_z^i}{mx^i my^i mz^i}\right]^{1/2}}{\left[\dfrac{mx + my + mz}{m_x m_y m_z}\right]^{1/2}}$$

[b]Denotes most intense peak for that vibrational mode.

In argon matrices, only the vibrational bands involving iron metal motions are intense enough to assign. Isotopic behavior and mode assignments of these bands are similar to the bands in methane matrices, although the frequencies are slightly shifted by the perturbation effects of different matrices.

Ammonia

When iron atoms are cocondensed with ammonia in argon, three sets of bands appear, labeled "a," "b," and "c" in Figs. 2A and 3A. Com-

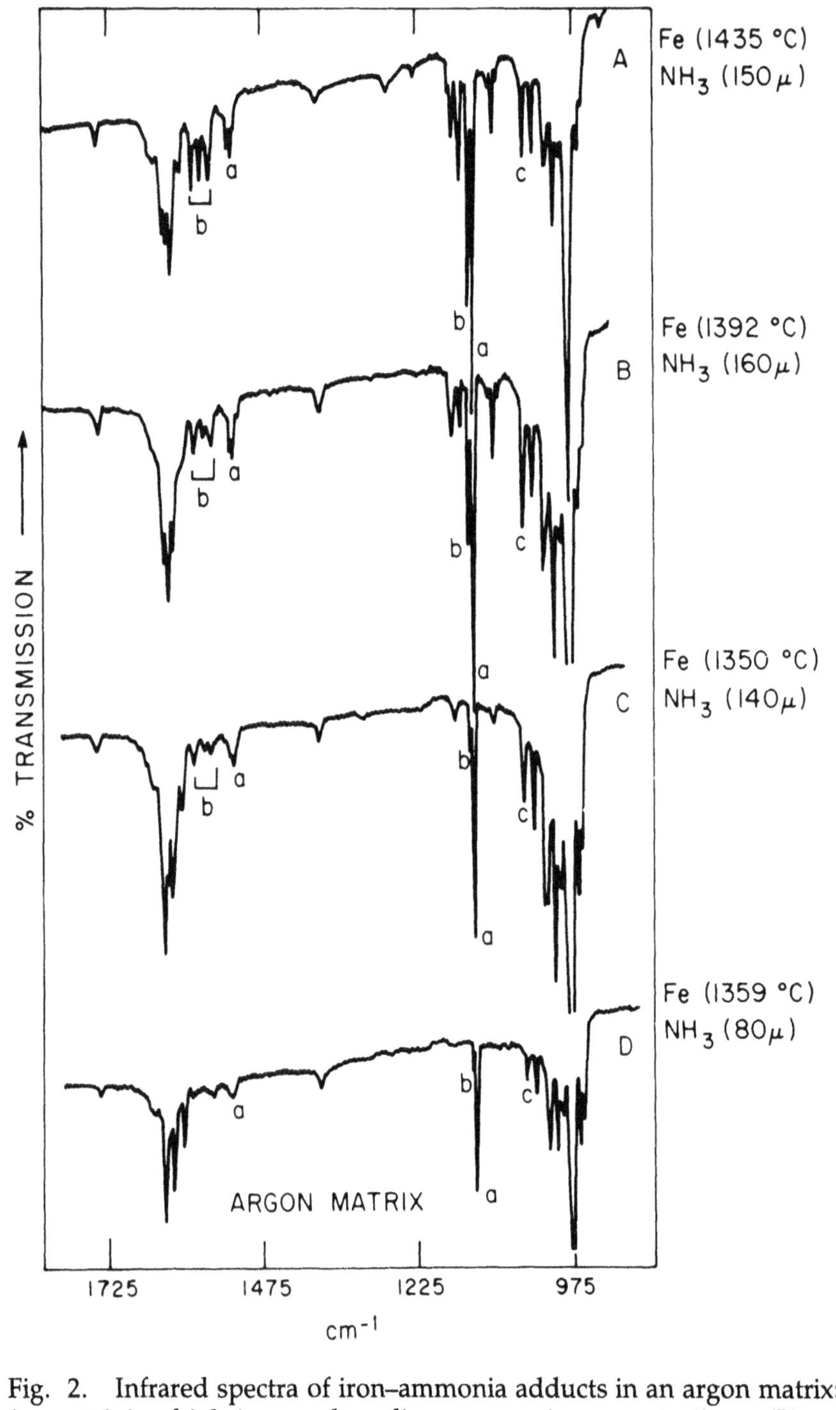

Fig. 2. Infrared spectra of iron–ammonia adducts in an argon matrix: (A) matrix containing high iron and medium ammonia concentrations; (B) matrix containing high to medium iron and medium ammonia concentrations; (C) matrix containing medium iron and ammonia concentrations; (D) matrix containing medium iron and low ammonia concentrations.

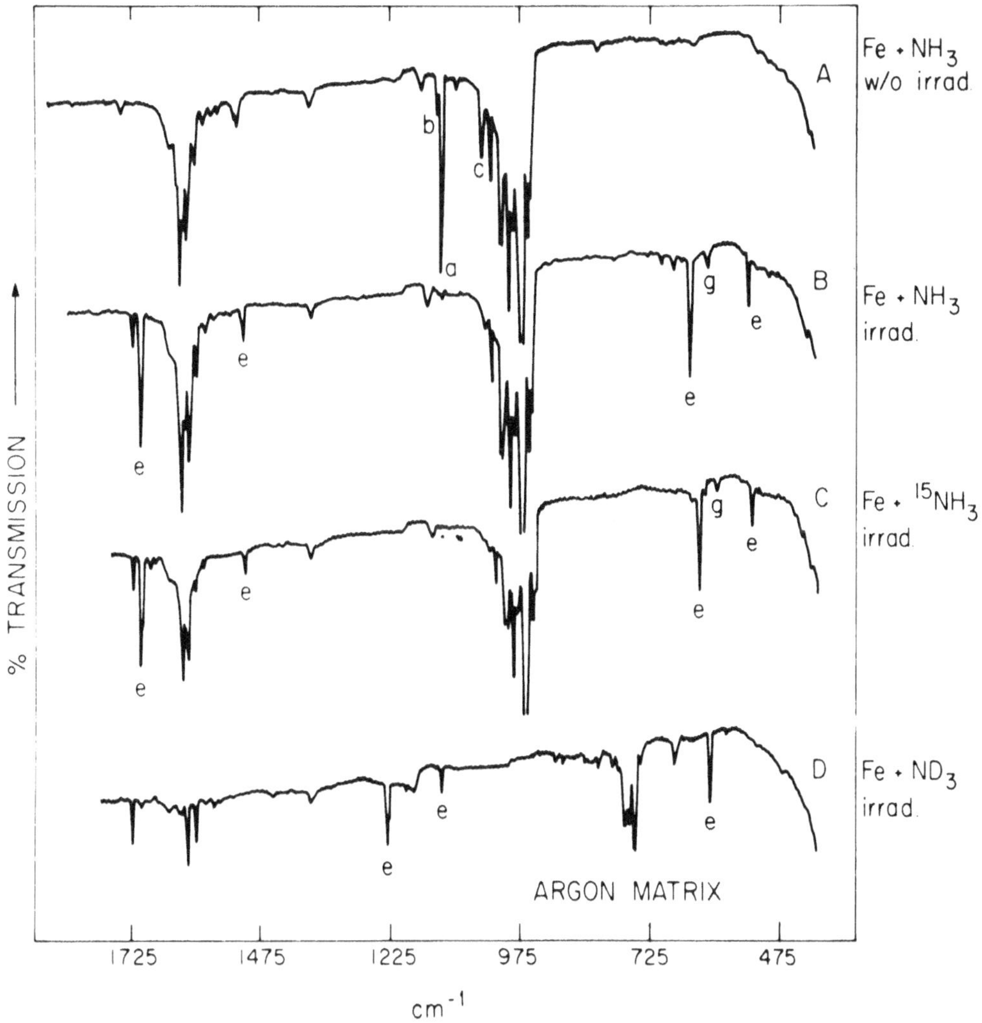

Fig. 3. Infrared spectra of iron–ammonia reaction products in an argon matrix: (A) matrix with medium iron and ammonia concentration immediately after cocondensation; (B) photolysis of matrix A with 340–300 nm light; (C) photolysis of matrix containing medium iron and $^{15}NH_3$ concentrations with 340–300 nm light; (D) photolysis of matrix containing medium iron and $^{14}ND_3$ concentrations with 340–300 nm light.

parison of the spectra in Fig. 2 shows the metal and ammonia concentration dependence of the "a," "b," and "c" products. The metal concentration is decreased (indicated by a decreasing furnace temperature) from spectrum A through C, while the ammonia concentration is held constant. The "a" bands are present at low metal concentrations and are assigned to the $Fe_1 \cdot NH_3$ adduct. The "a" adduct ammonia v_2 frequency increases $\sim$157 cm^{-1}, while the v_3 and v_4 modes decrease $\sim$50 and $\sim$98 cm^{-1}, respectively, from the free ammonia values listed in Table 3. Simi-

TABLE 3
Observed Frequencies and Vibrational Mode Assignments for Iron + Ammonia Adducts

	N—H stretch	Deg. deform.	Sym. deform
Free $^{14}NH_3$	3436	1624.8	974.7
"a" Adduct			
$Fe_1 \cdot {}^{14}NH_3$	3386	1527.1	1131.5
$Fe_1 \cdot {}^{15}NH_3$	3373	1525.9[a]	1160.3
		1521.0	1125.1[a]
$Fe_1 \cdot {}^{14}NH_2D$			1051.9
$Fe_1 \cdot {}^{14}NHD_2$			956.5
$Fe_1 \cdot {}^{14}ND_3$		1153.7	906.1
		1149.9[a]	877.8[a]
"b" Adduct			
$Fe_2 \cdot {}^{14}NH_3$		1577.4	1138.5[a]
			1100.7
$Fe_2 \cdot {}^{15}NH_3$			1131.7
			1095.1
$Fe_2 \cdot {}^{14}NHD_2$			965.0
$Fe_2 \cdot {}^{14}ND_3$		1164.0	884.1
"c" Adduct			
$Fe_1 \cdot ({}^{14}NH_3)_2$			1050.9
$Fe_1 \cdot ({}^{15}NH_3)_2$			1046.0
$Fe_1 \cdot ({}^{14}ND_3)_2$			814.0
"d" Adduct			
$Fe_3 \cdot {}^{14}NH_3$			1153.7
$Fe_3 \cdot {}^{15}NH_3$			1147.6
$Fe_3 \cdot {}^{14}ND_3$			895.9

[a]Denotes the most intense peak for that vibrational mode.

lar shifts have been observed for ammine complexes in the solid state, e.g., $[Mg(NH_3)_6]Cl_2$ (*10*), in which the metal bonds through the nitrogen. The "a" adduct frequencies with isotopically substituted ammonia $^{14}NH_3$, $^{15}NH_3$, $^{14}NH_2D$, $^{14}NHD_2$, and $^{14}ND_3$ are listed in Table 3.

The "b" bands in Fig. 2 are enhanced at higher metal concentrations, relative to the "a" bands. At higher metal concentrations, iron dimers are the first iron cluster most likely to form, so the "b" bands are assigned to the $Fe_2{\cdot}NH_3$ adduct. The "b" frequencies are listed in Table 3. Finally, at the highest iron concentrations, a new band appears at ~1154 cm^{-1}, labeled "d," possibly resulting from formation of an $Fe_3 \cdot NH_3$ adduct where X ⩾ 3. These adduct frequencies are listed in Table 3.

Spectra of matrices with a medium metal concentration and a high and low ammonia concentration are shown in Figs. 2C and 2D, respectively. The "c" band is more intense relative to the "a" bands in Fig. 2C. This suggests that the "c" adduct probably involves more than one ammonia molecule and so is tentatively assigned to the $Fe_1 \cdot (NH_3)_2$ adduct.

Spectrum A in Fig. 3 shows the "a," "b," and "c" adduct bands immediately after cocondensation. Photolysis of matrix A with 340–300 nm light causes the adducts to react, resulting in the disappearance of the adduct bands and the appearance of the "e," "f," and "g" bands. Only the "e" bands are shown in Fig. 3, since the "f" and "g" bands were generally very weak. The N—H stretching mode for the "e" product at 3381 cm^{-1} was not observed in $^{15}NH_3$ and $^{14}ND_3$ experiments for the same reason. The ~1717 cm^{-1} "e" band shows a large deuterium shift and a negligible nitrogen-15 shift, indicating assignment to an Fe—H stretching mode. The 650 cm^{-1} "e" band shows both a nitrogen-15 and deuterium shift, so it is assigned to an Fe—N stretching mode. The 537 cm^{-1} "e" band is tentatively assigned to an Fe—N—H mode. It shows a small nitrogen-15 dependence and is thought to shift below the 400 cm^{-1} spectrometer limit on deuterium substitution. The characteristics of the "e" bands can be explained by an insertion reaction of the metal atom into the N—H bond of ammonia to form $HFeNH_2$.

Photolysis of the $Fe_1 \cdot (NH_3)_2$ adduct causes it to react, forming the "f" product. The "f" product bands show red shifts comparable with the "e" bands on nitrogen-15 substitution for Fe—H and Fe—N stretching modes, and are assigned to the $HFeNH_2(NH_3)$ molecule. The "f" bands were not observed in the deuterium experiments. The "e" and "f" product bands are listed in Table 4.

Two weak "g" bands at 1687 and 581 cm^{-1} in the $^{15}NH_3$ studies are believed to arise from the reaction of the $Fe_2 \cdot NH_3$ adduct. The close proximity of these bands to the "e" product Fe—H and Fe—N stretching modes suggests like assignments. By analogy to the insertion reaction for the "e" and "f" products, the "g" bands are tentatively assigned to the metal dimer insertion product HFe_2NH_2. The "g" frequencies are listed in Table 4. Several unassigned metal dependent bands are also listed in Table 4.

Hydrogen Fluoride

When hydrogen fluoride is cocondensed with iron atoms in argon, the "a" and "f" bands appear in the spectrum, Fig. 4B, taken immediately after cocondensation. The "a" band at 3747 cm^{-1}, observed at relatively low metal concentrations, can obviously be assigned to an H—F

TABLE 4
Observed (obs.) and Calculated (calc.)[a] Frequencies (cm^{-1}) for Iron + Ammonia Reaction Products in an Argon Matrix

	N—H stretch, obs.	Fe—H stretch, obs.	H—N—H bend, obs.	Fe–N stretch obs.	Fe–N stretch calc.	Fe—N—H bend obs.
			"e" $HFeNH_2$			
$^{14}NH_3$	3381	1717.4[b]	1520.7	649.8[b]	649.8	536.8
		1714.2	1517.8[b]	649.0		
$^{15}NH_3$		1717.0[b]	1515.5	635.3	634.8	534.2
		1713.7	1512.8[b]			
$^{14}ND_3$		1236.4[b]	1131.7[b]	614.3[b]	614.1	—
		1233.8		613.4		

"f" $HFeNH_2(NH_3)$

	Fe—H stretch	Fe—N stretch	Fe—N—H bend,
$^{14}NH_3$	1698	615.0	498.6
$^{15}NH_3$	1696	601.5	496.6

"g" HFe_2NH_2

	Fe—H stretch, obs.	Fe—N stretch, obs.
$^{15}NH_3$	1687	581

Unassigned

$^{14}NH_3$	733.4	705.6
$^{15}NH_3$	715.3	694.0

[a]Calculated frequencies were obtained from the equation for an *xyz* molecule in Table 2.
[b]Denotes the most intense peak for that vibrational mode.

stretching mode, but it is red-shifted 172 cm^{-1} from the free H—F vibrational frequency. This band is assigned to the FeFH adduct. Photolysis of the matrix with 330—300 nm light causes the adduct to react, forming the "e" insertion product HFeF. The 1753 and 650 cm^{-1} "e" bands are close to the Fe—H and Fe—OH stretching frequencies found for the HFeOH molecule, which suggests assignment to the Fe—H and Fe—F stretching frequencies, respectively, of HFeF. The "a" and "e" frequencies are listed in Table 5.

The "f" frequencies are in the same region as the "e" bands, suggesting they might be assigned to Fe—H and Fe—F stretching modes, also. It is likely that the "f" bands result from insertion of iron dimers into hydrogen fluoride to give HFe_2F on cocondensation since this product is favored by high metal concentrations. The spontaneous reaction of Fe_2 with HF agrees with the absence of an $Fe_2 \cdot FH$ adduct band in any of the experiments. The HFe_2F frequencies are listed in Table 5.

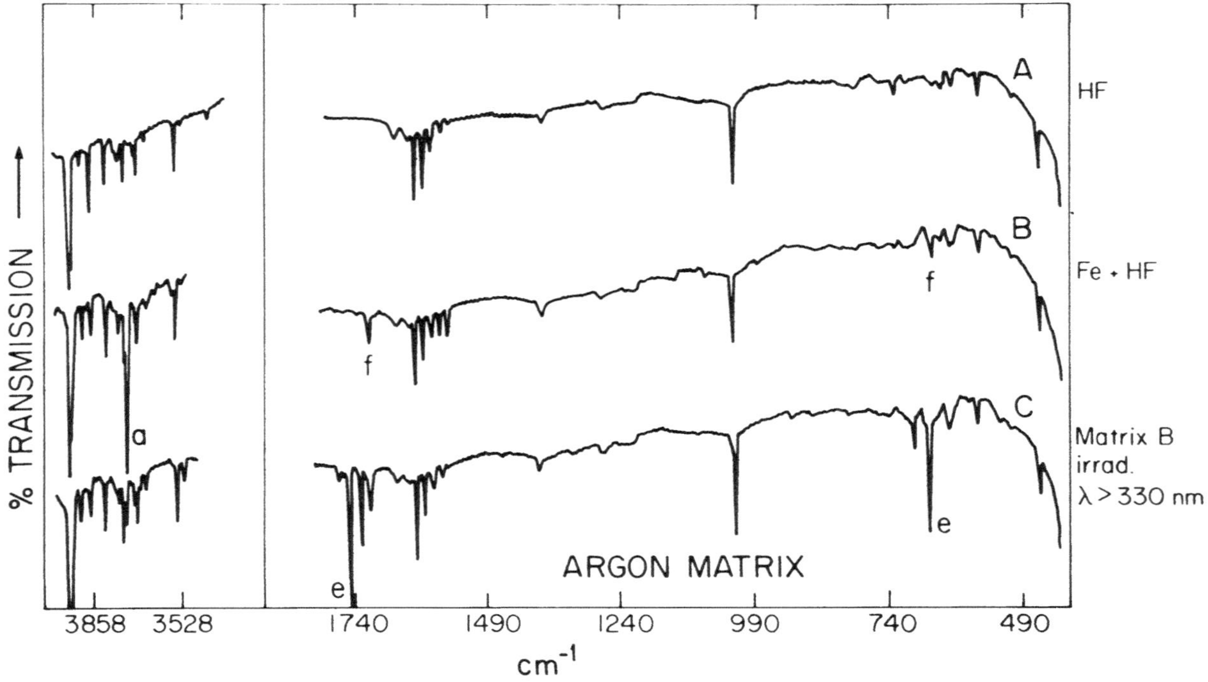

Fig. 4. Infrared spectra of iron + hydrogen fluoride reaction products in an argon matrix: (A) hydrogen fluoride in an argon matrix; (B) medium iron metal concentration + hydrogen fluoride; (C) photolysis of matrix B with 330–300 nm light.

TABLE 5
Observed Frequencies (cm^{-1}) of Iron + Hydrogen Fluoride Reaction Products

"a" Adduct			
			H—F stretch
H—F			3919.5
Fe · FH			3746.9
"e" HFeF			
	Fe—H stretch, obs.	Fe—F stretch, obs.	
	1774.0	652.4	
	1753.2[a]	650.0[a]	
"f" HFe_2F			
	Fe—H stretch, obs.	Fe—F stretch, obs.	Unassigned
	1713.9	651.6	609.2

[a]Denotes the most intense peak for that vibrational mode.

DISCUSSION

Iron metal atoms reacted with ammonia and hydrogen fluoride molecules on cocondensation at 15°K, forming the $Fe{\cdot}NH_3$ and Fe·FH adducts. Complexing with the metal caused the ammonia v_2 frequency to increase ~157 cm^{-1}, while the v_3 and v_4 frequencies decreased 50 and 98 cm^{-1}, respectively. Blue-shifting of the v_2 frequency is believed to result from restriction of the ammonia hydrogen atoms from moving out of plane by the iron. The decrease in the v_3 and v_4 frequencies is apparently caused by weakening of the N—H bond in the adduct, by analogy with the $Mg(NH_3)_6Cl_2$ complex studied by Fujita (*10*). Previous EPR matrix isolation experiments with lithium and ammonia show the metal atom bonding through the nitrogen similar to metal atom–water adducts (*2,3*). These results are supported by the theoretical calculations of Nicely and Dye (*11*) and Trenary (*5*), which predict the metal is bonded through the nitrogen with an Li—N bond strength of ~14 kcal/mol. Mossbauer studies by Barrett (*4*) also showed adduct formation between iron metal atoms and ammonia at low temperatures. These matrix isolation experiments also suggest that the $Fe_2{\cdot}NH_3$ and $Fe_1{\cdot}(NH_3)_2$ adducts are formed in analogy to the previously studied $M_2{\cdot}OH_2$ (*1*) and $Li{\cdot}(OH_2)_2$ adducts (*3*).

A metal atom–methane adduct was not detected, but photolysis of the matrix did lead to a metal atom insertion reaction, forming the $HFeCH_3$ molecule. It is likely that the metal atom and methane molecule react on photolysis by initially forming an excited-state intermediate adduct which then rearranges to form the insertion product.

The iron dimer apparently needs less energy to react with an HF molecule, inserting immediately on cocondensation, in contrast to the metal atom, which forms a stable adduct. This is in agreement with previous studies showing lower energy light is needed to induce the metal dimer adduct to react compared to the metal atom adduct (*1*). In contrast Ozin et al. (*7*) did not find evidence for Fe_2 reactivity with CH_4 at ~15°K, which would explain the absence of Fe_2-related bands in our methane experiments.

In summary, the following metal atom and dimer reactions with CH_4, NH_3, and HF have been observed:

$$X = NH_2, F \quad (1)\ HX + Fe \xrightarrow{15°K} Fe{\cdot}XH \xrightarrow{h\nu} HFeX$$

$$X = NH_2 \quad (2)\ 2HX + Fe \xrightarrow{15°K} Fe{\cdot}(XH)_2 \xrightarrow{h\nu} HFeX(HX)$$

$$X = CH_3 \quad (3)\ HX + Fe \xrightarrow{h\nu} Fe{\cdot}XH \rightarrow HFeX$$

$$X = NH_2 \quad (4)\ HX + Fe_2 \xrightarrow{15°K} Fe_2{\cdot}XH \xrightarrow{h\nu} HFe_2X$$

$$X = F \quad (5)\ HX + Fe_2 \xrightarrow{15°K} HFe_2X$$

ACKNOWLEDGMENTS

The authors are pleased to acknowledge support of this work by the Robert A. Welch Foundation and the National Science Foundation.

REFERENCES

1. Hauge, R. H., Kauffman, J. R., and Margrave, J. L., *J. Am. Chem. Soc.* **102,** 6005 (1980).
2. Hauge, R. H., Meier, P. F., and Margrave, J. L., *Ber. Bunsenges. Phys. Chem.* **82,** 102 (1978).
3. Meier, P. F., Hauge, R. H., and Margrave, J. L., *J. Am. Chem. Soc.* **100,** 2108 (1978).
4. Barrett, P. H., and Pasternak, M., *J. Chem. Phys.* **71,** 3837 (1979).
5. Trenary, M., and Schaefer, H. F., *J. Am. Chem. Soc.* **99,** 3885 (1970).
6. Billups, W. E., Konarski, M. M., Hauge, R. H., and Margrave, J. L., *J. Am. Chem. Soc.* **102,** 7393 (1980).
7. (a) Ozin, G. A., and McCaffrey, J. G., *J. Am. Chem. Soc.* **104,** 7351 (1982); (b) Ozin, G. A., and McCaffrey, J. G., *Inorg. Chem.* **22,** 1397 (1983).
8. Barrett, P. H., Pasternak, M., and Pearson, R. G., *J. Am. Chem. Soc.* **101,** 222 (1979).
9. Kauffman, J. W., Hauge, R. H., and Margrave, J. L., *High Temp. Sci.*, in press.
10. Fujita, J., Nakamoto, K., and Kobayashi, M., *J. Am. Chem. Soc.* **78,** 3295 (1956).
11. Nicely, V. A., and Dye, J. L., *J. Chem. Phys.* **52,** 4795 (1970).

0018–1536/84/1700–0251$03.00

Cadmium Activities of Silver–Cadmium Alloys Determined from Measurements on EMF Cells Involving Displacement Reactions

BARTON L. HOUSEMAN*

and

DONALD R. CONANT

Los Alamos Consultants, 133 La Senda Road, Los Alamos, NM 87544

Received September 23, 1983; Accepted December 31, 1984

ABSTRACT

Cadmium activities have been obtained for dilute solutions of silver in cadmium at 776 K from emf measurements using the cell: W-Cd|CdI_2 + Cd (a_{Cd} = 1)||CdI_2 + AgI + Cd (a_{Cd} < 1)|Cd, Ag–W. When the displacement reaction $2Ag + CdI_2 \rightleftarrows Cd + 2AgI$ is taken into account, the calculated cadmium activities follow the equation $a_{Cd} = 1 - N_{Ag} - 0.607\ N^2_{Ag}$ with a standard deviation of 0.00001 for those measurements (taken by the most precise method) used in the least squares fit. The largest deviation from the curve for all points measured was 0.0006 or 0.9 g/cal in $\bar{G}_{Cd}$.

Index Entries: Cadmium activities of Ag–Cd alloys; activities, of Cd in Ag–Cd alloys; silver–cadmium alloys, Cd activities in; alloys, Cd–Ag; emf cells, Cd activities in Ag–Cd from; displacement reactions, of Cd in Ag–Cd alloys.

*Author to whom all correspondence and reprint requests should be addressed.

INTRODUCTION

This work constitutes part of an investigation at this laboratory concerning the corrections that must be applied to emf measurements in order to obtain correct values for thermodynamic activities.

For cells containing cadmium alloys in the presence of molten CdI_2 electrolyte, the evaluation of cadmium activities from the emf data depends upon the value assigned to n in the equation $-nFdE = RT\, d \ln a_{Cd}$. The value of n, the average charge per mole of cadmium transferred, depends upon the solubility of cadmium in its iodide and upon the types, concentrations, and mobilities of the various chemical species present in the electrolyte.

When the solubility of cadmium in its iodide at 775 K is assumed to be 6.5 mol%, consistent with the cooling curve measurements of Topol and Landis (*17*), and when the mobilities of the various cadmium ion species are assumed similar, as is true for the chloride system investigated by Herzog and Klemm (*9*), emf data taken on the Ni–Cd system yield activities that agree with Raoult's limiting law (*2*). The value estimated for n from these assumptions is 1.87 at the limit of infinite dilution. This value agrees with a direct experimental determination of n by Houseman (*10*).

For Ag–Cd alloys a further complication arises because a significant amount of silver will displace the cadmium in the cadmium iodide electrolyte by the reaction

$$2Ag + CdI_2 \rightleftarrows Cd + 2AgI$$

EXPERIMENTAL

The lower half of an H-cell, as shown in Fig. 1, was surrounded by a furnace. The furnace contained two heaters outside three nesting nickel-plated copper cylinders separated by air gaps with the outer cylinder grounded to reduce stray emf effects.

Temperatures were measured directly with Pt/Pt–10% Rh thermocouple.

Cadmium (99.999% purity by manufacturer's analysis) was boiled with CdI_2 to free it from electrochemically important impurities and from oxide. Silver (99.9%) surfaces were cleaned by washing first with reagent-grade acetone, then with reagent-grade petroleum ether. The tungsten electrodes were cleaned electrolytically by using them to pass ac current through concentrated NaOH solution, and washing them with distilled water. Reagent CdI_2 was loaded into a side tube, boiled under vacuum, then cooled. The cell was back-filled with argon that had been purified by distillation. After closing the stopcock, the cell was removed from the vacuum system and loaded into the furnace. After the cadmium

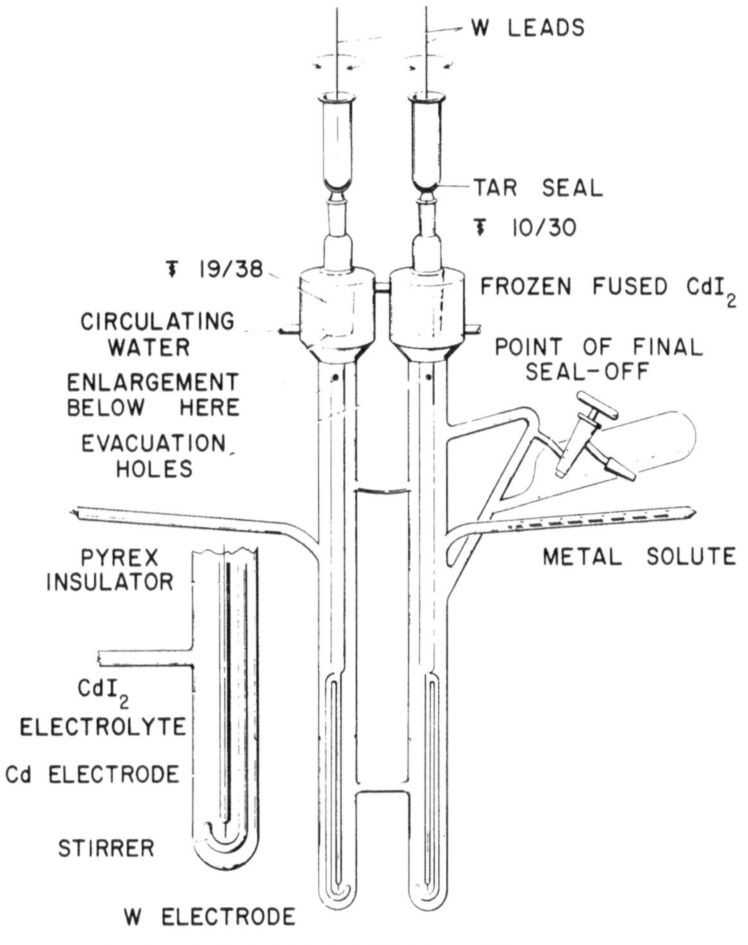

Fig. 1. Emf concentration cell.

in the legs melted, the CdI_2 was fused by torch and allowed to run into the H-cell legs.

The metals at the bottom of the tubes were relatively good conductors compared to the glass, and the conduction area of the metal relative to the cross-section of the tube walls was large. Therefore, the cadmium and the alloy should have been at a uniform temperaure throughout as well as the electrolyte in direct contact with the metal. However, a slight gradient across the electrolyte, especially across the crossarm containing a small amount of electrolyte would be expected, causing the permanent pure cadmium electrode to be at a slightly different temperature than the alloy electrode.

Thermodynamically, the emf arising from this cell may be considered the sum of two emf values, one arising from Cd(T_2) vs Cd(T_1) and the other from $Cd(T_1)$ vs alloy (T_1). The second emf is required for determining the activity of the cadmium at temperaure T_1. Since initially no solute had been added to the alloy side, the second emf was zero, and the first was just the initial emf measured and has been referred to as the bias. Since this value is essentially constant, it must be subtracted from emf readings obtained for the alloys to obtain the value that arises from the cell $Cd(T_1)$ vs alloy (T_1).

After the cell was allowed to equilibrate for 2 d with occasional stirring, an initial emf bias between the two pure cadmium legs was recorded.

For Run 1 the cell was then tipped so that when the sidearem was tapped, a piece of silver would slide down into the alloy leg. For Run 2 a ground glass joint allowed the leg to be rotated to obtain the angle necessary for tapping a solute piece into the alloy leg. During Run 2 the main cell body was fixed rigidly to the furnace so that the bias would remain more constant. The technique used in Run 2 markedly improves the precision of the data, as has been shown using a somewhat similar technique on the Ni–Cd system (2).

After each piece was added, both legs of the cell were stirred and the emf observed until no further change occurred. The thermocouple and cell emf values were then recorded.

Because of the improvement in experimental technique, the precision of the measurements of Run 2 was particularly good.

For Run 1, cell emfs were recorded to the nearest 0.1 μV and thermocouple emfs to the nearest μV. For Run 2, cell and thermocouple emfs were both recorded to the nearest 0.01 μV, although the uncertainty in absolute temperatures was probably as much as ±0.1°K. A correction factor of 1.37 μV was subtracted from the measured thermocouple emf to bring the equation [from ref. (*14*)]

$$E = -309.17 + 8.29558t + 0.00144103t^2 + 0.0000001634t^3$$

into agreement with more recent reference tables (*15*).

Cadmium was weighed to the nearest 0.1 mg and silver to the nearest 0.01 mg. The CdI_2 over each leg was estimated to be about 35 g. The uncertainty in cadmium added was less than one part in 100,000. For the most dilute silver piece added, the uncertainty was no better than one part in 3000. However, the uncertainty in total silver added decreased for each succeeding piece since the weighings were made on groups rather than on individual pieces after the first, i.e., after weighing the first piece, a second piece was placed on the balance pan together with the first piece, and the two were weighed together; then the third piece was added to the first two on the pan. The uncertainty in CdI_2 over each leg was perhaps as high as 10%.

RESULTS AND CALCULATIONS

Table 1 gives the results of measurements on the silver–cadmium system. The experimental values of cell emf, temperature, and material added are known to a high degree of accuracy as well as precision. The calculated value for alloy composition, electrolyte composition, and activity are subject to further uncertainties, however, that result from the significant solubility of cadmium in CdI_2 and from the significant displacement of cadmium by silver in the iodide electrolyte.

For the cell

Anode	A	B	Cathode
Cd	Cd (N^A_{Cd}) CdI_2 ($N^A_{CdI_2}$)	Cd (N^B_{Cd}) CdI_2 ($N^B_{CdI_2}$) AgI (N^B_{AgI})	Cd (N_{Cd}) Ag (N_{Ag})

the cadmium dissolved in the electrolyte may be represented as some general species Cd_j^{i+} where i represents the charge and j the number of cadmium nuclei associated with the dissolved species. the term N_x^M refers to the mole fraction of species x in solution M. The reactions for the above cell are:

Cathode

$$(1 - y)/(2j - i)\ Cd^{2+}\ (B) + (1 - y)e \rightarrow (1 - y)/(2j - i)\ Cd_j^{i+}\ (B)$$

$$yAg^+(B) + ye \rightarrow yAg(B)$$

Junction

$$(t_1/i)\ Cd_j^{i+}\ (A) \rightarrow (t_1/i)\ Cd_j^{i+}\ (B)$$

$$(t_2/2)\ Cd^{2+}\ (A) \rightarrow (t_2/2)\ Cd^{2+}\ (B)$$

$$t_3\ I^-(B) \rightarrow t_3\ I^-(A)$$

$$t_4\ Ag^+(A) \rightarrow t_4\ Ag^+(B)$$

Anode

$$[1/(2j - i)]\ Cd_j^{i+}\ (A) \rightarrow [j/(2j - i)]\ Cd^{2+}\ (A) + e$$

t_1, t_2, t_3, and t_4 being the transference numbers of Cd_j^{i+}, Cd^{2+}, I^-, and Ag^+, respectively, as defined relative to an arbitrary frame of reference.

Because cadmium dissolves in its halide to a significant extent, there is a continual irreversible flow of cadmium from the anode to the cathode. There is also an irreversible flow of Ag^+ ions. These irreversible fluxes cannot be handled by the thermodynamics of equilibrium, and

TABLE 1
Activity of Cadmium Alloyed with Silver: emf Measurements

Point no.	Silver, mol	Cell emf,[a] μV	Temp.,[b] K	Cadmium alloy mole fraction[c]	Cadmium activity	$\frac{1 - a_{Cd}}{N_{Ag}}$	Deviation of a_{Cd} from calculation[d]
			Run 1				
1	0.0004243	52.2	776.43	0.9985946	0.9985370	1.04101	
2	0.0008982	109.3	774.81	0.9970297	0.9969325	1.03273	
3	0.0015610	182.5	776.53	0.9948497	0.9948944	0.99131	
4	0.0022867	269.9	776.53	0.9924734	0.9924579	1.00206	
5	0.0030324	359.4	776.43	0.9900443	0.9899674	1.00772	
6	0.0038257	456.1	776.22	0.9874729	0.9872808	1.01533	
7	0.0046670	549.0	776.53	0.9847610	0.9847146	1.00305	
8	0.0055261	653.0	777.34	0.9820074	0.9818624	1.00806	
9	0.0066855	791.5	777.03	0.9783162	0.9780466	1.01244	
10	0.0078560	922.1	776.33	0.9746180	0.9744446	1.00683	
11	0.0099404	1167.3	777.34	0.9681030	0.9677937	1.00970	
12	0.0124488	1468.5	778.66	0.9603796	0.9597090	1.01693	
13	0.0150354	1767.2	777.34	0.9525459	0.9516216	1.01948	

14	0.0176957	2078.0	777.34	0.9446240	0.9433429	1.02314	
				Run 2			
15	0.0002611	30.16	776.16	0.9991447	0.9991542	0.98891	
16	0.0006359	73.40	776.12	0.9979193	0.9979426	0.98882	
17	0.0012080	140.81	776.41	0.9960546	0.9960580	0.99913	
18	0.0025951	303.55	776.44	0.9915635	0.9915204	1.00512	0.0000000
19	0.0034050	398.43	776.47	0.9889601	0.9888842	1.00688	−0.0000019
20	0.0041608	486.70	776.44	0.9865433	0.9864366	1.00793	0.0000032
21	0.0050811	594.53	776.43	0.9836168	0.9834547	1.00990	0.0000008
22	0.0062702	734.16	776.45	0.9798616	0.9796067	1.01266	−0.0000087
23	0.0074504	872.46	776.47	0.9761636	0.9758092	1.01487	−0.0000095
24	0.0099627	1165.66	776.46	0.9683850	0.9678027	1.01842	0.0000244
25	0.0125577	1471.23	776.50	0.9604825	0.9595263	1.02420	−0.0000083

[a] All cell emf values increased by constant 16.00 μV for Run 1 and by 6.44 μV for Run 2 because of slight temperature differences between two legs.

[b] Measured with Pt/Pt–10% Rh thermocouple against ice junction.

[c] 0.307636 mol of Cd in Run 1 and 0.311103 mol in Run 2 were added to alloy leg. 0.0956 mol of CdI_2 was estimated to be over the alloy leg in each run. The solubility of Cd in CdI_2 was assumed to be $N_{Cd}^{elect.} = 0.065\ N_{Cd}^{alloy}$.

[d] Calculated using $a_{Cd}^{calc.} = 1 - N_{Ag} - 0.607\ N_{Ag}^2$.

new assumptions are required to deal with the problem. For such steady-state irreversible processes, Thomson's hypothesis is assumed to apply. This hypothesis, which originated with W. Thomson (Lord Kelvin) (*16*) in his treatment of thermoelectricity, assumes that those fluxes that occur and cannot be "controlled" (i.e., stopped, allowed to occur, or reversed) under the experimental conditions of interest may be neglected under steady state conditions. Other fluxes, which can be "controlled," may be considered "reversible" and may be treated by the usual thermodynamic methods pertaining to reversible processes at true equilibrium. For the entropy changes associated with those processes that are considered "reversible," the equation

$$\sum_i dS_i = 0$$

is assumed to apply. The theory of Onsager (*12, 13*), which has been clarified by deGroot (*4–7*), is in agreement with Thomson's hypothesis for the steady state (*8*) and is the basis for the present thermodynamic theory of irreversible processes.

The primary fluxes that are present in the cells, and that are not "controllable" by external means once the cells have been set up, are those arising from the concentration gradients of Cd and Ag^+ in the electrolyte. The fluxes that we call "reversible" are those arising from the passage of electrical current through the electrodes. The "irreversible" fluxes occur as a result of diffusion at constant temperature and pressure.

Consider a series of cells varying continuously in composition from *A* to *B* as shown in Fig. 2. Each cell is assumed to be large enough so that the transfer of one mole does not significantly change the concentration.

If there is only an infinitesimal difference between the concentration of any given species in side *A'* of one cell and the concentration of the same species in side *B'*, then passing sufficient current to transfer one

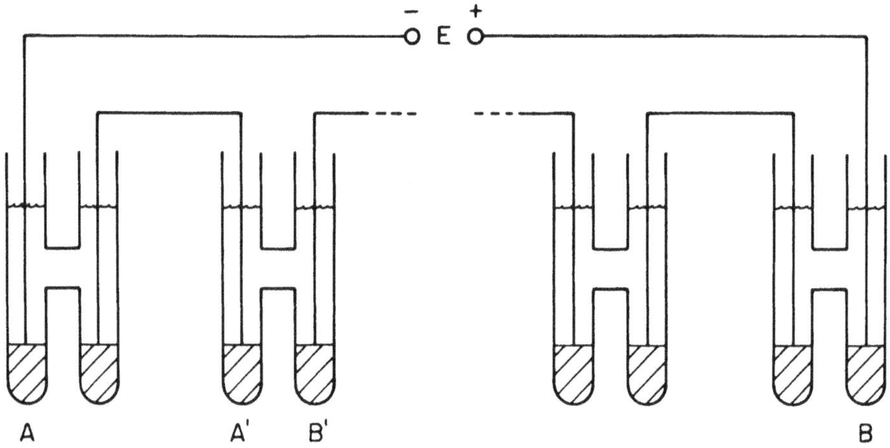

Fig. 2. A series of cells varying continuously in composition from *A* to *B*.

mole of cadmium from the more concentrated to the more dilute side will bring about a loss in free energy dG in the cell.

For any one of the cells, since Cd_j^{i+} *is* $(i/2)Cd^{2+} + [(2j - i)/2]Cd$, and $t_3 = 1 - t_1 - t_2 - t_4$, then the cell equations may be written as

$$1/2\ [1 - y + (2j - i/i)t_1]\ Cd(A') \rightarrow 1/2\ [1 - y + (2j - i/i)t_1]\ Cd(B')$$

$$[(1 - y - t_1 - t_2)/2]\ CdI_2(B') \rightarrow [(1 - y - t_1 - t_2)/2]\ CdI_2(A')$$

$$yAg(A') \rightarrow yAg(B')$$

$$(y - t_4)AgI(B') \rightarrow (y - t_4)AgI(A')$$

From thermodynamics,

$$dG = \Delta n_{Cd}\, d\bar{G}_{Cd} + \Delta n_{CdI_2}\, d\bar{G}_{CdI_2} + \Delta n_{AgI}\, d\bar{G}_{AgI} + \Delta n_{Ag}\, d\bar{G}_{Ag} \qquad (1)$$

However, because of equilibrium within the electrolyte, across the two-phase boundary, and within the alloy, three restrictions apply:

$$N'_{Cd}\, d\bar{G}_{Cd} + N'_{CdI_2}\, d\bar{G}_{CdI_2} + N'_{AgI}\, d\bar{G}_{AgI} = 0 \qquad (2)$$

$$d\bar{G}_{CdI_2} + 2d\bar{G}_{Ag} - d\bar{G}_{Cd} - 2d\bar{G}_{AgI} = 0 \qquad (3)$$

$$N_{Cd}\, d\bar{G}_{Cd} + N_{Ag}\, d\bar{G}_{Ag} = 0 \qquad (4)$$

where the primes refer the electrolyte. Thus,

$$dG = \frac{1}{2N'_{CdI_2} + N'_{AgI}} \Bigg[(2N'_{CdI_2} + N'_{AgI})\, \Delta n_{Cd} + \left(2N'_{AgI}\frac{N_{Cd}}{N_{Ag}} + N'_{AgI} - 2N'_{Cd}\right) \Delta n_{CdI_2} - \left(N'_{Cd} + 2N'_{CdI_2}\frac{N_{Cd}}{N_{Ag}} + N'_{CdI_2}\right) \Delta n_{AgI} - (2N'_{CdI_2} + N'_{AgI})\frac{N_{Cd}}{N_{Ag}}\, \Delta n_{Ag} \Bigg] d\bar{G}_{Cd} \qquad (5)$$

Since

$$\Delta n_{Cd} = \frac{1}{2}\left(1 - y + \frac{2j - i}{i}\, t_1\right) \qquad (6)$$

$$\Delta n_{CdI_2} = -\frac{1 - y - t_1 - t_2}{2} \qquad (7)$$

$$\Delta n_{AgI} = t_4 - y \qquad (8)$$

$$\Delta n_{Ag} = y \qquad (9)$$

we may write

$$dG = \frac{1}{2N'_{CdI_2} + N'_{AgI}} \left\{ N'_{CdI_2} + N'_{Cd} - N'_{AgI}\frac{N_{Cd}}{N_{Ag}} + \left[N'_{AgI}\frac{N_{Cd}}{N_{Ag}} + \frac{N'_{AgI}}{2} - N'_{Cd} + \frac{2j-i}{i}\left(N'_{CdI_2} - \frac{N'_{AgI}}{2}\right)\right]t_1 + \left(N'_{AgI}\frac{N_{Cd}}{N_{Ag}} + \frac{N'_{AgI}}{2} - N'_{Cd}\right)t_2 - \left(2N'_{CdI_2}\frac{N_{Cd}}{N_{Ag}} + N'_{CdI_2} + N'_{Cd}\right)t_4 \right\} d\bar{G}_{Cd} \tag{10}$$

Since $t_k = X_k Z_k u_k / \sum_l X_l Z_l u_l$, where X_l = ion fraction, Z_l = charge and u_l = mobility for species l, then

$$t_1 = 2iN'_{Cd}u_1/D \tag{11}$$

$$t_2 = 2[(2j -- i)N'_{CdI_2} - iN'_{Cd}]u_2/D \tag{12}$$

$$t_3 = -(2j - i)(N'_{AgI} + 2N'_{CdI_2})u_3/D \tag{13}$$

$$t_4 = (2j - i)N'_{AgI}u_4/D \tag{14}$$

where

$$D \equiv 2iN'_{Cd}\,u_1 + 2[(2j - i)N'_{CdI_2} - iN'_{Cd}]u_2 - (2j - i)\,[(N'_{AgI} + 2N'_{CdI_2})u_3 - N'_{AgI}u_4]$$

Substituting for the transference numbers and rearranging gives

$$dG = \frac{1}{2N'_{CdI_2} + N'_{AgI}} \left(N'_{CdI_2} + N'_{Cd} - \frac{N'_{AgI}}{N_{Ag}}N_{Cd} - \left\{ N'_{Cd}\left[i\left(\frac{N'_{AgI}}{N_{Ag}}N_{Cd} + \frac{N'_{AgI}}{2} - N'_{Cd}\right) + (2j - i)N'_{CdI_2}\right](u_2 - u_i) + N'_{CdI_2}\,(2 - i)\left(\frac{N'_{AgI}}{N_{Ag}}N_{Cd} + \frac{N'_{AgI}}{2}\right)(u_4 - u_2) + N'_{Cd}\,(2j - i)\frac{N'_{AgI}}{2}(u_4 - u_1)\right\} \Big/ \left[iN'_{Cd}\,(u_1 - u_2) + (2j - i)N'_{CdI_2}\,(u_2 - u_3) + (2j - i)\frac{N'_{AgI}}{2}(u_4 - u_3)\right]\right) d\bar{G}_{Cd} = -FdE \tag{15}$$

We see that the above expression, and hence the emf, is independent of the frame of reference chosen for the mobilities.

If we use the I^- ion for the frame of reference, assume $u_2 \approx u_1$ (see ref. 2), and neglect terms which become insignificant as infinite dilution is approached, we obtain for dilute alloy solutions

$$dG = 1/(2N'_{CdI_2})\,(1 - u_{Ag^+}\,N'_{AgI}/u_{Cd^{2+}}\,N_{Ag})d\bar{G}_{Cd}$$
$$= -FdE \qquad (16)$$

Finally, converting from free energy to activity and integrating

$$ln\ a_{Cd} = \frac{2F}{R[1 - u_{Ag^+}\,N'_{AgI}/u_{Cd^{2+}}\,N_{Ag}]} \int_A^B N'_{CdI_2}\, d\left(\frac{E}{T}\right) \qquad (17)$$

where the ratio $u_{Ag^+}N'_{AgI}/u_{Cd^{2+}}\ N_{Ag}$ is assumed to remain relatively constant over a small dilute solution range.

For further details concerning these equations, *see* ref. 3.

Since

$$a_{Cd}a^2_{AgI}/\ a_{CdI_2}a^2_{Ag} = e^{-\Delta G°/RT} \qquad (18)$$

the value for N'_{AgI}/N_{Ag} was obtained from

$$N'_{AgI}/N_{Ag} = \gamma_{Ag}\ N^{1/2}_{CdI_2}\ e^{-\Delta G°/2RT} \qquad (19)$$

where γ_i is the activity coefficient of component i, where γ_{CdI_2} was assumed to be unity because of approach to Raoult's limiting law for a_{CdI_2} in the electrolyte, where a_{Cd} was assumed to be approximately unity, and where γ_{AgI} was assumed equal to unity in the absence of experimental data for its determination. The values 0.181 for γ_{Ag} (*11*), 0.935 for N_{CdI_2} (*17*), −63.3 and −22.3 kcal/mol for the heats of formation, and −31.0 and −10.6 cal/K for the corresponding entropies of CdI_2 and AgI, respectively (*1*), were used. This gives $N'_{AgI}/N_{Ag} = 0.0048$.

Since Ag^+ has a charge that represents two thirds of the charge range from −1 of the iodide ion to +2 of the cadmium ion, the mobility of Ag^+ was assumed to be two-thirds that of Cd^{2+} in the absence of direct experimental measurements.

Consequently, the equation used for the integration was

$$\ln a_{Cd} = -0.02329 \int_A^B N'_{CdI_2}\ d(E/T) \qquad (20)$$

where E was in μV and T in degrees K.

The calculations were carried out on a CDC 6600 computer using iteration to obtain a self-consistent result for the various concentrations involved.

An initial concentration for N'_{Cd} was obtained by assuming that no AgI was formed. In this case we have

$$N'_{Cd} = n'_{Cd}/(n'_{Cd} + n'_{oCdI_2}) \qquad (21)$$

and

$$N'_{Cd} = s(n_{oCd} - n'_{Cd})/(n_{oCd} - n'_{Cd} + n_{oAg}) \quad (22)$$

where n_{oi} refers to the initial moles of component i introduced into the cell, s is the solubility of pure cadmium in its iodide, and it is assumed for dilute alloy solutions that $N'_{Cd} = sN_{Cd}$. From Eq. (21) we obtain

$$n'_{Cd} = [N'_{Cd}/(1 - N'_{Cd})]n'_{o\ CdI_2} \quad (23)$$

Using Eq. (23) to substitute for n'_{Cd} in Eq. (22) gives

$$(n_{o\ Cd} + n_{o\ Ag} + n'_{o\ CdI_2})N'^2_{Cd} - [n_{o\ Cd} + n_{o\ Ag} + s(n_{o\ Cd} + n'_{oCdI_2})]N'_{Cd} + sn_{o\ Cd} = 0 \quad (24)$$

Solution of this quadratic equation for N'_{Cd} gives the required result. The values $N_{Cd} = N'_{Cd}/s$, $N_{Ag} = 1 - N_{Cd}$, and from Eq. (23) the values for n'_{Cd} are calculated.

Iterations are carried out involving three nested loops. The outer loop adjusts N'_{AgI} and N_{Ag}; the middle loop adjusts n'_{AgI}, n^t_{Cd} (total cadmium), n_{Ag} and n'_{Cd}; the inner loop adjusts s', n_{Cd}, and N_{Cd}. First, N'_{AgI} is determined from Eq. (19). We may also write

$$N'_{AgI} = n'_{AgI}/(n'_{Cd} + n'_{CdI_2} + n'_{AgI})$$

The reaction $CdI_2 + 2Ag \rightleftarrows 2AgI + Cd$ removes some of the initial CdI_2 so that we have

$$n_{CdI_2} = n'_{o\ CdI_2} - (1/2)n_{AgI} \quad (26)$$

Substituting for n'_{CdI_2} from Eq. (26) in Eq. (25) and rearranging gives

$$n'_{AgI} = [2N'_{AgI}/(2 - N'_{AgI})\ (n'_{CdI_2} + n_{Cd}) \quad (27)$$

Equation (27) was used for calculating n'_{AgI}. Then $n^t_{Cd} = n_{o\ Cd} + (1/2)n'_{AgI}$, $n_{Ag} = n_{o\ Ag} - n'_{AgI}$, and $s' = sN'_{Cd}$ are calculated. With

$$s' \equiv n'_{Cd}/(n'_{Cd} + n'_{CdI_2} + n'_{AgI})$$

substitution for n'_{CdI_2} from Eq. (26) and rearranging gives

$$n'_{Cd} = [s'/(1 - s')]\ [n'_{oCdI_2} + (1/2)n'_{AgI}] \quad (28)$$

or

$$n_{Cd} = n^t_{Cd} - n'_{Cd} = n^t_{Cd} - [s'/(1 - s')]\ [n'_{o\ CdI_2} + (1/2)n_{AgI}] \quad (29)$$

Equation (29) was used to determine n_{Cd}; then new values were calculated for N_{Cd}, n'_{Cd}, and N_{Ag}.

The variation of N'_{CdI_2} with E/T is so slight over the range of the measurements that the trapezoid rule may be employed between the actual experimental points instead of drawing a smooth curve through the data. As a consequence, the experimental uncertainty is essentially retained in the calculated activities.

A plot of a_{Cd} vs N_{Cd} for both runs is shown in Fig. 3. A Dundee plot of the more precise Run 2 data is shown in Fig. 4. In the latter plot, all but the first three points appear to lie on a straight line. A small error in measuring the bias has the greatest effect on the most dilute points on such a plot and may be the cause of these deviations from linearity.

A least squares fit to the last eight points of Run 2 letting two parameters float gave

$$a_{Cd} = 1 - 1.00010\, N_{Ag} - 0.603 N_{Ag}^2 \quad (30)$$

The value 1.00010 obtained for the coefficient of N_{Ag} is in good agreement with the unit value expected from thermodynamic theory. Setting this parameter to unity and letting only the final parameter float gives

$$a_{Cd} = 1 - N_{Ag} - 0.607\, N_{Ag}^2 \quad (31)$$

with a standard deviation of 0.0000103 in the activity or 0.34 μV in the emf.

If cadmium activities are calculated ignoring the effects of the displacement reaction, a least squares fit to the last eight points of Run 2 gives

$$a_{Cd} = 1 - 0.99528\, N_{Ag} - 0.604\, N_{Ag}^2 \quad (32)$$

which disagrees with Raoult's limiting law slope by 0.00472. Consequently, the displacement reaction must be taken into account for this system in order to obtain satisfactory results.

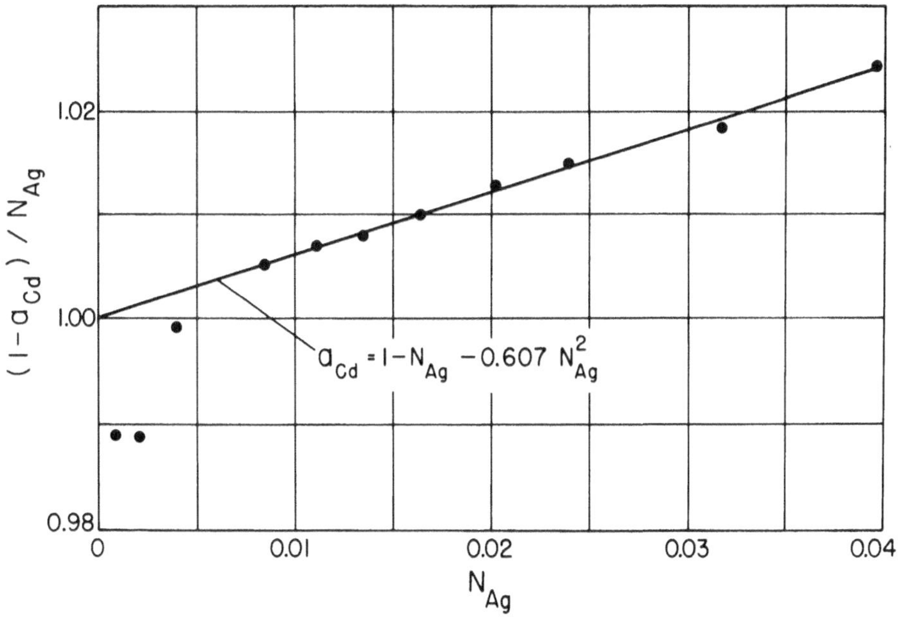

Fig. 3. Dundee plot of Run 2.

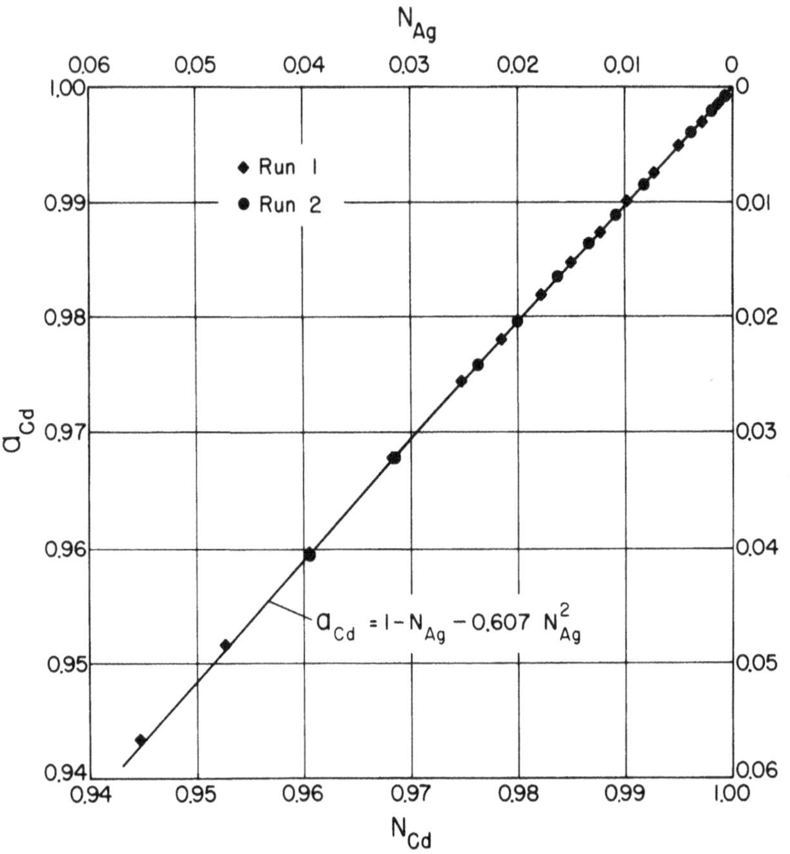

Fig. 4. Cadmium activities of Ag–Cd alloys.

DISCUSSION

Thermodynamic theory predicts that if the activity of a solute in a binary system has a finite slope at very great dilution, then the slope of the solvent activity will approach unity when referred to its mole fraction. Experimental tests of this limiting law are difficult and require measurements of the highest precision and accuracy. The precision of a given experimental method is usually discernible from the scatter in the data. The determination of accuracy is a more difficult problem, however. Few methods have even the precision required for testing the limiting law.

Emf measurements of cadmium activities over Ni–Cd (2) and Au–Cd alloys carried out at this laboratory are in excellent agreement with the limiting law. These solutes do not undergo a significant displacement reaction with CdI_2, so that cadmium activity calculations are simplified and subject to less uncertainty for these systems.

One would expect equally good agreement between Ag–Cd emf measurements and Raoult's limiting law since the method of measurement is the same. However, only by taking into account the displacement reaction can adequate agreement be obtained.

The experimental results obtained for the Ag–Cd system illustrate the fact that the calculation of activities from emf measurements can be complex.

ACKNOWLEDGMENT

Work done under the auspices of the US Department of Energy.

REFERENCES

1. Brewer, L., Bromley, L. A., Gilles, P., Lofgren, N. L., *National Nuclear Energy Series,* Vol. 19B, L. L. Quill, Ed., McGraw-Hill, New York, NY 1950.
2. Conant, D. R., *J. Chem. Eng. Data* **16,** 430 (1971).
3. Conant, D. R., Los Alamos Scientific Laboratory report LA-4102, 1969.
4. deGroot, S. R., "L'effect Soret; diffusion thermique dans les phases condensées" (Thesis), North-Holland Publishing Company, Amsterdam, 1945.
5. deGroot, S. R., *J. Phys. Radium* **8,** 188 (1947).
6. deGroot, S. R., Physica (The Hague) **13,** 555 (1947).
7. deGroot, S. R., *Thermodynamics of Irreversible Processes,* Interscience, New York, 1958, 1st ed., p. 5.
8. Denbigh, K. G., *The Thermodynamics of the Steady State,* Wiley, New York, 1951, p. 88.
9. Herzog, Von W., and Klemm, A., *Z. Naturforsch.* **15a,** 1100 (1960).
10. Houseman, B. L., *High Temp. Sci.* **8,** 25 (1976).
11. Hultgren, R., Orr, R. L., Anderson, P. D., and Kelley, K. K., *Selected Values of Thermodynamic Properties of Metals and Alloys,* Wiley, New York, 1963, p. 356.
12. Onsager, L., *Phys. Rev.* **37,** 405 (1931).
13. Onsager, L., *Phys. Rev.* **38,** 2265 (1931).
14. Roeser, W. F., and Wenzel, H. T., *J. Res. Nat. Bur. Stand.* **10,** 275 (1933).
15. Schenker, H., Lauritzen, J. I., Corruccini, R. J., and Lonberger, S. T., Nat. Bur. Stand. (US), Circ. 561 (April 27, 1955).
16. Thompson, W., *Mathematical and Physical Papers,* Cambridge University Press, 1882, Vol. I, p. 232.
17. Topol, L. E., and Landis, A. L., *J. Amer. Chem. Soc.* **82,** 6291 (1960).

0018–1536/84/1700–0267$04.20

Second Law and Solution Structure

GUY R. B. ELLIOTT,* DONALD R. CONANT,† AND
BARTON L. HOUSEMAN‡

*Los Alamos Consultants, 133 La Senda Road,
Los Alamos, NM 87544*

Received September 23, 1983; Accepted December 31, 1983

ABSTRACT

This summary paper starts from the fact that activity data are now published that give positive proof of the Second Law, i.e., highly precise experimental data show forms that would not occur unless the law held. Such positive proof contrasts with the essentially negative approach that "violations of the law have not been demonstrated" or with what Bridgman emphasized as an economic law controlling chemistry, i.e., "that you can't get something for nothing." In view of those activity data, we intend that all our present discussion shall be consistent with the Second Law. The paper deals with precise experimental data from nonrandom solutions and with our views as to (a) how the data should be taken and (b) how thermodynamics should be applied to such solutions, both solid and liquid. It deals with effects from numerous phenomena: nonstoichiometry and superlattices in solids; solid-like structures in liquids; metastability; hysteresis; persistent lattice stresses or slow diffusion; anomalies associated with one-component equilibrium in two-component systems; EMF changes caused by metal solubility in molten salts and the resultant mixed-valence electrolytes; and exchange reactions and electronic conduction in EMF cells. Proper plots for evaluating experimental data are considered. Effects of structure on partial molal entropies and enthalpies are discussed in relation

*Author to whom all correspondence and reprint requests should be addressed.

†Present address: 12903 Camphill Ct., Creve Coeur, Missouri 63141.

‡Present address: Chairman, Department of Chemistry, Goucher College, Towson, Maryland 21204.

to general behavioral trends of random solutions. For example, effects of solution ordering frequently look like fine structure imposed on the trends described by regular-solution theory (with its random-solution assumption).

Index Entries: Nonrandomness, in solutions; clusters, in solutions; precursors, in solutions; lattice species, in solutions; melting, and solution structure; one-component equilibrium, of solutions; entropy, and solution structure; solution structure, and the 2nd law.

I. INTRODUCTION

Several years ago we authors and our coworkers undertook an experimental and theoretical analysis of the Second Law. This work led us to develop experimental techniques of high precision and of apparently high accuracy that could be used for measurement of thermodynamic activity. Particular attention was given (a) to the solvent in moderately dilute solutions for which Raoult's law and Duhem integrations could be applied and (b) to metallic solutions with their complicated, long-ranging bonds. In turn, this work led us to analyze effects of solution order on solution thermodynamic relationships.

Experimental positive proof of the validity of the Second Law is offered by data for several chemical systems published within a series of papers (*1–9*) reporting the work. Thermodynamic anomalies for other systems are also discussed in these papers and are resolved.

Paper I (*1*), consistent with precise applicability of the Second Law, shows identical activities by vapor pressure and by emf, and it also shows asymptotic approach to Raoult's law for random solutions of molten gold in cadmium. Uncustomary correction must be made, however, for nonwhole-number valences for mixed Cd_2^{2+} and Cd^{2+}. This paper also shows activity effects associated with random and ordered alloys of the same compositions.

Paper II (*2*) shows precise Duhem correlation of activities for molten tin in mercury solutions, as required for consistency with the Second Law, even though the tin is highly nonrandom (only two-thirds of the random-solution value). Below 0.01 mol fraction, these solutions become random and Raoult's law is followed closely.

Paper III (*3*) deals with data clearly following Raoult's law or asymptotic approach to it (following regular-solution theory) for dilute alloys. However, some of these same systems show approach to limiting slopes very different from that of Raoult's law for all concentrations above 0.01 mol fraction. Precise vapor pressure, emf, and freezing point data are considered.

Paper IV (*4*) shows how we treat solution equilibria that create nonwhole-number average valences in electrochemical reactions, especially involving $Cd + Cd^{2+} = Cd_2^{2+}$. These equilibria lead to relationships such as $\Delta G = -1.86\mathcal{F}\mathcal{E}$ for cadmium iodide solutions that contact metallic cadmium at 500°C. When the correction is made, the systems discussed show asymptotic approach to the Raoult's law slope.

Paper V (*5*) relates experiments in which vapor pressures within a closed system appeared to approach different thermodynamic limits from those approached by emf within the same closed system at the same time. Resolution of the thermodynamic anomaly is discussed.

Paper VI (*6*) deals with largely unused equations that precisely describe midregion behavior in ordered alloys.

Paper VII (*7*) describes ionic constants applicable for aqueous freezing-point depressions from dilute to about 1 molal. The forms of these equation have been rejected in the past as a result of what we consider to be misapplications of thermodynamic theory.

Paper VIII (*8*) analyzes effects of oxide or other coats on so-called "inert" electrodes that allow ion movement and electrochemical reaction through the coat, but prevent direct chemical attack to equilibrium, thereby allowing approach to different equilibria for months at high temperature conditions within a single system.

Paper IX (*9*) gives the results of a computer program simulating microscopic reversibility for dilute solutions.

In this paper we generalize our conclusions from the preceding papers and from much other precise work, both ours and that from other laboratories. In particular, we attempt to communicate certain of our views about suitable ways to collect experimental data and to present and interpret these data.

For convenience in our presentation, we have divided this paper according to kinds of materials and methods of measurement.

II. SOLID BEHAVIOR

II.A. Equilibrium in Solids

With ever increasing clarity, developing from the early work of such people as Johansson and Linde, Brewer, Hagg and Magneli, J. S. Anderson, Haraldson, Jellinek, L. Eyring, Hyde, and their many coworkers, and too many others to name, it has become evident that inorganic compounds and alloys are neither simple Daltonides nor simple Bethollides—the ordering and disordering forces always compete. The enthalpy battles temperature times entropy, but neither totally carries the day.

We consider that equilibrium in such systems is seldom simple either to achieve or to understand in detail. In corollary with the comments above, inorganic compounds and alloys are all solutions both because ideal stoichiometry is statistically impossible to achieve or hold, and because thermally created defects would destroy any precise physical realism to ideal stoichiometry if it were achieved by accident. Likewise, random-solution concepts are generally highly idealized, and we believe care in their use is justified.

Even small departures in composition from that of the ideal parent lattice can sometimes produce complex superlattices that, though detectable only with most careful work, can measurably influence the thermodynamic properties of the system. An example that is now uncontestable, but used to cause considerable distress, is that shear planes, often misaligned with the parent unit-cell borders, can create crystal compositions different from that of the ideal parent structure. If the shear planes are of regular periodicity, a superlattice is created. Likewise, excess atoms or vacancies in solid alloys can order to form families of related superlattices. Because of the associated ranges of composition around the stoichiometry of ideal superlattices, "perfect" specimens of identical composition can have several different component activities. Of course, only one such structure can be thermodynamically most stable at any composition and temperature.

In practice, when a bulk material is being prepared from its chemical components, different regions may nucleate different lattices before the material has become homogeneous. Where only one component has effective mobility, e.g., O_2 via the gas as a metal film is oxidizing, the different crystallites (with their different lattices) will jointly seek equal activity of the mobile component. Because the existing lattices are different, the compositions produced in the crystallites will also be different. Such a bulk material becomes effectively locked into composition inhomogeneities. This material will satisfy many usual criteria of equilibrium and will *appear* to be in equilibrium, but actually it will be a one-component, metastable phenomenon with the other component unable to move effectively to new lattice sites and thus find its thermodynamically favored form.

Difficulties such as those described can lead to nonrandom scatter in experimental data, to hysteresis, to disagreements among data from different laboratories, and so on. Such difficulties are in addition to other widely recognized problems such as slow approach to equilibrium with need for annealing, the effects of impurities, and nonequilibrium produced as the sample is cooled after annealing.

II.B. Effects of Structural Complexities and Nonequilibrium in Solids Upon Thermodynamic Properties

In an earlier paper (*10*), we emphasized that (a) total free energies reflect primarily the bulk properties whereas (b) partial molal free energies often reflect nonstoichiometry, primarily through the creation or destruction of lattice defects. The partial and total properties are, however, interrelated through restrictions on the system that are described by the laws of thermodynamics.

II.B.1. Solid Oxide Thermodynamics

One of the best single sources of information concerning the influence of structural complexity upon thermodynamic properties comes from measurements (*11*) of oxygen activities in alpha-praseodymium oxide (PrO_{2-x}). Here the oxygen activities for metastable equilibrium and for true equilibrium (or as close to true as one is likely to get) were established as functions of both temperature and composition to determine partial molal enthalpies and entropies for the oxygen.

Three fields of distinctly different behavior within the alpha-phase region were detected. The oxygen activity–composition relationships within and between these fields behaved like the relationships one would expect for a series of distinct compounds with the partial molal entropies and enthalpies varying smoothly within each field, but abruptly between fields. In contrast, because of approximately cancelling effects of changes in the partial molal entropies and enthalpies, the partial molal free energy for the oxygen varied much more slowly, and the total free energy varied very little across the alpha-phase region. Specifically, between the behavioral fields the oxygen partials shift abruptly by about 20 kcal/mol O_2 and about 20 cal/deg-mol O_2, but across a single behavioral field the shifts are continuous and amount to about 15 kcal/mol O_2 and about 15 cal/deg-mol O_2 (1 cal = 4.187 J). Meanwhile, variation in the free energy partial is only about 3 kcal/mol O_2 across the whole portion studied within the alpha-phase region, and Duhem integration to give the total free-energy change would give a much smaller value (in the calorie range). The variations of partial molal entropy and enthalpy follow linear equations (*11*).

From an atomistic viewpoint, these values make qualitative sense—very little change has taken place in the total lattice, but the stresses upon the oxygen ions that exchange most easily with the gas have been altered significantly (changing their partial enthalpy), and their fractional occupancy of available sites has changed (changing their entropy partial). These comments do not imply that the systems are out of equilibrium for any composition.

For the more general case of solid oxides, enthalpy changes are usually established by calorimetry. Comparison of different calorimetric methods (*12*) suggests that the lattice nonequilibrium in bomb combustions usually does not alter heats by more than 0.5%.

It is possible that the ordering of excess atoms and defects, with resultant superlattices, is the rule rather than the exception for oxides—evidence demonstrating or suggesting such complexity has been given for lanthanides, actinides, and elements in Groups IVB, VB, VIB, VIIB, and VIII of the periodic chart.

From consideration of oxide data and related alloy data, we believe the effects of lattice imperfections and superlattices on the thermodynamic properties of solid oxides can be summarized as follows: Total enthalpies, entropies, and free energies probably will change only gradually within a nominal solid solution region, even if superlattices are present. On the other hand, the partial molal free energy of a component will change more rapidly, because it reflects the effects of a relatively small percentage of the atoms, i.e., those that are involved in nonstoichiometry or in superlattices. Here the superlattice effects on the partial molal free energy are apt to look like a fine structure that is imposed on the trends of behavior that could be associated with nonstoichiometry and random occupancy of off-lattice positions of a particular type. Finally, the effects of forming superlattices reduce randomness and strengthen bonding, correspondingly affecting (often strongly) the partial molal entropies and enthalpies.

Historically, it has proved difficult to establish reliable values of changes of partial molal enthalpies from bulk measurements of total changes of enthalpy for various compositions. In view of the foregoing discussion, the difficulties are not surprising. The practical hazard is that an improperly evaluated partial molal enthalpy change may be combined with an improperly estimated partial molal entropy change to predict a wrong partial molal free energy change. The saving aspect of the problem is that trends of changes of partial molal free energy with composition tend to be gradual, and the values for unordered defects and for superlattices are not greatly different.

II.B.2. Solid Alloy Thermodynamics

In the case of $CeCd_{\sim 4.5}$ "solid-solution" region, some 70 superlattices (microphases that are actually distinct compounds) and a more broad single-phase solid solutions were observed by vapor pressure studies (*6, 14*) within a composition range of only 0.01 mol fraction units.

Unlike the one-component equilibria that oxides frequently exhibit (only oxygen moves), these alloy data reflect true two-component equilibria in which a single superlattice structure propagates observably (in 10–15 min, once nucleated at 600°C) through a porous, one-piece, 3 g sample—the regularity with composition within microphase families, and the supporting superlattice X-ray patterns (*15*) can have no other plausible explanation. However, it does not imply that the most stable phase necessarily has nucleated and, in fact, equilibrium for several compositions and several metastable phases can be observed at different times (different hysteresis) for a single cadmium vapor pressure.

In this case a rough Duhem integration (*14*) was used to establish the free energy difference between the broad solid solution range (random occupancy of acceptable lattice positions) and the most stable microphase (largely ordered occupancy) of identical composition. Here the change in total free energy is only about 10 cal/mol of cadmium; in contrast, the change in the partial molal free energy of the cadmium is about 2 kcal/mol of cadmium. Very similar trends in the activity–composition relationships were followed both by the broad solid-solution range and through the groups of microphases in the same composition range.

Thus, in terms of the partial molal free energy, the variations in properties associated with the microphases look like a fine structure that has been imposed on the trends with composition that would be observed in the absence of superlattices in these alloys.

It was not possible to establish temperature coefficients for these microphase vapor pressures (it was too easy to convert to different superlattices as conditions changed), but theoretical calculation indicates that about 16 kcal/mol of defects would be required to overcome the temperature–entropy loss on ordering.

Again, as with the oxide case, the temperature–entropy variations largely balance the enthalpy effects. By corollary, the trends of free energy partials for the superlattices appear as such small steps (two phase, one phase) that they will look gradual and continuous (like a broader solid-solution range) through the composition region if only a relatively few data points are taken or if the data are not precise.

The microphase activity–composition data all plot linearly, within small experimental scatter, and one can develop simple atomistic models that predict corresponding behavior. For the more broad solid-solution ranges of $CeCd_{\sim 4.5}$ and for similar data for $CeCd_{\sim 6}$, the activity–composition plots show curvatures that require slightly more complicated atomistic models. Whether or not these atomistic models have factual validity, they offer useful and accurate means for presenting the experimental data:

Start from the experimental fact that there can be equilibrium at 600°C between cadmium vapor and a cerium–cadmium microphase. Probably there is an intermediate surface region containing migrating and bouncing cadmium atoms that establishes this equilibrium while retaining microscopic reversibility and satisfying sticking coefficients.

Consider that the microphase has slightly more cadmium atoms than would correspond to ideal stoichiometry, giving a cadmium species fraction of X^*_{excess}. Most cadmium atoms are in ideal lattice sites, but the excess cadmium atoms will be in less-than-perfect lattice sites, and they will be bound less completely and less tightly than the atoms at ideal sites. The buildup of excess atoms will (while retaining microscopic reversibility through the surface layer) add relatively rapidly to the vapor pressure over the microphase. If the effect is linear (as with Henry's law for solutes), there will be a cadmium activity–composition effect that plots with the form, $y = mx + b$, as follows:

$$a_{Cd} = gX^*_{excess} + B \tag{1}$$

where B is the cadmium activity for the pure microphase of ideal stoichiometry, and g is an essentially constant species fraction coefficient (like an activity coefficient) that corrects for the effect of cadmium atoms at nonideal sites in this microphase.

This equation form is widely applicable for alloys, solid or (as will be discussed) liquid. All 70 $CeCd_{\sim 4.5}$ microphases show the form. Furthermore, among the related microphases in a family of superlattices, the value of g often applies throughout and the value of B changes systematically.

For the broader, random, solid-solution regions of $CeCd_{\sim 4.5}$ and $CeCd_{\sim 6}$, other kinds of defects must be considered, but the descriptive relationships remain simple. Here the interested reader is referred to the original articles (*10, 16*).

II.B.3. Bringing Solids to Equilibrium

We do not know any really satisfactory ways to bring complex solids to their most stable structures—the principal problems lie with nucleation and diffusion. For now, at least, we chemists seem to be immutably faced with uncertainties like those discussed earlier, though understanding of the problem frequently can help.

What are the forces leading to complex order? Clearly, the microphases require communication of information over long distances, in spite of strong thermal vibrations—there are at least 8000 atoms per unit cell with ordering-cycle distances of at least 126 Å in two directions, as measured or produced at 600–700°C. The ordering forces, although not large in total, do individually have to overcome much temperature–entropy loss per excess atom forming the superlattice (amounting to about 16 kcal/mol of defects, as noted earlier).

We presume the microphase forces arise from (a) the depression of metal energy levels that occurs at Brillouin-zone boundaries (*17*) plus (b) the lattice stabilization that occurs when excess atoms in the lattice, acting as defects, are removed to ordered positions (*14, 15*). The stability, in the face of the strong thermal vibrations mentioned, is ascribed to the low-lying, long-wavelength nature of the electron waves—they lie too far below the Fermi surface to allow thermal excitation of the electrons to that surface, and the electronic transition is effectively prohibited.

Many homologous oxides also exhibit electron conduction and metal-like interatomic distances; thus, ordering associated with metallic resonance is also not excluded as the source of ordering information (*14, 15*) for some oxides. Such bonding is not the total answer (*15*), however, because it does not explain the long-range order of rare-earth oxides (e.g., ref. *13*) in which thousands of atoms per unit cell are involved. Focusing of vibrations, as suggested elsewhere (*18*), may also be part of the explanation.

This view of the ordering forces is important both to why nucleation can be so difficult and to how it can occur, as is discussed in the next paragraph. For $CeCd_{\sim 4.5}$, the random solutions of excess atoms in the ordered parent lattice (in porous, single-piece, uniformly heated samples of a few grams) could be studied at numerous compositions for days or weeks before nucleation happened to occur. In contrast, if any new microphase formed within the parent lattice, it became a nucleus whose structure propagated through the sample, changing the sample weight for a few minutes as the new structure adjusted to the ambient cadmium-vapor activity set in the isopiestic balance (*16*). Also, the nucleation of new microphases within a family was relatively easy as compositions shifted further off the ideal composition.

During the random motions of the excess atoms in solution in certain sites of an ordered lattice, there is little likelihood, at any moment or in any local region of the crystal, that these random atoms (whose free energy, however, could be reduced if they were ordered) will create a pattern of site occupancy that could become a nucleus for microphase growth. Therefore, there is a good chance that a random solution can be measured for long periods before chance nucleation creates a microphase. But once the random motion has been replaced by higher order, the statistical barrier to forming other microphases is much reduced, and the barriers to formation of different microphases within a single family of microphases are even lower.

Nucleation within microphase families can be assisted by temperature or concentration gradients, and these same variables can assist the formation of new microphase families. If a microphase composition is shifted enough off its ideal composition to make the random solution more stable, then the random solution forms at once—there is experimentally (*14*) no barrier to the disordering reaction in that case, consistent with theoretical expectations. Perhaps mechanical stresses, e.g., by ultrasonics, might help—we did not try that. But, to repeat, we have no proven way to hasten the ordering of a random solution, whether oxide or alloy.

What we have just said is consistent with thermodynamics and the Second Law.

II.B.4. Component Activity Measurements on Solids or Using Solids

As long as a whole solid phase is being created, modified, or destroyed, and equilibrium is being sought, the problems will be like those already described. Here we speak of distribution between immiscible phases, freezing point depressions, or isopiestic vapor equilibrations, none of which add a new class of problems.

But a new class of problems and uncertainties can arise when significant quantities of one component must flow out of the solid solution during the measurement. Here we are considering Langmuir evaporations, large-hole Knudsen-cell measurements, or emf determinations with low-resistance equipment. In such cases, if the bulk does not reach and maintain near-true equilibrium with the surface of the solid solution, a loss of meaning of the measurements will be corollary.

On the other hand, if the outflow is small, other problems can arise. For example, small amounts of impurities can produce large voltages or can represent large proportions of the evaporating gas. Even with radioactive tracers, the vapor pressure determinations have sometimes been questionable, perhaps because of surface problems such as surface oxidation.

Measurements using solid electrolytes have been very useful, but the measurements frequently involve millivolt precision—molten-salt electrolytes frequently offer microvolt precision. Thus the solid electrolytes are useful in determining free energies of formation of compounds, but they are less useful in evaluating the variation of activity during solution formation. Also, when solid electrolytes are used with high-impedance voltmeters, there is danger that the voltage may reflect only a minor impurity rather than the constituent of interest. On the other hand, solid electrolytes effectively eliminate exchange reactions and the formation of mixed-valence electrolytes such as must be considered with molten salts and are discussed later.

III. LIQUID BEHAVIOR

III.A. Equilibrium in Liquids

Like solids, liquids also can be very complex in their variation of chemical and physical properties with composition and temperature. Precise data on activities as measured by emf, by vapor pressure, and by freezing studies show the complexity; hysteresis and metastability are demonstrated experimentally in both physical and chemical properties; X-ray studies show nonrandom atomic groupings in molten alloys. Splat cooling shows two-liquid regions and precursor structures near phase boundaries. The list is not complete.

These liquid complexities, particularly in molten alloys, can lead to low-level metastability even for equilibrations of several days. On the other hand, locked-in compositional inhomogeneities on a bulk scale, as observed for solids, are not a problem—given a well-mixed solution and nucleation of a more stable microstructure, often that structure will grow through the liquid solution and dominate it in a few minutes.

III.A.1. Nonmetallic Liquids

It is convenient to distinguish all other liquid solutions from molten alloys because of unique effects associated with the metallic bond. Therefore, metallic solutions will be discussed in a separate section.

III.A.1.a. AQUEOUS FREEZING POINT EQUATIONS. A widely used plot for freezing-point depressions of ionic, aqueous solutions (e.g., 19) is $\theta - \theta^{st}$ vs m^2. The symbol θ represents the freezing point depression that is demonstrated experimentally, and θ^{st} is the freezing point depression that would result from Debye-Hückel (ion–ion) interactions for an otherwise ideal solution of the same composition. With θ^{st}, a standard distance of closest ionic approach (e.g., 3.085Å) is often used for all ions because it simplifies the calculations, and the exact value is not known anyway. The term m is the ionic solute's molality. Linearity on the plot corresponds to

$$\theta - \theta^{st} = k_1 m + k_2 m^2 + k_3 m^3 \quad (2)$$

with $k_1 = 0$. This type of plot has two major disadvantages: (a) It unnecessarily assumes $k_1 = 0$, and (b) squaring m results in effectively losing data for mole fractions that are large (off scale) or small (bunched around the origin through which, by definition, all data descriptions must pass anyway.)

A better plot (*7, 20*) uses $(\theta - \theta^{st})/m$ vs. m. This distinction is important because frequently k_1 is nonzero and $k_1 m$ should not be neglected; also, in practice k_3 is often near zero with the nonzero constant k_2 applying over much of the composition range measured for most systems: The plot often is essentially linear beyond great dilution, and this linear portion extrapolates to a nonzero intercept. Calculating θ^{st} on a standard 3.085Å distance of closest ionic approach can produce curvature at low concentrations in these plots—this low-concentration curvature can sometimes be eliminated by evaluating an apparent distance of closest approach (*20, 21*), which removes curvature, but such evaluated distances are usually different from those calculated by other means.

A thorough study of the best aqueous freezing-point depression data (*20*) showed that the actual Debye-Hückel corrections have practical effectiveness to larger concentrations than has been recognized generally. This statement is true in practice even though it has been asserted many times that Debye-Hückel should not be valid beyond about 0.1 molal. Specifically, one can use the equation form

$$\theta - \theta^{st} = k_1 m + k_2 m^2 \quad (3)$$

with k_1 nonzero to describe almost all the best aqueous freezing point data out to the largest concentrations measured, namely about 1 molal.

For plots of all these aqueous freezing point depression data, see ref. *20*. In these plots, there is low-concentration curvature for most systems, and this curvature is stronger for large ions (formate, acetate, iodide): Presumably, much of this curvature results from the inaccurate distance of closest approach (3.085Å) used in the conventional calculations. The use of other distances of closest approach are discussed in the original papers (*20, 21*).

Five bivalent sulfates whose freezing points had been measured (*19*) were plotted (*7, 20*) to test for correspondence to Eq. (3). In each case a structural change showed in the range 0.4–0.6 molal, i.e., Eq. (3) was obeyed both above and below the transition, but there was an abrupt shift in the values of k_1 and k_2, as plotted.

III.A.1.b. "PREMONITORY PHENOMENA" AND PHASE CHANGES IN NONMETALS. Many liquid phenomena, particularly transport properties such as viscosity, show changes in their trends of behavior as the freezing point is closely approached (e.g., ref. 22). Such phenomena (though not really premonitory) are much too pronounced to be denied, and they are often reproducible, though supposedly identical experiments sometimes show different behavior. Hysteresis in the phenomena may also be observed—the rising-temperature phenomenon may not match that for falling temperatures. The clustering of molecules into coherent groups with structure different from that of the usual liquid is well recognized as the source of premonitory phenomena (e.g., ref. 22) near phase boundaries.

Clustering can only occur, of course, if the product solution has lower free energy than the corresponding solution without clusters. Therefore, a solution with clusters will depart from the previous trends of behavior, and it will be detectably different from the metastable extension of the previous trends. This departure will take place before the phase boundary is reached as the liquid cools. Such differences have been measured experimentally, and (contrary to some published assertions) the liquid structure that could supercool past a phase boundary often is not the thermodynamically most stable liquid structure at temperatures just above that of the phase boundary.

III.A.2. Molten Alloys

As will be elaborated, the metallic bond produces unique features in molten alloys compared with other liquid solutions. The fundamental difference is that metallic resonance is optimized by identical atoms in perfect array—nonstoichiometry, lattice defects, and unusual groupings all serve to weaken the metallic resonance. Thus, with molten alloys, as it also true with complexity in solid alloys, there are strong forces tending to order the defects; by corollary, liquid alloy interactions are very complex, almost as complex as those for solid alloys.

III.A.2.a. CLUSTERS IN MOLTEN ALLOYS. First, some comments on terminology. Many words are used to describe a group of related phenomena. Early work in Eastern Europe spoke (in translation) of cybotaxes or microheterogeneous regions. In our own writings, as we searched for communication, we have used interaction units and clusters. Pseudomolecules, ionic assemblies, chemical or covalent bonding, and groups are terms that have been used to describe identical or closely related phenomena. We now speak of clusters because that term seems most widely accepted.

However, in our discussion with a group of physicists at a recent Gordon Conference, it developed that some of them object to the concept of clusters because the atomic positions and atomic exchange do not appear to correspond with a rigid cluster shape. On the other hand, our consideration of the thermodynamics and nature of allowable bonding as suggested below leads us to emphasize the cluster aspects of the solution behavior. We do not consider that the two points of view are mutually exclusive.

It is now becoming more widely accepted (*23–26* and several papers in ref. *27*), that a metallic solute may lose its metallic character when it forms a dilute solution in another metal: In a dilute, random solution, the solutes do not often even touch each other, so they cannot resonate together; normally, they also cannot resonate well with the solvent metal either, because their electronic states will usually be too different from those of the solvent host.

The more usual case is a strong covalent or ionic interaction with the host metal to form a nonmetallic cluster involving the solute atom and a number of solvent atoms. In addition, arrays of the clusters appear to form in many cases in the solvent metal, as will be discussed further later. The energy of bonding in the clusters must be large: For example, when gold dissolves in cadmium to form a liquid solution at 600 K, the gold loses its very strong metallic resonance, and the conditions for resonance in the cadmium are made less favorable. The nonmetallic bonds in the cluster must be at least strong enough to overcome the 87 kcal/mol heat of vaporization of gold and to reduce the cadmium liquid activity of the solutions much below the Raoult's law values.

III.A.2.b. STRUCTURE AND LIQUID–LIQUID IMMISCIBILITY IN MOLTEN ALLOYS. In liquid metals and alloys there is enough lattice breakdown so that the materials pour, and the usual X-ray diffraction techniques are inadequate. However, the small heats of fusion and the small changes of electrical conductivity on melting for metals indicate that the solids and liquids are very similar in their bonding mechanisms. Therefore, it should not surprise scientists that molten alloys often resemble solid alloys far more than they resemble the random solutions that have so often been assumed in the past, largely for want of better methods.

It can be inferred from the results of careful experiments that considerable order does indeed exist in many liquid alloys. From X-ray studies it is established that alloy near neighbors do not randomly occupy solution positions but, rather, they seek preferred groupings. From extensive and detailed activity measurements at different temperatures and at different compositions, it is established that frequently the systems show distinct, single-phase, liquid regions with two-liquid regions of phase immiscibility between. Examples of two-liquid regions are found in Au–Cd, Au–Sn, Cu–Cd, and Cu–Sn systems, plus others (*1, 6, 23, 24, 28–32*).

III.A.2.C. STRUCTURAL FIELDS IN A SINGLE-PHASE REGION IN MOLTEN ALLOYS. Like the structural fields in a single-phase region of PrO_{2-x} discussed earlier, there can also be structural fields in liquid alloys. For example, in Au–Cd molten alloys (*6*), note the fields G, I, and K; also see refs. *23, 24, 26, 29*.

The distinction between structural changes via fields in a single-phase region and immiscibility with two-phase regions seems to lie with the host liquid structure: If there is incompatibility in the host structures of two stable structural regions (for example, bcc) structures cannot go continuously to fcc), then a two-liquid region must form, thus satisfying the phase rule. If the host structure merely changes the form of its clusters, or if for other reasons there can be continuous changes from one structure to the next in composition, then the single phase will be maintained, but behavioral fields will develop.

The abruptness of the change from one cluster type to another strongly suggests that even fairly dilute clusters can order in the metallic–solvent host, e.g., see Pb–Cd (*3, 26*), thereby creating behavioral fields.

From an experimental viewpoint with molten alloys, it is important to recognize that structures frequently shift abruptly, not gradually, from one to another—it is not like a gradual shift from Al^{3+} through a series of species to AlF_6^{3-}, as would occur in water solutions. Rather, the solution may look on a mixture of cluster species as unordered defects that weaken resonance and are to be rejected if possible, to be replaced by an ordered array of clusters of one particular type. In thermodynamic terms, the activity coefficient will not follow a simple variation with composition.

III.A.2.d. Precursors Near Phase Boundaries and Hysteresis in Molten Alloys. For nonmetallic liquids we discussed premonitory phenomena and clusters; here structures that are precursors of solids were implied, but not explicitly stated. Very similar structures often will also be present and may be detectable if two alloy phases, both liquid, are together at equilibrium. If the liquids are separated, then are changed slightly off the compositions for two-phase equilibrium, the precursors will still remain. These precursors have the possibility of forming each phase because the entropy of mixing often will favor at least some mixing of the two types of structures.

The formation of these precursors (as with the earlier discussion under the common, but objectionable, term, premonitory phenomena), produces a deviation in the activity–composition relationships in a phase—the precursors could not form unless their formation reduced the system free energy relative to that in their absence. These activity deviations show experimentally at liquid–liquid transformation regions sometimes, but often more strongly in the liquid near a liquidus boundary (*see* refs. *6, 23–25, 29*).

Depending upon whether or not the precursors happen to have nucleated, there may be hysteresis in the molten-alloy behavior.

III.A.2.e. Clusters at Great Dilution. The metallic bonding of individual clusters must be retained even at near-infinite dilution. The cluster order, which involves many clusters, however, must finally convert from order to independent clusters that move without coordinated positions or orientations. Such a conversion to random position has been observed for Sn–Hg (*2, 33, 34*) below 0.01 mol fraction of tin at 525 K and for Au–Sn (*3, 35*) below 0.01 mol fraction at 540 K.

In a similar vein, it has been concluded that ordered clusters in pure metals (e.g., ref. *36*) and alloys (*34, 37*) break up at higher temperatures to produce more random solutions. The clusters around individual solute atoms will remain, however.

III.A.2.f. PRODUCING STABLE ALLOYS FROM METASTABLE STRUCTURES. The formation of random Cu or Fe clusters in small amounts in liquid tin is reported to promote the production of the tin clusters that are present at equilibrium in liquid tin at lower temperatures (*38*), thus reducing the chance of supercooling the tin. In line with this reported effect, we would anticipate the addition of foreign clusters in small amounts in a tin melt would promote the formation of tin clusters by locally breaking up the metallic bonding of the tin.

Stirring of alloys that have been well mixed already appears to do very little to convert metastable alloys to the stable structures (e.g., refs. *30, 39*). Heating and composition changes do sometimes nucleate more stable structures (e.g., ref. *39*).

It appears for liquid alloys that some solutions may exhibit metastability amounting to a few calories uncertainty in the partial molal free energy, depending upon what structure has nucleated. Furthermore, we do not yet know how to predict whether liquid alloys will be ordered or random—we have seen much evidence of liquid order, but we have also seen several cases of random solutions out to around 0.1 mol fraction or more of solute (e.g., ref. *40*).

III.A.2.g. ACTIVITY–RELATIONSHIP EQUATIONS FOR LIQUID ALLOYS. In describing the activities of dilute alloys, we have often used Dundee plots of $(1 - a_1)/X_2$ vs X_2. Linearity on this plot corresponds to

$$a_1 = 1 + k_1X_2 + k_2X_2^2 \qquad (4)$$

By using this plot, we often get very good representations of the data as we establish the effect of each solute addition in reducing the activity of the pure solvent.

As the solution moves into new phase regions or new structural fields, it will usually be necessary to replace the 1 in Eq. (4) by a particular k_0 that applies uniquely for the particular structure being studied

$$a_1 = k_0 + k_1X_2 + k_2X_2^2 \qquad (5)$$

In effect, k_0 retains the thermodynamic standard state of pure solvent, but acknowledges that there are now compounds in solution that alter the solvent behavior. [By comparison, Raoult's law modified by square terms corresponds to Eq. (5) with $k_0 = 1$ and $k_1 = -1$.]

On rare occasions $k_3X_2^3$ may be invoked for very precise data.

The Houseman equation gives very similar results in dilute solutions, but fits better than Eq. (5) for some concentrated solutions, i.e.,

$$\log \gamma_1 = \alpha_1X_2 + \alpha_2X_2^2 \qquad (6)$$

Again on rare occasions with very precise data, an $\alpha_3 X_2^3$ term may be invoked to accommodate curvature in plots of (log γ_1)/X_2 vs X_2. By comparison, linear α-function corresponds with

$$\log \gamma_1 = \alpha_1 X_2 + \alpha_2 X_2^2 + \alpha_3 X_2^3 \tag{7}$$

with $\alpha_1 = 0$, and (log γ_1)/X_2^2 vs X_2 is plotted.

Possibly the most surprising development for us in our whole research program on liquid metal solutions was to find that trends of behavior over broad regions of composition may be described by one equation form whereas the details follow a very different equation (e.g., ref. *29*): The broad-range behavior over several tenths in the solution mole fraction might be observed to follow linear α-function rather closely [Eq. (7) with $\alpha_1 = 0$ and with α_2 and α_3 nonzero]; yet the exact description of the experiments within their uncertainty require linear γ-function relationships [Eq. (7) with α_1 and α_2 nonzero and α_3 indistinguishable from zero]. Similarly, Raoult's law plus square terms may be followed approximately [Eq. (5) with $k_0 = 1$, $k_1 = -1$, and k_2 nonzero) when the more exact equations over the individual fields or phases may require equations of another form [Eq. (5) with k_0 not 1, k_1 not -1, and k_2 indistinguishable from zero].

The equation form

$$\gamma_1 = a_1/X_1 = k_1 + k_2 X_1 \tag{8}$$

can also be used with plots of γ_1 vs X_1.

As was suggested earlier, it is experimentally difficult to distinguish the validity of the activity and activity coefficient forms of these equations for metals, but determining which constants should be used with precise data often is straight-forward once an equation form has been selected. For nonmetals, where the transitions from one solution species to another are not sharp, one will not expect necessarily to be able to sort out the component equation forms as well.

III.B. Evaluation of Data Involving Liquids

III.B.1. Liquid-Alloy Vapor Pressures.

For fractional-atmosphere pressures and long equilibrations, the equilibrium vapor activities are generally unambiguous for the liquid structure that happens to have nucleated. One of the few uncertainties lies with the conversion of pressure to fugacity, and here the activity is very insensitive to whether fugacity or pressure ratios have been used. For the Ga–Cd case (*3, 41*)

$$\begin{aligned} a_{\mathrm{Cd}} &= f_{\mathrm{Cd\ alloy}}/f_{\mathrm{Cd\ pure}} \\ &\approx 1.000001\ p_{\mathrm{Cd\ alloy}}/p_{\mathrm{Cd\ pure}} \end{aligned} \tag{9}$$

and the activity was experimentally indistinguishable from the pressure ratio in the reported experiments.

There are some complications relating to peculiar heat flow that can cause minor problems in isopiestic balances if they are pushed to their limits but, in general, isopiestic balances such as we use are a highly reliable source of vapor pressure data.

III.B.2. EMF Measurements

III.B.2.a. AQUEOUS EMF VALUES. Aqueous standard cells are, of course, chosen for their high reproducibility in behavior. Weston standard cells in good condition give voltages roughly in the range 1018.3–1019.6 mV, and the cells are stable to about ±0.1 mV for long periods, so long as ambient conditions are kept constant; transient variations in ambient conditions can produce permanent changes in cell voltages of several tenths of a millivolt.

If one is to have a liquid metallic electrode in contact with aqueous electrolyte, one usually is limited to amalgams. Some such amalgams are stable as used, but many solute metals are measured with dripping amalgams because the solute reacts with water—in the analysis of such measurements, one usually makes the dubious assumption that the side reactions do not influence the emf being measured. Aqueous emf measurements must usually be checked against other methods of determining activities, as a consequence (e.g., *see* ref. *42*).

Concentration cells with transference can be measured without the amalgam side reaction. Here an inert electrode such as platinum will be used in the measurements. There is some irreproducibility caused by strains and impurities in such electrodes. Also, tests of various noble-metal electrodes have shown differences of up to 3 mV in the voltages associated with such "inert" materials (*43*).

We authors have generally more confidence in the emfs based on molten-salt electrolytes than we do in those from aqueous systems, but molten salts also have their problems, as will be discussed.

III.B.2.b. MOLTEN-SALT EMF VALUES. Molten electrolytes can be brought into equilibrium with the metallic phase in contact with them. In this sense the molten salts have major advantages over the aqueous electrolytes that often are unable to come to equilibrium with the desired concentration of the reactive amalgam, as was just discussed. Solid salts, also discussed earlier, are difficult to bring to equilibrium in many cases.

Bringing about equilibration between molten metal and molten salt requires overcoming physical restrictions, and chemical reactions may occur for which correction may be required. Physically, one wants to provide for the stirring of both metal and electrolyte, in order to allow material transfer across the interface, and one should isolate the two sections of the single electrolyte phase by a capillary salt bridge of some form to prevent cross stirring between the two sections. The analysis of the voltages of emf cells of this type by irreversible thermodynamics (*44*) shows that the cell emf should not be sensitive to the fact that the electrolyte in the salt bridge contains a concentration gradient.

Also in a physical sense, one should insulate electrical leads where they pass through the electrolyte: Leads that are supposedly inert, such as Mo or W, may react with components of the electrolyte and develop a chemically unreactive coating that prevents further chemical reaction with the underlying metal. However, when the cell voltage is to be measured, the coating may not bar electrochemical reaction with the substrate (*8, 25*), with the consequence that spurious voltages may be measured.

Chemically speaking, at least three types of side reactions may become involved. First, many molten salts dissolve their parent metals. Where the dissolution produces a species of lower valence (e.g., Cd_2^{2+} in Cd^{2+} electrolyte), a correction of the electrolyte valence to a nonwhole number for emf-free energy calculation will be required; this calculation involves both the species concentrations and their mobilities (*1, 40, 45*). For CdI_2 electrolyte in contact with nearly pure cadmium at 773 K, the proper value for the effective valence is 1.86.

(Contrary to some opinion, one does not avoid the problem of a mixed-valence electrolyte by dissolving the reactant in an inert solvent, e.g., $CdCl_2$ in LiCl–KCl—the equilibrium constant for the reaction with the cadmium metal still holds, but now one has further uncertainties in activity coefficients and mobilities.)

The value of the effective valence is not easy to establish because the apparent value can change with current density in an electrolysis experiment (*1, 45, 46*). The problem has been solved experimentally by carrying out a series of electrolysis experiments that could be extrapolated to zero current (*45*).

In the absence of experimental data for the effective valence, one can correct by using measured solubility data for the metal in the electrolyte plus assumptions about mobilities and activity coefficients (*1, 40*).

The second type of reaction indicated above involves displacement between solutes in the more reactive metal and the electrolyte containing ions of the reactive metal. Such reactions proceed to form dilute products in the electrolyte, and they change the alloy composition also.

In certain cases the net effect of the displacement reactions on the emf can be evaluated by simultaneous vapor and emf equilibrations (*5, 47*). The residual emf at vapor equilibrium indicates the extent of the combined displacement–mobility effects.

If one wishes to assume mobility values and activity coefficients, he can use data from standard free energy tables to approximate the effects of displacement reactions upon his case of interest.

The third type of reaction involves effective electronic conduction in the molten salt. Where species such as Fe^{2+} and Fe^{3+} are in solution, the simple exchange of an electron when the species collide during thermal agitation can create the effect of electronic conduction that can internally drain an emf cell. In contrast, the collision of Cd_2^{2+} normally will not lead to significant electronic conduction because here a whole atom of cadmium must move rather than just an electron.

One way to handle this problem approximately is to pass a countercurrent to opppose the internal drain (48).

IV. SUMMARY

The collection and analysis of precise thermodynamic data are not simple, but the work can be done successfully and accurately. Thermodynamic theory must be applied with care to distinguish which aspects of a "thermodynamic" analysis are true consequences of inviolable laws and which are based on models that reflect intuition about what the systems "ought" to do. There is nothing wrong with using models, so long as the models are consistent with thermodynamics and reflect the data—in fact, most solution thermodynamics could not be applied without the use of models to join the data points and smooth their scatter.

ACKNOWLEDGMENTS

This work was performed in part under the auspices of the US Department of Energy.

REFERENCES

1. Conant, D. R., and Elliott, G. R. B., *High Temp. Sci.* **14,** 215 (1981).
2. Elliott, G. R. B., and Conant, D. R., *High Temp. Sci.* **14,** 231 (1981).
3. Elliott, G. R. B., Conant, D. R., and Swofford, H. S., Jr., Second Law Studies: III Confirmation of Thermodynamic Theory for Dilute, Molten Alloys, Both Ordered and Random, Los Alamos Consultants Report No. 80-3, Jan. 5 (1980). Available for $6.00 from address on title page.
4. Conant, D. R., Houseman, B. L., and Elliott, G. R. B., Second Law Studies: IV Confirmation of Thermodynamic Theory by EMF with Mixed-Valence Electrolytes, Los Alamos Consultants Report No. 80-4, Jan. 5 (1980). Available for $6.00 from address on title page.
5. Conant, D. R., and Elliott, G. R. B., Second Law Studies: V The Kinetics and Thermodynamics of EMF-Cells with Vapor Equilibration: Confirmation of the Laws of Irreversible and Equilibrium Thermodynamics, Los Alamos Consultants Report No. 80-5, Jan. 5 (1980). Available for $6.00 from address on title page.
6. Elliott, G. R. B., Conant, D. R., Robinson, P. L., and Wang, D. N-K., Second Law Studies: VI Molten Alloy Behavior at Midregions of Composition, Los Alamos Consultants Report No. 80-6, Sept. 15 (1980). Available for $6.00 from address on title page.
7. Elliott, G. R. B., and Conant, D. R., Second Law Studies: VII Thermodynamic and Behavioral Correlations for Ionized Aqueous Solutions Through Universal Constants for Individual Ions, Los Alamos Consultants Report No. 80-7, Sept. 15 (1980). Available for $6.00 from address on title page.
8. Elliott, G. R. B., Robinson, P. L., and Conant, D. R., Second Law Studies: VIII Molybdenum and Tungsten Bare Electrodes in Stannous Chloride and

Cadmium Iodide Molten Salts, Los Alamos Consultants Report No. 80-8, Sept. 10 (1980). Available for $6.00 from address on title page.
9. Conant, D. R., and Elliott, G. R. B., Second Law Studies: IX Computer Models of Solvent-Solute Interactions in Dilute Solutions, Los Alamos Consultants Report No. 80-9, Sept. 10 (1980). Available for $6.00 from address on title page.
10. Elliott, G. R. B., and Lemons, J. F., Nonstoichiometry in Intermetallic Compounds, in *Nonstoichiometric Compounds, Advances in Chemistry Series No. 39,* Gould, R. F., ed., American Chemical Society, 1963, p. 144.
11. Jenkins, M. S., and Eyring, L., *Proceedings of the Seventh Rare Earth Research Conference, Oct. 28–30, 1968,* National Science Foundation, p. 163.
12. Holley, C. E., *Pure Appl. Chem.* **8,** 131 (1964).
13. Hyde, B. G., and Eyring, L., *Proceedings of the Fourth Conference on Rare Earth Research, April 22–25, 1964,* Eyring L., ed., Gordon and Breach, New York, 1966, p. 623.
14. Elliott, G. R. B., and Lemons, J. F., Nonstoichiometry, Order, and Microphases in $CeCd_{\sim 4.5}$ Solid Solutions, in *Nonstoichiometric Compounds,* Advances in Chemistry Series No. 39, Gould, R. F., ed., American Chemical Society, 1963, p. 153.
15. Roof, R. B., and Elliott, G. R. B. *Inorg. Chem.* **4,** 691 (1965).
16. Elliott, G. R. B., and Lemons, J. F., *J. Phys. Chem.* **64,** 137 (1960).
17. Slater, J. C., *Phys. Rev.* **84,** 179 (1951).
18. Ubbelohde, A. R., book review in ref. *10* (ref. not relocated).
19. Brown, P. G. M., and Prue, J. E., *Proc. Royal. Soc.* **232A,** 320 (1950).
20. Elliott, G. R. B., Some Individual Ion Correlations Among Aqueous Freezing Point Measurements, Los Alamos Scientific Laboratory Report, No. LA-3891 (1968).
21. Conant, D. R., The Relationship of Certain Measurements on Aqueous Electrolyte Systems to the Debye-Huckel Theory to Raoult's Limiting Law, and to Thermodynamics, Los Alamos Scientific Laboratory Report No. LA-4428 (1970).
22. Ubbelohde, A. R., *J. Chim. Phys.* **61,** 58 (1964).
23. Elliott, G. R. B., Herrick, C. C., Lemons, J. F., and Nordine, P. C., Structure in Liquid Au-Cd and Ce-Cd Solutions. Vapor Pressure and Electrical Resistivity. Liquid Compounds, Two-Liquid Regions, Premonitory Phenomena, and Freezing, Los Alamos Scientific Laboratory Report No. LA-3526 (1966).
24. Elliott, G. R. B., Herrick C. C., Lemons, J. F., and Nordine, P. C., *High Temp. Sci.* **1,** 58 (1969).
25. Wang, D. N-K., Elliott, G. R. B., and Pitt, C. H., *J. Electrochem. Soc.* **122,** 905 (1975).
26. Swofford, H. S., and Elliott, G. R. B., *Inorg. Chem.* **7,** 666 (1968).
27. Takeuchi, S., *The Properties of Liquid Metals, Proceedings of the Second International Conference, Tokyo, Japan, 3–8 Sept. 1972,* Taylor and Francis, London, 1973.
28. Conant, D. R., and Swofford, H. S., Jr., *J. Chem. Eng. Data* **15,** 539 (1970).
29. Robinson, P. L., and Elliott, G. R. B., *High Temp. Sci.* **4,** 1 (1972).
30. Sutliff, K. E., Ruhlander, B. R., Pohlman, S. L., Jackson, J. S., and Elliott, G. R. B., Structural Complexity and Quasi-Solid Behavior in Molten Cu-Sn Alloys by Stirred H-Cell Measurements, Los Alamos Consultants Report No. 80-11, Dec. 1, (1980).
31. Hopkins, D. N., Metallographic Study of Copper-Tin Alloys Splat-Cooled from a Two-Phase Liquid Region, Thesis, Department of Mining, Metallurgical, and Fuels Engineering, University of Utah, Salt Lake City, 1973.

32. Hopkins, D. N., Elliott, G. R. B., and Pitt, C. H., "Splat-Cooling Studies of Two-Liquid Regions and Possible Precursor Phenomena in Copper-Tin Alloys," Los Alamos Consultants Report No. 81-1, Jan. 2 (1981). Available for $6.00 from address on title page.
33. Elliott, G. R. B., *J. Electrochem. Soc.* **115,** 1143 (1968).
34. Weaver, C. S., Whitelatch, R. C., Jerome, D. J., and Elliott, G. R. B., *High Temp. Sci.* **5,** 155 (1973).
35. Elliott, G. R. B., and Lemons, J. F., *J. Electrochem. Soc.* **114,** 935 (1967).
36. Lad'yanov, V. I., Arkharov, V. I., and Novokhatskiy, I. A., and Kisun'ko, V. Z., *Fiz. metal. metalloved., (Eng.)* **34,** 150 (1972).
37. Hopkins, D. N., and Elliott, G. R. B., Splat-Cooled, Two-Liquid Solutions of Gold in Cadmium as Studied by Metallography, Los Alamos Consultants Report No. 81-2, Jan. 2 (1981). Available for $6.00 from address on title page.
38. Braun, G., and Mannchen, W., *Z. phys. Chem. Leipzig* **252,** 7 (1973).
39. Griffiths, D. A., Braithwaite, J., Beckstead, L., and Elliott, G. R. B., *J. Electrochem. Soc.* **120,** 301 (1973).
40. Conant, D. R., *J. Chem. Eng. Data* **16,** 430 (1971).
41. Elliott, G. R. B., Lemons, J. F., and Swofford, H. S., Jr., *J. Phys. Chem.* **69,** 933 (1965).
42. Harned, H. S., and Owen, B. B., *The Physical Chemistry of Electrolytic Solutions,* Reinhold, New York, 1950, pp. 302–303.
43. Willard, H. H., and Fenwick, F., *J. Amer. Chem. Soc.* **44,** 2504, 2516 (1922).
44. Conant, D. R., A Derivation of Alloy Activity Equations for EMF Cells Involving Displacement Reactions, Los Alamos Scientific Laboratory Report No. LA-4102 (1969).
45. Houseman, B. L., *High Temp. Sci.* **8,** 25 (1976).
46. Houseman, B. L., and Elliott, G. R. B., *J. Electrochem. Soc.* **116,** 1363 (1969).
47. Conant, D. R., and Elliott, G. R. B., Test of Thermodynamic Theory Using EMF Cells Undergoing Vapor Equilibration, Los Alamos Scientific Laboratory Report No. LA-4791 (1972).
48. Morris D. R., Aksaranan, C., Waldron, B. S., and White, S. H., *J. Electrochem Soc.* **120,** 570 (1973).

0018–1536/84/1700–0289$02.60

Diatomic Partition Functions from Classical and Semiclassical Phase Integrals

JOEL TELLINGHUISEN

Department of Chemistry, Vanderbilt University, Nashville, Tennessee, 37235

Received September 23, 1983; Accepted December 31, 1983

ABSTRACT

Two methods for calculating diatomic partition functions directly from potential curves are examined with regard to their computational efficiency and accuracy. The first utilizes semiclassical phase integrals to estimate the rotation–vibration eigenvalues and may be considered exact. The second method entails numerical integration of the classical phase integral. The two methods are compared in test calculations of the bound and metastable contributions to the partition functions q_{vr} for Ar_2 and HgBr. These calculations show that the classical method, which is typically one to two orders of magnitude faster than the semiclassical method, gives results for the bound contributions to q_{vr} within 1% of exact when kT is at least twice the vibrational energy. For the metastable contributions, two alternative modifications of the classical expression perform less well, and it appears that these contributions can best be estimated by the semiclassical method.

The calculations for HgBr employ the latest spectroscopic constants for this molecule. The resulting values of q_{vr} are represented within 0.2% over the range 200–3000 K by an empirical expression containing six parameters, with the latter determined by the method of least squares.

Index Entries: Partition functions, of diatomic molecules; classical phase integrals, for diatomic partition functions; semiclassical phase integrals, for diatomic partition functions; HgBr, diatomic partition functions for; diatomic partition functions.

INTRODUCTION

Although Leo Brewer is very much "the Compleate Scientiste," he has always been, first and foremost, a thermodynamicist. Yet in the late 60s, when I was a member of his research group, there was a strong spectroscopic thrust to his research efforts, centered mainly around work on radiative lifetimes and the characterization of species isolated in low-temperature matrices. The problems that occupied most of my time and effort and became my thesis project were not directly concerned with thermodynamics or even with high temperatures. Rather, they involved the spectroscopy and photophysics of the I_2 molecule, and they developed almost "involuntarily," as a spinoff of an unexpected v'-dependence in decay rates measured on exciting I_2 to the B state at various wavelengths. (I_2 was originally to have been the simple test case for checking out the lifetime equipment prior to conducting experiments on high-temperature diatomics; it ended up as at least a major component of ~10 PhD theses and publications from Brewer's group.) In any event, with my preoccupation with I_2 (which continues today), I must confess that I completed my tenure with Leo without ever removing an uneasy feeling from the pit of my stomach . . . the fear that I didn't really understand thermodynamics!

Years of teaching thermo and statistical mechanics have done much to remedy that malaise. Now, although I am not active in thermodynamics on the research level, I feel at least conversant with the subject. Recently I have come to reconsider one of Leo's favorite thermodynamic aphorisms: "Thermodynamics doesn't care what you call a material—atoms, molecules, or whatever; as long as you treat the system consistently, you will obtain correct results." When I first heard him say this, I had a gut understanding that the truth of this statement was wrapped up in those fudge factors we know as activity and fugacity coefficients. But I think I might have been hard pressed to verify the point through numerical calculations. Now I believe I can do the latter, at least for the simplest kind of system—gaseous dimers in equilibrium with their parent atoms.

Although thermodynamics doesn't care about the formal definition of a system, the thermodynamicist can simplify his numerical calculations by choosing a definition which makes the system as close to "ideal" as possible. For example, we can call a container of xenon gas "Xe" or "Xe_2"; but under most conditions (except possibly at very low temperature and high pressure) the latter definition will show deviations roughly a factor of two from ideal in PVT behavior, and will require anomalous fugacity coefficients far from unity. On the other hand, by defining the system as a mixture of Xe and Xe_2 in equilibrium at all times, we can force the thermodynamics toward ideality. If we are careful how we define "Xe" and "Xe_2" in this system, we can be sure that any further deviations from ideality can be attributed to "polyatomic" contributions (e.g., Xe_3, Xe + Xe_2, Xe + Xe + Xe, Xe_2 + Xe_2, Xe_4, etc.). Then, in principle at least, we could redefine our system once again to include such species, and thereby force the system into still closer compliance with ideality.

In a recent paper (*1*) Mies and Julienne have shown very clearly how the partitioning of the system, dimers ↔ atoms, is to be carried out on a statistical mechanical level, in order to achieve the desired rigorously ideal behavior (neglecting higher-order interactions, of course, but including atom–atom collisions). Among their recommendations were the following: (1) To achieve ideal *PVT* behavior for this system, one should include the molecular continuum and metastable levels in the partition function q_{vr} for the dimer. (2) The safest procedure for evaluating q_{vr} is to begin with the potential curve or curves which define the molecular states of significance. (3) At high temperatures the classical method is both very accurate and by far the easiest procedure for evaluating q_{vr} (*2*). It is worth noting (as Mies and Julienne do) that similar recommendations were put forward long ago [e.g., refs. (*3*) and (*4*)], but were never fully developed. It should also be mentioned that for weakly attracting species such as the rare gases, this approach may lead to *negative* concentrations of "molecules" over much of the temperature range! This is because excluded volume effects dominate and contribute negatively to the continuum part of q_{vr}.

I can add nothing to the formal treatment by Mies and Julienne and can only recommend that their paper receive close scrutiny. However, I can offer some information and suggestions on the mechanics of calculating the molecular partition functions. In a recent paper I described a very efficient computational procedure for evaluating the "exact" bound and metastable contributions to the diatomic partition function for any potential at any temperature (*5*). This method utilizes semiclassical phase integrals to obtain the eigenvalues. Its accuracy and efficiency are attributable to the following observations: (1) For "realistic" diatomic potentials, the first-order semiclassical eigenvaluese are very close to the quantum mechanical eigenvalues (typically within 1 cm^{-1}); for the RKR potentials commonly used by diatomic spectroscopists, they are exact by definition. (2) The semiclassical eigenvalues can be evaluated very quickly and accurately by means of the Gauss-Mehler quadrature (*6*), coupled with interpolation. As a consequence, partition functions can be calculated by this method about as quickly as through the usual sum over states using spectroscopic constants to calculate the latter, and without certain problems concerning the range of validity of these spectroscopic constants (*5*).

In spite of its accuracy and efficiency, the semiclassical procedure is still slower and more intricate to code for machine computations than is the classical method. The latter is also particularly easy to apply when the molecular continuum is included in the phase space of the dimers (*1*, *2*). In the present paper I have carried out numerical calculations designed to examine the validity of the classical method, using the semiclassical method as my benchmark. The test cases are a weakly bound dimer (Ar_2) approximated by a 6–12 potential, and a more realistic "high-temperature" species (HgBr) having an RKR-based potential defined by recent spectroscopic work. These calculations indicate that the classical method is virtually exact at high temperatures and surprisingly accurate even at temperatures low enough to make it suspect (i.e., where the vibrational energy interval is roughly equal to kT). Although the calculations focus on just the bound and metastable contributions to q_{vr}, the near exact performance of the classical method for these parts makes it plausible that similar reliability will hold for the continuum region, in agreement with the suggestions of Mies and Julienne (*1, 2*).

THEORY

The usual starting approximation for evaluating the rotation–vibration partition function q_{vr} of a diatomic molecule is the rigid rotor/harmonic oscillator (RR/HO) model (*7*). At not-too-high temperatures the RR/HO partition function may be corrected for effects of anharmonicity, rotation-vibration interaction and centrifugal distortion (*8*). With increasing T these approximations must eventually fail, and the only safe procedure is the direct sum over states of the Boltzmann factors. In the usual implementation of this method, one uses spectroscopic constants to represent the energies $E(v, J)$ as functions of v and J, the vibrational and rotational quantum numbers. Whereas this procedure is computationally easy to perform using modern computers, one must nevertheless exercise caution in extrapolating the expressions for the spectroscopic constants beyond their range of validity, i.e., the range of the spectroscopic observations. It is for this reason that methods that focus on the molecular potential rather than the spectroscopic constants are preferred at high T, because the potential can be extrapolated to dissociation more easily than can the spectroscopic constants. [However, the recently developed methods that represent the spectroscopic constants by expressions using the dissociation limit as reference (*9–11*) make the use of spectroscopic constants much more straightforward for levels near dissociation.]

Apart from fine effects related to electron spin interactions, which for thermodynamics calculations can usually be handled through a simple degeneracy factor, the eigenvalues $E(v, J)$ for a given electronic state of a diatomic molecule are the solutions of the Schrödinger equation for the effective potential.

$$U_J(R) = U_0(R) + \beta J(J + 1)/R^2 \tag{1}$$

where the constant β is inversely proportional to the molecular reduced mass μ. (In cm^{-1} units, $\beta = B_e R_e^2$, where B_e and R_e are the equilibrium rotational constant and internuclear distance, respectively.) Thus if the rotationless potential $U_0(R)$ is known, one can systematically solve for all the eigenvalues of each effective potential $U_J(R)$, $J = 0, 1, 2, \ldots$ To do so by quantum mechanical methods, however, is tedious and expensive in CPU time, except for very light molecules or shallow potentials. As an alternative, I have employed the semiclassical phase integral (5),

$$h(\upsilon + 1/2) = (8\mu)^{1/2} \int_{R_1}^{R_2} [E - U_J(R)]^{1/2} dR \tag{2}$$

where h is Planck's constant and R_1 and R_2 are the classical turning points for energy E. The method I have devised based on this approach is very efficient; in test calculations I was able to evaluate 5×10^4 eigenvalues and ~10 partition functions in several minutes of CPU time on a DEC-1099 (not a "supercomputer").

Assuming that the potential $U_0(R)$ represents the ground electronic state of the diatomic and there are no other low-lying electronic states, the molecular partition function is just

$$q_{\upsilon r} = \sum_{J,\, \upsilon} (2J + 1)e^{-E(\upsilon,\, J)/kT} \tag{3}$$

The energy levels are designated as bound, metastable, or continuum levels, depending on the relationship of $E(\upsilon, J)$ to the effective potential $U_J(R)$. For reference, I show in Fig. 1 effective potentials for $J = 0$, 200, 350, and 500 for a Morse curve approximating HgBr (see below). All levels having $E < \mathfrak{D}_e$ are bound. All levels lying above the rotational barrier $E_{b,\,J}$ for that J belong to the continuum. Those levels inside the well for which $\mathfrak{D}_e \leq E \leq E_{b,\,J}$ are designated as metastable. Note that for $J > 350$ in Fig. 1, only metastable and continuum levels exist, whereas for $J > 500$ only continuum levels occur. In the present test calculations I am concerned with only the bound and metastable contributions to $q_{\upsilon r}$, which always dominate at sufficiently low T. For the continuum contributions one must be careful to include just the "molecular" part of the continuum, i.e., to exclude contributions from colliding noninteracting (ideal) atoms (*1, 2, 12*). (Note that for noninteracting atoms, $U_0 = 0$, but $U_J \neq 0$ for $J \neq 0$).

The classical partition function q_{cl} is given by the classical phase integral (*7*).

$$q_{cl} = h^{-n} \int exp[-H_{cl}(q, p)/kT]dq_1dp_1 \ldots dq_ndp_n \tag{4}$$

where n is the number of degrees of freedom, and q_i and p_i are conjugate coordinate and momentum variables. In the present case, $n = 3$, and we can conveniently express the element of volume in phase space in Cartesian or spherical coordinates:

$$\begin{aligned} d\tau &= dq_1dp_1 \ldots dq_3dp_3 = dxdp_xdydp_ydzdp_z \\ &= dRdp_Rd\theta dp_\theta d\varphi dp_\varphi \end{aligned} \tag{5}$$

The classical Hamiltonian is the sum of kinetic and potential energies. Since the latter is a function of internuclear separation only, it is more appropriate to use spherical coordinates:

$$H_{cl} = (1/2\mu)\ [p_R^2 + (p_\theta/R)^2 + (p_\varphi/R\ sin\ \theta)^2] + U_0(R) \tag{6}$$

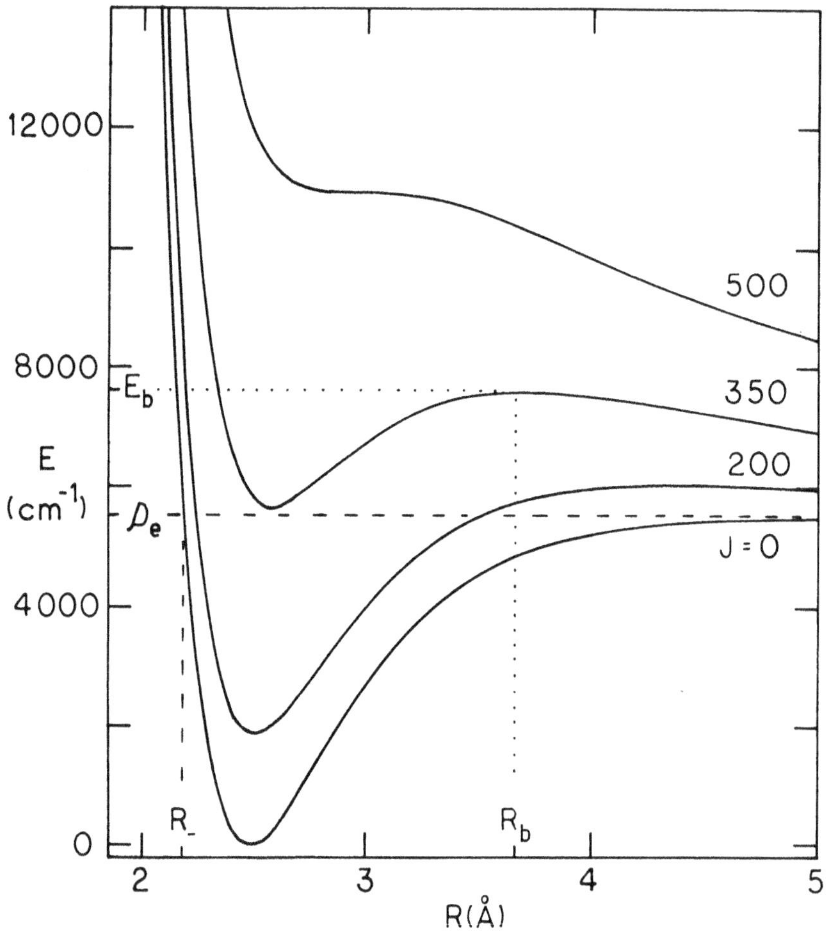

Fig. 1. Effective potentials, $U_J(R)$, for HgBr at selected values of the rotational quantum number J. Also shown are the quantities $\mathfrak{D}_e$ and R_- for the rotationless curve, and the location of the barrier (R_b, E_b) for $J = 350$.

For the bound part of q_{cl} we must include just that part of phase space that restricts the particle to be within the well. This we do by limiting the total kinetic energy to the value, $\mathfrak{D}_e - U_0(R)$. For this purpose it is only the total magnitude of the momentum that is important, and it is convenient to rewrite Eqs. (5) and (6) as

$$d\tau = (4\pi p^2 dp)\,(4\pi R^2 dR) \tag{7}$$

and

$$H_{cl} = (p^2/2\mu) + U_0(R) \tag{8}$$

Now the momentum integral in Eq. (4) is limited to the range, $-p_{max}(R) \leq p \leq p_{max}(R)$, where

$$p_{max}^2 = 2\mu[\mathfrak{D}_e - U_0(R)] \tag{9}$$

Correspondingly, the coordinate integral is limited to the range of R where $\mathfrak{D}_e \geq U_0(R)$, $R_- \leq R \leq \infty$ (see Fig. 1). The result is (13)

$$q_{cl,B} = (2\pi\mu kT/h^2)^{3/2} \times 4\pi\int_{R-}^{\infty} R^2 dR \, [e^{-U_0(R)/kT} f(A)] \tag{10}$$

where

$$f(A) = \int_0^A (4y/\pi)^{1/2} e^{-y} dy = \mathrm{erf}(A^{1/2}) - (4A/\pi)^{1/2} e^{-A} \tag{11}$$

In these equations erf is the error function and $A = [\mathfrak{D}_e - U_0(R)]/kT$. Note that although the exponential term in brackets in Eq. (10) becomes constant as $R \rightarrow \infty$, $f(A)$ goes to zero faster than R^{-3} for most potentials (see below), so the integral in Eq. (10) remains bounded.

The classical treatment outlined above requires the potential $U_0(R)$, but nowhere involves the effective potential $U_J(R)$. Thus it is not clear how the metastable contributions to q_{cl} might be calculated, since they are defined specifically in terms of the effective potentials. One could return to an explicit sum over J, with the (one-dimensional) vibrational spectrum treated classically, and presumably regain thereby an explicit evaluation of the metastable contributions. However, this seems pointless, since (a) rotation is generally more classical than vibration, and (b) the calculations are anyway likely to be more time-consuming than the full semiclassical calculations outlined above. Instead, I have examined two other ways of limiting the range of the momentum integration in Eq. (4), to see how closely these account for the metastable contributions. In the first case the three components of momentum are *individually* limited by the maximum allowed kinetic energy. In the second the rotational and vibrational energies are separately limited to $\mathfrak{D}_e - U_0(R)$. For these treatments the quantity $f(A)$ becomes

$$\begin{aligned} &(\text{For 1}): \quad [\mathrm{erf}(A^{1/2})]^3 \\ &(\text{For 2}): \quad (1 - e^{-A})\,\mathrm{erf}(A^{1/2}) \end{aligned} \tag{12}$$

I will call the corresponding partition functions $q_{cl,1}$ and $q_{cl,2}$. Note that both must be larger than $q_{cl,B}$, since the momentum space is less restricted. For the same reason $q_{cl,1}$ must be greater than $q_{cl,2}$. However, there is still no direct involvement of the effective potential $U_J(R)$, so no obvious relation to the exact $(q_B + q_M)$. This connection can only be investigated numerically, and is done so in calculations discussed below.

COMPUTATIONAL METHODS

The semiclassical computations have been described in detail in ref. (5). Briefly they involve evaluating the phase integral of Eq. (2) at specified, equidistant intervals of E between the minimum, $E_{m,J}$, and the barrier, $E_{b,J}$, for each effective potential $U_J(R)$. The eigenvalues are then estimated by interpolation over the resulting grid of E and (noninteger) v values. A surprisingly small number of E intervals (6–10), together with low-order Lagarangian interpolation, serve to estimate the important low υ levels with spectroscopic accuracy (~0.01 cm^{-1}). (This is because the relationship between E and υ is nearly linear at low υ). Levels near the top of the well may be considerably more in error (~10 cm^{-1}), but for these levels such errors have little effect on $q_{\upsilon r}$. In any event, their effect can be checked easily by increasing the number of points in the interpolation grid and the order of the interpolation.

To evaluate the classical partition function I have used a Simpson-type integration with higher-order corrections. Since this method involves a grid of equidistant points on the R-axis, it is necessary to specify the upper limit of the integration, R_+. In test calculations I found that the range beyond R = 10 Å made small but systematic contributions, and I elected to estimate these large-R contributions by assuming the potential behaves as $-C_n/R^n$ for $R \geq R_+$, with C_n determined by $\mathfrak{D}_e$ and $U_0(R_+)$. With this assumption the integrand in Eq. (10) takes the limiting form.

$$(4/3\sqrt{\pi})\ (C_n/kT)^{3/2} \exp\ (-\mathfrak{D}_e/kT)\ R^{(2\ -\ 3n/2)} \tag{13}$$

As long as $n > 2$ (which covers all but ionic interactions: these require special attention in any treatment), the integral of the quantity in Eq. (13) from R_+ to infinity is finite. For the commonly occurring case, $n = 6$, the large-R contribution to the integral in Eq. (10) is

$$(2/9\sqrt{\pi})\ (C_6/kT)^{3/2} \exp(-\mathfrak{D}_e/kT)\ R_+^{-6} \tag{14}$$

In the corresponding calculations for $q_{cl,1}$ and $q_{cl,2}$ [Eq. (12)], the first factor in Eq. (14) becomes $(4/3)\pi^{-3/2}$ and $(1/3\sqrt{\pi})$, respectively. Since $n = 6$ is appropriate in my test cases, I have used these expressions.

For the Ar_2 calculations I have used the same 6-12 potential employed in Ref. (*5*).

$$U_0(R) = 7.98 \times 10^8/R^{12} - 5.65 \times 10^5/R^6 \tag{15}$$

where the units of energy and distance are cm^{-1} and Å, respectively. For this potential $\mathfrak{D}_e$ = 100.01 cm^{-1} and R_e = 3.76 Å, which approximate the current best spectroscopic estimates (*14*).

For HgBr, recent spectroscopic work on the single isotopic species $^{200}Hg^{79}Br$ has given an improved description of the vibrational structure in the ground state for υ = 0–34 (*15, 16*), and from a partial rotational analysis, the estimate $R_e \approx 2.49$ Å. Without a complete rotation analysis, I am unable to calculate the full RKR curve. Hence I have used a Morse-RKR potential (*17*), for which the repulsive branch is defined by a Morse curve based on the experimental estimates of R_e and the first two vibrational constants, ω_e and $\omega_e x_e$. The RKR f-integral is then used to evaluate the width of the potential as a function of υ, from which the attractive branch is constructed.

The vibrational constants in ref. (*16*) were obtained from a direct fit to a mixed representation for the ground state—the usual polynominal in

(υ + *1*/2) for low υ levels, a near-dissociation expansion for high υ—with smoothness constraints as the switchover point. For the latter we chose $\upsilon = 25$, and to insure smoothness we required that both representations agree at $\upsilon = 25$ and 26. Although these constraints insure continuity of the function, the first derivative is only *approximately* continuous, as was evident from slow convergence of the RKR calculations for levels just above $\upsilon = 25$. Thus I have concluded that it is better to explicitly constrain the function and its first derivative to be continuous at a specified value υ. Accordingly I have modified the constants of ref. (*16*) very slightly to achieve this smoothness. The changes are completely insignificant for the present purposes and are mentioned here purely as a technical point. The constants are likely to change somewhat anyway as we extend the assignments to higher υ. In this last context it should be noted that although I am using the vibrational constants of ref. (*16*) to define the entire potential curve, the upper region of the curve remains experimentally uncertain, with $\mathfrak{D}_e$ uncertain by ± 75 cm^{-1}. For thermodynamic applications the latter uncertainty remains the limiting factor.

With the above mentioned changes in the smoothness constraints, the RKR calculations were extended to $\upsilon = 62$ with no convergence problems. This level is only three quanta below υ_D (the vibrational quantum number at dissociation) and 0.5 cm^{-1} below $\mathfrak{D}_e$ (= 5527.2 cm^{-1}), and has a right-hand turning point near 11 Å. The mixed representation for the vibrational energy, $G(\upsilon)$, is readily handled by the method I have described previously (*18, 19*), which employs another version of the Gauss-Mehler quadrature (*20*).

In the Simpson's method estimation of the classical phase integral (Eq. 10), the use of a step size $\Delta R = 0.01$ Å agreed with results for $\Delta R =$ 0.02 Å to better than one part in 10^4, so no further decrease in step size was needed. For the evaluation of $q_{cl,B}$, $q_{cl,1}$, and $q_{cl,2}$, at ten temperatures, the classical calculations (which could be done in single precision) took about 5 *s* of CPU time, which was about two orders of magnitude faster than the corresponding semiclassical calculations in the case of HgBr, or one order of magnitude in the case of Ar_2.

RESULTS AND DISCUSSION

The results of the Ar_2 calculations are illustrated in Fig. 2. The semiclassical and classical estimates of the bound contribution to q_{vr} agree too closely to be distinguished in the scale of this plot. (The classical value was 1.1% high at 100 K, decreasing to 0.6% at 800 K.) On the other hand the classical functions, $q_{cl,1}$ and $q_{cl,2}$, sandwich the exact (semiclassical) values for the combined bound and metastable contributions. In fact the average of $q_{cl,1}$ and $q_{cl,2}$ is within 1.5% of the exact value at all T, hence is a reasonably good approximation of the latter.

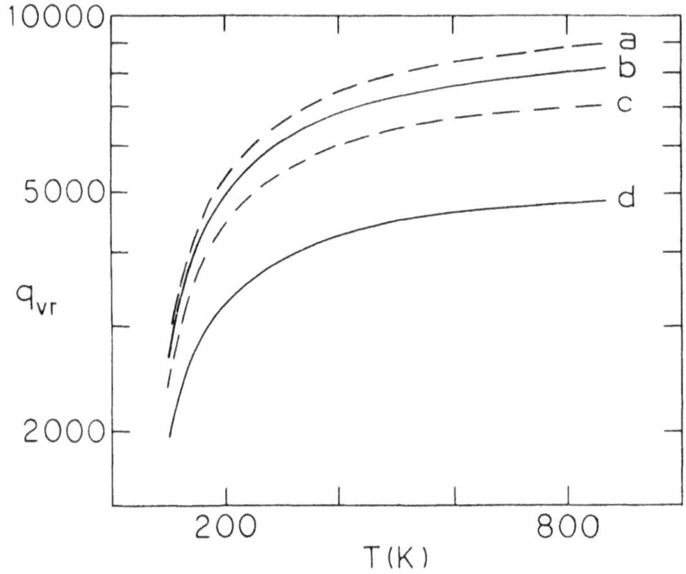

Fig. 2. Comparison of classical and exact partition functions for Ar_2 as a function of T. The plotted quantities are as follows: (a) $q_{cl,1}$, (b) the exact (q_B + q_M), (c) $q_{cl,2}$, and (d) $q_{cl,B}$ ($\approx$ the exact q_B).

Unfortunately the last result is not general, as is evident from the HgBr results, shown in Fig. 3. Here $q_{cl,1}$ is reasonably close to ($q_B + q_M$), but it deviates progressively at high T (see curve b). The function $q_{cl,2}$ (not shown) is about halfway between $q_{cl,B}$ and $q_{cl,1}$. Note that again $q_{cl,B}$ (curve b) is within 0.5% of exact except at the low-T end. However, even at 200 K, where kT is only 74% of the vibrational energy, the error is only 4%. From simple considerations of zero-point energies, one might expect the classical result to be in error by as much as a factor of two. The present calculations make it clear that the classical method is in fact much better than has generally been recognized. There is one slightly surprising result in the calculations displayed in Fig. 3: Although the error in the classical approximation to q_B is only 0.35% at 3000 K, it is actually *increasing* with T. I was suspicious of this result, so I checked it by repeating the calculations using a Morse approximation to the HgBr potential. The agreement was better in this case but still not exact, and again the deviation was increasing at high T, but in the opposite direction. This effect may have to do with the manner in which the phase space near the dissociation limit is sampled discretely by the exact partition function.

For completeness I have included in Fig. 3 the comparisons for the exact evaluation of ($q_B + q_M$) for the Morse approximation to the ground state (curve a) and the RR/HO model (curve d). The Morse results are high, because the attractive branch of the potential lies below the "true" (Morse-RKR) curve in its upper region, making more phase space available to bound molecules. The RR/HO partition function drops below the

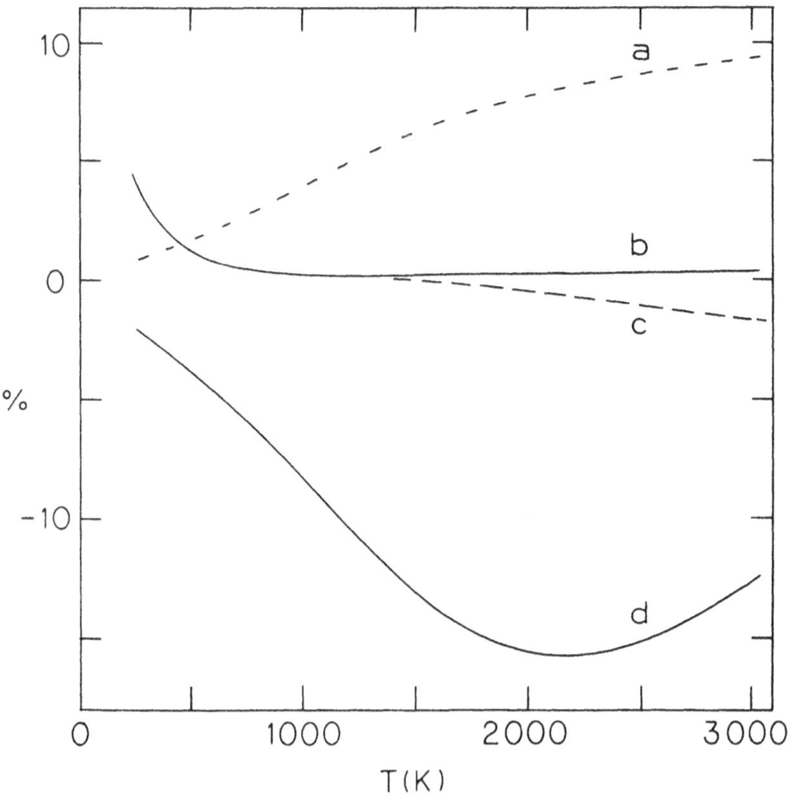

Fig. 3. Percent deviation from exact for various partition functions of HgBr. For curves (a), (c), and (d), the reference is the exact ($q_B + q_M$) calculated for the Morse-RKR curve. The quantities compared are: (a) the exact ($q_B + q_M$) for the Morse approximation to the ground state potential, (c) the classical function, $q_{cl,1}$, and (d) the RR/HO partition function. Curve (b) shows the comparison between the classical and exact values of q_B for the Morse-RKR curve.

exact values in this temperature range: however, it is rising at high T, as it must, since the RR/HO model assumes an infinite rotational and vibrational spectrum, and consequently the RR/HO q_{vr} must exceed the exact q_{vr} at high T (5).

For thermodynamic applications I have experimented with several ways of representing q_{vr} for HgBr and have settled on the expression,

$$\ln (q_{vr}) = 7.485 + 1.615 \ln(y) - 0.877/y + 6.78 \times 10^{-2}y - 1.406 \times 10^{-3}y^2 + 6.96 \times 10^{-6}y^3, \quad (15)$$

where $y = (T/100)$. The constants in this expression were obtained from a least-squares fit of the semiclassical ($q_B + q_M$) values for $^{200}Hg^{79}Br$ at T = 200–1200 K in steps of 100 K, extending to 3000 K in steps of 300 K thereafter. The coefficients have been rounded with care, and Eq. (15) reproduces the original q_{vr} values with a maximum deviation of 0.2%. I have verified that this function and its first derivative are "smooth" functions of T over the stated range and somewhat beyond. Note that the reference of energy is the minimum of the potential and that $\mathfrak{D}_e$ is 5527.2 cm^{-1} for this curve. The first constant in Eq. (15) includes ln 2 for the spin degeneracy. For other isotopomers of HgBr this constant can be altered in an obvious way to reflect the $\mu^{3/2}$ dependence in q_{vr} [Eq. (10)].

CONCLUSION

In this work I have examined the numerical accuracy of the classical method of calculating diatomic rotation–vibration partition functions: As a reference I have used an exact sum over states, with the latter estimated by a semiclassical technique. Test calculations for Ar_2 and HgBr indicate that the classical estimate of the bound partition function, $q_{vr,B}$, is correct within 1% when $kT > 2hc\omega_e$. This performance is much better than is generally recognized. For the estimation of the metastable contribution to q_{vr}, I have tried two other ways of defining the available momentum space in the classical method; but neither has proved fully satisfactory, and it appears that q_M can best be assessed by the semiclassical procedure.

The present calculations have dealt with just the bound and metastable contributions to q_{vr}. However, it seems likely that the classical expression for the full q_{vr}, including the continuum, will be equally reliable and by far the easiest way to evaluate q_{vr} when the molecular continuum is included within the phase space allocated to the molecule (*1*, *2*). For the calculation of concentrations of bound and metastable molecules in equilibrium with the parent atoms, however, the partition functions, q_B and q_M, are still of value and can be calculated easily by the methods I have outlined.

For thermodynamic applications I have expressed my results for $^{200}Hg^{79}Br$ in terms of a simple empirical equation, with coefficients determined by the method of least squares. The results in this case are still limited by experimental uncertainties, primarily in the dissociation energy and the sketchy rotational analysis. We are continuing work on the spectroscopy of this molecule and the other HgX diatomics and hope to reduce the remaining uncertainties.

ACKNOWLEDGMENTS

I want to thank Fred Mies and Paul Julienne for several helpful discussions and for sending me a copy of ref. (*1*) prior to its publication. This work was supported by the Office of Naval Research.

REFERENCES

1. Mies, F. H., and Julienne, P. S., *J. Chem. Phys.* **77,** 6162 (1982).
2. Mies, F. H., and Julienne, P. S., to be published.
3. Beckett, C. W., and Haar, L., *Proc. of the Joint Conference on Thermodynamic and Transport Properties of Fluids,* Institute of Mechanical Engineers, London, p. 27 (1957).
4. Haar, L., *Kinetics, Equilibria and Performance of High Temperature Systems,* Proc. of the First Conference of the Western States Section of the Combustion Institute, Butterworth, London, 1959.
5. Tellinghuisen, J., *Chem. Phys. Lett.* **102,** 4 (1983).
6. Tellinghuisen, J., *Chem. Phys. Lett.* **18,** 544 (1973).
7. Hill, T. L., *An Introduction to Statistical Thermodynamics,* Addison-Wesley, Reading, Mass., 1960.
8. Pitzer, K. S., and Brewer, L., *Thermodynamics* (revised from original by G. N. Lewis and M. Randall), McGraw-Hill, New York, 1961.
9. Le Roy, R. J., and Lam, W.-H., *Chem. Phys. Lett.* **71,** 544 (1980).
10. Tromp, J. W., and Le Roy, R. J., *Can. J. Phys.* **60,** 26 (1982).
11. Tellinghuisen, J., *J. Chem. Phys.* **78,** 2374 (1983).
12. Hirschfelder, J. O., Curtiss, C. F., and Bird, R. B., *Molecular Theory of Gases and Liquids,* Wiley, New York, 1954.
13. Hedges, R. E. M., Drummond, D. L., and Gallagher, A., *Phys. Rev. A.* **6,** 1519 (1972).
14. Colbourn, E. A., and Douglas, A. E., *J. Chem. Phys.* **65,** 1741 (1976).
15. Tellinghuisen, J., and Ashmore, J. G., *Appl. Phys. Lett.* **40,** 867 (1982).
16. Tellinghuisen, J., and Ashmore, J. G., *Chem. Phys. Lett.* **102,** 10 (1983).
17. Tellinghuisen, J., and Henderson, S. D., *Chem. Phys. Lett.* **91,** 447 (1982).
18. Tellinghuisen, J., *J. Mol. Spectr.* **44,** 194 (1972).
19. Tellinghuisen, J., *Computer Phys. Commun.* **6,** 221 (1974).
20. Kopal, Z., *Numerical Analysis,* Chapman and Hall, London, 1961.

0018–1536/84/1700–0303$06.00

Surface/Interfacial Free Energies and the Surface Tension of Uranium Dioxide

M. S. DESHPANDE AND P. D. DESAI*

Center for Information and Numerical Data Analysis and Synthesis, Purdue University, 2595 Yeager Road, West Lafayette, Indiana 47906

and

A. A. SOLOMON

School of Nuclear Engineering, Purdue University, West Lafayette, Indiana 47907

Received September 23, 1983; Accepted December 31, 1983

ABSTRACT

The purpose of this study is to review literature on surface/interfacial free energies and surface tension of $UO_{2\pm x}$. The data available in the literature are reviewed and critical evaluation and analyses of the available data are made by comparing them not only with each other, but also with the estimated values based on the available theoretical models.

In light of the complexity of the material and the problems associated with the available literature data, no recommendations of surface/interfacial free energies and surface tension values are possible at this time. However, an attempt is made to point out problems associated with the data in general and also to develop procedures that can be used to analyze surface energies.

Index Entries: Surface/interfacial energies, of UO_2; surface tension, of UO_2; uranium dioxide, thermodynamic properties of; interfacial energies, of UO_2 surfaces.

*Author to whom all correspondence and reprint requests should be addressed.

1. INTRODUCTION

The safety of nuclear power is directly linked to the reliable and predictable behavior of the UO_2 fuel. Thus, better understanding of this fuel's behavior is necessary. The surface and interfacial free energies are keys to understanding this behavior. Knowledge of the surface energies of nuclear fuels is also important to understand, predict, and quantitatively evaluate a number of diffusion-controlled processes as well as the swelling phenomena in nuclear fuels. The surface energy is a significant parameter in the growth, migration, and coalescence of gas bubbles. Its role in determining equilibrium size and pressure in gas bubbles is well-known. Similarly, the surface free energy can play an important role in the microcracking and failure of brittle materials. Thus, reliable fuel performance codes require accurate and dependable surface free energy values as an input.

In accident situations, the wetting behavior between oxide phases and liquid metals and solids is equally important; this is also true for cermet technology and phase boundary strength.

2. REVIEW OF MEASUREMENT METHODS FOR SURFACE/INTERFACE FREE ENERGIES AND SURFACE TENSIONS

2.1. Surface Tension Measurements of Liquids

Several reviews of surface tension measurement techniques have been reported in the literature [e.g., Adam (*1*), Bikerman (2), Davies and Rideal (*3*), Semechenko (*4*), White (*5*, *6*), and most recently Murr (*7*)]. The purpose of this presentation is to review briefly those measurement techniques that are important and/or have been used for $UO_{2\pm x}$. The detailed classification and discussion of methods can be found in Deshpande et al. (*8*), along with a consideration of assumptions made, requirements for optimal experiments, and expected accuracy for various experimental methods. Therefore, only those methods that have been commonly used for $UO_{2\pm x}$ or are potentially useful because of their accuracy are discussed here.

2.1.1. Sessile Drop Method

A schematic of the sessile drop is illustrated in Fig. 1. The shape of a liquid drop at equilibrium on a horizontal solid surface is determined by minimization of the summation of the potential energy caused by the gravitational field and the interfacial energy contributions from liquid–vapor, solid–vapor, and solid–liquid interfaces. A mathematical description of the profile of such a drop (Fig. 1) can be obtained by applying the Laplace Equation. A derivation of the Laplace equation as one

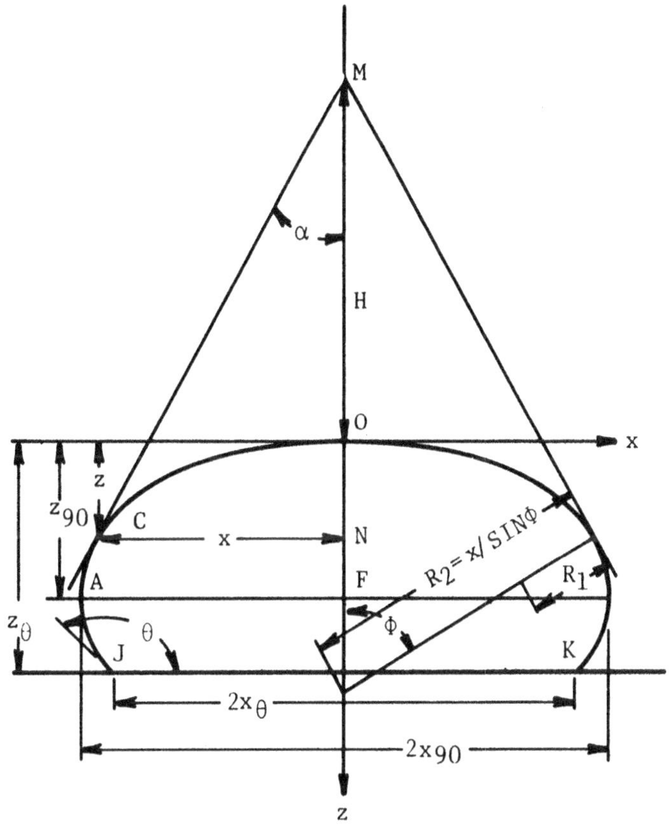

Fig. 1. Profile of a sessile drop showing elements of measurements.

of the conditions of equilibrium of a sessile drop was carried out by Keskar [9], who showed that

$$\Delta P = \gamma_{LV} \, [1/R_1 + 1/R_2] \tag{1}$$

where ΔP is the pressure difference across a general curved interface at any point; R_1 and R_2 are the principal radii of curvature, and γ_{LV} is the liquid–vapor surface tension.

At any point on the interface between the liquid and vapor, the ΔP term is given by

$$\Delta P = gz(\rho_L - \rho_V) + c \tag{2}$$

where ρ_L and ρ_V are bulk densities of liquid and vapor phases, respectively, c is a constant, g is the gravitational acceleration, and z is the vertical coordinate of the point with respect to the origin, selected at the topmost point of the profile, as shown in Fig. 1.

From Eqs. (1) and (2) and neglecting ρ_V (since it is about four orders of magnitude less than ρ_L),

$$\gamma_{LV}[1/R_1 + 1/R_2] = gz p_L + c \tag{3}$$

At the origin $z = 0$, $R_1 = R_2$ (= R say), and $c = 2\gamma_{LV}/R$. Therefore,

$$\gamma_{LV}[1/R_1 + 1/R_2] = gz\rho_L + 2\gamma_{LV}/R \tag{4}$$

Since $R_2 = x/\sin\phi$ (Fig. 1), Eq. (4) can be rewritten as

$$R/R_1 + \sin\phi/(x/R) = 2 + g\rho_L R^2/\gamma_{LV}\,[z/R] \tag{5}$$

The last term in Eq. (5), $g\rho_L R^2/\gamma_{LV}$ is a dimensionless parameter referred to as β. It is constant for a given drop at a given temperature and must be determined experimentally in order to calculate the surface tension, γ_{LV}.

From geometrical considerations of Fig. 1, we know

$$R_1 = \{1 + [\partial z/\partial x]^2\}^{3/2} \Big/ d^2z/dx^2 \tag{6}$$

and

$$\sin\phi = (dz/dx)\,[1 + (dz/dx)^2]^{1/2} \tag{7}$$

Substituting Eqs. (6) and (7) into Eq. (5) we get

$$\frac{d^2z}{dx^2} + \left\{1 + \left(\frac{dz}{dx}\right)^2\right\}\frac{1}{x}\frac{dz}{dx} = \left(2 + \frac{g\rho_L}{\gamma_{LV}}\cdot z\right)\left\{1 + \left(\frac{dz}{dx}\right)^2\right\}^{3/2} \tag{8}$$

and the boundary conditions are: at $x = 0$, $z = 0$, and $1/x\,(dz/dx) = 1$, an analytical solution to Eq. (8) is not known. Extensive effort has been made to develop approximate solutions, but the validity of these solutions is limited to very large sessile drops that are either hard to obtain or to maintain experimentally. These solutions have been reviewed by Porter (*10*) and are not discussed here because of their limited applicability. Bashforth and Adams (*11*) proposed a numerical solution and reported tabulated values of β and x/z for $\phi = 90°$, along with separate tables of z/R and x/R at various β values as function of ϕ. From measurements of x_{90} and z_{90} (Fig. 1), β and R can be determined by using these tabulated values. Paddy (*12*) has revised these tabulated values so that γ_{LV} values can be calculated more accurately using the simple relation:

$$\gamma_{LV} = g\rho_L R^2/\beta \tag{9}$$

A relatively convenient approach was taken by Koshevnik et al. (*13*) for the calculation of surface tension by using direct measurement of the drop profile. For small drops of very close to spherical geometry, they obtained a simpler relation:

$$\gamma_{LV} = D(2x_{90})^2\,\rho_L g \tag{10}$$

where D is a constant. Bikerman (2) has also tabulated D as a function of $(2x_{90}/z_{90})$. It was observed that, since D increases with decreasing $(2x_{90}/z_{90})$, sessile drops become more spherical as they decrease in size.

For an infinitely large sessile drop, which has effectively no curvature at the summit of the drop, Eq. (10) can be rewritten as [Bikerman (2)].

$$\gamma_{LV} = \rho_L g(z_{90})^2/2 \tag{11}$$

Some of the most careful measurements of surface tensions of liquid metals and alloys were reported by Humenik and Kingery (*14*). They used the criteria of Eq. (9) and an alternative method of calculation suggested by Dorsey (*15*) to calculate surface tension from a sessile drop profile, using the empirical relation

$$\gamma_{LV} = [g\rho_L(x_{90})^2][(0.0520/f) - 0.1227 + 0.0481f] \quad (12)$$

where

$$f = (z_{90}/x_{90}) - 0.4142 \quad (13)$$

The γ_{LV} value obtained from Eq. (12) depends on profile measurements from the top of the drop to the intersection of the axis with a 45° tangent to the drop (see Fig. 2.1) and does not rely on interpolations or calculations based on tabulated D versus (x_{90}/z_{90}) values by Bikerman (*2*). According to Bonfield (*16*), Eq. (12) represents perhaps the most accurate method of measuring surface free energies of liquid metals and alloys, particularly when the sessile drop is small in diameter (<4 cm). The method is attractive since the amount of UO_2 required to melt and to form sessile drops is much less than that required in maintaining a larger pool of liquid UO_2 as required in other techniques. With the aid of photographs of the sessile drop profile, the method is particularly suitable for investigations of isotropic surfaces and symmetric drops. Photographs can be taken at any desired and well-defined interval without disturbing the surface in any way. The measurements of drop profiles from photographs then can be made at a later time as desired. The major requirement for the use of the sessile drop technique is that the contact an-

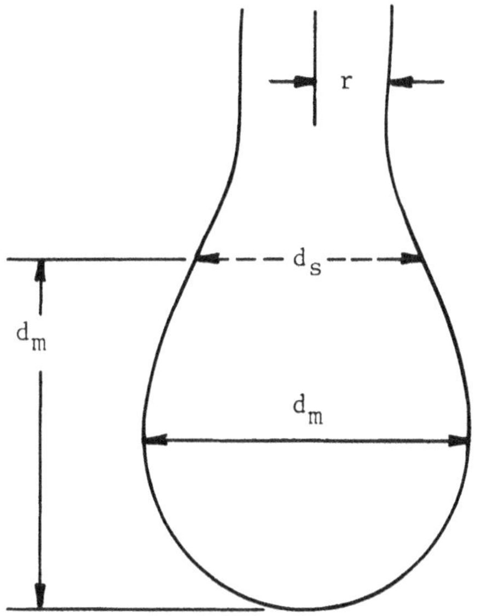

Fig. 2. Pendant drop profile and associated geometric parameters.

gle should be >90° in order to obtain symmetrical drops and for better precision of the measurements. Usually this requirement is satisfied with proper choice of the substrate. Butler and Bloom (*17*), Maze and Burnett (*18*), and Keskar (*9*) developed computer programs to analyze sessile drop profiles, and such analysis has become routine rather than use of contact angle measurements. Keskar (*9*) presented an extensive error analysis of the technique, where he pointed out the requirements for better accuracy and noted that for $\phi < 30°$ (see Fig. 1) excessively large errors could result and suggested that these points should not be included as input data in profile analysis. To get a better profile, Keskar (*9*) also pointed out that larger sessile drops with symmetry are useful and recommended use of cup-shaped substrates. However, their use precludes any measurement of contact angle or density. He also added that the method was the most suitable when the surface tension value was not very low in magnitude. Use of this method is not recommended for liquids with very low surface tension and/or high volatility. In such cases, Keskar (*9*) concluded that an alternate method in which measurements are made on fresh surfaces (e.g., the maximum bubble pressure method) appear suitable.

The accuracy of sessile drop measurements, therefore, depends heavily on the geometric details of the profile. Since the shape of the profile is controlled by the weight of the sessile drop, the recent study of Sangiorgi et al. (*19*) can be taken to represent conditions for an optimal experiment, since they considered the effect of sessile drop weight on measured γ_{LV} values.

Keskar (*9*) also suggested a procedure for objective comparison of various data obtained by the sessile drop technique. He emphasized inclusion of density $\rho = \rho(T)$ data to support surface tension measurements, and suggested comparison of γ_{LV}/ρ values.

2.1.2. The Drop Weight Method

When a falling drop (Fig. 2) detaches itself from the bulk, it could be reasoned that the weight of such a drop balances the surface tension at the point of detachment. The relationship between the drop weight, the detachment force, and the surface tension is given by (ref. *20*)

$$\gamma_{LV} = m_o g / 2\pi r f \tag{14}$$

where m_o is the weight of the static drop hanging from a wetted vertical rod of radius r and f defines a correction factor for the true surface tension of drop separating from the rod (usually the end of the rod is melted to form the drop). The quantity f is a single-valued function of $r/(m_o/\rho_L)^{1/3}$, where ρ_L is the liquid density, with values usually in the range 0.6–0.8 [*see* Harkins and Brown (*20*)]. The term m_o/ρ_L represents the volume, V, of the detached drop. Equation (14) is usually rewritten as

$$\gamma_{LV} = (m_o g / r) \cdot \phi \tag{15}$$

where ϕ is the drop weight correction factor, which is related to f by a simple relation: $\phi = 2\pi f$. Tables relating ϕ to $r/(m_o/\rho_L)^{1/3}$ have been made available by Lando and Oakley (*21*).

This method was recently used by Schins (22) to measure γ_{LV} of UO_2. He carried out experiments for optimal drop weight. His results are presented in Section 4 of the numerical data.

2.2. Surface/Interfacial Free Energy Measurements for Solids

Some of the popular methods used for the measurement of surface free energies of solids are reviewed by Inman and Tipler (23), Greenough (24), Bikerman (25), Blackley and Maiya (26), Hondros (27), Levy and Regnier (*28*), Linford (*29*), Murr (*7*), and very recently by Regnier (*30*). Regnier (*30*) based the evaluation of methods on a thermodynamic approach, but evaluation of these methods for different classes of materials and sources of errors is not completely realized. The commonly used methods are reviewed very briefly here to highlight the current state-of-the-art.

2.2.1. Cleavage Method

The cleavage method involves measurements of the work needed to create new single crystal surfaces. Since cleavage is a dynamic irreversible process and not a quasistatic one as required by the condition of reversibility, the data produced are not in accordance with definition. The reversibility cannot allow use of growing cracks. Even for nongrowing cracks, dissipative processes operate at the crack tip as instability is approached. This yields large errors even at low temperatures and especially for brittle materials. Because of these limitations, the methods like 'Hertzian Indentation' as devised by Matzke et al. (*31*) are merely useful to evaluate relative behavior and cannot be used for quantitative measurement of the intrinsic surface energy of the material. Regnier (*30*) concluded that the γ values obtained by this method are 'fracture surface energies' that are upper limits of surface free energy.

2.2.2. Calorimetric Method

The principle of the calorimetric method is to convert the total surface energy, γA into heat. However, for experimental simplicity, the difference between the heats of solution of equal masses of powdered and lump forms of a sample are measured, and not the difference between the heats of sublimation [see, Lipsett et al. (*30A*)]. Thus, the method gives surface tension of the solid-solvent interface rather than surface free energy.

2.2.3. Zero-Creep Method

The zero-creep method for determining surface free energy of solids has been in use for the last 50 yr. The first measurement of surface energy is dated back to 1930 when Sawai and Nishida (*32*) used this technique on metals. Since then, the method has been modified and used even for alloys. The technique is based on the fact that very fine wires and foils contract when heated to temperatures close to the melting temperatures. For a large surface-to-volume ratio, the surface tension (or shrinkage force) will exceed the static stress associated with the weight of wire or foil sample. It is generally accepted that when a thin wire or foil specimen is heated to a high enough temperature it contracts due to the surface-tension-driven diffusion of atoms, in order to minimize its surface energy. On the other hand, when it is slightly loaded, it deforms. The initial change in length of such specimens is governed mainly by a redistribution of dislocations, offsetting inclined grain boundaries [Greenough (*33, 34*)] and orientation changes which reduce the specific grain boundary energies [Herring (*35*)]. The linear expansion or contraction of the specimen is generally assumed to be caused by the Nabarro-Herring mechanism, where material transport is affected by a flow of vacancies from the grain boundaries to the external surface (solid–vapor interface) or vice versa. Thus, there exists a critical load for which the creep rate vanishes, given by the generalized relation

$$F_c \, dl = \gamma_s \, dA_s + \gamma_b \, dA_b \tag{16}$$

where F_c is the critical load, dl the virtual infinitesimal change in length, $\gamma_s \, dA_s$ the infinitesimal work of surface expansion, and $\gamma_b \, dA_b$ the corresponding term for boundary expansion. Assuming volume is constant, the variation of length can be related to the variation of area if the geometry of the surfaces and boundaries are known. This can be easily done for a bamboo-structure wire, where grain boundaries are perpendicular to wire axis. The expression for surface free energy is given by the equation

$$\gamma_{SV} = W_o g/\pi r\{1 - 2(r/l)\cos(\theta_s/2)\} \tag{17}$$

where W_o is zero-creep load, r is the radius of the wire, l is the average grain length, g is the gravitation constant, and θ_s is the dihedral angle. A similar relationship holds for a sheet specimen [see Murr (*7*) for details]. The advantage of wire over sheet is the probability of producing a bamboo-structure wire specimen, although very fine wire is weaker than thin sheet. The experiment is usually carried out with a set of samples from which different small loads are hung. The positive or negative creep of each sample and plots of the steady-state creep rate versus load are observed. From this plot the critical load is interpolated, and Eq. (17) is used to obtain the surface tension. Regnier (*30*) pointed out that the use of the principle of virtual work to write Eq. (16) is not absolutely rigorous. This is because the wire with zero-creep load is not in complete equilibrium since the lateral surface of this wire can still move by surface-tension motivated diffusion of atoms. Regnier (*30*) suggested that the condition of zero-creep rate for a bamboo-structure wire is that the sum of the divergence of the flux of atoms along the grain boundary and the flux of atoms normal to the boundary vanishes, then

$$\mathrm{div}(J_b) + J_m = 0 \tag{18}$$

During elongation of the wire, part of the work caused by applied load is reversibly converted into surface energy and the rest is dissipated by the diffusion current. Also, part of the work of the surface tension is dissipated by diffusion processes. It is the total energy dissipated per second that controls the rate of the shape evolution of the grain. A full derivation of the relationship between the surface free energy and the zero-creep load is not fully known.

A further discrepancy between the parameters measured and the surface tension is that the wire is usually surrounded by an inert gas or by hydrogen to prevent evaporation or oxidation, and thus only the interfacial tension between the gas and the solid can be measured. The relative errors reported for zero creep vary with authors which ranges from 10 to 20%. On the other hand Kumikov and Khokonov (*36*) reported an accuracy of 1–3%.

From evaluation of Eq. (*17*), one can see that systematic errors are also possible, which are mainly due to error in measurement of dihedral angles. The aim is to choose a sample and a grain size such that the contribution of γ_{gb} to zero creep load is a minimum during γ_{SV} measurements. This forces us to review equilibration of grain boundaries with free surfaces.

There are four possible mechanisms controlling the rate of equilibration of grain boundaries with free surfaces. These can be summarized as (1) evaporation–condensation, (2) volume diffusion in the solid, (3) diffusion in the vapor phase, and (4) surface diffusion.

These mechanisms have been studied carefully to predict the groove profiles. The details are summarized by Deshpande et al. (*8*) in a detailed report. It is known that groove width for various mechanisms is proportional to various powers of time, *t*; e.g., for evaporation-condensation mechanisms it is proportional to $t^{1/2}$, for volume diffusional mechanisms to $t^{1/3}$, whereas for surface diffusion mechanisms it is proportional to $t^{1/4}$.

Thus, θ_S values are time dependent and the use of proper θ_S values should yield reliable γ_{SV} values. It was generally argued that for the wire samples with very small *r*/*l* ratios negligible grooving takes place and the $\cos(\theta_S/2)$ term in Eq. (17) is very small. Regnier and Felsen (*37*) derived a correction to γ_{SV} values assuming grain boundary work was not negligible and reported net errors of up to 50–150% in the γ_{SV} values for copper wires reported in the literature. The feasibility of this method for ceramics like UO_2 is also not yet established although tensile creep experiments have recently been carried out on polycrystalline ceramics like CoO and UO_2 (*38, 38A*).

2.2.4. Multiphase Equilibrium

The equilibrium of the stress vectors at two or three meeting interfaces leads to the following relation among the components (ref. *39*):

$$\sum_i (\gamma_i + \partial\gamma_i/\partial\psi_i) = 0 \tag{19}$$

where the γ_i are surface tensions and the $d\gamma_i/d\psi_i$ are torques at corresponding interfaces, as shown in Fig. 3(a). When the torque terms are neglected and the junction is that of a boundary (solid–solid interface) with a surface (solid–gas interface), this equation reduces to:

$$\gamma_{SS}/\gamma_{SV} \text{ or } \gamma_{gb}/\gamma_{SV} = 2 \cos \theta_{SV}/2 \tag{20}$$

whereas if the atmosphere is gas + metal vapor, Eq. (2) is written as

$$\gamma_{SS}/\gamma^*_{SV} \text{ or } \gamma_{gb}/\gamma^*_{SV} = 2 \cos \theta^*_{SV}/2 \tag{21}$$

A similar relation holds when a drop of a nonreacting liquid fills the junction (see Fig. 3c). Thus,

$$\gamma_{SS}/\gamma_{SL} = 2 \cos \theta_{SL}/2 \tag{22}$$

In addition, if the liquid does not wet the solid, an additional condition of equilibrium is:

$$(\gamma^*_{SV} - \gamma_{SL})/\gamma_{LV} = \cos \theta_{LV} \tag{23}$$

Thus, the four tensions [γ_{SS} (γ_{gb}), γ_{SV} or γ^*_{SV}, γ_{SL}, and γ_{LV}] are related by Eqs. (20–23), and four equilibrium angles are experimentally measurable. Therefore, if the surface tension of the neutral liquid is known, the other tensions can be determined as

$$\gamma_{SV} = \gamma_{LV} \frac{\cos \theta_{SV} \cdot \cos \dfrac{\theta_{SL}}{2} \cdot \cos \dfrac{\theta_{SV}}{2}}{\cos \dfrac{\theta_{SV}}{2} \left(\cos \dfrac{\theta_{SL}}{2} - \cos \dfrac{\theta^*_{SV}}{2} \right)} \tag{24}$$

It should be noted that this method gives the relationships among tensions because the junction is in equilibrium with respect to the transport of atoms. This method as reported by Allen (*39*) gives only a crude measurement of the surface tension of solids because torque terms are neglected, and in practice according to Regnier (*30*) there is always some liquid segregated in the grain boundary and some vapor adsorbed on the surface. Mykura (*40*) has derived the anisotropy of the surface tension from the equilibrium angles of the two inverted grooves of a twin boundary by considering torque terms. The data generated using this technique are presented in Section 4.2.

2.2.5. Multiphase Equilibrium with Interphase Control of Stoichiometry

Hodkin and Nicholas (*41*) specially devised this modification of multiphase equilibrium method to study effect of O/M ratios on surface free

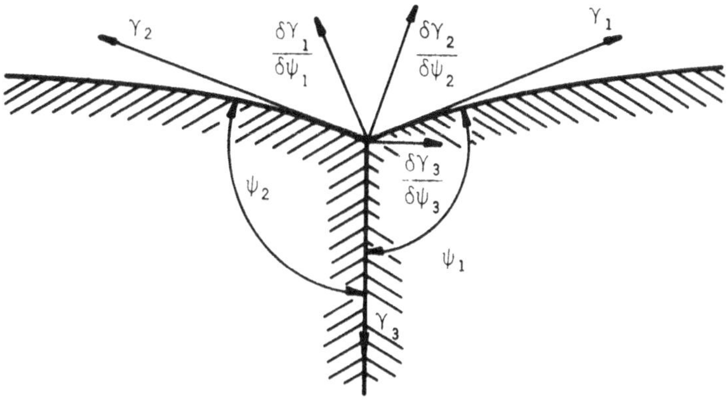

Fig. 3a. Equilibrium of surface tensions at the intersection of a grain boundary with a free surface.

energy of UO_{2+x} and mixed oxide fuels. They used copper–1 at% thorium alloy as reactive fluid in measuring contact angles at high temperatures. According to Hodkins and Nicholas (*41*), the multiphase equilibrium approach can not be adopted to nonstoichiometric UO_2 because of the compositional changes that could be produced during the prolonged heat-treatments involved in some of the equilibrations. Thus, the interfacial control technique, as suggested by Martimore and Nicholas (*41A*) is usually used to derive surface energies. The use of short term grooving experiments can give grain boundary energy along with the mass transport data.

The interfacial technique basically involves measuring contact angles of a reactive alloy on two substrates. If reactive elements in the alloy control the chemistry of the interfaces, their energy will be identical for both substrates and hence (ref. *41*)

$$\gamma^*_{SV_1} - \gamma^*_{SV_2} = \gamma_{LV}(\cos\theta_1 - \cos\theta_2) \tag{25}$$

where $\gamma^*_{SV_1} - \gamma^*_{SV_2}$ are the surface free energies of two substrates in presence of vapor from liquid alloy in stagnant argon atmosphere and θ_1

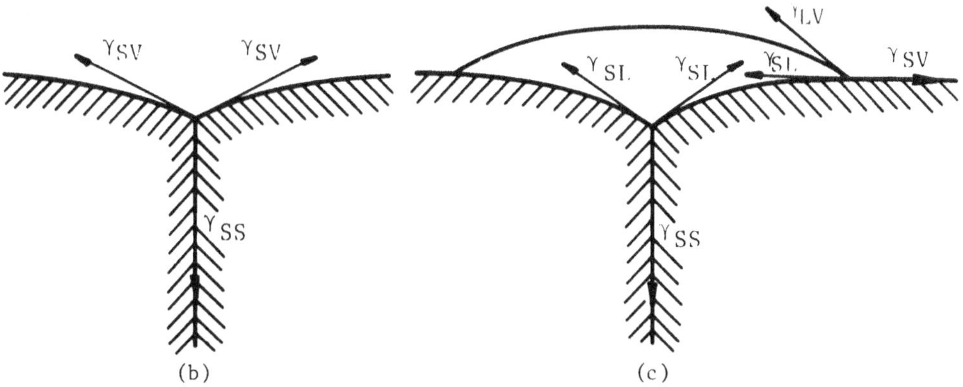

Fig. 3b, c. The multiphase equilibrium method.

and θ_2 are contact angles measured for drops on the two substrates. The data produced by Hodkin and Nicholas (*41*) using this technique are included in Section 4.2.

2.2.6. Other Methods

Additionally, there are indirect measurement methods that fall into two major categories, equilibrium and nonequilibrium. These methods have not been used to generate data for $UO_{2\pm x}$ and, hence, are not discussed here. The details and related discussion are surveyed carefully by Murr (*7*) and Regnier (*30*).

Maiya (*45*) was the only one to take an indirect approach to study surface free energies of UO_2; namely, measurement of the product $D \cdot \gamma$ in the case of thermal grooves, where D is the diffusion coefficient and γ is the surface energy. He has concluded that the accuracy of γ depends significantly on the D values which are used to calculate γ. His data, reported in Section 4.2, show an erroneous trend, which was attributed to orientation of grain boundaries to the crystallographic directions.

An interesting variation of the zero creep method is the zero-swelling method used by Hsu and Solomon (*42A*) to measure the surface free energy of ZnO. The method equilibrates pressurized pores with external pressure and surface tension. However, the required independent knowledge of pore pressure was obtained from their experiments of sintering under known ambient pressures.

3. METHODOLOGY OF CHARACTERIZATION OF NUMERICAL DATA

Numerical data on surface/interfacial free energies and surface tensions are tabulated, along with complete sample characterization and methods used for objective comparison of various data available in the literature. The methodology adopted to characterize numerical data emphasizes complete description of specifications such as the technique used, sample details, accuracy of method, and reported uncertainty of the measurements. Along with these specifications, the tabulated data includes remarks pertinent to these specifications and the validity of the data. The remarks are usually pertinent to the atmosphere used, the precautions taken, the care taken to reduce systematic error, or the attempts made to optimize the experiment. The second important aspect of characterizing the data set is 'Sample Details.' The specification tables of numerical data report chemical composition (purity), manufacturing details of the sample, and the stoichiometry of $UO_{2\pm x}$. In some of the recent studies, surface contamination and complex compound formation on surfaces have been studied to explain abnormal variations in surface free energies. Such details are also provided along with the data. It is noted that very few investigators have reported these.

A complete characterization of numerical data was not always possible because of the lack of such information and because the techniques used are not critically evaluated to report accuracies and the precautions needed to minimize systematic errors.

4. NUMERICAL DATA AND ANALYSIS

4.1. Surface Tension of Liquid $UO_{2\pm x}$ (γ_{LV})

There are six experimental data sets available for the surface tension of liquid UO_2. These data sets are presented in Fig. 4. The stoichiometric analyses of the samples are not reported. As pointed out earlier in Section 2.1, the value of γ_{LV} is very sensitive to contact angle, which is very difficult to measure. The data reported are also subject to various sources of systematic errors as reported by Keskar (*9*) and Sangiorgi et al. (*19*) for the sessile drop method, and by Murr (*7*) for various other techniques. The requirement of optimal experiments is pointed out in Section 2.1 for each method. Most of the reported data are rather old, and efforts to carry out optimal experiments, if any, are not thoroughly reported. Schins (*22*) attempted an optimal experiment by the drop weight method, and observed much higher values than reported by others. He did not give any details of the test atmosphere, stoichiometry, and/or surface contamination. For the same reason the data sets from secondary sources (data sets 5′ and 6′) cannot be analyzed. Thus, it can be concluded that there is a lack of reliable data and wide disagreement in the available data sets. This disagreement can very well be attributed to stoichiometry, measurement environment (which provides information on surface contamination), and non-optimized experiments.

The data of Nikolopoulos and Schulz (*43*) (data set 1′) are calculated values based on the theory of physical behavior of ionic liquids, as developed by Furth (*44*) and experimentally verified by Bockris and Richards (*45*). Nikolopoulos and Schulz (*43*) used experimental compressibility at the melting point in solid and liquid state obtained by Slagle and Nelson (*46*) and Young's modulus and poisson's ratio reported by Schmidt (*47*). Since Young's modulus and Poisson's ratio are functions of stoichiometry and porosity [see Matzke (*31*)] the calculated values are thus useful to understand the trend of $\partial\gamma_{LV}/\partial T$ near the melting point, and are applicable only for given constant stoichiometry and density and porosity. The stoichiometry reported by Nikolopoulos and Schulz (*43*) (data set 1′) is assumed to be 2.000. Their relation for γ_{LV} in terms of T(K),

$$\gamma_{LV} = 522.0 - 0.19(T - 3125)\ \text{mJ/m}^2 \tag{26}$$

depends significantly on the reliability of elastic constant data for a given stoichiometry. Fink et al. (*48*) recommended the value of 494 mJ/m^2 as a

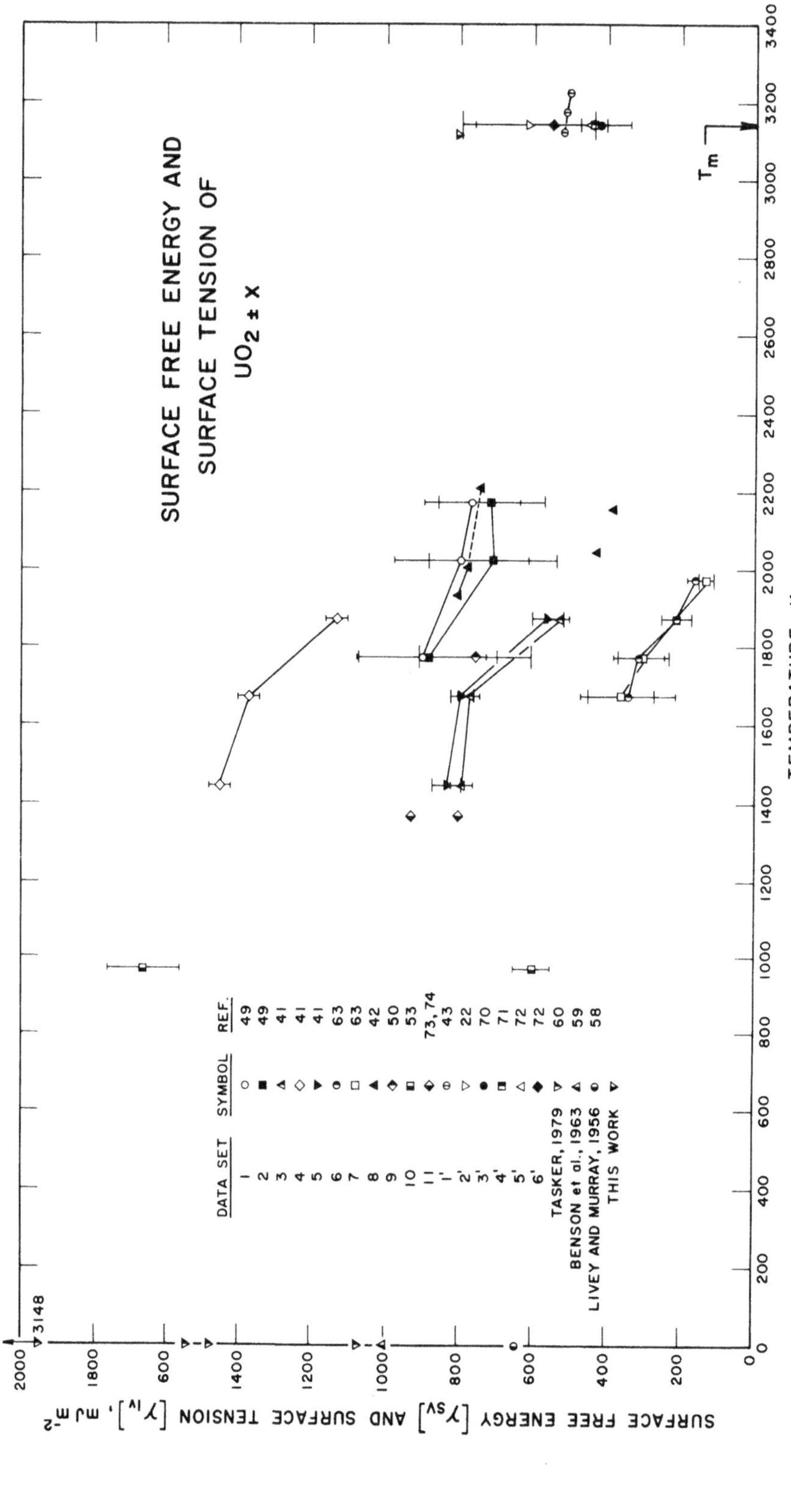

Fig. 4. Surface free energy and surface tension of $UO_{2\pm x}$.

mean γ_{SL} value with estimated error as ±15% at melting point. They also suggested an expression for γ_{LV} as

$$\gamma_{LV} = 494 - 0.19(T - 3120)\ \text{mJ/m}^2 \tag{27}$$

It can be seen that they used 494 mJ/m^2 for the γ_{LV} value at the melting temperature, which is an arithmetic mean of the reported data and used the estimated $d\gamma_{LV}/dT$ value of Nikolopoulos and Schulz (*43*) to develop this relation. Thus, their claim of estimated error ±6–15% is only based on very limited data (6 data points). Only measurement of γ_{LV} by various methods can reduce this limit of estimated error, and efforts to carry out optimal experiments are needed.

4.2. Surface Free Energy of Solid $UO_{2\pm x}$ (γ_{SV})

Because of the importance of surface free energy of UO_2, numerous measurements have been made. There are 12 experimental data sets available; these are presented in Table 1 and shown in Fig. 4. Most of these data sets are obtained by the multiphase equilibrium method or its modification, the interfacial control technique used by Hodkin and Nicholas (*41*). Only the data set provided by Maiya (*42*) is obtained by diffusion-controlled thermal grooving. The wide disagreement observed in the data is not surprising and might occur for the following reasons.

The methods used to obtain data probably do not maintain constant stoichiometry of UO_2 during the measurement. Only Nikolopoulos et al. (*49*) (data sets 1 and 2) reported stoichiometry before and after testing. Changes were also found as a result of holding the sample at very high temperatures for a long period of time in a stagnant atmosphere during various angle measurements. Hodkin and Nicholas (*41*) (data sets 3–6) allowed longer periods for equilibration of the thermal groove angle at high temperature in a stagnant atmosphere, and Bratton and Beck (*50*) also allowed up to 24 h for equilibration to demonstrate this effect. Thus, the equilibration of the composition and careful angle measurement are in question. It is argued by Nikolopoulos et al. (*52*) that the changes in observed surface free energy with composition are very small, whereas Hodkin and Nicholas (*41*) and very recently Matzke et al. (*31*) observed wide variations for similar stoichiometric changes. Therefore, the importance of stoichiometry cannot be ignored.

The proposition that stoichiometric changes have minor effects as compared to changes in temperature needs further evaluation, and it is necessary to quantify these effects separately. Matzke et al. (*31*) attempted measurements of fracture surface energy by the Hertzian indentation technique. As pointed out earlier, this is not an absolute measure of surface free energy, but it may provide a way to compare relative changes. They noted an increase in fracture surface energy with an in-

TABLE 1. SURFACE FREE ENERGY OF SOLID UO_{2+x}

Data Set No.	Author(s), Year [Ref. No.]	Method of Measurement	Sample Characterizations: Details	Sample Characterizations: Analysis	Stoichiometry	Temp. T, K	Surface Free Energy, γ_{sv} mJ m^{-2}	Remarks
1	Nikolopoulos, P., Nazare, S., and Thümmler, F., 1977 [49]	Multiphase equilibrium [Ni drop on UO_2]	Disc shape sample of 8 mm thick, 16 mm diam; initial stoichiometry 2.005; grain size 10 μm; 94-95% theoretical density.	element ppm Al 8-10 Cr 5-10 Cu 1 Fe 20-30 Mn 5-10 Mo 2-3 Ni 5(max) Si 50 Sn 10(max) W,Mg 5	2.025 2.006 1.999	1773 2023 2173	899 ± 174 793 ± 182 765 ± 131	Flowing argon atm was used ($H_2 \leq 20$ vpm, $O_2 \leq 20$ vpm, $N_2 \leq 20$ vpm); pure nickel used for dihedral angle measurements and contact angle; stagnant argon with no metal vapor was used to measure thermal groove angle.
2	Nikolopoulos, P., et al., 1977 [49]	Same as above	Same as above	Same as above	2.025 2.006 1.999	1773 2023 2173	883 ± 188 708 ± 173 712 ± 144	Same as above except stagnant argon atm with Ni vapor was used for thermal groove angle measurement.
3	Hodkin, E.N. and Nicholas, M.G., 1977 [41]	Interfacial control	2-3 mm thick, 10.0 mm diam disc; 2.084 stoichiometric single phase ingot sample was heated in argon at 1700°C for 20 h; polished to 1/4 μm finish.	element ppm Al 300 C 200 Cr 20 Fe 70 N 10 other <10	2.053	1448 1673 1873	790 ± 30 770 ± 30 530 ± 30	Measurements in pure stagnant argon at pressure of 50 cm of mercury; experiments conducted in presence of and in absence of Cu vapor which results in same γ_{sv} values; ultra-pure copper and Cu-1 at.% Th reactive alloy were used to measure contact angles; starting material of stoichiometry 2.084.
4	Hodkin, E.N. and Nicholas, M.G., 1977 [41]	Interfacial control	Disc shape sample 2-3 mm thick, 10.0 mm diam; polished to 1/4 μm finish.	Same as above	2.084	1448 1673 1873	1450 ± 30 1370 ± 30 1130 ± 30	Similar to the above.
5	Hodkin, E.N. and Nicholas, M.G., 1977	Interfacial control	Disc shape sample 2-3 mm thick, 10.0 mm diam; stoichiometric ingot sample with starting 1/4 = 2.001 was heated for 3 h at 1740°C in 10^{-6} torr vacuum to get stoichiometry of 1.998; some evidence of uranium-precipitation was observed; sample was polished to 1/4 μm.	element ppm Al 500 C 200 Cr 20 Fe 70 N 10 other <10	1.998	1448 1673 1873	830 ± 40 790 ± 30 560 ± 40	Measurements in pure stagnant argon at pressure of 50 cm of mercury; experiments were conducted in presence of and in absence of Cu vapor which results in same γ_{sv} values; ultra pure copper and Cu-1.0 at.% Th reactive alloy used to measure contact angles; chemical analysis of UO_{2+x} given in the respective column is of starting material of 2.001 stoichiometry.
6	Hodkin, E.N. and Nicholas, M.G., 1973 [63]	Multiphase equilibrium [Cu drop on UO_2]	Disc shape of 2.3 mm thick 10 mm diam; cut from hot-pressed and sintered material; oxide is single phase and has no interconnected porosity; annealed at 1700°C for 20 h.	element ppm Al 300 C 200 Cr 20 Fe 70 N 10 Ni 5	2.001	1673 1773 1873 1973	340 ± 130 310 ± 70 210 ± 40 160 ± 20	The thermal groove angle measurements were carried after 45-175 h by allowing equilibration; the time given for contact angle equilibration varies from 90-240 h.

TABLE 1. SURFACE FREE ENERGY OF SOLID UO_{2+x} (continued)

Data Set No.	Author(s), Year [Ref. No.]	Method of Measurement	Sample Characterizations: Details	Sample Characterizations: Analysis	Stoichiometry	Temp. T,K	Surface Free Energy, γ_{sv} mJ m^{-2}	Remarks
6 (cont.)				element ppm Mn 2 Pb,Sn <2 Pd,Sb, Ti,Zn <20 Cd,Mg <10 Co,Cu, Mo,H <5 Ca,V <2 Bi,Ge, In <1 Be <0.1				
7	Hodkin, E.N. and Nicholas, M.G., 1973 [63]	Same as above	Same as above	Same as above	2.001	1673 1773 1873 1973	360 ± 90 300 ± 70 210 ± 40 130 ± 20	Similar to the above except measured in purified stagnant argon at 1 atm along with Cu vapor.
8	Maiya, P.S., 1971 [42]	Thermal grooving	Annealed at respective temperatures for 8 h for thermal grooving.	element ppm H_2 19 C 12	2.000	1933 2003 2048 2158 2213	806 775 425 385 741	Specimen contained in high-purity tungsten crucible which controls stoichiometry during experiment and not necessarily as claimed; values of γ_{sv} reported here are calculated from the diffusion coefficient, diffusion of UO_2 vapor in helium in which measurements were carried out from kinetic theory of gases and product of $\gamma_{sv} \cdot D$ obtained from kinetics of grain boundary grooving; measured D smaller than the calculated; drop in γ_{sv} values at 2048 and 2158 K was associated to a probable bias towards grain boundaries selection having orientation towards index planes; this trend observed here appears to be erroneous.
9	Bratton, R.J. and Beck, C.W., 1971 [50]	Multiphase equilibrium [Ni drop on UO_2]	Solid-solid vapor equilibrium determined by grain boundary; groove angle was obtained by thermal grooving of a high-density polycrystalline sample prepared by hot-pressing reactor grade depleted UO_2; to establish stoichiometry of 1.997 of starting material; sample was annealed at 1600°C in high purity argon; specimen was thermally grooved at 1500°C for 24 h; solid-solid-liquid equilibrium was		2.010	1773	754 ± 150	High purity flowing argon atm; the gas chromotography of atm reported 0.10 ppm of O_2 in argon; method takes direct measurement approach which is thermodynamically reliable technique; standard deviation on thermal groove angle and dihedral angle was ±2°; the stoichiometry reported is based on 10^{-7} O_2 potential.

TABLE 1. SURFACE FREE ENERGY OF SOLID UO_{2+x} (continued)

Data Set No.	Author(s), Year [Ref. No.]	Method of Measurement	Sample Characterizations: Details	Sample Characterizations: Analysis	Sample Characterizations: Stoichiometry	Temp. T,K	Surface Free Energy, γ_{sv} mJ m^{-2}	Remarks
9 (cont.)			1500°C for 24 h; solid-solid-liquid equilibrium was established to get dihedral angle using nuclear grade depleted UO_2 and 10 wt.% 325 Ni powder of high purity; the impacts were sintered at 1500°C for 8 h.					
10	Eberhart, J.G., 1968 [53]	Wetting of Pb-Bi	Not available.	Not available	~2.000 ~2.000	970 970	600 ± 50 1650 ± 100	The first data point was measured in argon atm with probable oxygen content and the second one in argon + (unknown%) hydrogen atm; nature of atm (stagnant or flowing not reported).
11	Rhee, S.K., 1973 [73]	Multiphase equilibrium [(Cu-Al) alloy drop on UO_2]	Not available.	Not available	~2.000	1370	931	Atm for experiment was argon with (unknown%) hydrogen; probable presence of oxygen was suspected.
12	Rhee, S.K., 1974 [74]	Sessile drop (tin drop on UO_2)	Not available.	Not available	~2.000	1370	838	Same as above.

crease in oxygen content in UO_2. In summary, it can be said that the reported variation in the data is a combined representation of stoichiometric and temperature changes, and the reported stoichiometry information is questionable.

The wide disagreement among data is also attributed to use of unoptimized methods which inherently have many assumptions. Furthermore, complete evaluation of the limitations of these methods as applied to UO_2 have not been carried out, and it can surely be said that there is need to evaluate the available methods as applied to ceramics. This should reveal the requirements of optimal experiments necessary to generate data when the method of measurement and sample requirements are in conflict with the processes controlling the material behavior.

Maiya (*42*) also suggested that all reported data are subject to a bias in orientation of grain boundary chosen for experiment. Wide fluctuations in observed values are expected for the diffusion kinetic approach to get surface energy. It is very difficult to obtain any quantitative evaluation of the trend of temperature dependence from the available data.

Hodkin and Nicholas (*55*) criticized the data of Nikolopoulos et al. (*49*) and argued that the data of Nikolopoulos et al. (*49*) may not be at constant stoichiometry. Nikolopoulos et al. (*52*) later suggested that the dependence of γ_{SV} or the sensitivity of γ_{SV} values to stoichiometry can be obtained by extrapolation of γ_{SV} values to the melting point and 0 K, and also by comparing them with available *estimated* values in the literature. They concluded that extrapolation of their values is close to an estimate of γ_{SV} at the melting point by Eberhart (*53*) (545 mJ m^{-2}), and argued that possible disagreement cannot be explained by variations in stoichiometry, but only by probable errors in measurement of the groove angle.

However, Eberhart (*53*) used the relation

$$\gamma_{SV} = 1.1\ \gamma_{LV}, \tag{28}$$

derived by Allen (*54*) for metals. Alternatively, the drop in surface free energy during melting can be approximately written (ref. *7*) as

$$\Delta\gamma \mid_{T_m} = \gamma_{SV} \mid_{T_m} - \gamma_{LV} \mid_{T_m} = \Delta H_m / A_o \tag{29}$$

where ΔH_m is latent heat of fusion and A_o is area per molecule on the surface of the compound. This gives solid surface free energy γ_{SV} at the melting point (T_m) as

$$\gamma_{SV} \mid_{T_m} = \Delta\gamma \mid_{T_m} + \gamma_{LV} \mid_{T_m} \tag{30}$$

where $\gamma_{LV} \mid_{Tm}$ is the liquid surface free energy (surface tension) at the melting point (T_m). Skapski (*55*) devised a more rigorous formulation of the above equation, which considers the coordination number in bulk and on the surface as well as density variation at the melting point. The resulting empirical expression is

$$\gamma_{SV} = (Z - Z_s)/Z\ [\Delta H_m / A_S] + [\rho_S/\rho_L]^{2/3}\ (\gamma_{LV})_m \tag{31}$$

where Z and Z_s are liquid bulk, and surface atom coordination numbers, A_S is the molar solid surface area, and ρ_S and ρ_L are solid and liquid densities at the melting temperature (T_m), respectively.

Oriani (*56*) has shown that $(Z - Z_s)/Z \cong 0.25$ for fcc and is $\cong 0.36$ for bcc. If we assume a value of 0.3 for UO_2 and use Skapski's relation for molar surface area, i.e.,

$$A_S = 1.12 N_A^{1/3} (M/\rho_S)^{2/3} \tag{32}$$

where N_A is Avogadro's number and M is molecular weight, then $\gamma_{SV} =$ 805 mJ/m^2 at the melting temperature. [The data required for these calculations are taken from Fink et al. (*48*).] Thus, if it is assumed that Skapski's approach is as applicable to oxides as it is to metals, the estimate of Eberhart (*53*) of 545 mJ/m^2 are grossly in error. It is important to note that this estimate assumes a certain coordination number at the surface and in bulk, which is very difficult to justify. In fact, very little is known about liquid–vapor interface structure and there are no means to illustrate such interfaces [see Murr (*7*), p. 167]. But Skapski's theory has been successfully used on metals and alloys by Allen and Kingery (*57*). The application of Skapski's theory is usually supported by indirect evidence that the modulus of elasticity (Young's modulus) at 0.9 T_m is proportional to surface free energy (γ_{SV}) for metals and alloys [see Murr (*7*), pp. 126–127]. In fact, a similar observation holds true for ceramics (ref. *8*), and thus a more rigorous approach for calculations of γ_{SV} by Eq. (31) is used here for UO_2 for the first time.

Theoretical calculations for the (111) and (110) surfaces for fluorite structures after relaxation is another means to obtain insight of the surface free energy γ_{SV}. Such calculations are reported in the literature by Livey and Murray (*58*), Benson et al. (*59*), and recently by Tasker (*60*) for fluorite-type structures. Livey and Murray (*58*) reported a value of 640 mJ/m^2, whereas Benson et al. (*59*) reported a value of $\geq$1000 mJ/m^2 and suggested a value of 1030 mJ/m^2. Tasker (*60*) used an ionic model with short-range potentials and allowed the surface to relax to equilibrium. The relaxation he considered approximated details of surface structure from a relaxed ion position, and the surface energy was computed after relaxation. He used parameters for the interionic potential and the shell model for UO_2 of Catlow (*61, 62*) and assumed no cation–cation short-range interactions. The assumed lattice parameter is 2.734 Å. The resulting values for relaxed (111) and (110) surfaces are 1069 and 1539 mJ/m^2, which represents minimum and maximum, respectively, whereas for the same unrelaxed surfaces they are given as 1479 and 3148 mJ/m^2, respectively. These values are shown in Fig. 4 at 0 K. Now, assuming γ_{SV} is a linear function of temperature, we get $\partial\gamma_{SV}/\partial T$ [computation based on relaxed values and our estimate at the melting point obtained by Eq. (31)] as -0.085 and -0.235 mJ/m^{2}K. In actual cases, if surfaces studied do not have bias towards similar indices planes, one could observe different slopes, which may be totally different and can be out of range of −0.235 to −0.085 mJ/m^{2}K. The reported data are conflicting, and the uncertainty limits are sometimes so high that $\partial\gamma_{SV}/\partial T$ estimates become meaningless [e.g., data sets 1 and 2 of Nikolopoulos et al. (*49*)] in Fig. 4. The uncertainty limits for Hodkin and Nicholas (*63*) are reasonably small, but the $\partial\gamma_{SV}/\partial T$ for these data sets (data sets 6 and 7) turn out to be quite below the range mentioned above. Hodkin and Nicholas (*41*) (data sets 3–5) have three data points for each data set that generally lie within the range. Maiya (*42*) (data set 8) did not report uncertainty limits of the data, but, with exclusion of two data points (with a higher probability of systematic error because of the orientation of grain boundaries to low index planes) found a value of −0.164 mJ/m^{2}K for $\partial\gamma_{SV}/\partial T$ that is within the above-mentioned range. Therefore, $\partial\gamma_{SV}/\partial T$ at constant composition may be a weak function of stoichiometry, however, the surface energy calculations have not been made for nonstoichiometric compositions.

Finally, to put the status of available data in perspective, the above discussion can be summarized as follows:

(1) The reported data are not sufficiently reliable and extensive to specify the dependence of γ_{SV} on stoichiometry and temperature.
(2) $\partial\gamma_{SV}/\partial T$ appears to be a very weak function of composition.
(3) The reported γ_{SV} values are often questionable, especially when γ_{SV} at the melting temperature is less than γ_{LV} reported to date.
(4) There is a complete lack of data between 2100 K and the melting temperature.
(5) The surface free energy γ_{SV} at the melting point can be calculated applying Skapski's theory and the resulting approximate γ_{SV} value at the melting point for $UO_{2.000}$ is ≅805 mJ/m^2.

4.3. The Grain-Boundary Free Energy of $UO_{2\pm x}$ (γ_{gb} or γ_{SS})

There are a total of six data sets available for the grain boundary free energy of UO_2. They are presented in Fig. 5. The data sets show quite significant disagreement and even indicate different trends in the same regime of temperature. Most of the data reported are from 1200 to 2200 K, and are measured by either the multiphase equilibrium technique or its modification, the interfacial control technique. Only the data of Maiya (*42*) (data set 8) are obtained by the kinetic method. In the multiphase equilibrium or interfacial control method, the γ_{gb} value is obtained by knowing the dihedral angle that quantifies the ratio γ_{gb}/γ_{SV}. Thus, the disagreement in the γ_{SV} values is reflected in the γ_{gb} values along with the probable uncertainties and systematic errors in the measurement of the dihedral angles. Hence, the analysis presented in Section 4.2 is helpful in understanding the data presented here and warrants no further

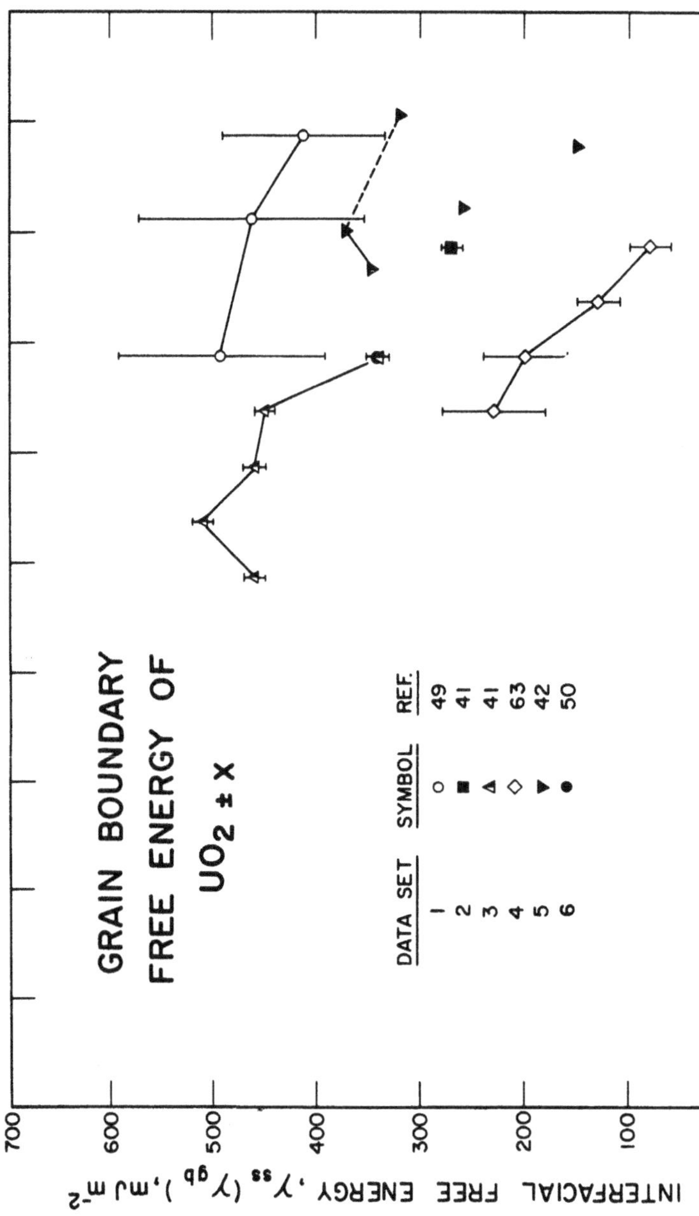

Fig. 5. Grain boundary free energy of $UO_{2\pm x}$.

comment. Only the better measurement of γ_{SV} would help to resolve the discrepancies observed.

4.4. Interfacial Free Energies of the $UO_{2\pm x}$–Liquid Metal System ($\gamma_{\alpha\beta}$)

There are over 30 data sets available for the interfacial free energies of $UO_{2\pm x}$–liquid metal systems. The interest in $UO_{2\pm x}$–liquid metals (or alloys) arose because of their technological importance in nuclear safety calculations and the need to calculate the γS_V value of $UO_{2\pm x}$, as explained earlier (Section 2.2.4). The data on the interfacial free energies of $UO_{2\pm x}$–liquid metals ($\gamma_{\alpha\beta}$) are presented in Fig. 6. The data are presented here as a function of homologous temperature, T/T_m, where T_m is the melting point of metal and T is the temperature of the system. The homologous temperature was used as a parameter for ease of comparison with the data of Nikolopoulos and Ondracek (*64*), who reported $\gamma_{\alpha\beta}$ values for 13 different elements (data sets 1–13). The $\gamma_{\alpha\beta}$ values for UO_2–liquid metal systems such as Cu, Ni, and Stainless Steel 1.4970 (close to 300 type nuclear grade SS) have been investigated by many researchers, but these studies disagree with each other. The prime reason for such disagreement is possibly surface contamination. No attempt has been made along this line with the exception of Nikolopoulos et al. (*65*). They used Auger spectroscopy to characterize the surface of $UO_{2\pm x}$ and then to evaluate the effect of the presence of sodium vapor on the $\gamma_{\alpha\beta}$ value of UO_2–SS 1.4970 system. These results are presented in Fig. 6 (data sets 19, 20). It was noted that the effective change in $\gamma_{\alpha\beta}$ in the presence of Na vapor possibly arises from the formation of Na—O—Ti and Na—O—U compounds. Nikolopoulous et al. (*65*) were the only ones to detect these compounds and to find this change, within the uncertainty limit of the measurement, as negligible. Thus, the following general observations can be made from the available data:

(1) No liquid metal wets unirradiated UO_2 ($\theta_c > 90°C$) at the melting temperature of the metal.
(2) From studies of the wetting of irradiated UO_2 by molten Cu it was qualitatively observed that, as burnup increases, the liquid Cu approaches wetting (i.e., $\theta_c \rightarrow 90°$). This provides indirect evidence that irradiation induced defects, stoichiometric changes of UO_2, and fission-product contamination all affect the surface free energy.
(3) For very low burnups (~1.2%), irradiated UO_2 has surface energy a similar to that of unirradiated UO_2. However, exceptions are available in the literature [see Hodkin (*66*)].
(4) The contact angle θ_c generally decreases with increasing liquid metal melting temperature.
(5) $d\gamma_{\alpha\beta}/dT$ increases as the melting temperature of the metal increases.

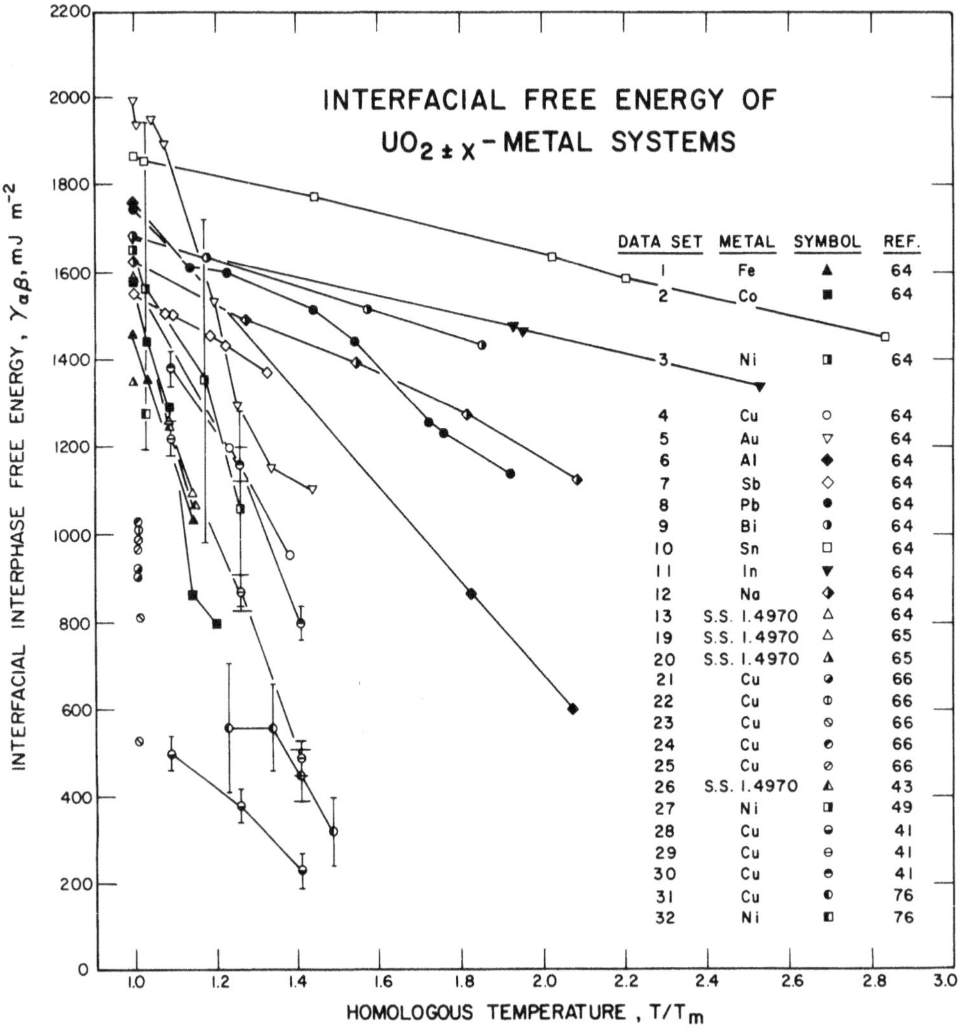

Fig. 6. Interfacial free energies of $UO_{2\pm x}$—liquid metal systems.

(6) The interfacial free energies of unirradiated UO_2–liquid metal systems at the melting points of the metals fall within the range from 1200 to 2000 mJ/m^2.

Nikolopoulous and Ondracek (*64*) (data sets 1–13) concluded that the average mean value of $\gamma_{\alpha\beta}$ at the melting temperature of any metallic system is about 1676 ± 142 mJ/m^2 for unirradiated UO_2. This value appears to be in good agreement with all other available data in the literature (see Fig. 6). Furthermore, they suggested the estimate for $\gamma_{\alpha\beta}$ for UO_2–liquid metal systems at the melting point by the equation

$$\gamma_{UO_2\text{–liquid metal}} = KT_m/A^n \tag{33}$$

where T_m is the melting temperature of metal, K is a correlation factor that can be obtained by experimental data on interfacial free energies, and A^n is the molar surface area of metal. A similar empirical relation obtained by Nikolopolous and Ondracek (64) for unirradiated UO_2 is

$$\gamma_{UO_2\text{-liquid metal}} = 1679 + 0.062 \times 10^{-6}\, T_m/A^n \ (\text{mJ/m}^2) \qquad (34)$$

Lastly, it is important to note from the available data that for wetting of unirradiated UO_2 (i.e., to get $\theta_c > 90°$), the system temperature must be far above the melting temperature of metal, so that the temperature effects reduce the $\gamma_{\alpha\beta}$ and consequently facilitate wetting. This implies, for instance, that the liquid-phase sintering of oxide cermets is possible only at very high temperatures.

4.5. Surface Free-Energy Ratio (γ_{gb}/γ_{SV}) of $UO_{2\pm x}$

There are 15 data sets available for the ratio of grain-boundary free energy to surface free energy. These are presented in Fig. 7. The data reported at high temperatures are obtained by the multiphase equilibrium method, which predicts a surface-energy ratio in the range 0.25 to 0.80 over the temperature range 1360 to 2200 K. The remaining data are obtained by pore geometry measurement involving the dihedral angle, θ_{SV}, such that

$$\gamma_{gb}/\gamma_{SV} = 2 \cos\, [\theta_{SV}/2] \qquad (35)$$

The more than twofold discrepancy in the data obtained from these two methods is rather surprising to all fuel performance evaluators. Reynolds et al. (67) suggested that the chemical changes, particularly fission-product contamination at the bubble surface, could markedly lower γ_{SV} values; this would account for the higher ratios, although there is no direct proof that fission product contamination would lower the surface free energy. Kingery and Francois (68) assert that the apparent high ratios are not caused by contamination, but by the pressure exerted by entrapped gases. Hodkin (69) attempted an evaluation of this discrepancy and reported the data for unirradiated and irradiated fuel elements. He found similar ratios for both and could not explain the discrepancy. He argued that the lack of change in pore angle upon irradiation does not necessarily mean that the surface free energy is not changing. It may simply mean that two counteracting mechanisms, such as structural changes and stoichiometric changes, are cancelling and hence that there is a need for more elaborate experimental and theoretical considerations to explain the phenomenon.

5. CONCLUSIONS

The surface/interfacial free energies and surface tension of $UO_{2\pm x}$ have been reviewed here along with the experimental methods that are important and useful in the measurements of data. The data on these surface properties is, for the first time, reviewed systematically and com-

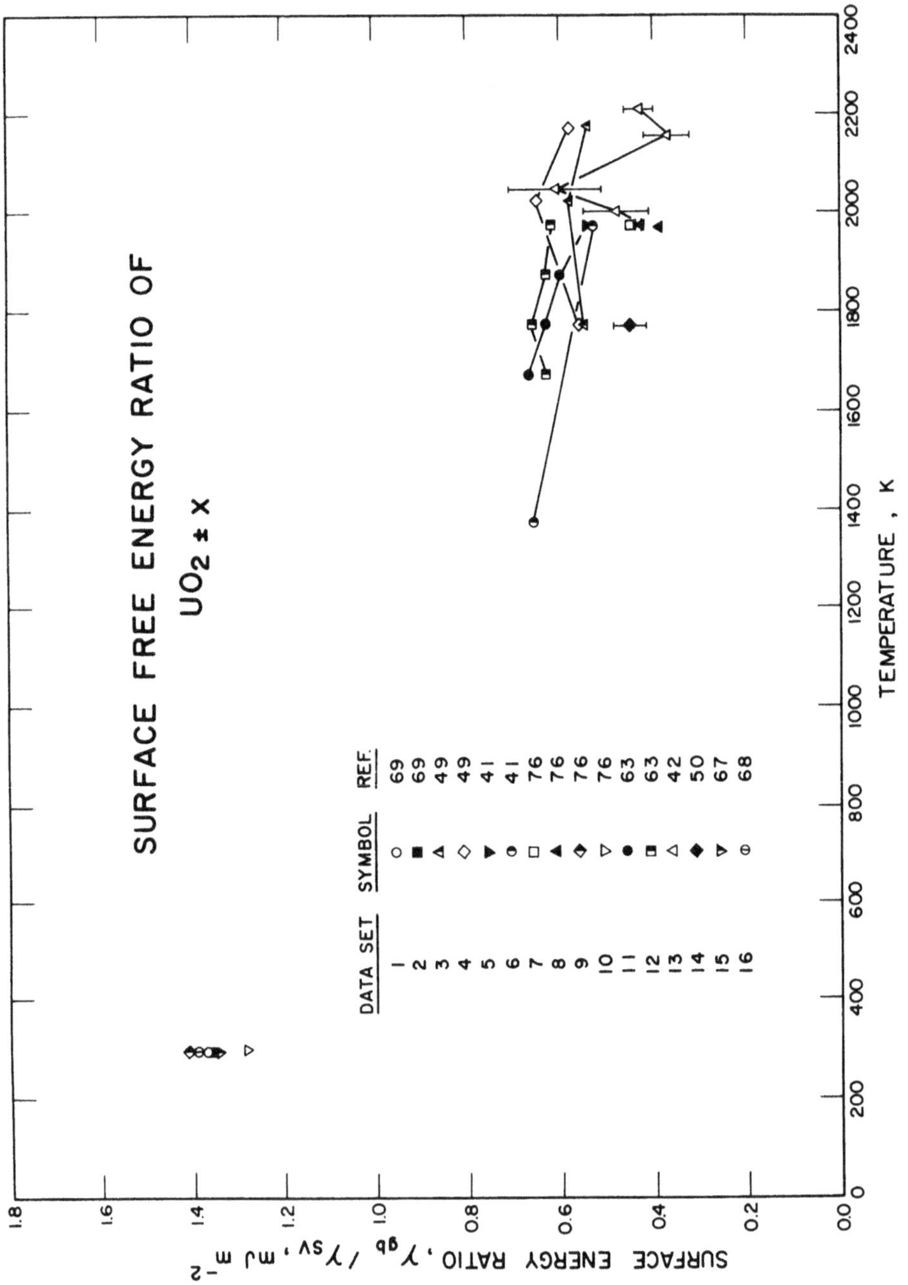

Fig. 7. Surface energy ratio (γ_{gb}/γ_{sv}) of $UO_{2\pm x}$.

pared objectively. The following important conclusions can be drawn from the present investigation:

(1) There is a lack of reliable γ_{LV} data. The available information does not indicate that optimal care was taken during the experiments to reduce systematic errors. The reported information is difficult to analyze because of a lack of completeness in reporting the data. In the absence of such complete characterization, no recommendations for γ_{LV} values were possible.
(2) No attempt has been made to measure $d\gamma_{LV}/dT$ of $UO_{2\pm x}$. The only available estimate is by Nikolopoulos and Schulz (*43*), based on the theory of the physical behavior of ionic liquids. Their estimated value for $d\gamma_{LV}/dT$ is -0.19 mJ/m^{2}K.
(3) There is wide disagreement among available data on γ_{SV} values and a complete lack of data between 2100 K and the melting point. Based on the reported data, it is not possible to give a quantitative picture of the dependence of γ_{SV} on stoichiometry and temperature.
(4) From the available limited information, it can be tentatively inferred that $d\gamma_{SV}/dT$ may be a very weak function of stoichiometry, but the actual dependence of γ_{SV} on stoichiometry has not been derived.
(5) The surface free energy γ_{SV} at the melting temperature can be estimated from Skapski's theory. Thus, the estimated γ_{SV} value at the melting temperature for $UO_{2.000}$ is approximately 805 mJ/m^2. However, based on this information, the reported trends in $\gamma_{SV}(T)$ values may be often erroneous, such as implying that at the melting point $\gamma_{SV} < \gamma_{LV}$.
(6) The grain-boundary free energy is usually obtained by measuring the γ_{gb}/γ_{SV} ratio. Thus, the reported γ_{gb} values are scattered considerably. Only a better measurement of γ_{SV} will help resolve the discrepancies observed.
(7) From the studies of wetting behavior carried out to measure interfacial free energy ($\gamma_{\alpha\beta}$) of both unirradiated and irradiated $UO_{2\pm x}$–liquid metal systems, the following general observations can be made:

(a) At the melting temperature of the metal, no liquid metal wets unirradiated UO_2 (i.e., $\theta_c > 90°$).
(b) Wetting of irradiated commercial fuels was studied with liquid Cu. It was observed that up to burnups of ~1.2% both irradiated and unirradiated UO_2 have similar surface free energies. An exception to this observation is available in the data for 0.5% burnup [see Hodkin (*66*)].
(c) The contact angle θ_c generally decreases as the melting temperature of metals increases.
(d) The slope of $d\gamma_{\alpha\beta}/dT$ increases as the melting temperature of the metal increases.

(e) The $\gamma_{\alpha\beta}$ of unirradiated UO_2–liquid metal systems measured at the melting point of the metal usually falls in the range of 1200 to 2000 mJ/m^{-2}.

(f) The required engineering approximation according to which γ_{UO_2}–liquid metal may be obtained for any metal is that of Nikolopoulos and Ondracek (62); it is given by the equation

$$\gamma_{\alpha\beta} = 1679 + 0.062 \times 10^{-6}\ T_m/A^n \quad mJ/m^2$$

where T_m is the melting point and A is the molar surface area of the metal in question.

(8) There is a two- to three-fold discrepancy in the measurement of the γ_{gb}/γ_{SV} ratio. The measurements carried out by pore dihedral angle measurements are usually threefold higher than those obtained by the multiphase equilibrium. This discrepancy is not yet resolved. Efforts to obtain γ_{SV} by other measurement techniques and as well as efforts to reduce systematic errors will help to gain insight into the discrepancy.

(9) From the review of measurement techniques (Section 2), it is clear that there are several methods that are not presently used, but do have potential to give reliable γ_{SV} or γ_{LV} values. Efforts to determine the feasibility of these measuring techniques are needed and would be of great help to enhance the knowledge about surface/interfacial free energies in general.

ACKNOWLEDGMENT

The support of A. A. Solomon by the Office of Basic Energy Sciences, US Department of Energy, under contract no. DE-AC02-80ER10591, is gratefully acknowledged.

REFERENCES

1. Adam, N. K., in *The Physics and Chemistry of Surfaces,* Oxford University Press, London, 1941, 363–88.
2. Bikerman, J. J., in *Surface Chemistry,* Academic Press, New York, 1958, pp. 4–23.
3. Davies, J. T., and Rideal, E. K., in *Interfacial Phenomena,* Academic Press, New York, 1961, pp. 42–47.
4. Semechenko, V. K., in *Surface Phenomena in Metals and Alloys* (translated from Russian), Pergamon Press, London, pp. 304–312.
5. White, D. W. G., *Trans. Am. Soc. Metals* **55,** 757 (1962).
6. White, D. W. G., *Met. Mater.* **2**(7), 73 (1968).
7. Murr, L. E., in *Interfacial Phenomena in Metals and Alloys,* Addison-Wesley, Reading, MA, 1976, pp. 87–126, 116–24.

8. Deshpande, M. S., Desai, P. D., and Solomon, A. A., CINDAS Special Rept., in preparation, Aug. 1983.
9. Keskar, A. R., Liquid–Gas Interfacial Energies of Copper and Be–Ti Alloys, Purdue Univ. PhD Thesis, 1971, p. 146.
10. Porter, A. W., *Phil. Mag.* **25,** 752 (1938).
11. Bashforth, F., and Adams, J. C., in *An Attempt to Test the Theory of Capillary Action,* Cambridge University Press, London (1892).
12. Paddy, J. F., in *Surface and Colloidal Science,* Matijevic, E., ed., Wiley-Interscience, New York, 1969.
13. Koshevnik, A. Yu., Kusakov, M. M., and Lubman, N. M., *Zh. Fiz. Khim.* **27,** 1887 (1953).
14. Humenik, M., and Kingery, W. D., *J. Am. Ceram. Soc.* **37,** 18 (1954).
15. Dorsey, N. E., *J. Wash. Acad. Sci.* **18,** 505 (1928).
16. Bonfield, W., *J. Mater. Sci.* **7,** 148 (1972).
17. Butler, J. N., and Bloom, B. H., *Surf. Sci.* **4,** 1 (1966).
18. Maze, C., and Burnett, G., *Surf. Sci.* **13,** 451 (1969).
19. Sangiorgi, R., Caracciolo, G., and Passerone, A., *J. Mater. Sci.* **17**(10), 2895 (1982).
20. Harkins, W. D., and Brown, F. E., *J. Am. Chem. Soc.* **4,** 499 (1919).
21. Lando, J. L., and Oakley, H. T., *J. Colloid Interface Sci.* **25,** 526 (1967).
22. Schins, H., *J. Nucl. Mater.* **78,** 215 (1978).
23. Inman, M. C., and Tipler, H. R., *Met. Rev.* **8,** 105 (1963).
24. Greenough, A. P., *Appl. Mater. Res.* **4,** 25 (1965).
25. Bikerman, J. J., *Phys. Status Solidi* **10,** 3 (1965).
26. Blackley, J. M., and Maiya, P. S., *J. Appl. Phys.* **38,** 698 (1967).
27. Hondros, E. D., in *Techniques of Metal Research,* Bunshah, R., ed., Vol. IV(A), Chapt. 8A, Wiley, New York, 1970.
28. Levy, V., and Regnier, P., *J. Phys. (Paris)* **31,** Suppl. 4, C1-159 (1970).
29. Linford, R. G., in *Solid State Surface Science,* vol. 2, Dekker, 1 (1973).
30. Regnier, P., in *Handbook of Surfaces and Interfaces,* vol. 2, Chapt. 1, Garland STPM Press, New York-London, 1980.

30A. Lipsett, S. G., Johnson, F. M., and Mass, O., *J. Am. Chem. Soc.* **49,** 925 (1927).

31. Matzke, H., Inoue, T., and Warren, R., *J. Nucl. Mater.* **91,** 205 (1980).
32. Sawai, I., and Nishida, M., *Z. Anorg. Allg. Chem.* **190,** 375 (1930).
33. Greenough, A. P., *Phil. Mag. Ser. 7,* **43,** 1075 (1952).
34. Greenough, A. P., *Phil. Mag. Ser. 8,* **3,** 1032 (1958).
35. Herring, C., *J. Appl. Phys.* **21,** 459 (1950).
36. Kumikov, V. K., and Khokonov, Kh.B., *J. Appl. Phys. 54*(3), 1346 (1983).
37. Regnier, P., and Felsen, M. F., *Phil. Mag.* **43,** 11 (1981).
38. Marioton, B., and Solomon, A. A., *Bull Amer. Cer. Soc.* **62**(3), 388 (1983).

38A. Solomon, A. A., private communication (1982).

39. Allen, B. C., Univ. Microfilm No. AD 673 793 (1968).
40. Mykura, H., *Acta Metall.* **5,** 346 (1957).
41. Hodkin, E. N., and Nicholas, M. G., *J. Nucl. Mater.* **67,** 171 (1977).

41A. Mortimer, D. A., and Nicholas, M. G., *J. Mater Sci.* **8,** 640 (1973).

42. Maiya, P. S., *J. Nucl. Mater.* **40**(1), 57 (1971).

42A. Solomon, A. A., and Hsu, F., *J. Am. Cer. Soc.* **63,** 7, 467 (1980).

43. Nikolopoulos, P., and Schulz, B., *J. Nucl. Mater.* **82,** 172 (1979).
44. Furth, R., *Proc. Cambridge Phil. Soc.* **37,** 252 (1941).
45. Bockris, J. O. M., and Richards, N. F., *Proc. Roy. Soc. London* **241,** 44 (1957).

46. Slagle, O. D., and Nelson, R. F., *J. Nucl. Mater.* **40,** 349 (1971).
47. Schmidt, H. E., *High Temp.-High Press.* **3,** 345 (1971).
48. Fink, J. K., Chasanov, M. G., and Leibowitz, L., in *Thermodynamic Properties of UO_2,* ANL-CEN-RSD-80-3 (1981).
49. Nikolopoulos, P., Nazare, S., and Thummler, F., *J. Nucl. Mater,* **71,** 89 (1977).
50. Bratton, R. J., and Beck, C. W., *J. Am. Ceram. Soc.* **54**(8), 379 (1971).
51. Hodkin, E. N., and Nicholas, M. G., *J. Nucl. Mater.* **74,** 178 (1978).
52. Nikolopoulos, P., Nazare, S., and Thumler, F., *J. Nucl. Mater.* **78,** 213 (1978).
53. Eberhart, J. G., *J. Nucl. Mater.* **25,** 103 (1968).
54. Allen, B. C., *Trans. AIME* **236,** 903 (1966).
55. Skapski, A. S., *Acta Metall.* **4,** 576 (1956).
56. Oriani, R. A., *J. Chem. Phys.* **18,** 575 (1950).
57. Allen, B. C., and Kingery, W. D., *Trans. AIME* **215,** 30 (1959).
58. Livey, D. T., and Murray, P., *J. Am. Ceram. Soc.* **39,** 363 (1956).
59. Benson, G. C., Freeman, E. I., and Dempsey, E., *J. Am. Ceram. Soc.* **46,** 43 (1963).
60. Tasker, P. W., *Surf. Sci.* **79,** 315 (1979).
61. Catlow, C. R. A., *Proc. Roy. Soc. London, Ser. A* **353,** 533 (1977).
62. Catlow, C. R. A., *Proc. Roy. Soc. London, Ser. A* **364,** 473 (1978).
63. Hodkin, E. N., and Nicholas, M. G., *J. Nucl. Mater.* **47,** 23 (1973).
64. Nikolopoulos, P., and Ondracek, G., *J. Nucl. Mater.* **98,** 306 (1981).
65. Nikolopoulos, P., Nold, E., Schneider, H., and Schulz, B., *J. Nucl. Mater.* **101,** 277 (1981).
66. Hodkin, E. N., *J. Nucl. Mater.* **97,** 27 (1981).
67. Reynolds, G. L., Beere, W. B., and Sawbridge, P. T., *J. Nucl. Mater.* **41,** 112 (1971).
68. Kingery, W. D., and Francois, B., *Proceedings of the Notre Dame Conference on Sintering,* 471 (1965).
69. Hodkin, E. N., *J. Nucl. Mater.* **88,** 7 (1980).
70. Chasnov, M. G., Leibowitz, L., and Gabelnick, S. D., *J. Nucl. Mater.* **49,** 129 (1974).
71. Bates, J. L., McNeilly, C. E., and Rasmussen, J. J., *Mater. Sci. Res.* **5**(11) (1970).
72. Christensen, J. A., BNWL-SA-584A (1966).
73. Rhee, S. K., *Mater. Sci. Eng.* **11**(6), 311 (1973).
74. Rhee, S. K., *Mater. Sci. Eng.* **16,** 45 (1974).
75. Pastushkov, V. G., Novoselov, G. P., and Borisov, G. B., *At. Energ.* **34**(5), 345 (1973).
76. Wollfrey, J. L., and Wright, S. J., *J. Aust. Ceram. Soc.* **12**(2), 21 (1976).

0018–1536/84/1700–0333$06.40

Transpiration Mass Spectrometric Analysis of Liquid KCl and KOH Vaporization

J. W. HASTIE,* K. F. ZMBOV,† AND D. W. BONNELL

Inorganic Materials Division, Center for Materials Science, National Measurement Laboratory, National Bureau of Standards, Washington, DC 20234

Received December 22, 1983; Accepted December 31, 1983

ABSTRACT

Existing thermodynamic functions for the equilibrium vapor species over liquid KCl and KOH are based largely on an extrapolation of data for the lower temperature solid systems together with estimated spectroscopic constants. The degree of importance of dimeric or more complex vapor species is especially uncertain, with a wide disparity indicated for the various literature studies. Using a transpiration mass spectrometric method, we have determined the equilibrium vapor composition in the presence of liquid over a wide range of temperature and pressure. The results for KCl are in very good agreement with the JANAF evaluation of previous work. In addition, thermodynamic data are given for the $(KCl)_3$ trimer species for the first time. For the KOH system, the dimer species $(KOH)_2$ is much more important than suggested by the JANAF evaluation of previous work. Second law thermodynamic data, at 1000 K, with current JANAF values given in parentheses are: for vaporization to yield KOH(g), ΔH_v = 164 ± 4 (159 ± 17) kJ/mol, ΔS_v = 109 ± 4 (102 ± 4) J/K mol; for vaporization to yield $(KOH)_2(g)$, ΔH_v = 159 ± 5 (123 ± 17) kJ/mol, and ΔS_v = 95.0 ± 4 (50.0 ± 13) J/K mol. Thermochemical data are also reported for the KO_2 species produced by the reaction:

$$2KOH(l) + 1.5\ O_2 = 2KO_2 + H_2O$$

*Author to whom all correspondence and reprint requests should be addressed.
†NBS Guest Worker; permanent address: Boris Kidrich Institute, Belgrade, Yugoslavia.

from which we derive at 1000 K, ΔG_f (KO_2) = −169 ± 12 kJ/mol and ΔH_f (KO_2) = −73.6 ± 13 kJ/mol.

Bond dissociation energies and entropies for the various potassium halide and hydroxide species are found to correlate well with other alkali halide systems. Evidence of temperature dependent electron impact ionization is also indicated in these studies.

Index Entries: Transpiration mass spectrometry, of KCl and KOH vaporization; thermodynamics, of KCl and KOH vaporization; KO_2; ionization cross-sections; vaporization of KCl and KOH, mass spectrometric analysis of; KCl, mass spectrometric analysis of the vaporization of; KOH, mass spectrometric analysis of the vaporization of; mass spectrometry, of KCl and KOH vaporization.

1. INTRODUCTION

Vapors containing alkali metal species have diverse implications for high temperature processes, as we have discussed in detail elsewhere (*1*). With the need for improved efficiency in materials and energy management, specifications for alkali control are becoming more stringent. Also, the expanded use of computer modeling for process design, with a thermodynamic data base, requires more accurate thermodynamic functions than are presently available. Key data-base species in this category include the alkali metal halides, hydroxides, and oxides, all of which are believed to be significant in various industrial vapor transport processes. The present study deals with potassium halide, hydroxide and oxide systems.

1.1. Status of KCl System

The JANAF (*2,3*) thermochemical tables provide a critical evaluation of most of the available data for the KCl solid, liquid, vapor system. Sufficient vaporization, calorimetric, and spectroscopic data exist to assign KCl monomer and $(KCl)_2$ dimer as the major vapor species. Excellent second and third law agreement and other thermochemical tests have accurately established the vaporization thermodynamics for the monomer. The dimer is less accurately established, with key uncertainties arising from estimates of vibration frequencies used for entropy and free energy function calculations (*3*) and from ionization cross-section assumptions (*4*). Following the JANAF (*3*) estimation of $(KCl)_2$ vibrational frequencies, experimental values have been obtained (*5*). The B_{1u} mode frequencies are somewhat less than the JANAF (*3*) estimates and hence the dimer entropies should be greater than the JANAF values at typical vaporization temperatures.

Various studies indicate the dimer/monomer proportion to vary from being negligible to about 30 %. The only direct evidence for dimer formation is semiquantitative and is limited to the vapor over solid KCl, e.g., see Milne and Klein (*6*) and Grimley and Joyce (*4*). In these two

Knudsen effusion-mass spectrometric studies, the dimer sublimation enthalpy was reasonably well-established by the second law treatment of data. Grimley and Joyce (*4*) obtained partial pressure data by estimation of ionization cross-sections. Their estimation was based on the often used additivity of atomic cross sections assumption, which remains questionable for high temperature species. These studies also indicated, qualitatively, the presence of small amounts of $(KCl)_3$ trimer over solid KCl. However, since the partial pressure and sublimation enthalpy data were not determined, the possible significance of this species at higher temperatures/pressures could not be established.

The present study attempts to reduce these uncertainties, and to extend measurements to the liquid state, through use of the Transpiration Mass Spectrometric (TMS) method (*7*). The data will also be compared with the analogous KOH system for further development and testing of empirical correlations between halides and hydroxides (*8*).

1.2. Status of KOH System

The JANAF (2) thermochemical tables provide a critical evaluation of all pertinent data for the KOH solid, liquid, vapor system. However, there are a number of key assumptions required in their analysis and many of the literature data inconsistencies remained unresolved. The greatest uncertainties occur in the ΔH_f and ΔS_f data for the dimer and monomer species of molecular KOH. JANAF's (2) most recent critical evaluation uses as a starting point the third law heat of dimerization based on the mass spectrometric data of Porter and Schoonmaker (*9*) and of Gusarov and Gorokhov (*10*). This analysis appears to have favored the data of the former workers even though Gusarov and Gorokhov (G&G) also provide precise second law data. The third law analysis is particularly unreliable for this system owing to problems of ion precursor assignment, ionization cross-section estimates, impure samples leading to low dimer activity, and the tenuous estimates of vibrational frequencies. We argue that preference should be given to the second law data of G&G since it bypasses practically all of the problems associated with the third law data. The general reliability of their experimental approach is evident from the good agreement obtained for K_2SO_4 vaporization among G&G (*10*), Lau et al. (*11*), and Plante et al. (*12*).

In the JANAF (2) analysis the dimer becomes much less significant with increasing temperature and is considered negligible at the boiling point. This is contrary to the second law results of G&G (*10*), which suggest high dimer concentrations, even at the boiling point. JANAF (2) also apparently prefers the early boiling-point data of Wartenberg and Albrecht (*13*) even though their second law enthalpies differ by ~10 kcal/mol from the JANAF (2) analysis†. The more recent transpiration data of

†cal = 4.184 J.

Dubois and Millet (*14*) are in much closer agreement with JANAF (2) even though the dimer is necessarily neglected in the data analysis. Also, the single temperature transpiration data point of Jackson and Morgan (*15*), which was used by G&G to calibrate their mass spectrometric ion intensities, is in good agreement with the results of Dubois and Millet (*14*). We consider the KOH data of Jackson and Morgan (*15*) to be credible since their KCl data (corrected for dimer) agree well with the accepted JANAF (2) analysis. G&G (*10*) have shown that oxide impurities in KOH lead to low pressures and sublimation enthalpies for the dimer species and the data of Porter and Schoonmaker (*9*) may suffer from this problem.

With regard to the entropy functions for gaseous KOH and $(KOH)_2$, we estimate that at typical experimental temperatures (average 1000 K) the vibrational contributions are uncertain by 2 and 6 cal/K mol, respectivaly. The JANAF (2) vibrational frequencies are estimates based on an analogy with the KF system, which is itself relatively uncertain. In fact, there is still insufficient vibrational data for any of the dimers of the alkali salt species to make accurate entropy calculations. The entropy functions for these species are usually derived by forcing a second–third law agreement in reaction enthalpies. And, the third law data are sensitive to the absolute pressures of dimer species that often rely on mass spectrometric assumptions about ionization processes. A further difficulty with these alkali salt systems is the closeness of the vaporization enthalpies of monomer and dimer species. This effect complicates the experimental separation of dimer and monomer contributions to mass spectra or to total or apparent (e.g., transpiration) pressure measurements. To summarize the status of the KOH system, the following thermochemical data are highly uncertain:

(1) Absolute partial pressures of monomer and dimer—uncertain by factors of 2 and 10, respectively (e.g., at 1000 K).
(2) Dimerization enthalpy and entropy—uncertain by ±5 kcal/mol and ~10 cal/K mol, respectively.

The present study attempts to reduce these uncertainties, through application of the TMS method, which provides an accurate second law analysis over a relatively large temperature range (*7*). Also, the high pressure capability of this technique allows for measurements over the liquid that were not possible with the Knudsen effusion techniques of G&G (*10*) and Porter and Schoonmaker (*9*). Such an extension in pressure range is also desirable as higher polymers such as $(KOH)_3$ could form under relatively high pressure conditions. The ability to externally adjust the carrier gas composition in transpiration mass spectrometry also allows us to suppress the tendency of KOH(l) to decompose to yield K and H_2O vapor species. Thus experiments were carried out both in inert N_2 at-

and in H_2O–O_2 atmospheres. Under high O_2 pressure conditions the production of KO_2 vapor is possible, i.e.,

$$2KOH(l) + 1.5\ O_2 = 2KO_2 + H_2O$$

and results were obtained for this process. No thermochemical data exist for this species, though its presence in lean flames has been inferred (*16*), as has the similar species NaO_2 (*17*).

2. EXPERIMENTAL APPROACH

The transpiration mass spectrometric (TMS) apparatus used has been described in detail elsewhere (*7*). With this apparatus and experimental procedure, accurate thermodynamic vaporization data have been obtained for systems including: NaCl and Na_2SO_4 (*7*), complex slag (*1*), and glass systems (*18*). The TMS apparatus consists of a chamber for transpiration and molecular beam production and a second chamber in which modulation of the molecular beam and mass spectrometric analysis are performed. The experimental procedure involves heating the sample in a carrier gas stream (N_2 in the present study) and measuring the intensities of positive ions produced by electron impact from molecules present in the molecular beam. In addition, the mass of sample transported (m) is determined gravimetrically and the volume (V) of carrier gas is calculated from the gas flow rate and transport time. Total gas pressures used typically fall in the range 0.1 to 1.0 atm†.

The measured ion intensities (I_i) are converted into partial pressures (P_i) of the molecular species (i) through the basic relation

$$P_i = k_1 I_i T \tag{1}$$

where T is the temperature and k_i is the instrument and chemical system sensitivity constant.

The sensitivity constant k_i is obtained by several methods [see ref. (*7*)] that use the following basic relations:

$$k_i = R(n/V)(\Delta t/\Sigma I \Delta t) \tag{2}$$

and

$$k_i = k_j(\sigma_j/\sigma_i)(S_j/S_i)\ (1/S_g) \tag{3}$$

where (for species i and carrier gas j) R is the gas constant, n the number of mols of salt transported (m/M), V the volume of gas transported, t the transport time, I_i the species ion intensity (corrected for isotope abundance), σ the species ionization cross-section, S the beam expansion and quadrupole analyzer mass discrimination factor (species i,j), and S_g an

†atm = 101,325 N/m^2 (or Pascal).

instrument transport gas scattering correction [=0.6 (7)]. The reference or transport gas sensitivity factor k_j is obtained from Eq. (1), since the pressure of the transport gas is known from an external manometric determination.

In practice, a value of k_j and hence k_i is determined for each data point to minimize the effects of changing beam dynamics. This approach assumes that the ratio σ_j/σ_i remains constant with varying T and fixed $P(N_2)$. Hence

$$P_i \propto P(N_2)\, I_i\, [I\ (N_2)]^{-1} \tag{4}$$

In the present study, where the vapors are comprised mainly of monomer and dimer species, the transpiration equation can also be used for pressure determination, i.e.:

$$P_i = n_i[P(N_2) + P_i][n(N_2) + \sum_i n_i]^{-1} \cong n_i P(N_2)[n(N_2)]^{-1} \tag{5}$$

where $P(N_2) >> P_i$ and $n(N_2) >> n_i$ and T is constant. Calculation of n_i from the total transport mass requires a knowledge of P_d/P_m or of P_m. Values of P_i determined by this approach can be used together with Eq. (1) and (3) for measurement of σ_i.

3. RESULTS FOR THE KCl SYSTEM

3.1. Ion Precursor Assignments

Vaporization of liquid KCl with electron impact ionization of the expansion-cooled molecular beam leads to the following positive ions: $K^+(KCl)$, $K_2Cl^+(K_2Cl_2)$, $KCl^+(KCl)$, $K_3Cl_2^+(K_3Cl_3)$, in decreasing order of ion intensity. The assignment of molecular presursors is indicated in parenthesis. Observation of the correct Cl^{35}, Cl^{37} isotopic mass and intensity distribution for these ions provided unambiguous verification of ion assignments. The following appearance potentials also support the various ion-precursor assignments: $K^+(9.1)$, $K_2Cl^+(10.2)$, and $KCl^+(9.6)$, in electron volt units (corrected using N_2^+) with ±0.5 eV uncertainty. In addition, the temperature coefficients (see Tables 2 and 3 below) also indicate K^+ and KCl^+ to result primarily from a single precursor and K_2Cl^+ and $K_3Cl_2^+$ each arise from other distinct precursors. Total vaporization isotherms also agree with these observations. These results and interpretations are similar to those of earlier work using Knudsen effusion mass spectrometry [e.g., see ref. (4)]. Possible small (~10%) contributions of K_2Cl_2 and K_3Cl_3 to K^+, KCl^+, and K_2Cl^+ (from the latter species) are not ruled out by any of the above arguments. However, errors from this uncertainty will be less than from other sources.

3.2. Molecular Beam Characteristics

With the TMS approach, it is desirable to test for the possible effects of gas unsaturation, beam scatter, Mach focusing, beam cooling, temperature-dependent electron impact ionization, and cluster formation. These effects are conveniently revealed, when present, by the mass spectral dependence on N_2 carrier gas pressure (and flow rate) for a fixed sample pressure, i.e., temperature. In these studies, the gas flow rate increases with the applied N_2 pressure.

As we have discussed elsewhere (*19*), the final beam temperature (T_b) depends on orifice size, source temperature (T_o) and pressure (P), the local Mach number, and the gas heat capacity ratio (C_p/C_v), and collision cross-section and effectiveness terms. For the two orifice sizes of the present study, and a typical source temperature, the beam (or expansion) temperature is calculated to depend on source pressure [$\sim P(N_2)$] as shown in Fig. 1.

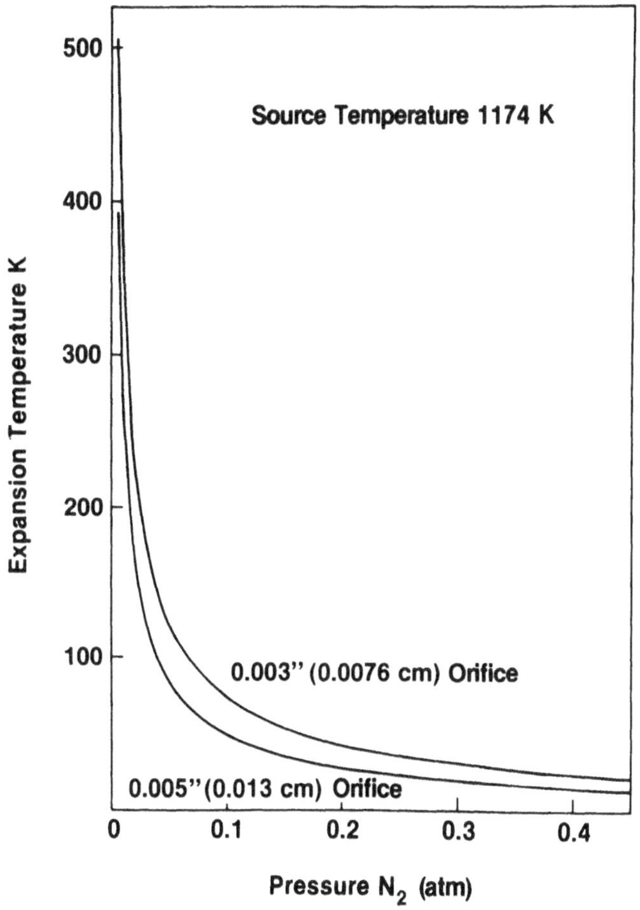

Fig. 1. Effect of N_2 source pressure and sampling probe orifice size on beam or expansion temperature.

Effects of Mach focusing and quadrupole mass discrimination are monitored using a gas mixture of known composition, as discussed earlier (7). These combined effects are accounted for as S_i-factors in Eq. (3). Typical S_i values for the conditions of the present study are indicated in parentheses, as follows: N_2^+ (1.0), K^+ (0.9), KOH^+ (0.7), KCl^+ (0.6), K_2OH^+ (0.4), K_2Cl^+ (0.25), and $K_3Cl_2^+$ (0.15).

The variation of mass spectral ion intensities and ratios with source pressure, and hence T_b, for the two orifices used is shown in Figs. 2 and 3. For the larger orifice, the invariance of dimer to monomer partial pressures [$\propto K_2Cl^+/(K^+ + KCl^+)$] with N_2 pressure indicates first, no ten-

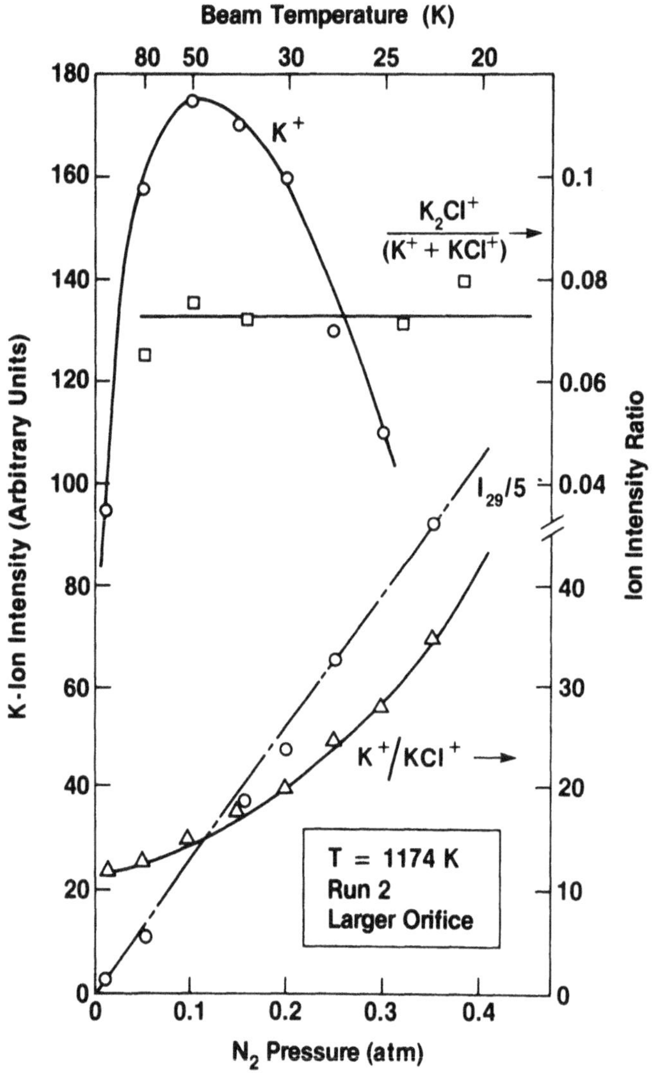

Fig. 2. Variation of KCl system ion intensities and ratios with N_2 source pressure at a fixed source temperature (1174 K), using a capillary probe (0.005 in. orifice). $I_{29} = I^+(N_{14}N_{15})$ isotopic signal.

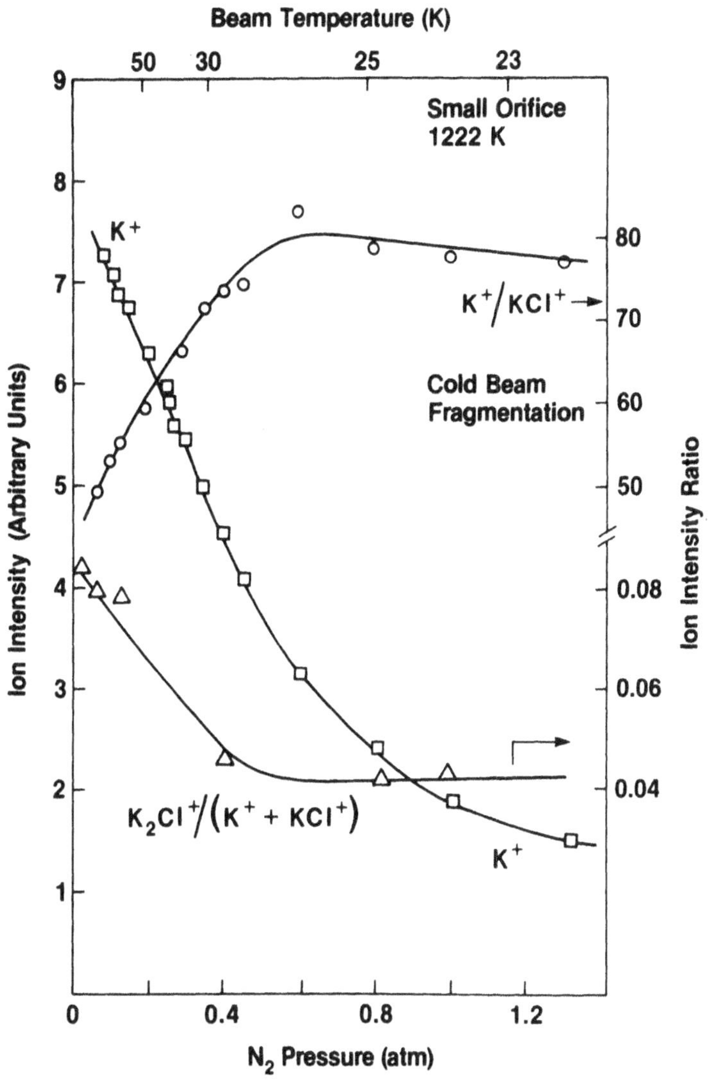

Fig. 3. Variation of KCl system ion intensities and ratios with N_2 source pressure and a fixed source temperature (1222 K), using a conical probe (0.003 in. orifice).

dency for cluster formation under these conditions and second, no unsaturation (*see* Fig. 2). For the smaller orifice (lower T_b), the $K_2Cl^+/(K^+ + KCl^+)$ ion intensity ratio decreases with increasing N_2 pressure, and decreasing T_b, indicating no clustering (*see* Fig. 3). This ratio decrease could be interpreted as evidence of gas unsaturation that generally affects the dimer more than the monomer. However, this effect was absent for the higher flow rate conditions used with the larger orifice, e.g., see Fig. 2, and can be discounted for the conditions of Fig. 3. The variation in the K^+/KCl^+ ion intensity ratio with $P(N_2)$, and T_b, is attributed to temperature-dependent electron impact fragmentation. A similar effect was noted in earlier studies on NaCl (*7*). The effect results from a relatively poor Franck-Condon overlap from the lower vibrational levels favored by beam cooling. We have developed a quantitative anharmonic oscillator model of the effect for alkali halides, as discussed elsewhere (*20*).

The decreased K^+ ion intensity at pressures greater than about 0.15 atm could result from a reduced efficiency in beam formation from gas collisions in the free jet. However, the N_2 ion intensity (I_{29}) is proportional to N_2 pressure over the whole pressure range, e.g., see Fig. 2. This proportionality is indicative of good beam-forming conditions rather than collision dominated behavior [see ref. (*7*)]. We therefore attribute the reduced production of K^+ at higher N_2 pressures to a changing cross-section (σ) for K^+ production (from KCl). That is, $\sigma_m(K^+)$ decreases with decreasing T_b. Likewise, the $K_2Cl^+/(K^+ + KCl^+)$ dependence on $P(N_2)$ (See Fig. 3) is interpreted as a decreasing dimer to monomer cross section ratio σ_d/σ_m with decreasing T_b. It follows from Eqs. (1) and (3) that for the present condition of $I(N_2^+) \propto P(N_2)$, and T_o constant

$$\sigma_m/\sigma(N_2) \propto I_K^+$$

$$\sigma_d/\sigma_m \propto I(K_2Cl^+)/[I(K^+) + I(KCl^+)]$$

and

$$\sigma_m(K^+)/\sigma_m(KCl^+) \propto I(K^+)/I(KCl^+).$$

Hence the profiles of ion intensities and ratios shown in figs. 2 and 3 reflect, directly, the cross-section ratios given above. In view of this changing cross-section behavior, we should not expect TMS cross-sections to *necessarily* agree with those for Knudsen effusion mass spectrometric conditions, nor follow any empiricisms based on classical high temperature mass spectrometry. It is pertinent to note that the NaCl system reported on earlier (*7*) shows similar behavior with changing T_b.

Another noteworthy beam-forming effect is the variation of K^+/KCl^+ ion intensity ratio with source temperature (see Tables 2 and 3 below). Knudsen effusion mass spectrometric (KMS) studies on KCl and our earlier TMS measurements on NaCl did not show this effect, but KMS studies on CsCl did so (*21*). This effect is attributed to the same temperature-dependent electron impact fragmentation process so evident in the variable $P(N_2)$ [i.e., T_b] experiments (see Figs. 2 and 3). Because of this effect we have used the intensity of the predominant ion, K^+ rather than KCl^+, as a measure of monomer partial pressure. Use of KCl^+ yields an incorrect (too low) vaporization enthalpy.

Consideration of these various beam-formation phenomena indicates the optimum conditions for thermodynamic studies, as follows. These potentially perturbing phenomena can be factored out by obtaining data at a relatively low, fixed value of P (N_2) and referencing the

KCl ion intensities to N_2^+. Also, K^+, K_2Cl^+, and $K_3Cl_2^+$ are considered the optimum ions for partial pressure determinations of monomer, dimer, and trimer, respectively.

3.3. Partial Pressure Calibration

Conversion of the classical transpiration data to partial pressures can be made [using Eq. (4)] if the ratios of dimer to monomer partial pressures (P_d/P_m) are known. The JANAF (3) data for P_m are sufficiently well established that we can assume only those ratios that yield P_m data close to the JANAF (3) values are valid. The key thermodynamic quantities upon which JANAF monomer pressures are based include the KCl molecule bond energy and entropy and the corresponding condensed phase data, all of which are well-established by methods other than vaporization. For the dimer, the literature (3) uncertainty is greater. However, the P_d/P_m ratios obtained from the present transpiration experiments (e.g., see Table 1), with the restriction that $P_m \sim P_m$ [JANAF], are in very good agreement with the corresponding ratios from JANAF (3). From the P_d and P_m data, obtained at several transpiration temperatures (given in Table 1) values of k_m and k_d were determined for use in converting ion intensities to partial pressures at all temperatures. For the relatively insignificant trimer species, k_t values were obtained from k (N_2) using an estimated ionization cross section for $K_3Cl_2^+$ production of 6.1, relative to the known value (in our instrumentation) of 1.48 for N_2. This estimate is based on $\sigma_t/\sigma_m \cong 1.12$, which in turn is derived from the empiricism $\sigma_t/\sigma_m = 1.5$ and a molecular weight scaling factor. The resulting partial pressure-ion intensity relationships are summarized in Tables 2 and 3.

Ionization cross-sections can be derived for the monomer and dimer species from these relationships, e.g., for the capillary data $\sigma_m = 7.4$ and

TABLE 1
Pressure Calibration for the KCl System

Sampling probe	Conical	Capillary
Temperature	1261	1116 K
N_2 pressure	0.255	0.37 atm
N_2 flow rate	3	20 cm^3/min (273 K, 1 atm)
Transport time	215	275 min
Transport mass of KCl	0.252	0.13 g
P_d/P_m (assumed)	0.43	0.45
P_m	1.2×10^{-2}	1.4×10^{-3} atm
P_d	8.9×10^{-3}	6.26×10^{-4} atm
P_m (JANAF, 3)	1.2×10^{-2}	1.38×10^{-4} atm
P_d (JANAF, 3)	5.2×10^{-3}	6.2×10^{-4} atm

TABLE 2
Mass Spectrometric and Partial Pressure Data for KCl(l) (Run 3, capillary)[a]

	Ion intensities (arbitrary units)					Pressures, 10^{-3} atm units		
T, K	$I(29)$	$I(K^+)$	$I(KCl^+)$	$I(K_2Cl^+)$	$I(K_3Cl_2^+)$	P_m	P_d	P_t
987	24.5	2.8	0.04			0.10		
1012	23.0	4.5	0.15	0.23		0.17	0.04	
1012	23.0	4.8	0.16	0.24		0.18	0.05	
1040	24.4	8.3	0.31	0.55		0.29	0.10	
1042	24.5	9.2	0.32	0.59		0.33	0.11	
1075	25.0	19.3	0.70	1.42		0.68	0.25	
1076	25.0	19.9	0.70	1.44		0.71	0.26	
1104	26.0	33.5	1.26	2.90	0.03	1.15	0.50	0.01
1128	28.0	51.2	1.93	4.60	0.04	1.63	0.74	0.012
1151	28.5		2.87	7.25			1.14	
1149	28.5	72.5	2.75	6.95		2.26	1.1	
1148	28.5	70.2	2.72	6.90		2.19	1.09	
1183	29.0	121.7	4.58	13.20	0.11	3.73	2.05	0.033
1183	29.0	124.2	4.40	13.30	0.11	3.81	2.07	0.033
1181	29.0	124.0	4.00	14.38		3.81	2.23	
1181	29.0		4.10	14.06			2.2	
1204	29.0	173.2	5.27	22.00	0.14	5.3	3.4	0.042
1204	29.0	173.8	5.50	20.80		5.3	3.2	
1204	29.0		5.65	20.76			3.2	
1236	28.0	241.5	7.65	35.00	0.19	7.7	5.6	0.059
1236	28.0	255.0	7.90	34.00		8.1	5.5	
1259	28.0	357.0	9.50	52.00	0.23	11.3	8.4	0.071
1259	28.0	348.0	9.60	49.20	0.25	11.1	7.9	0.077
1289	25.7	470.0	11.90	78.00	0.27	16.3	13.7	0.091
1289	25.7	465.0	12.30	72.00	0.27	16.1	12.6	0.091

[a]Data are listed in chronological order of experiment.
$P(N_2) = 0.37$ atm.
$I(29)$ is $N^{14}N^{15}$ measured ion intensity ($\equiv 0.0072 \times N^{14}N^{14}$).
$P_m = 8.9 \times 10^{-4} \times I(K^+)/I(29)$ atm.
$P_d = 4.51 \times 10^{-3} \times I(K_2Cl^+)/I(29)$ atm; largest isotope.
$P_t = 8.64 \times 10^{-3} \times I(K_3Cl_2^+)/I(29)$ atm; largest isotope.
$\Delta H_{T,\,m} = 40.7 \pm 0.27$ (JANAF, 41.8) kcal/mol; $\Delta S_{T,\,m} = 23.3 \pm 0.23$ (JANAF, 24.4) cal/K mol.
$\Delta H_{T,\,t} = 34.8 \pm 1.5$ kcal/mol ($= 145.6 \pm 6.3$ kJ/mol); $\Delta S_{T,\,t} = 8.7 \pm 1.2$ cal/K mol ($= 36.4 \pm 5.0$ J/K mol).

$\sigma_d = 5.5$. These values are somewhat higher than we would estimate from analogy with other alkali halides, e.g., $\sigma(NaCl) = 1.0$, $\sigma(NaCl)_2 = 1.5$, σ (CsCl) = 3.1, all at 30 eV. Earlier TMS work established the cross-section for NaCl as 1.0 ± 0.3, relative to 1.48 for N_2 (*7*). From molecular orbital arguments (22), ionization of NaCl or KCl involves primarily removal of an electron from Cl^- in the molecule. The larger separation of K^+, compared with Na^+, and Cl^- suggests a higher cross-section for KCl by a factor of 1.13 (based on internuclear separations in NaCl and

TABLE 3
Mass Spectrometric and Partial Pressure Data for KCl(l)
(Run 4, Conical Probe)[a]

	Ion intensities (arbitrary units)				Pressures, 10^{-3} atm units	
T, K	$I(29)$	$I(K^+)$	$I(KCl^+)$	$I(K_2Cl^+)$	P_m	P_d
1085	4.3	4.2	0.08	0.09	0.85	0.33
1087	3.9	4.5	0.09	0.10	1.0	0.4
1087	3.9	4.7	0.09	0.11	1.05	0.44
1120	3.7	7.5	0.13	0.19	1.76	0.81
1120	3.7	7.6	0.14	0.20	1.78	0.85
1132	3.8	8.9	0.16	0.23	2.0	0.95
1134	3.7	9.1	0.16	0.23	2.1	0.98
1152	3.5	11.7	0.19	0.30	2.9	1.3
1152	3.4	11.5	0.20	0.31	2.9	1.4
1170	3.5	14.2	0.23	0.42	3.5	1.9
1170	3.3	14.2	0.23	0.43	3.7	2.0
1186	3.3	17.1	0.27	0.48	4.5	2.3
1193	3.2	18.9	0.30	0.58	5.1	2.8
1193	3.2	19.4	0.31	0.58	5.3	2.8
1219	3.0	25.6	0.40	0.88	7.4	4.6
1217	3.1	25.0	0.38	0.83	7.0	4.2
1217	3.2	25.1	0.38	0.85	6.8	4.2
1235	3.3	32.0	0.48	1.17	8.4	5.6
1235	3.2	32.0	0.48	1.18	8.7	5.8
1257	2.9	40.5	0.62	1.66	12.1	9.0
1256	2.9	40.0	0.62	1.63	12.0	8.8
1283	2.8	49.5	0.74	2.37	15.3	13.3
1285	2.7	51.0	0.77	2.37	16.4	13.8
1305	2.5	56.5	0.84	2.94	19.6	18.5
1305	2.4	57.0	0.84	2.94	20.6	19.2

[a]Data are listed in chronological order of experiment.
$P(N_2) = 0.25$ atm.
$P_m = 8.68 \times 10^{-4} \times I(K^+)/I(29)$ atm.
$P_d = 1.57 \times 10^{-2} \times I(K_2Cl^+)/I(29)$ atm;
$\Delta H_{T,\,m} = 39.07 \pm 0.37$ (JANAF, 41.8) kcal/mol; $\Delta S_{T,\,m} = 22.23 \pm 0.31$ (JANAF, 24.4) cal/K mol.

KCl). Thus the observed σ(KCl) of 7.4 appears to be anomalously high. One possible explanation is the production of excess K^+ by electron impact fragmentation of $(KCl)_2$. However, the near-constancy of K^+/KCl^+ over a wide temperature range and the relatively low concentration of $(KCl)_2$ species suggest another explanation. The high σ values for KCl and $(KCl)_2$ are plausible given the possibility of cross-section enhancement from beam cooling. These values are still significantly less than those based on the widely used additivity-of-atomic-cross-sections approximation. As we have noted above, the KCl cross sections appear to depend on T_b and they are not particularly useful for other beam conditions.

3.4. Vapor Pressure Versus Temperature Determination

Four separate experiments were performed on KCl vaporization as a function of temperature with fixed N_2 pressure. This condition of a fixed N_2 pressure leads to a decreased N_2^+ intensity with increased temperature, in accordance with the gas law. Runs 1–3 used a capillary orifice of 0.005 in. diameter and run 4 used a conical orifice with 0.003 in. diameter. Tables 2 and 3 show typical data, which is then summarized as Clausius-Clapeyron plots in Figs. 4 and 5. As shown in Fig. 4, the KCl

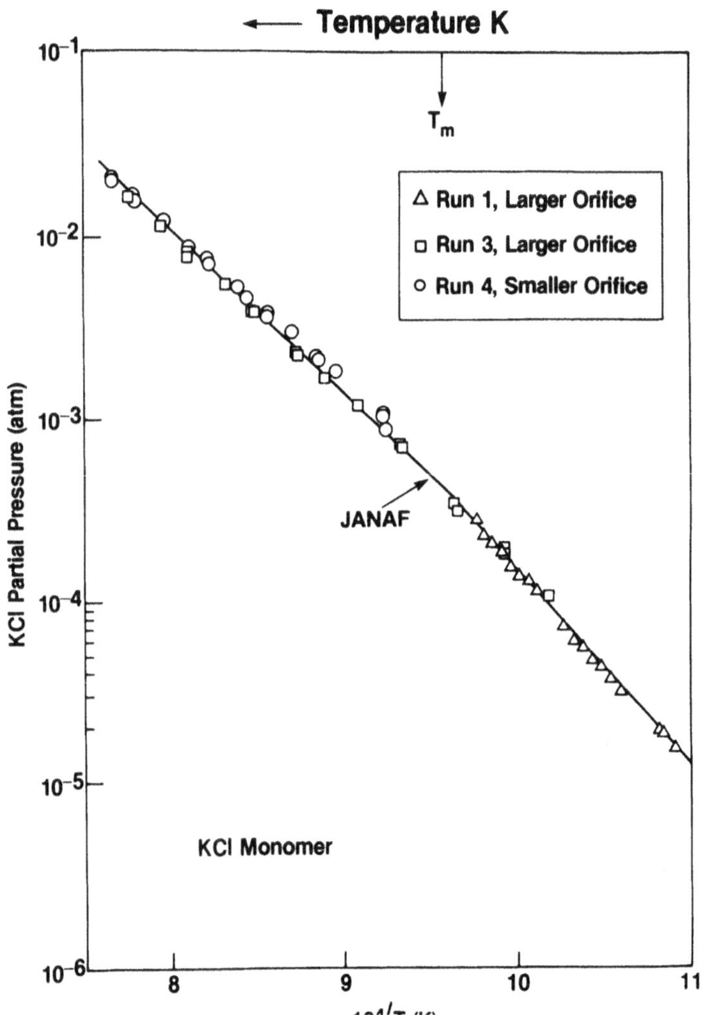

Fig. 4. Variation of KCl monomer partial pressure with reciprocal temperature based, in part, on data given in Tables 2 and 3. Solid line is the JANAF (1971) curve and T_m is the literature melting point.

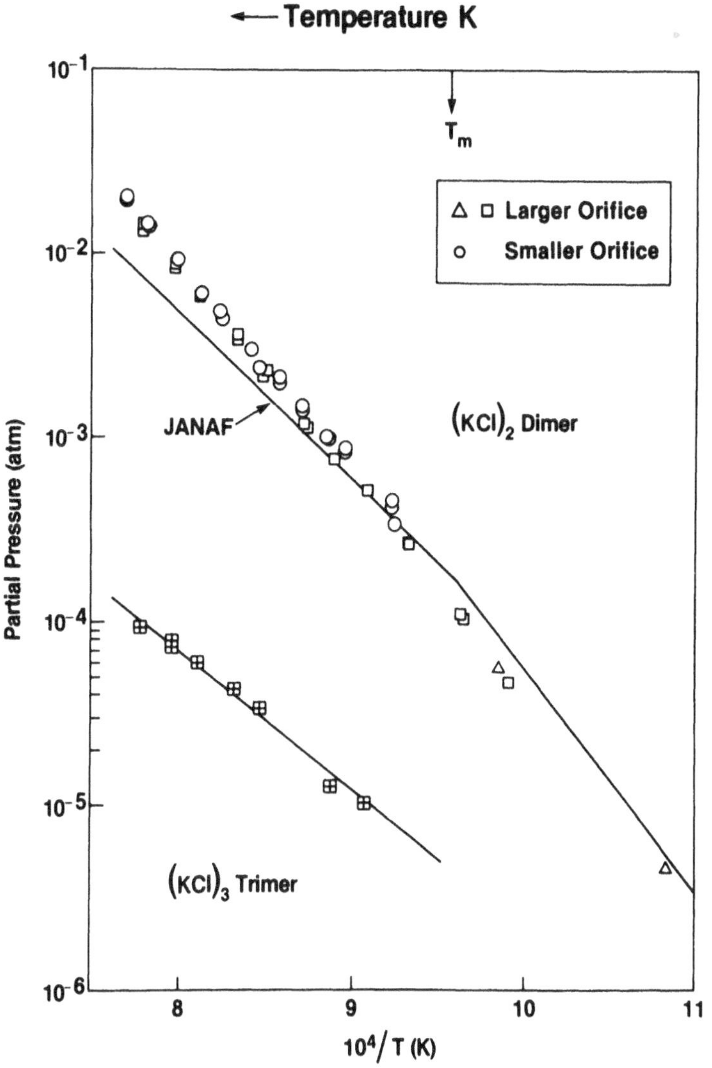

Fig. 5. Variation of KCl dimer and trimer partial pressures with reciprocal temperature based on data given in Tables 2 and 3. Solid dimer curve is from JANAF (3).

monomer partial pressures over solid and liquid KCl are in excellent agreement for each run and probe, as well as with the JANAF (3) data. Evidence of a slight systematic error (< 1 kcal/mol) is given by the lack of agreement, within the statistical uncertainties, between Runs 3 and 4 (see footnotes in Tables 2 and 3). It is likely that this error results from the $\sigma_m/\sigma(N_2)$ temperature dependence. The slope discontinuity agrees well with the JANAF (3) melting temperature (T_m in Fig. 4). In Fig. 5 the dimer data also agree well with JANAF (3) for partial pressures up to about 10^{-3} atm. The breakdown of this agreement at higher pressures is attributed either, to $\sigma_d/\sigma(N_2)$ increasing with T_o, or (less likely) to $(KCl)_2$ cluster formation, as already discussed for the variable $P(N_2)$ experiments. We conclude that the present results and the JANAF (3) data for KCl monomer and dimer are basically in agreement, and that no modification of the current JANAF tables is warranted for these species. It may be argued that our use of JANAF data in our calibration procedure (see Table 1) forces an agreement between our data and that of JANAF. However, this is not the case, for the following reasons. For a particular experimental run, the JANAF value of P_m at a single temperature was assumed in converting the transpiration (non-mass spectrometric) data to partial pressures, P_d and P_m. Thus P_d is independent of the use of JANAF P_m data. Also, the vaporization enthalpies of both monomer and dimer are independent of the JANAF data. Only the vaporization entropy of monomer is biased by the use of JANAF data for calibration and such data is considered reliable. The use of JANAF data could have been avoided completely if the ionization cross-sections could be reliably estimated. However, as already discussed, KCl appears to behave in an unpredictable, anomalous manner on electron impact.

No literature data exist for $(KCl)_3$ and the present results are considered a reliable indication of the equilibrium trimer partial pressures. The extent of cluster formation, if present, for the dimer is sufficiently small that, based on collision frequency arguments, it can certainly be neglected for the trimer. Also, the possible effect on P_t of σ_t varying with T_o is considered small even if it is similar to the dimer behavior.

4. RESULTS FOR THE KOH SYSTEM

4.1. Ion Precursor Assignments

Vaporization of liquid KOH with electron impact ionization of the cooled molecular beam leads to the following positive ions: K^+, KOH^+, K_2OH^+, and K_2O^+, in decreasing order of ion intensity. The first two ions are assigned to KOH and the latter two, the $(KOH)_2$ molecular species. As Gusarov and Gorokhov [G&G, ref. (*10*)] noted, some part of the K^+ signal could also be attributed to the dimer precursor, though we have no means of establishing the magnitude of this contribution, if any. G&G have convincingly established KOH as the only significant precursor to KOH^+, and we have no evidence to the contrary. Also, the electric field deflection of polar species technique indicates, for the analogous NaOH system, that $NaOH^+$ and Na_2OH^+ form from monomer (polar) and dimer (nonpolar) species, respectively (*23*). Thus, in relating mass spectral ion intensities to species partial pressures, we utilize KOH^+ and K_2OH^+ for the monomer and dimer species, respectively.

The K_2O^+ ion has two possible molecular precursors, $(KOH)_2$ and K_2O. In the study of G&G (*10*), carried out at much lower pressures of KOH and $(KOH)_2$, they concluded that K_2O^+ was a fragment of the dimer electron impact process. They cite a ratio of $K_2O^+/K_2OH^+ = 0.1$, compared with our values in the range 0.13–0.2 (at 30 vs 50 eV for G&G). The following vapor phase equilibrium reaction between K_2O and $(KOH)_2$ is possible:

$$(KOH)_2 = K_2O + H_2O \qquad [a]$$

and similarly,

$$2KOH = K_2O + H_2O \qquad [b]$$

Hence, since the dimer pressures are considerably higher in the present study than for that of G&G (because of the higher temperatures used in our study), it is possible that K_2O could be present. However, we can assign most or all of the K_2O^+ ion signal to $(KOH)_2$ for the following reasons. The ion intensity ratio K_2O^+/K_2OH^+ is invariant with the experimental conditions of temperature, total pressure, gas flow rate, and time even though the ion intensity ratio K_2OH^+/KOH^+ depends on these variables. Also, when additional H_2O was added no change in the ion intensity ratio K_2O^+/K_2OH^+ was observed. Assuming as an upper limit that 10% of K_2O^+ could be attributed to K_2O then an upper limit K_p for reaction [b] is,

$$K_p(K_2O) \leqq 10^{-2}$$

from which ΔG_{950}(Reaction [b]) $\geqq$ 8.7 kcal/mol and $\Delta G_{f,950}$ K_2O > −55.3 kcal/mol. Also, since ΔS for the Reaction [b] is estimated to be negligible because no change in the number of species occurs then ΔH(Reaction [b]) > 8.7 kcal/mol and $\Delta H_{f,950}$ K_2O > −49.4 kcal/mol. Thus the average K—O bond energy in K_2O is less than 74 kcal/mol. This result may be compared with a corresponding value of 89 kcal/mol for Li_2O. A lower stability for K_2O, compared with Li_2O, is consistent with the trend expected from ionic bonding arguments.

Small ion signals corresponding to $K_3(OH)_2^+$ may be attributed to the trimer species $(KOH)_3$. At 1000 K the partial pressure ratio for trimer to dimer was ~0.05, neglecting small cross-section and sensitivity differences. This ratio is notably similar to a value of 0.04 for the analogous KCl system.

Earlier mass spectrometric studies have noted the presence of additional K^+ arising from K atoms, thereby indicating sample decomposition (*9,10*). In the initial heating stage of the present experiments, using N_2 carrier gas, atomic K was also observed. A high K^+/KOH^+ ratio, low appearance potential (4.5 eV), and time-dependent behavior were observed during this K-forming process. Figure 6 shows that the high K^+/KOH^+ ratio decays with time to a value characteristic of KOH ionization. Higher than normal (i.e., for H_2O desorption from the sample) H_2O^+ signals were also present when excess K^+ was observed. A potential concomitant source of K and H_2O is the reaction:

$$2KOH(l) = H_2O + 2K + \tfrac{1}{2} O_2$$

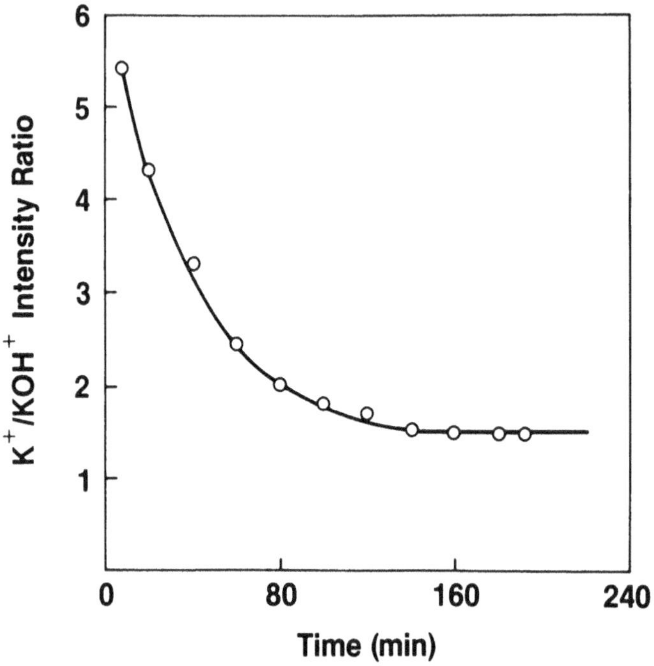

Fig. 6. Variation of K^+/KOH^+ ion intensity ratio with time.

From the known Gibbs energies of these reactants, K pressures in the range 10^{-5} to 10^{-4} atm would be formed under our experimental temperature conditions (1000–1200 K). However, we observed maximum K pressures at least an order of magnitude greater than predicted by this reaction. These higher initial K pressures are attributed to the presence of Pt(s) which, we hypothesize, incorporates O_2 as an adsorbed, dissolved, or oxide intermediate product. The energetics of this process would require a Gibbs energy change of only about −10 kcal/mol K formed to yield the observed K pressures. This process effects only the K^+ signals and, as only the KOH^+ and K_2OH^+ ions are used in the thermochemical analysis, the reaction does not affect the partial pressures of KOH monomer and dimer, provided the KOH(l) activity remains unity.

Further evidence for ion-precursor assignments is provided by the effect of O_2 and H_2O additions to the carrier gas stream. When the H_2O partial pressure was varied over the range 10^{-4}–10^{-1} atm, the ion intensity ratio K^+/KOH^+ decreased from 2.5 to 1.7. Also, when O_2 was present (at 10^{-3}–10^{-1} atm), in addition to H_2O, the ratio K^+/KOH^+ was further reduced to 1.3. These observations are consistent with the reduction of free K in accordance with the reaction

$$2K + H_2O + \tfrac{1}{2}O_2 \rightarrow 2KOH(l)$$

The presence of additional H_2O and O_2 had no effect in the ion intensity ratio KOH^+/K_2OH^+. Hence the activity of liquid KOH(l) is not affected by the presence of free K and the observed KOH^+ and K_2OH^+ ions can be used as measures of the monomer and dimer partial pressures, respectively, for unit activity KOH(l).

An additional ion KO_2^+ was observed when high levels of O_2 (~ 0.1 atm) were added to the carrier gas. As the intensity of this ion showed no correlation with the intensities of K_2O^+ and K_2OH^+, the dimer $(KOH)_2$ can be eliminated as a potential molecular precursor. Under the present experimental conditions, the most reasonable precursor assignment is KO_2.

4.2. Total Vaporization Isotherms

Impurity effects have proven to be a significant problem in earlier KOH vaporization studies, as noted by G&G. These authors found the activity of KOH solid to decrease with time owing to the build up of less volatile $K_2O(s)$. This problem was not present in the present study as the ion intensity ratio of dimer to monomer remained constant during isothermal total vaporization calibration experiments. Also, when the sample was depleted all ion intensities dropped to zero concomitantly (e.g., see Fig. 7). Thus the KOH liquid activity remained at unity throughout the experiment.

4.3. Molecular Beam Characteristics

Beam-formation phenomena were monitored in a similar manner to that for the KCl system as discussed above. Typical data for the experi-

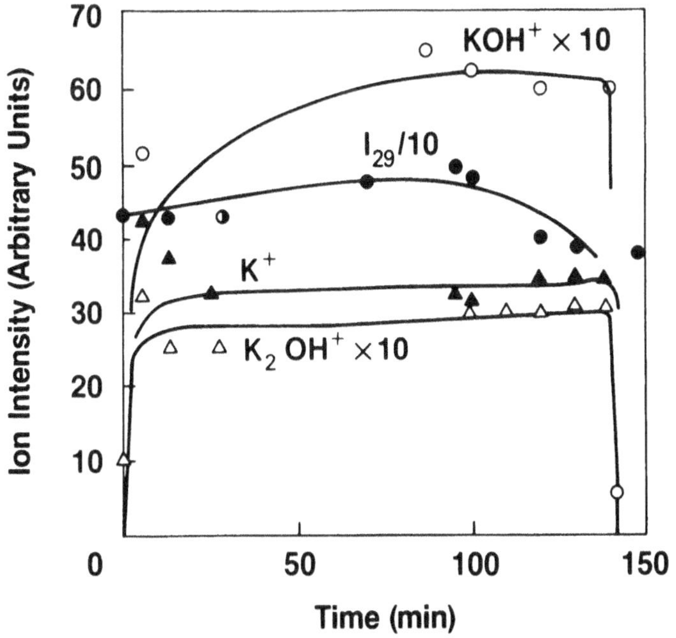

Fig. 7. Ion intensity versus time data at 1024 K and an N_2 pressure of 0.16 atm.

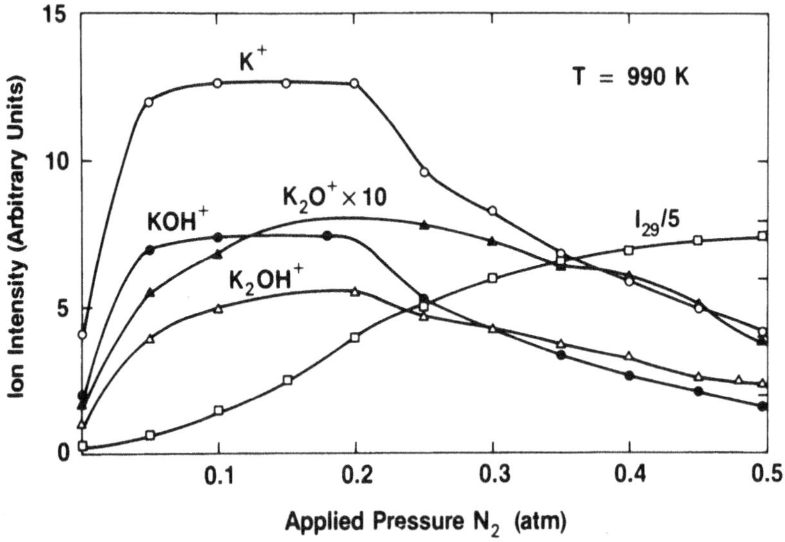

Fig. 8. Profiles of ion intensity versus N_2 pressure at 990 K.

mental run 1 are shown in Figs. 8 and 9. Note in Fig. 9 that the ion intensity ratio K_2OH^+/KOH^+ increases slightly with $P(N_2)$, and hence gas flow rate, thereby indicating carrier gas saturation with KOH vapor. The similar intensity-pressure profiles of K^+ and KOH^+ are indicative of a common precursor, namely, KOH. The N_2 isotopic ion intensity signal, I_{29} increases less with N_2 pressure above 0.2 atm. This is indicative of significant beam scattering above this pressure owing to the relatively large orifice used. Hence, the thermochemical experiments were con-

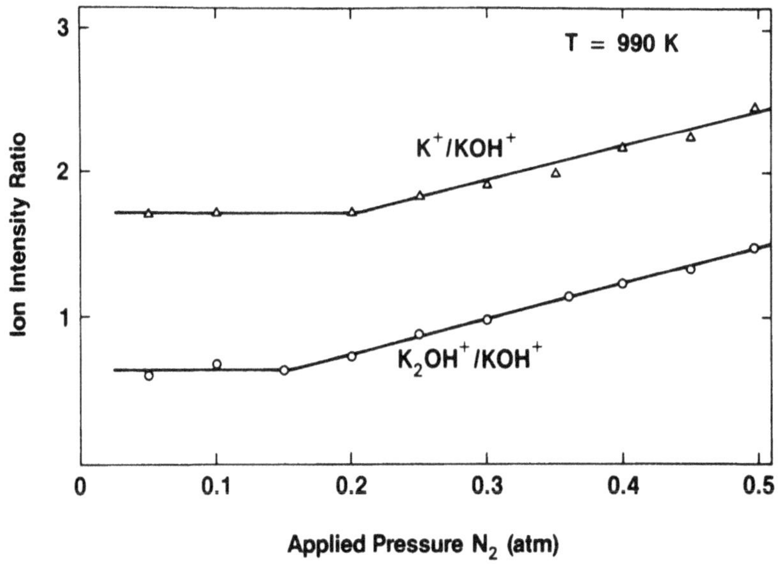

Fig. 9. Ion intensity ratios versus N_2 pressure at 990 K.

ducted at pressures in the range 0.15–0.2 atm. The ion intensity ratio data in Fig. 9 reveal that above 0.2 atm the ionization process increasingly favors K^+ over KOH^+. This effect is analogous to that observed by us previously for NaCl (7) and KCl (above), though very much less pronounced. As can be seen in Fig. 1, the beam temperature decreases markedly with increased N_2 pressure. This reduces the population of excited vibrational levels in KOH and alters the *partial* ionization cross sections for the processes yielding K^+ and KOH^+.

The increased intensity of dimer to monomer ions with N_2 pressure (above 0.15 atm) can be attributed either to the formation of additional dimer species by clustering during the free jet expansion process, or, more likely, to an increasing σ_d/σ_m ratio. It is well known that cluster formation is proportional to gas pressure. In the present case the effect can be related to the increased number of third body (N_2) collisions and decreased temperature that stabilizes the exothermic clustering process. Note in Fig. 9 that this process, if present, is only significant at pressures in excess of 0.2 atm. For the conditions of run 1 this possible clustering effect contributes less than 15% to the total dimer pressure. For the conditions of run 2, this effect should be even less. As the magnitude of this effect is smaller than the pressure calibration uncertainty, or the experimental precision, we neglect this possible contribution to the dimer signal. For the case where the effect results from σ_d/σ_m varying with $P(N_2)$, the thermodynamic data remain unaltered as approximately constant beam dynamics are maintained during these measurements.

A separate experimental run was made using a small-orifice conical sampler, at pressures of ~ 0.2 atm, thereby eliminating the possibility of $(KOH)_2$ cluster formation or changing σ_d/σ_m. About 10^{-3} atm of H_2O vapor was also added to the N_2 carrier gas for this experiment. The ion intensity ratio KOH^+/K^+ and K_2OH^+/KOH^+ were similar to those listed for the capillary sampler runs 1 and 2. These comparisons indicate that runs 1 and 2 were essentially free of cluster effects and also of significant decomposition to produce K and H_2O species.

4.4. Partial Pressure Calibration

The key factor in relating ion intensity to partial pressure data via Eqs. (2) and (1) is the value of the dimer to monomer partial pressure ratio, P_d/P_m, at the calibration temperature. This value is needed to partition the total amount of salt transported into monomer and dimer components. Alternatively, with the independent use of Eq. (3) one requires the dimer to monomer ionization cross-section ratio, σ_d/σ_m. The pressure calibration procedure that provides the best agreement between the results of methods (2) and (3) is as follows.

The extrapolated data of Gusarov and Gorokhov (*10*) give a value of $P_d/P_m = 0.42$ at 1024 K. With this value, our weight loss measurement gives P_d and P_m values similar to those of G&G. Other ratios give P_d

and P_m values well outside of an acceptable range. This range of possible upper and lower limit partial pressures can be set from the transpiration data of Dubois and Millet (*14*) and Jackson and Morgan (*15*). Thus our weight loss data are consistent with the *relative* pressure data of G&G. Using the values of k_m and k_d thus obtained, we derive from Equation (3) the following cross-sections: $\sigma_m(KOH^+) = 1.1$, $\sigma_m(KOH^+ + K^+) \leqq 2.7$, $\sigma_d(K_2OH^+) = 3.6$; i.e., $\sigma_d/\sigma_m \leq 1.4$. These cross-sections are close to what we would estimate based on comparisons with the NaCl monomer–dimer system and the expected changes with increased alkali size. Farber et al. (*24*) in their Knudsen effusion mass spectrometric study estimated a value of $\sigma_m(KOH^+) = 4.79$. The only quantity needed for method (3) calibration is σ_d/σ_m and previous mass spectrometric work has shown this value to lie in the range 1.5 ± 0.5 for many systems (*25*). Thus the two calibration approaches are self-consistent. In practice, $k(N_2, 28\text{ amu})$ values are determined for each data point. Hence, these calibration results are more conveniently expressed in the form:

$$k(KOH^+) = 2.5\ k(N_2, 28\text{ amu})$$

and

$$K(K_2OH^+) = 1.65\ k(N_2, 28\text{ amu})$$

Table 4 summarizes the pressure calibration data obtained using the classical transpiration approach with the P_d/P_m ratio determined from the extrapolated mass spectrometric data of G&G.

4.5. Vapor Pressure Versus Temperature Determination

Two separate experiments were performed on KOH vaporization as a function of temperature with fixed N_2 pressure. A separate transpiration experiment at constant temperature was used for pressure calibration, as described in the previous section. The temperature-dependent ion intensity and partial pressure data are summarized in Tables 5 and 6. Figure 10 shows typical Clausius-Clapeyron plots from which second law vaporization enthalpy and entropy data are derived. A comparison of the

TABLE 4
Pressure Calibration for the KOH system

Temperature	1024 K
N_2 pressure	0.155 atm
N_2 flow rate	12 cm^3/min (273 K, 1 atm)
Transport time	250 min
Transport mass of KOH	0.253 g
P_d/P_m (G&G)	0.42
P_m	2.8×10^{-3} atm
P_d	1.2×10^{-3} atm

TABLE 5
Mass Spectrometric and Partial Pressure Data for KOH (*l*) (Run 1)[a]

	Ion intensities (relative units)					Pressures, 10^{-3} atm units	
T, K	$I(29)$	$I(K^+)$	$I(KOH^+)$	$I(K_2OH^+)$	$I(K_2O^+)$	P_m	P_d
870	18.0	1.6	3.30	0.24		0.06	0.03
903	18.0	2.4	8.50	0.35		0.10	0.05
929	15.8	4.8	12.00	1.20		0.27	0.18
933	12.0	3.6	1.07	0.52		0.32	0.10
971	13.3	8.1	2.40	1.23		0.65	0.22
971	14.0	8.0	2.60	1.33	0.27	0.67	0.23
993	16.0	14.2	5.50	2.70	0.60	1.24	0.24
996	16.7	14.2	6.25	3.30	0.71	1.35	0.47
1023	17.4	21.8	10.50	5.80	1.15	2.17	0.79
1024	17.5	21.5	10.40	5.80	1.23	2.14	0.79
1059	16.0	36.0	18.30	10.10	1.86	4.12	1.50
1060	16.0	35.0	18.30	10.10	1.86	4.12	1.50
1044	15.8	25.0	14.20	7.75	1.50	3.23	1.16
1043	15.5	25.0	13.60	7.60	1.45	3.15	1.16
1014	16.1	14.7	8.50	4.80	0.95	1.90	0.71
1013	16.4	14.6	8.60	4.90	0.95	1.89	0.71
982	10.4	4.9	2.90	2.90	0.27	1.01	0.33
982	9.8	4.9	2.90	1.45	0.27	1.06	0.35
959	9.2	3.2	2.05	1.07	0.15	0.80	0.28
958	9.1	2.7	1.80	0.73		0.73	0.19
958	9.3	2.9	1.80	0.76		0.70	0.19
922	11.1	1.4	0.92	0.50		0.30	0.11
889	12.3	0.7	0.48	0.28		0.14	0.05
859	10.1	0.3	0.17	0.12		0.06	0.03
840	10.0	0.2	0.10	0.06		0.04	0.01

[a] Data are listed in chronological order of experiment.
$P(N_2) = 0.2$ atm.
$k(N_2) \cdot T = 1.44 \times 10^{-3}/I(29)$.
$P_m = 3.6 \times 10^{-3} \times I(KOH^+)/I(29)$.
$P_d = 2.38 \times 10^{-3} \times I(K_2OH^+)/I(29)$.
$I(29)$ is $N^{14}N^{15}$ measured ion intensity ($\equiv 0.0072 \times N^{14}N^{14}$).

least squares curves from these experiments with those of earlier work is given in Figs. 11 and 12 for monomer and dimer species, respectively. The transpiration results of Dubois and Millet (*14*) assumed KOH as the sole vapor species and these results have been corrected for dimer as follows. It follows from the transpiration equation that

$$P_m = P(1 + 2P_d/P_m)^{-1}$$

and

$$P_d = P(P_d/P_m)(1 + 2P_d/P_m)^{-1}$$

TABLE 6
Mass Spectrometric and Partial Pressure Data for KOH(*l*) (Run 2)[a]

	Ion intensities (arbitrary units)					Pressures, 10^{-3} atm units	
T, K	*I*(29)	$I(K^+)$	$I(KOH^+)$	$I(K_2OH^+)$	$I(K_2O^+)$	P_m	P_d
955	19.3	7.5	4.0	3.3		0.58	0.32
957	18.3	7.3	4.0	3.0		0.61	0.30
976	17.0	10.6	6.0	3.8		1.01	0.43
978	17.1	11.0	6.3	4.0	0.50	1.06	0.45
979	17.4	12.2	6.7	4.5		1.11	0.49
1002	17.7	17.7	9.5	7.0	0.98	1.59	0.78
1002	17.1	16.1	9.1	4.4		1.58	0.50
1001	17.0	16.4	9.2	6.0		1.61	0.69
1001	14.0	15.5	8.8	5.4		1.62	0.76
1012	17.1	21.0	11.7	7.8	1.04	1.99	0.88
1011	15.5	20.5	11.8	7.2	0.85	2.22	0.89
1024	9.8	18.0	9.3	4.5		2.70	0.87
1024	7.7	15.5	9.0	3.6	0.44	3.20	0.89
1024	7.3	17.0	9.5	4.0		3.87	1.07
1041	9.2	25.6	13.8	5.8		4.11	1.14
1042	8.8	26.0	14.2	5.8		4.42	1.19
1066	9.5	38.0	19.9	9.2	1.33	5.99	1.83
1066	9.9	38.0	19.8	9.3	1.33	5.73	1.77
1049	10.4	29.7	15.5	8.8	1.10	4.21	1.60
1049	10.1	29.7	15.5	8.0		4.34	1.48
1028	10.0	21.2	11.6	5.7	0.73	3.16	1.03
1027	10.6	20.9	11.5	5.8		2.99	0.99
1005	10.0	14.4	7.9	3.4	0.45	2.16	0.61
1004	8.5	14.5	8.5	2.9		2.76	0.62
975	7.0	6.1	3.3	8.5		1.31	0.22
974	5.5	2.7	0.6	—		1.36	0.20

[a]Data are given in chronological order of experiment.
$P(N_2) = 0.16$ atm.
$k(N_2) \cdot T = 1.15 \times 10^{-3}/I(29)$.
$P_m = 2.88 \times 10^{-3}\,[I(KOH^+)/I(29)]$.
$P_d = 1.9 \times 10^{-3}\,[I(K_2OH^+)/I(29)]$.

where *P* is the pressure calculated assuming KOH as the only vapor transport species, e.g., as was done by Dubois and Millet (*14*). Thus the effect of dimer species is to reduce the total pressure as calculated from transpiration data. The curves of Gurvich et al. (*26*) in Figs. 11 and 12, are based on recently published evaluated thermochemical tables. The curve of Farber et al. (*24*) in Fig. 11 is an extrapolation of their lower temperature Knudsen effusion mass spectrometric results. These authors did not report dimer data, except to note that its partial pressure was less than 10% of the monomer's.

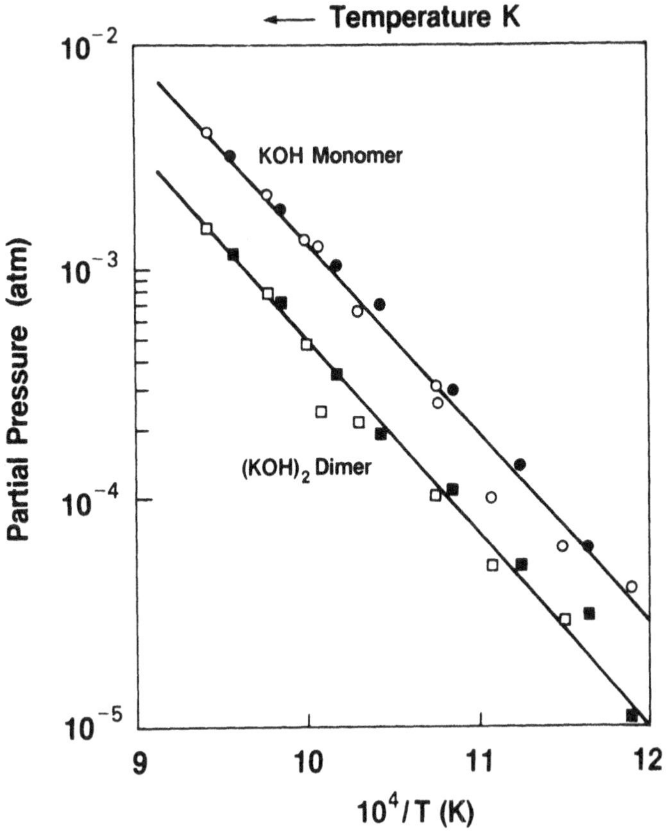

Fig. 10. Variation of partial pressure with reciprocal temperature (Run 1); open and closed symbols represent increasing and decreasing temperature run chronology, respectively.

4.6. KO_2 Formation

With the substitution of O_2 for N_2 in the carrier gas, a new ion was noted with mass corresponding to KO_2^+. Typical ion intensity data are given in Table 7. For the vapor phase reaction:

$$2KOH + 1.5\ O_2 = 2KO_2 + H_2O$$

$$K_p(T) = \frac{[P(KO_2)]^2 P(H_2O)}{[P(KOH)]^2[P(O_2)]^{1.5}} \sim \frac{[I(KO_2)]^2 I(H_2O)}{[I(KOH)]^2[I(O_2)]^{1.5}[k(KOH) \cdot T]^{0.5}}$$

This K_p relationship assumes a cancellation of cross-section and sensitivity factors. The constant $k(KOH) \cdot T$ is determined from the KOH pressure and ion intensity data. With these considerations the following values of K_p result,

$$K_p\ (1013) = 7.9 \times 10^{-2}$$

$$K_p\ (987) = 1.8 \times 10^{-3}$$

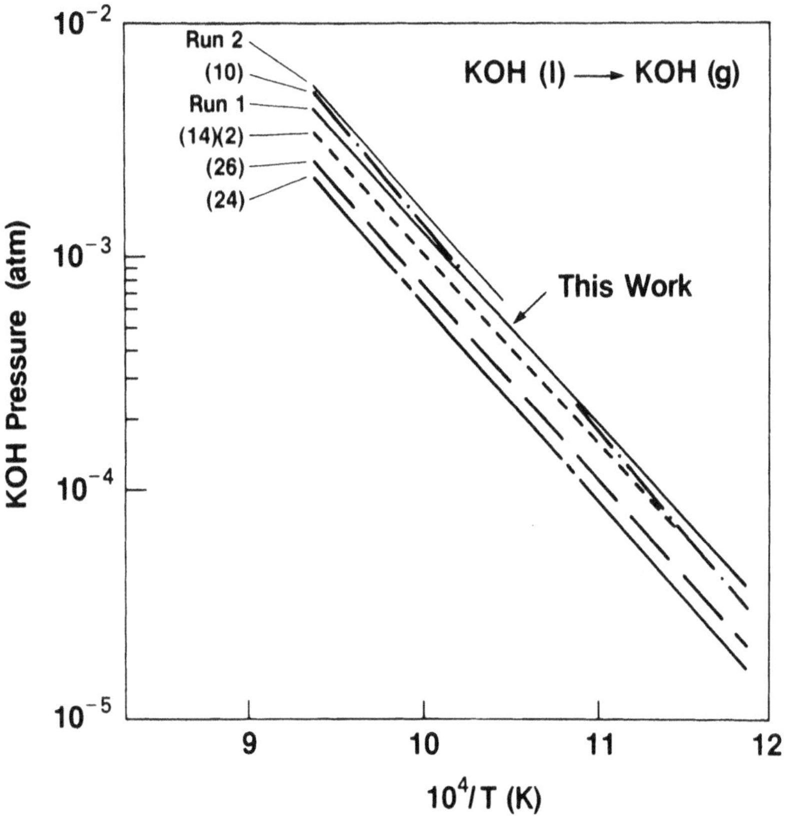

Fig. 11. Comparison of KOH partial pressure versus reciprocal temperature curves with literature data. The numbers in parentheses indicate literature references. Curves based on the work reported in refs. (*2*) and (*14*) are indistinguishable in this figure. The curve labeled "This Work" represents our recommended result.

from which the reaction Gibbs energies are

$$\Delta G(1013) = 5.1 \text{ kcal/mol}$$

$$\Delta G(987) = 12.4 \text{ kcal/mol}$$

As the KO_2-forming reaction involves a change of only one-half in mol number, the entropy change will be small and is estimated to be $\Delta S(T) = -12 \pm 2$ cal/K mol. Hence the two values of ΔG should have a negligible temperature dependence and an average value is

$$\Delta G(1000) = 8.8 \pm 3 \text{ kcal/mol}$$

From these values of ΔS, and ΔG, the reaction enthalpy change is

$$\Delta H(1000) = 20.8 \text{ kcal/mol}$$

from which we obtain

$$\Delta H_f\,(1000)KO_2 = -17.6 \pm 3 \text{ kcal/mol}$$

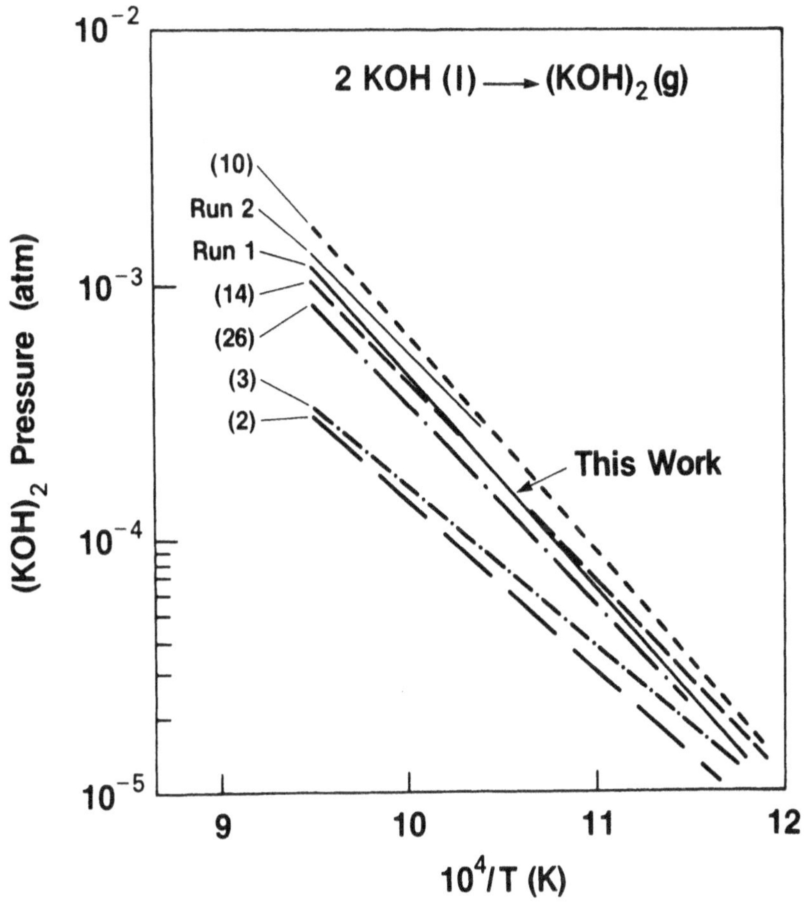

Fig. 12. Comparison of $(KOH)_2$ partial pressure versus reciprocal temperature curves with literature data. The numbers in parentheses indicate literature references.

and

$$D(1000)K - O_2 = 36.7 \pm 3 \text{ kcal/mol}$$

for the dissociation enthalpy of KO_2 to yield K and O_2. This value seems reasonable when compared with the analogous values for HO_2 and NaO_2 of 47 and 35 ± 5 kcal/mol (17), respectively. These KO_2 stability data indicate that under lean flame conditions, such as those used by Kaskan (*16*), a measureable loss of K atoms should occur because of KO_2 formation. This effect was in fact noted by Kaskan (*16*).

TABLE 7
KO_2 Relative Ion Intensity Data

P, atm	*T*, K	H_2O	O_2	K	KOH	K_2OH	KO_2
0.86	1013	470	700	18	17	1.7	0.03
0.8	987	144	57	17	—	13	0.04

5. CONCLUSIONS

5.1. KCl System

The present results support the JANAF (3) analysis and the relative significance of dimer and trimer to monomer is reasonably well-established at sufficiently high temperatures (and partial pressures) to allow a more reasonable extrapolation to the boiling point. Thermodynamic data have also been obtained for the trimer species for the first time. Evidence is given for jet expansion phenomena, including temperature-dependent electron impact ionization. Ionization cross-sections determined for KCl and $(KCl)_2$ are somewhat anomalous in their absolute and relative magnitudes and this is attributed to temperature dependent Franck-Condon effects. The extent to which these complicating cross section effects occur in other systems remains in question. However, among the many gaseous and vapor systems studied by TMS to date, only NaCl and KCl have shown striking cross-section dependences on beam conditions. The possibility of cluster formation, which is either small or non-existent, is obscured by the dependence of σ_m and σ_d on T_b.

5.2. KOH System

The vapor pressure and vaporization enthalpy data for KOH monomer compare reasonably with the literature data, as shown in Fig. 11. A notable exception is the curve of Farber et al. (*24*), which is about a factor of two below our curves and is well outside the combined uncertainties between the two studies. A likely explanation for their low pressures is reaction with the alumina cell to form stable potassium aluminate. Such a reaction would lower the KOH(l) activity, yielding lower values of P_m and particularly P_d. Their P_d values based on their statement of 10% of P_m are typically a factor of seven less than those of the present study. It appears certain that their conditions do not provide a saturated KOH vapor.

As shown in fig. 12, the JANAF (2) recommended dimer pressures differ markedly from those of the present work and the results of G&G (*10*). The results of Dubois and Millet (*14*) and Jackson and Morgan (*15*), when corrected for dimer, yield data in excellent support of the present dimer data (see Fig. 12).

Table 8 summarizes the second law enthalpy and entropy data from this and previous work, as analyzed by JANAF (2), together with the recent results of Farber et al. (24). As mentioned earlier, a third law enthalpy analysis is not considered reliable in view of the highly uncertain vibrational frequencies and other molecular constants. These data indicate that the KOH dimer is a much more significant species over the whole temperature range, i.e., to the boiling point, than has been previ-

TABLE 8
Second Law Thermodynamic Functions for Liquid KOH Vaporization at 1000 K

Reaction	ΔH_v, kcal/mol		ΔS_v, cal/K mol	
KOH(l) = KOH(g)	39.2 ± 1	(Run 1)[a]	26.1 ± 1	(Run 1)[a]
	40.7 ± 2	(Run 2)	28.1 ± 2	(Run 2)
	37.9 ± 4	(2)	24.4 ± 1	(2)
	40.8 ± 1	(24)[b]	26.6 ± 1	(24)[b]
2KOH(l) = $(KOH)_2$(g)	37.9 ± 1.3	(Run 1)[c]	22.7 ± 1.3	(Run 1)[c]
	36.9 ± 2.7	(Run 2)	22.1 ± 2.7	(Run 2)
	29.5 ± 4	(2)	11.9 ± 3	(2)

[a]Recommended values are 1640 ± 4 kJ/mol and 109.2 ± 4 J/K mol.
[b]Values listed from ref. (24) are for 770 K, also see text.
[c]Recommended values are 158.6 ± 5.4 kJ/mol and 95.0 ± 5.4 J/K mol.

ously recognized. Both the enthalpy and entropy functions for this species require considerable revision and a future experimental study of the vibrational spectrum is most desirable. In the interim, new vibrational frequency estimates can be made using the second law entropy data of the present study. One can infer from these results that the NaOH vaporization thermochemistry also warrants further investigation and studies are underway in this laboratory.

From the second law vaporization enthalpy for monomeric KOH, the zero point dissociation energy to yield K + OH is D_0° = 84.1 ± 1 kcal/mol. The JANAF (2) values of $H_T^\circ - H_0^\circ$ were used in this calculation. this value of the dissociation energy is consistent with the JANAF (2) value of 85.4 ± 2.6 kcal/mol, Farber et al.'s (24) value of 84.2 ± 0.5 kcal/mol, and an empirical value of 88.2 ± 5 kcal/mol. The empirical value was taken as 0.86 D_0°(KCl) using the average ratio of hydroxide to halide bond energies as discussed elsewhere (*8*). Another empiricism suggested by Krikorian (27) of 1.05 D_0°(KCl) yields 107.7 kcal/mol, which is clearly much too large. The agreement with Farber et al. (24) seems fortuitous since their partial pressure data were not consistent with those of the present study. Gole (28) has spectroscopic evidence for a lower bound value of 88.0 kcal/mol, though this could be invalid if other chemiluminescence pathways than those considered by Gole are present.

5.3. Thermodynamic Correlations

As has been argued elsewhere, there exist systematic trends and correlations for metal halides and hydroxides (*8*). For instance, gas-phase reaction entropies, to produce dimer and trimer from monomer species, generally fall in the range of −30 ± 5 and −60 ± 5 cal/K mol, respectively. Also, the ratios of dissociation energies for dimer or trimer bridge bonds (D_b) to that for the monomer (or terminal bond) species (D_t) generally fall in the range of 0.6 ± 0.04 and 0.65 ± 0.04, respectively. These empirical correlations are sufficient to allow prediction of reaction equilibrium constants and partial pressures for species where thermodynamic data is lacking. The extent to which the present results for KCl and KOH follow these empiricisms is indicated in Table 9. Note that the KCl trimer data are consistent with the correlations, as are the NaCl data calculated from our earlier vaporization data (*7*). The data of Gurvich et al. (*26*) for $(NaCl)_3$ are probably based on the experimental data of Feather and Searcy (*29*) and Akiskin et al. (*30*). Both the present data and the tables of Gurvich et al. (*26*) for KOH agree with the expected values of ΔS and D_b/D_t, contrary to the JANAF results. However, the JANAF data for KCl

TABLE 9
Thermodynamic Correlations Between KCl, KOH, and Other Alkali Species

	KCl	KOH	NaCl	LiCl
		reaction 2 monomer → dimer		
$-\Delta H_r$, kcal/mol[a]	43.4 (3)[b] 33.4[d]	40.5 42.2 (26) 49.9 (2)	47.4 (3)	48.3 (3)
$-\Delta S_r$, cal/K mol	27.5 (3) 28.4 (26) 36.5[d]	29.5 29.7 (26) 36.9 (2)	29.9 29.4 (26)	30.6 (3)
D_b/D_t	0.606 (3) 0.58[d]	0.62	0.62	0.605 (3)
		reaction 3 monomer → trimer		
$-\Delta H_r$, kcal/mol	90.6	90[e]	90.0 (7)[c] 88 (26) 92.6 (29)	97.6 (3)
$-\Delta S_r$, cal/K mol	64.5	65[e]	62.7 (7) 57.2 (26) 66.4 (29)	68.9 (3)
D_b/D_t	0.65		0.66 (7) 0.65 (26) 0.66 (29)	0.64 (3)

[a]All thermodynamic quantities are for 1000 K, with typical uncertainties of ±3 kcal/mol and ±3 cal/K mol.
[b]Numbers in parentheses indicate reference source.
[c]Based on ΔH_{sub} (1000 K) = 63.9 kcal/mol and ΔS_{sub} (1000 K) = 34.2 cal/K mol.
[d]Value obtained by assuming (incorrectly, we believe) that our dimer data at the higher temperatures are thermodynamically significant.
[e]Our estimate, based on observation of P_t (KOH) ~ P_t (KCl).

dimer are more consistent with the correlations than the present results which, as discussed elsewhere, appear to be anomalous at the higher temperatures, possibly because of temperature-dependent cross-section effects. The perennial high temperature mass spectrometric problem of unknown ionization cross-sections can be exacerbated in cool beams and further study of this behavior is warranted.

The present confirmation of the JANAF (3) evaluation of the KCl vaporization thermodynamics is also consistent with the dimensional analysis of Frurip and Blander (*31*). Their analysis essentially indicates self consistency in the JANAF (3) evaluation of entropy data for the alkali halide monomer and dimer species. The dimensional analysis correlations argue strongly against entropy errors of more than 2 cal/K mol in the JANAF (3) values for $(KCl)_2$. Likewise, the exchange perturbation theory used by Brumer and Karplus (*32*) gives good agreement with the JANAF (3), $(KCl)_2$ dissociation energy. Additional recent support for the experimental alkali halide dimer binding energies is given by the good agreement between Rittner model calculations and experimental data (*33*). For the cases of $(NaCl)_2$ and $(KCl)_2$, where we have additional new experimental data, a comparison between our zero-point dimerization energies and the prediction of Gowda and Benson (*33*) indicates 49.3 ± 2 (obs), 50.8 (prediction), and 45.1 ± 1 (obs), and 45.0 (prediction), kcal/mol, respectively.

Evidence of self-consistency among the MOH alkali metal hydroxide entropies is given by the dimensional model correlations of Frurip et al. (*34*). The non-electronic entropy [JANAF (2) data] follows the dimensional correlation for M = Li, Na, K, and Cs. Also, for the $(KOH)_2$ dimer species, the dimensional model yields entropies which are more consistent with the present second law experimental results than the JANAF (2) values. For instance, S (1000) $(KOH)_2$ = 117.8 ± 1, 114, and 106.1 cal/K mol for the present work, the dimensional model, and the JANAF (2) evaluation, respectively.

ACKNOWLEDGMENTS

Valuable technical assistance was provided by Mr. Art Sessoms. Partial financial support was provided by AFOSR for the KCl study and by ARO for the KOH and KO_2 experiments.

REFERENCES

1. Hastie, J. W., Plante, E. R., and Bonnell, D. W., in *Metal Bonding and Interactions in High Temperature Systems,* Gole, J. L., and Stwalley, W. C., eds., ACS Symp. Series **179,** 543 (1982).
2. *JANAF Thermochemical Tables,* Chase, M., Curnutt, J. L., Hu, A. T., Prophet, H., Syverud, A. N., and Walker, L. C., 1974 Supplement, *J. Phys. Chem. Ref. Data* **3**(2), (1974).

3. *JANAF Thermochemical Tables*, Stull, D. R., and Prophet, H., NSRDS-NBS 37 (1971).
4. Grimley, R. T., and Joyce, T. E., *J. Phys. Chem.* **73,** 3047 (1969).
5. Schaber, H., and Martin, T. P., *J. Chem. Phys.* **70,** 2029 (1979).
6. Milne, T. A., and Klein, H. M., *J. Chem. Phys.* **33,** 1628 (1960).
7. Bonnell, D. W., and Hastie, J. W., *Proceedings of the 10th Materials Research Symposium*, NBS SP-561, 1979, p. 357.
8. Hastie, J. W., *High Temperature Vapors: Science and Technology,* Academic Press, New York (1975).
9. Porter, R. F., and Schoonmaker, R. C., *J. Phys. Chem.* **62,** 486 (1958).
10. Gusarov, A. V., and Gorokhov, L. N., *Russian J. Phys. Chem.* **42,** 449 (1968).
11. Lau, H. K., Cubicciotti, D., and Hildenbrand, D. L., *Electrochem. Soc. Solid State Sci. Technol.* **126,** 490 (1979).
12. Plante, E. R., Olson, C. D., and Negas, T., Vapor Pressure Studies of K_2SO_4, to be published; see also ref. (*1*).
13. Wartenberg, H. von, and Albrecht, P., *Z. Elektrochem.* **27,** 162 (1921).
14. Dubois, J., and Millet, J., *Compt. Rend.* **269,** 1336 (1969).
15. Jackson, D. D., and Morgan, J. J., *J. Ind. Eng. Chem.* **13,** 110 (1921).
16. Kaskan, W. E., *Tenth International Symposium Combustion,* 1965, p. 41.
17. Hynes, A. J., Steinberg, M., and Schofield, K., "The detailed chemistry and thermodynamics of sodium in oxygen-rich flames", paper presented at Fall Meeting West. States Comb. Inst., Sandia, Livermore, CA (1982).
18. Hastie, J. W., Plante, E. R., and Bonnell, D. W., "Vaporization of Simulated Nuclear Waste Glass". NBSIR 83-2731. Available NTIS (1983).
19. Hastie, J. W., and Bonnell, D. W., in *Thermochemistry Today and Its Role in the Immediate Future,* Ribeiro da Silva, M. A. V., ed., Reidel, Boston (1983).
20. Bonnell, D. W., and Hastie, J. W., "A Theoretical Analysis of Temperature Dependent Electron Impact Fragmentation", 31st AAMS Meeting, Boston, MA (1983); also to be published.
21. Dronin, A. A., and Gorokhov, L. N., *Zh. Fiz. Khim.* **49,** 798 (1975).
22. Hastie, J. W., and Margrave, J. L., *High Temp. Sci.* **1,** 481 (1969).
23. Buchler, A., and Berkowitz-Mattuck, J. B., in *Adv. High Temp. Chem.,* Eyring, L., ed. **1,** Academic Press, NY, 1967, p. 105.
24. Farber, M., Srivastava, R. D., and Moyer, J. W., *J. Chem. Thermodynam.* **14,** 1103 (1982).
25. Meyer, R. T., and Lynch, A. W., *High Temp. Sci.* **5,** 192 (1973).
26. Gurvich, L. V., Veits, I. V., Medvedev, V. A., Khachkuruzov, G. A., Yungman, V. S., Bergman, G. A., et al. "Termodinamicheskie Svoistva Individual'nykh Veshchestv," (*Thermodynamic Properties of Individual Substances*); Glushko, V. P., gen. ed., Vol. 4, parts 1 and 2, 1982.
27. Krikorian, O. H., *High Temp.-High Press.* **14,** 387 (1982).
28. Gole, J., *J. Optical Eng.* **40,** 546 (1981).
29. Feather, D. H., and Searcy, A. W., *High Temp. Sci.* **3,,** 155 (1971).
30. Akiskin, P. A., Gorokhov, L. N., and Sidorov, L. N., *Zh. Fiz. Khim.* **33,** 2822 (1959).
31. Frurip, D. J., and Blander, M., in *Characterization of High Temperature Vapors and Gases,* NBS SP 561/2, Hastie, J. W., ed., US Government Printing Office, Washington, 1979, p. 1597.
32. Brumer, P., and Karplus, M., *J. Chem. Phys.* **64,** 5165 (1976).
33. Gowda, B. T., and Benson, S. W., *J. Chem. Soc. Faraday Trans. 2,* **79,** 663 (1983).
34. Frurip, D. J., Chatillon, C., and Blander, M., *J. Phys. Chem.* **86,** 647 (1982).

0018–1536/84/1700–0365$02.00

A New Gaseous Iridium Oxide, Ir_2O_3

S. N. TRIPATHI AND M. S. CHANDRASEKHARAIAH*

Water Chemistry Section, Chemical Group, Bhabha Atomic Research Centre, Bombay 400 085, India

Received September 23, 1983; Accepted December 31, 1983

ABSTRACT

A careful reappraisal of the available data of oxidation of iridium in light of the recent measurements of the pressures of iridium oxides in this laboratory has revealed the existence of a new gaseous iridium oxide Ir_2O_3. The results presented here support this conclusion and a structural formula is suggested.

Index Entries: Gaseous iridium oxides; Ir_2O_3; entropy of formation, of Ir_2O_3.

INTRODUCTION

Recently, in this laboratory, iridium transport in a flowing atmosphere of dry oxygen was remeasured from 1435 to 1628 K (*1*). The experimentally determined iridium transport, especially above about 1500 K, could not be satisfactorily accounted for in terms of the mass spectrometrically reported (*2,3*) gaseous iridium oxides, namely IrO_3, IrO_2, and IrO. The observed increase in the iridium transport could be reasonably explained by the presence of a new iridium oxide, Ir_2O_3 (g), having the same oxygen pressure dependence as IrO_3 (*4,5*). In this communication we are presenting further evidence in favor of the formation of Ir_2O_3(g) species.

*Author to whom all the correspondence and reprint requests should be addressed.

RESULTS AND DISCUSSION

Out of nine studies reporting the enthalpy and entropy of formation of $IrO_3(g)$ (ref. *1*, Table 2), there is a good agreement among six, and hence one could write the following expression for the Gibbs enthalpy of formation of $IrO_3(g)$.

$$\Delta G_f^\circ (IrO_3, g)/J\ mol^{-1} = (19{,}040 \pm 5350) + (43.93 \pm 3.77)T \qquad (1)$$

As shown recently (*1*), the total iridium transport in flowing oxygen at 1 atm was significantly higher than the calculated values from Eq. (1), especially between 1500 and 2000 K. It was proposed that another species, $Ir_2O_3(g)$, exists and was responsible for the observed increase in iridium transport. This oxide species, Ir_2O_3, would not only contribute to the iridium loss in oxidizing environment under equilibrium conditions, but also under dynamic non-equilibrium situations such as oxidation rate studies. Here a reexamination of the kinetic studies reported in the literature is made.

From the experimental results of Chandrasekharaiah et al. (*1*), the Gibbs free enthalpy of formation of Ir_2O_3 (g) was calculated as

$$\Delta G_f(Ir_2O_3, g) = (303{,}800 \pm 37{,}200) - (113.8 \pm 23.8)T\ J/mol \qquad (2)$$

Combining Eqs. (1) and (2) gives the change in the Gibbs enthalpy for the equilibrium, Eq. (3).

$$Ir(s) + IrO_3(g) = Ir_2O_3(g) \qquad (3)$$

$$\Delta G_T^\circ = (284{,}760 \pm 42{,}550) - (157.7 \pm 27.6)T\quad J/mol \qquad (4)$$

Reaction (3) shows that the relative abundance of IrO_3 and Ir_2O_3 is independent of oxygen pressure and Eq. (4) shows that the $Ir_2O_3(g)$ species becomes increasingly important with the rise of temperature. In fact, its partial pressure equals that of IrO_3 around 1800 K.

Recently, Jehn et al. (*6*) reported oxidation rates of iridium in 0.1–100 Pa oxygen from 1500 to 2150°C. Assuming the Langmuir evaporation mode, and that gaseous species contain one iridium atom per molecule, i.e., IrO_x, their weight loss data was converted to the pressures of iridium oxides. From the data of Chandrasekharaiah et al. (*1*) and Peterson (*2*), the equilibrium partial pressures of all the iridium oxides as well as Ir(g) were calculated under the same oxygen pressures. The results are shown in Fig. 1.

The observed iridium loss at 10 Pa is higher than the calculated total iridium oxide pressures including Ir(g), but excluding $Ir_2O_3(g)$ (lower curve). But the calculated iridium transport under equilibrium conditions becomes larger than the measured values of Jehn et al. (*6*) if $Ir_2O_3(g)$ species are also taken into account (dashed curves in Fig. 1). This is especially so at higher temperatures. This trend is observable even at 30 Pa, at which pressure other mass transport mechanisms besides the heteroge-

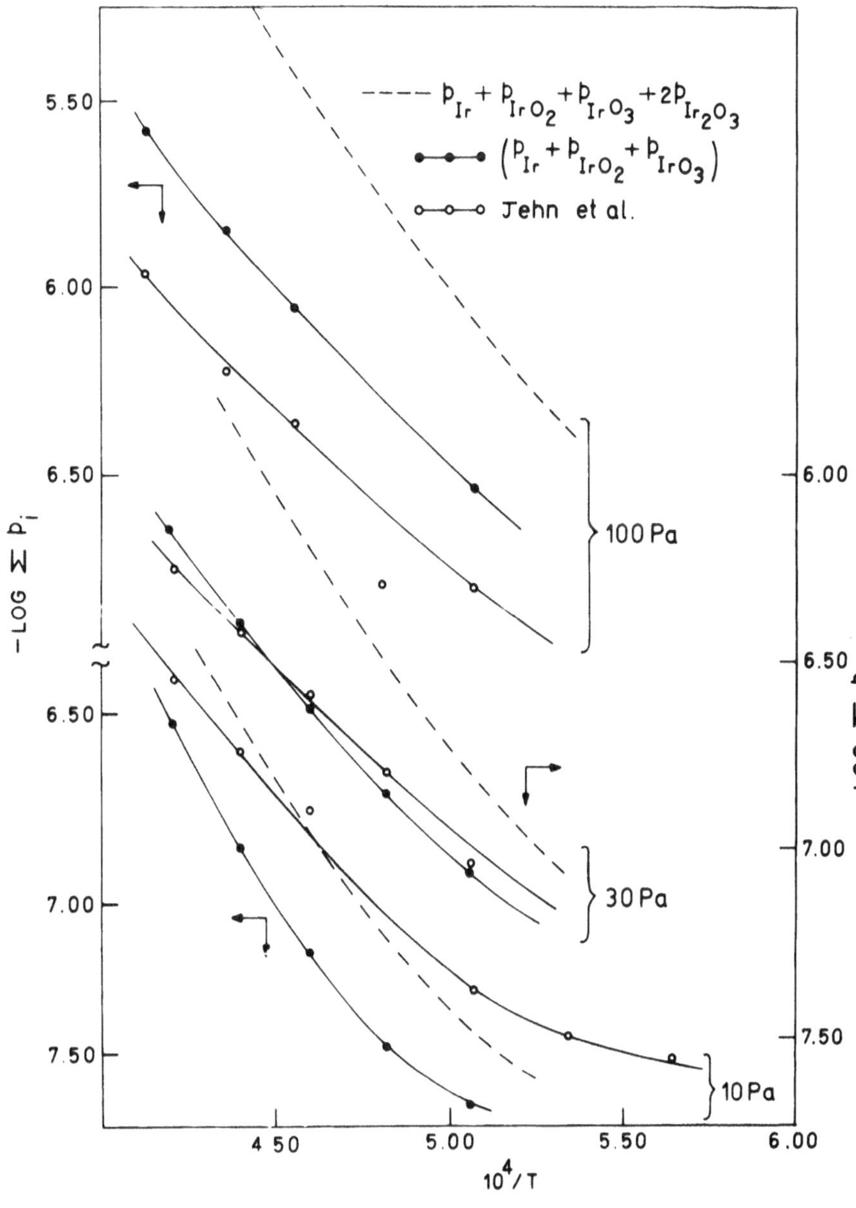

Fig. 1. Total pressure of iridium-bearing species vs 1/*T*.

neous surface reactions may become rate-determining steps. Thus at 100 Pa, Jehn et al. (*6*) reported an iridium loss rate lower than both the calculated curves.

Wimber et al. (*7*) also have reported oxidation rates of iridium wires in different oxygen pressures 0.0046 to 0.84 atm oxygen at 1948 and 2238 K. From their reported data (Table IV, ref. *7*) one could calculate the iridium mass loss rate from the surface of the wire. From the available thermodynamic data on Ir(g), IrO_2(g), IrO_3(g), and Ir_2O_3(g) (*1,2*), the

TABLE 1
Oxidation Rates of Iridium under Equilibrium and Dynamic Flow Conditions

		Iridium loss, g/cm^2 s, $\times 10^6$		
Temp, K	Oxygen pressure, atm.	Wimber et al.	Calculated not including Ir_2O_3	Calculated including Ir_2O_3
	0.0046	15.2	7.6	80
1948	0.066	33.2	412	4150
	0.0076	82.5	17.5	1800
2238	0.0132	65.0	40.0	4.120
	0.132	113 to 132	1270	13000

maximum iridium loss from the surface of the hot wire under equilibrium conditions, such as in a transpiration experiment at the same temperature and oxygen pressures as those of Wimber et al. (7) was calculated. In Table 1, two sets of calculated mass losses are shown, one of which does not consider the iridium loss caused by $Ir_2O_3(g)$ while the other includes this species.

It can be seen that under the oxidation rate measurements, where the ambient gas pressures are such that gas phase mass transport is not the predominant rate-determining step, the measured rate of iridium loss by Wimber et al. (7) is higher than that calculated on the basis of $Ir(g)$, $IrO_2(g)$, and $IrO_3(g)$, but lower than that found when $Ir_3O_3(g)$ is also included.

It is well known that the maximum vapor phase loss occurs under equilibrium conditions and is always less under any non-equilibrium conditions. Thus the data in Table 1 also support the existence of $Ir_2O_3(g)$ at higher temperatures.

Thus, this analysis of iridium oxidation data further supports our earlier proposal of $Ir_2O_3(g)$ as contributing significantly to iridium loss in oxidizing atmospheres, especially above 1500 K and in air or oxygen.

Three possible structures can be written for $Ir_2O_3(g)$ molecule that are consistent with the known chemical behavior of iridium. One is the linear (A) molecule with $D_{\infty h}$ symmetry, the second is a bent (C_{2v}) structure, and the third is the trigonal bipyramid (B) molecule with D_{3h} symmetry

$$O = Ir - O - Ir = O$$

A

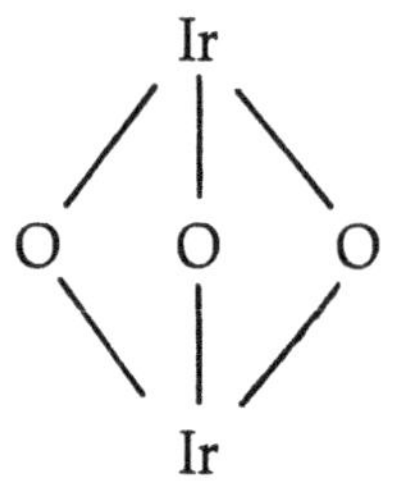

B

Chandrasekharaiah et al. (*1*), have reported an entropy of formation of (114 ± 24) J K^{-1} mol^{-1}, which indicates the structure B to be more probable. Further it is hard to conceive the linear $Ir_2O_3^+$ ion to be so unstable as to escape mass spectrometric observation under the experimental conditions, under which the equally sparse IrO_2^+ was observed. On the other hand, ionization of Ir_2O_3 of structure B may result in fragmentation of the $Ir_2O_3^+$ ion. There are very few gaseous metallic oxides of the formula M_2O_3 and Ir_2O_3 would be a good candidate to be investigated by matrix isolation spectroscopy.

ACKNOWLEDGMENT

The authors would express their thanks to Dr. H. A. Jehn, Max Planck Institute, Stuttgart, FRG, who provided the original data for their oxidation studies.

REFERENCES

1. Chandrasekharaiah, M. S., Karkhanavala, M. D., and Tripathi, S. N., *J. Less Common Met.* **80,** 9 (1981).
2. Peterson, D. E., LA-Report 6352 (1976).
3. Norman, J. H., Stanley, H. G., and Bell, W. E., *J. Chem. Phys.* **42,** 1123 (1965).
4. Alcock, C. B., and Hooper, G. W., *Proc. Roy. Soc., London Ser A.* **254,** 551 (1960).
5. Bell, W. E., and Tagami, M., *J. Phys. Chem.* **70,** 640 (1966).
6. Jehn, H. A., Volker, R., and Ismail, M. I., *Platinum Met. Rev.* **22,** 92 (1978).
7. Wimber, R. T., Hills, S. W., Wall, N. K., and Tempero, C. R., *Metal. Trans. A,* **8,** 193 (1977).

0018–1536/84/1700–0371$02.00

Gibbs Free Energies of Formation for Intermetallic Compounds Involving Transition Elements, Lanthanides, and Actinides

JOHN K. GIBSON[1] AND PAUL R. WENGERT*,[2]

[1]*Chemistry Department, Oak Ridge National Laboratory, Oak Ridge, TN 37830 and* [2]*Materials Engineering Dept., Owens-Illinois, Inc., Toledo, OH 43666*

Received September 23, 1983; Accepted December 31, 1983

ABSTRACT

Values or limits are provided for the Gibbs free energies of formation of "A_nB_m" intermetallic compounds for 84 binary systems. "A" elements are: Li, Be, Mg, Ca, Sr, Ba, Al, Si, Sc, Ti, V, Mn, Fe, Co, Ni, Y, Zr, Nb, Hf, Ta, La, Ce, Pr, Nd, Sm, Eu, Gd, Tb, Dy, Ho, Er, Tm, Yb, Lu, Th, U, and Pu. "B" elements are Fe, Co, Ni, Ru, Rh, Pd, Ag, Re, Os, Ir, Pt, and Au. The significance and applications of this information are discussed.

Index Entries: Gibbs free energies of formation, of intermetallic compounds; free energies, of formation for intermetallics; formation free energies, for intermetallics; intermetallic compounds, Gibbs free energies of formation of; transition elements, free energies of formation of; lanthanides, free energy of formation of; actinides, free energy of formation of; platinum group elements, free energies of formation of.

INTRODUCTION

John Gibson is the most recent student of Leo Brewer's to have developed limits on the Gibbs free energies of formation for intermetallic

*Author to whom all correspondence and reprint requests should be addressed. Current address: Technical Staff, Sandia National Laboratories, Albuquerque, NM 87185.

compounds. Paul Wengert is Leo's first student to have developed such limits in 1969. It is our privilege to present a summary of free energies in honor of Leo Brewer.

The reaction:

$$ZrC + 3Pt \rightarrow ZrPt_3 + C$$

is significant for several reasons:

First, both ZrC and Pt generally are considered refractory, nonreactive materials in most applications. Yet, they react when placed in contact with each other at high temperatures.

Second, since the reaction proceeds, the Gibbs free energy of formation (ΔG_f^0) of $ZrPt_3$ is less than ΔG_f^0 (ZrC). Simple reactions have been useful in setting the limits on many of the ΔG_f^0 values presented.

Third, with excess ZrC, the reaction becomes:

$$2ZrC + 3Pt \rightarrow ZrPt_3 + C + ZrC$$

The region of the Zr–C–Pt ternary phase diagram in which the three phases are in equilibrium becomes identifiable.

Fourth, determining the extreme stability of $ZrPt_3$ provided the empirical evidence that the *d*-orbital electrons were participating in a Lewis-Acid-Base type of bonding.

VALUES AND LIMITS OF ΔG_f°

Simple reactions provide a method for determining limits on ΔG_f° values for many compounds in a relatively short period of time. Since earlier limits were set, others have developed actual values by using galvanic cells, oxide reductions, vapor pressure measurements, and calorimetry.

Table 1 presents values and limits for compounds in 84 binary systems.

The temperature of the observed ΔG_f° also is given in Table 1. However, ΔG_f° does not vary significantly with temperature since neither products nor reactants are in their gaseous standard state. Assuming the values that are presented in Table 1 over several hundred Kelvin degrees is reasonable.

APPLICATIONS

The information presented in Table 1 has been applied in several ways. Six applications are presented.

ZrC and Pt should not be used in assemblies that will see high temperatures unless their reaction is desired. This precaution is relevant to all "A" carbides with B elements where "A" is Y, Zr, Hf, a lanthanide, or an actinide and "B" elements are Rh, Pd, Ir, and Pt.

TABLE 1
Gibbs Free Energies of Formulation for "A*n*B$_m$"
(kcal/g-at)

Compound	ΔG_f°	*T*, K	Ref.
"A" = Li			
$LiPt_7$	≤− 2.5	1500	*1*
"A" = Be			
$BePt_{15}$	≤− 2.4	1500	*1*
"A" = Mg			
$MgPt_3$	≤−10.1	1500	*1*
$MgPt_7$	≤− 5.0	1500	*1*
"A" = Ca			
$CaPt_2$	≤−16.6	1500	*1*
Ca_2Pt_7	≤−12.4	1500	*1*
$CaPt_5$	≤− 8.3	1500	*1*
"A" = Sr			
$SrPt_2$	≤−13.3	1500	*1*
$SrPt_3$	≤−10.0	1500	*1*
$SrPt_5$	≤− 6.6	1500	*1*
"A" = Ba			
$BaPt_3$	≤− 5.8	1500	*1*
$BaPt_5$	≤−11.5	1500	*1*
"A" = Al			
AlRh	≤−21.3	1500	*1*
AlPd	≤−29.3	1500	*1*
Al_3Pd_5	≤−22.0	1500	*1*
$AlPd_2$	≤−19.5	1500	*1*
AlIr	≤−17.1	1500	*1*
Al_3Pt_2	≤−27.5	1500	*1*
AlPt	≤−24.6	1500	*1*
Al_3Pt_5	≤−18.5	1500	*1*
$AlPt_3$	≤−12.4	1500	*1*
Al_3Pt_{13}	≤− 9.3	1500	*1*
"A" = Si			
$SiFe_3$	≤− 4.8	1500	*1*
$SiCo_2$	≤−11.4	1500	*1*
$SiNi_2$	≤− 8.6	1500	*1*
$SiNi_5$	≤− 7.3	1500	*1*
SiRu	≤−12.0	1500	*1*
$SiRh_2$	≤− 4.4	1500	*1*
$SiPd_2$	≤−11.9	1500	*1*
SiIr	≤− 9.3	1500	*1*
"A" = Sc			
$ScPt_3$	≤−19.3	1500	*1*

(continued)

TABLE 1 (continued)

Compound	ΔG_f°	T, K	Ref.
"A" = Ti			
TiRu	≤ -19.4	1800	2
$TiRh_3$	≤ -4.4	1500	1
TiPd	-16.0	1873	3
$TiPd_3$	-13.5	1873	3
Ti_3Ir	≤ -6.63	2000	4
TiIr	-28.8	1623	5
$TiIr_3$	-17.2	1623	5
TiPt	≤ -9.0	1500	1
$TiPt_3$	-17.8	1300	6
"A" = V			
VPt_3	≤ -2.8	1500	1
"A" = Mn			
$MnPt_3$	-9.80	1773	7
"A" = Fe			
FePd	-0.956	1123	8
FePt	-7.29	1123	8
"A" = Co			
CoPt	-5.62	1573	9
"A" = Ni			
NiPt	-4.30	1625	10
"A" = Y			
YIr_2	≤ -11	1500	1
YPt_5	≤ -12.5	1500	1
"A" = Zr			
ZrRu	-21.5 ± 1.5	1570	11
$ZrRu_2$	-14.2 ± 0.9	1800	11
Zr_2Rh	≤ -5	1800	11
$ZrRh_{1-}$	≤ -7	1800	11
$ZrRh_{1+}$	≤ -7	1800	11
Zr_3Rh_5	≤ -9	1800	11
$ZrRh_3$	≤ -11	1850	11
Zr_2Pd	≤ -5.1	1360	11
ZrPd	≤ -7	1650	11
$Zr_{44}Pd_{56}$	≤ -8	1650	11
$ZrPd_2$	≤ -10	1650	11
$ZrPd_3$	≤ -11	1650	11
$ZrPd_4$	≤ -12	1650	11
Zr_2Ag	≥ -20	1228	11
ZrAg	≥ -15	1228	11
Zr_2Re	≥ -29	1800	11
$ZrRe_2$	≥ -15	1800	11
Zr_5Re_{24}	≥ -7	1800	11
Zr_7Os_3	≥ -16	1800	11

TABLE 1 (*continued*)

Compound	ΔG_f°	*T*, K	Ref.
ZrOs	≥−11	1800	*11*
$ZrOs_2$	≥− 8	1800	*11*
Zr_3Ir	<− 4	1800	*11*
Zr_2Ir	<− 5	1800	*11*
Zr_5Ir_3	<− 6	1800	*11*
ZrIr	<− 7	1800	*11*
$ZrIr_2$	<−10	1800	*11*
$ZrIr_3$	<−11	1800	*11*
Zr_5Pt_3	<−15	1800	*11*
ZrPt	<−20	1800	*11*
$ZrPt_{14}$	<−20	1800	*11*
Zr_4Pt_5	−37 ± 15	1800	*11*
Zr_2Pt_{3-}	−35.5 ± 11.5	1800	*11*
$ZrPt_3$	−25 ± 5	1300	*12,13*
$ZrPt_{3+x}$	−27 ± 2	1800	*11*
$ZrPt_8$	≤−13	1800	*11*
Zr_3Au	<− 4	1370	*11*
Zr_2Au	<− 5	1370	*11*
Zr_5Au_4	<− 7	1400	*11*
Zr_7Au_{10}	<− 9	1400	*11*
$ZrAu_2$	−13.5± 2.5	1400	*11*
$ZrAu_3$	−12 ± 1	1400	*11*
$ZrAu_5$	−10 ± 0.5	1340	*11*
"A" = Nb			
$NbRh_3$	−11.8	1300	*14*
NbPd	≤− 6.6	1515	4
$NbPd_3$	≤− 5.9	1700	4
$NbIr_3$	≤− 8	1770	*11*
$NbPt_3$	≤− 8	1500	*11*
"A" = Hf			
$HfRh_3$	≤−14	1800	*11*
$HfPd_3$	≤−11.1	1900	4
Hf_5Re_{24}	≤− 9	1800	*11*
$HfIr_3$	≤−14	2300	*11*
$HfPt_3$	−24 ± 4	2400	*11,12*
"A" = Ta			
$TaPd_3$	≤− 8.5	2000	4
Ta_2Re_3	− 5.5 ± 2.5	2730	*11*
$TaRe_{1.85}$	− 5 ± 2	2700	*11*
$TaRe_{3.6}$	− 4.5 ± 1	2700	*11*
$TaIr_3$	≤− 8.5	2154	4
$TaPt_3$	≤− 9	1800	*11*
"A" = lanthamide			
$LaPt_2$	−24.9	1373	*15*

(*continued*)

TABLE 1 (*continued*)

Compound	ΔG_f°	T, K	Ref.
$LaPt_5$	−15.0	1373	*15*
$CePt_2$	−24.6	1373	*15*
$CePt_5$	−14.5	1273	*15*
$PrPt_2$	−25.5	1373	*15*
$PrPt_5$	−15.1	1273	*15*
$NdPt_2$	−25.2	1373	*15*
Nd_2Pt_7	−18.2	1373	*15*
$NdPt_5$	−14.8	1273	*15*
$SmPt_2$	−25.6	1373	*15*
Sm_2Pt_7	−18.5	1373	*15*
$SmPt_5$	−14.9	1273	*15*
$EuPt_2$	−17.8	1373	*15*
Eu_2Pt_7	−13.0	1373	*15*
$EuPt_5$	−10.7	1273	*15*
$GdPt_2$	−26.3	1373	*15*
Gd_2Pt_7	−18.6	1373	*15*
$GdPt_5$	−14.9	1273	*15*
$TbPt_3$	−21.5	1373	*15*
$TbPt_5$	−15.4	1273	*15*
$DyPt_3$	−21.1	1373	*15*
$DyPt_5$	−15.0	1273	*15*
$HoPt_3$	−21.9	1373	*15*
$HoPt_5$	−15.3	1273	*15*
$ErPt_3$	−22.3	1373	*15*
$ErPt_5$	−15.4	1273	*15*
$TmPt_3$	−21.9	1373	*15*
$TmPt_5$	−14.9	1273	*15*
$YbPt_3$	−20.1	1373	*15*
$LuPt_3$	−22.0	1373	*15*
"A" = Th		1100	*16*
Th_7Ru_3	− 0.932	1100	*16*
Th_3Ru_2	−10.8	1100	*16*
ThRu	−11.2	1100	*16*
$ThRu_2$	− 8.28	1100	*16*
Th_7Rh_3	−15.4	1100	*17*
ThRh	−24.5	1100	*17*
Th_3Rh_4	−24.0	1100	*17*
Th_3Rh_5	−21.8	1100	*17*
$ThRh_2$	−20.2	1100	*17*
$ThRh_3$	−16.8	1100	*17*
$ThRh_5$	−11.2	1100	*17*
Th_7Os_3	− 7.29	1200	*18*
Th_3Os_2	− 7.98	1200	*18*
$ThOs_2$	− 8.04	1200	*18*
$ThIr_3$	−13.2	1200	*18*
$ThIr_5$	−12.0	1200	*18*

TABLE 1 (*continued*)

Compound	ΔG_f°	*T*, K	Ref.
"A" = U			
URu_3	-10 ± 1	1100	*19, 20*
U_2Rh	$-$ 8.44	900	*21*
URh	$-$ 8.60	1200	*21*
U_3Rh_4	$-$ 9.42	1200	*21*
U_3Rh_5	$-$ 9.59	1200	*21*
URh_3	-14 ± 5	1200	*19, 21,22*
UPd_3	-15.5	1673	*23*
UOs_2	-10.7	1100	*24*
UIr_2	-17.4	1100	*24*
UIr_3	-15.4	1100	*24*
"A" = Pu			
$PuRu_2$	$-$ 7.25	1640	*25*
$PuRh_2$	-10.4	1519	*26*
$PuRh_3$	$-$ 7.17	1725	*26*
$PuOs_2$	$-$ 6.37	1553	*26*
$PuIr_2$	-15.5	1929	*26*
$PuPt_2$	-10.8	1629	*26*
$PuPt_3$	$-$ 8.96	1579	*26*
$PuPt_4$	$-$ 5.64	1596	*26*
$PuPt_5$	$-$ 6.33	1614	*26*

ZrO_2 and Pt react in reduced atmospheres. Since ZrO_2 is a common sheathing material for Rt/Pt–Rh thermocouples, erroneous temperature readings will occur in reducing atmospheres (*27*). Erroneous galvanic cell readings also have occurred when Pt electrodes have been in contact with ZrO_2 under reducing conditions (*28*).

Limits on the partial pressure of oxygen (P_{O_2}), which will not cause container attack, can be set in glass melting processes where Pt, Pt–Rh alloys, and similar precious metals are used as the container materials. It is sometimes desirable to use reducing melting conditions in order to control the solubility of precious metals into the glass and to control the corresponding reprecipitation of the precious metal in the form of "sparklers," i.e., small crystals floating in the glass. Reducing melting conditions also may be used to control the valence state of multivalent dopants (*29*).

High purity actinide elements can be isolated by reduction of their very stable oxides in the presence of platinum group metals and their subsequent vaporization from the alloy and condensation onto the desired substrate (*33*). An example is the formation of pure Am by the reactions:

$$AmO_2 + 2H_2O + 5Pt \rightarrow AmPt_5 + H_2O \text{ (ref. } 30\text{)}$$

and

$$AmPt_5 \rightarrow 5Pt + Am(g)$$

Reactions have been reported between platinum group metals and TiO_2 support metals in catalyst assemblies (*31*).

Reactions have been reported between fission products with nuclear fuels to form intermetallic alloys, such as URh_3 (*32*).

Only one of the above examples demonstrates a beneficial use for the intermetallic compounds by themselves. However, knowledge of their stability aids in avoiding their formation when such formation is not desired. In the case of Am purification, the intermetallic compound $AmPt_5$ was useful as an intermediate species.

REFERENCES

1. Wengert, P. R., and Spanoudis, L., *J. Amer. Cer. Soc.* **57** (2), 94 (1974).
2. Raub, E., and Falkenburg, G., *Metallwiss.* Technik **27**, 669 (1973).
3. Choudary, U. V., Gingerich, K. A., and Cornwell, L. R., *Met. Trans.* **8A,** 1487 (1977).
4. Gibson, J., PhD thesis, Univ. of California, Berkeley, 1983.
5. Choudary, U. V., Gingerich, K. A., and Cornwell, L. R., *J. Less-Common Met.* **50,** 201 (1976).
6. Meschter, P. J., and Worrell, W. L., *Met. Trans.* **7A,** 299 (1976).
7. Rao, B. K. D. P., and Gaskell, D. R., *Met. Trans.* **12A,** 207 (1981).
8. Alcock, C. B., and Kubik, A., *Acta Met.* **17,** 437 (1969).
9. Schwerdtfeger, K., and Muan, A., *Acta Met.* **12,** 905 (1964).
10. Walker, R. A., and Darby, J. B., Jr., *Acta Met.* **18,** 1261 (1970).
11. Brewer, L., and Wengert, P. R., *Met. Trans.* **4,** 83 (1973).
12. Meschter, P. J., and Worrell, W. L., *Met Trans.* **8A,** 503 (1977).
13. Carbonra, R. S., and Blue, G. D., *High Temp. Sci.* **3,** 225 (1971).
14. Kleykamp, H., *J. Less-Common Met.* **83,** 105 (1982).
15. Kemmler, G., "Untersuchung zur Stabilitat von Lanthaniden-Platin Verbindungen," Kernforschungszenstrum Karlsruhe Report KFK 2452, 1977.
16. Kleykamp, H., and Murabayashi, M., *J. Less-Common Met.* **35,** 227 (1974).
17. Marabayashi, M., and Kleykamp, H., *J. Less-Common Met.* **39,** 235 (1975).
18. Kleykamp, H., *J. Less-Common Met.* **63,** P25 (1979).
19. Wijbenga, G., and Cordfunke, E. H. P., *J. Chem. Thermodynam.* **14** 409 (1982).
20. Holleck, H., and Kleykamp, H., *J. Nucl. Mater.* **35,** 158 (1970).
21. Naraine, M. G., and Bell, H. B., *J. Nucl. Mater.* **50,** 83 (1974).
22. Holleck, H., and Kleykamp, H., *J. Nucl. Mater.* **45,** 47 (1972/73).
23. Lorenzelli, N., and Marcon, J. P., *J. Nucl. Mater.* **44,** 57 (1972).
24. Holleck, H., Kleykamp, H., and Franco, J. I., *Z. Metallk.* **66,** 298 (1975).
25. Peterson, D. E., *J. Nucl. Mat.* **91,** 306 (1980).
26. Peterson, D. E., "Thermodynamics of Plutonium-Noble Metal Compounds," L.A.S.L. Report #LA-UR 82-2580, 1982.
27. Schaller, H. J., *Ber. Bunsen-Gesellschaft* **80,** 999 (1976).
28. Weppner, W., *J. Electroanal. Chem.* **84,** 339 (1977).

29. Boling, N. L., Spanoudis, L., and Wengert, P. R., *Damage Threshold Studies of Glass Laser Materials,* DOE Contract No. DAHC 15-69-C-0303, Semi-Annual Report, July 31, 1971.
30. Erdmann, B., and Keller, C., *Inorg. Nucl. Chem. Lett.* **7,** 675 (1971).
31. Tauster, S. J., Fung, S. C., and Garten, R. L., *J. Am. Chem. Soc.* **100,** 170 (1978).
32. Holleck, H., and Kleykamp, H., Kernforschungszenstrum Karlsruhe Report KFK 1271-1, 1971.
33. Muller, W., Reul, J., and Spirlet, J. C., *Atomwirtschaft* **17,** 415 (1972).

0018–1536/84/1700–0381$02.80

Prediction of High Temperature Metallic Phase Diagrams

PAUL R. WENGERT*

Materials Engineer, Owens-Illinois, Inc., Toledo, OH 43666

Received September 23, 1983; Accepted December 31, 1983

ABSTRACT

The Engel Correlation and the Brewer Predictions are reviewed. Multicomponent phase diagrams for the 30 metallic elements of K through Ni, Rb through Pd, and Cs through Pt are discussed.

Index Entries: Engel–Brewer theory; phase diagrams; transition elements, phase diagrams of; Pt-group elements, phase diagrams of.

INTRODUCTION

The following article is a review of Leo Brewer's publication: "Prediction of High Temperature Metallic Phase Diagrams," that appeared in *High-Strength Materials*(*1*). The publication predicts the multicomponent phase diagrams for the the 30 metals of the three transition series from K through Ni, Rb through Pd, and Ce through Pt. These metals in various combinations require over 2 billion multicomponent phase diagrams. Less complex systems either with total solubility or with limited solubility are discussed without the illustration of phase diagrams. Phase diagrams are presented for combinations of metals of the fourth to sixth groups with metals of the seventh to the tenth groups. A total of 1500 multicomponent systems are illustrated in 30 diagrams.

Phase stability is most heavily influenced by electronic structure. As an aid to understanding the predictions, Leo Brewer thoroughly describes the Engel Correlation and confirms it with spectroscopic data of

*Current address: Technical Staff/7472, Sandia National Laboratories, Abuquerque, NM 87185.

the gaseous elements and with thermodynamic data. Important secondary effects are also considered.

In the following article, the Engel Correlation is reviewed first. The predicted phase diagrams are presented next.

ENGEL CORRELATION

The Engel Correlation simply states that the crystal structures of transition elements and their intermetallic compounds correlate to the concentration of electrons in the "*s,p*" valence atomic orbital as follows: body centered cubic (bcc) correlates with one electron, hexagonal closed pack (hcp) correlates with two electrons, and cubic closed pack (ccp) correlates with three electrons.

Review of existing phase diagrams demonstrates that the electron concentration range is broader than originally postulated and that other crystal structures are influenced by electron concentration. Specifically, the crystal structure of the transition elements and their intermetallic compounds correlates to the concentration of electrons in the *s,p* valence atomic orbitals as follows: bcc correlates to 1.0 to 1.5 electrons, hcp correlates to 1.7 to 2.1 electrons, and ccp correlates to 2.5 to 3.0 electrons. When other, more complex crystal structures are stabilized by size considerations, then their presence typically appears at increasing concentrations of electrons in the *s,p* valence atomic orbitals in the following sequence: bcc, A15 (Cr_3Si), σ (β-U) and related ϕ, δ, *P*, *R*, and *D*8 structures, χ (α-Mn), β (β-Mn), hcp, ccp.

The correlation of crystal structure to the concentration of electrons in the *s,p* valence atomic orbitals does not ignore the bonding character of electrons in "*d*" valence atomic orbitals. The *d* orbital electrons bond to nearest neighbors and do not control the crystal structure as do the *s,p* orbital electrons that extend beyond nearest neighbors. Indirectly, the *d* orbital electrons influence crystal structure because they offer a competing energy state for electrons. Within the Brewer Predictions, *d* orbital electrons play a significant role and are discussed on an element-by-element basis.

BREWER PREDICTIONS

For the 30 metallic elements of K through Ni, Rb through Pd, and Cs through Pt, 435 binary phase diagrams are possible. Including all multicomponent systems, there are over 2 billion possible diagrams.

Fortunately, most of these systems are simple. One phase dominates for many systems. For other systems, the secondary factors of size ratios, electronegativity, or solubility parameters allow calculation of solubility ranges of phases. The various combinations of the 18 metals of the first six groups fall into this category, as do the metals of the seventh, eighth, ninth, and tenth groups. These systems are discussed in detail by Brewer without the aid of phase diagrams.

Of the remaining systems, phase diagrams have been presented for combinations of metals of the fourth to sixth groups with metals of the seventh to tenth groups. They are presented in Figs. 1 through 30. Figure 31 offers a key to identify the phase designations. Figure 32 gives the phases for the pure elements as a reference.

Each phase diagram with six elements, such as Fig. 1, is a compilation of 57 phase diagrams that includes 15 binary diagrams.

All diagrams are projections of phase regions over the entire temperature region up to the solidus temperatures. The composition ranges given represent the maximum ranges at the optimum temperature for each phase. The phases shown do not necessarily exist at a common temperature. When the maximum composition ranges of adjoining phases overlap to a large extent, the limiting composition range of the low temperature form is indicated as a dotted line. In some instances where the maximum ranges just barely overlap, the drawings do not show any overlap to avoid any confusion in regard to identification of the phases.

A simplified method of representing multicomponent phase diagrams is used. Figure 1, the diagram for alloys of Hf with W, Re, Os, Ir, and Pt, can be used to illustrate the method. The horizontal scale gives the composition in atomic percent with 100% Hf on the left and 100% of the alloying metal on the right. The top horizontal line represents the binary Hf–W system. A horizontal line at the level of the symbol Re represents the binary Hf–Re system. A horizontal line half-way between Re and Os, for example, would represent an equimolal mixture of Re and Os mixed with Hf with 100% Hf at the left and 0% Hf at the right. The various horizontal lines that can be drawn between Re and Os correspond to various ratios of Re and Os in alloys with Hf. For most of the systems chosen, the internal pressures and sizes of the atoms are close enough together so that different combinations of metals yielding the same electron concentration show approximately the same phase behavior. Thus the horizontal line half-way between Re and Os intersects the phase boundaries at compositions corresponding to the behavior of a system of Hf mixed with an equimolal mixture of Re and Os. This same horizontal line also represents the mixing of Hf with an equimolal mixture of W and Ir, the mixing of Hf with an alloy containing 25% each of W, Re, Os, and Ir, or any mixture of W, Re, Os, Ir, and Pt, which averages out to 7.5 valence electrons per atom. The major difference between binary systems and multicomponent systems of the same electron concentration will be an increased stabilization of phases with nonequivalent sites, such as σ and χ phases compared to phases with equivalent sites such as fcc, for example. With this use of the diagrams, each diagram corresponds to 57 multicomponent diagrams. This use of the diagrams for multicomponent systems works best for metals of the third transition series and poorest for metals of the first transition series.

Fig. 2

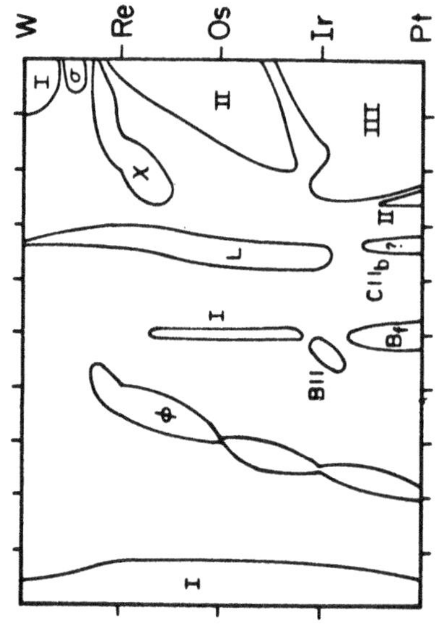

Fig. 4

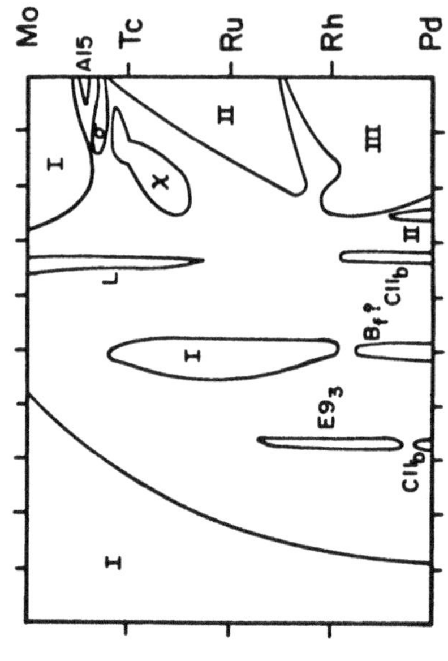

Fig. 1

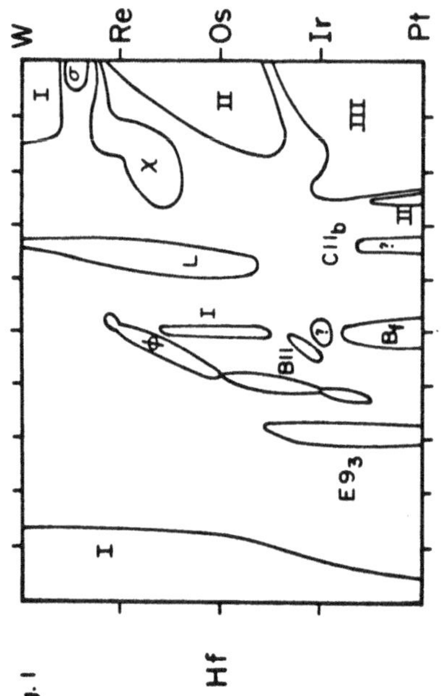

Fig. 3

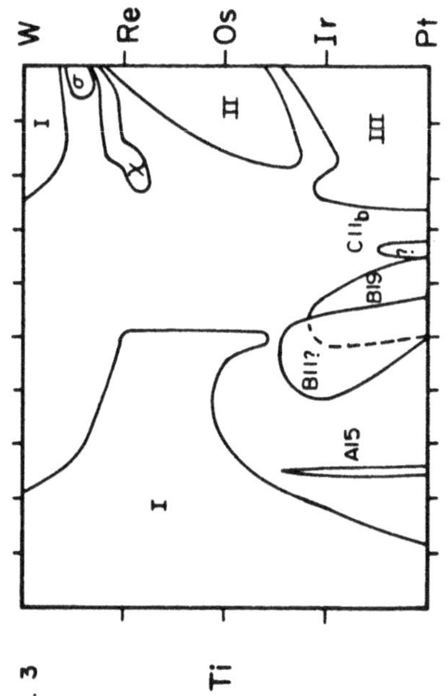

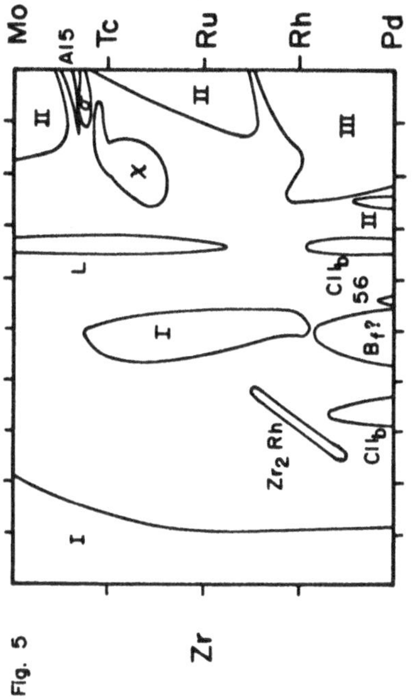

Fig. 5

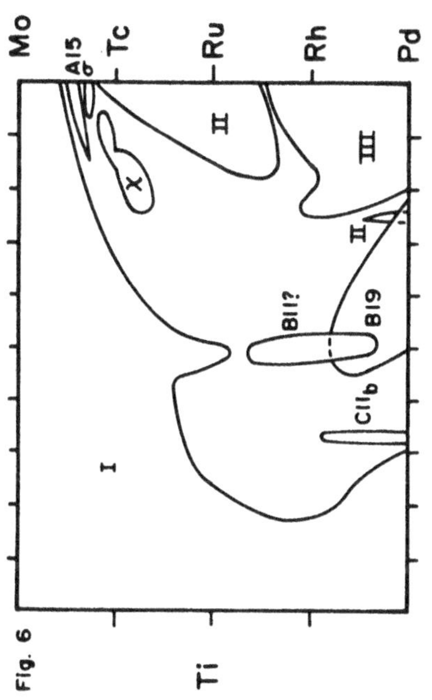

Fig. 6

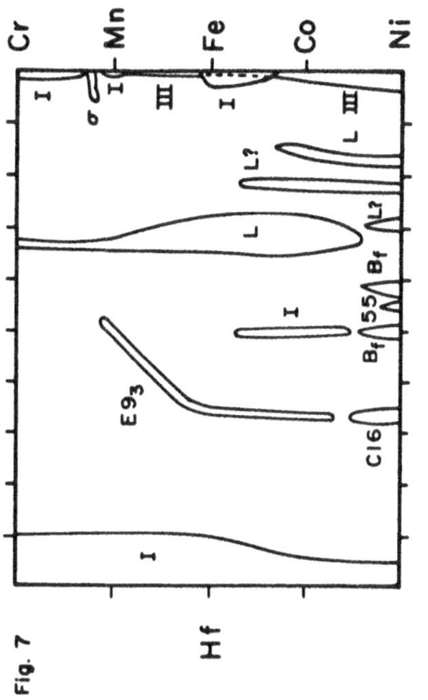

Fig. 7

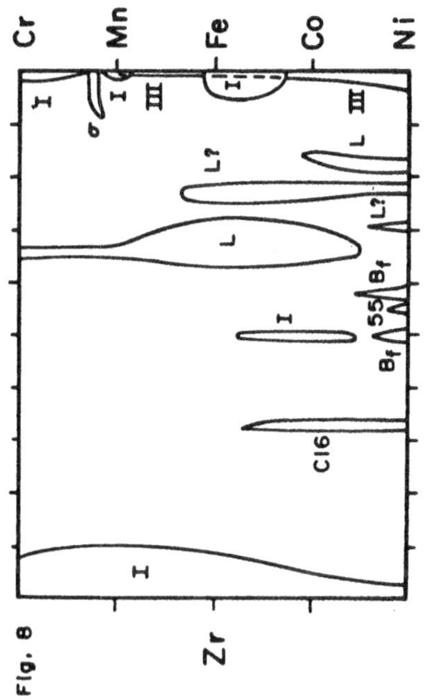

Fig. 8

Fig. 9

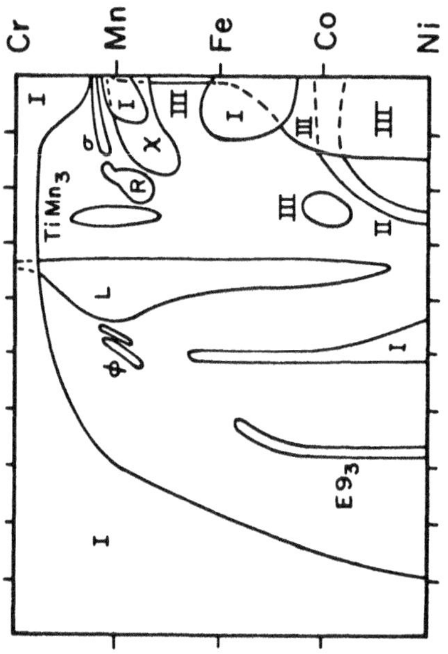

Fig. 10

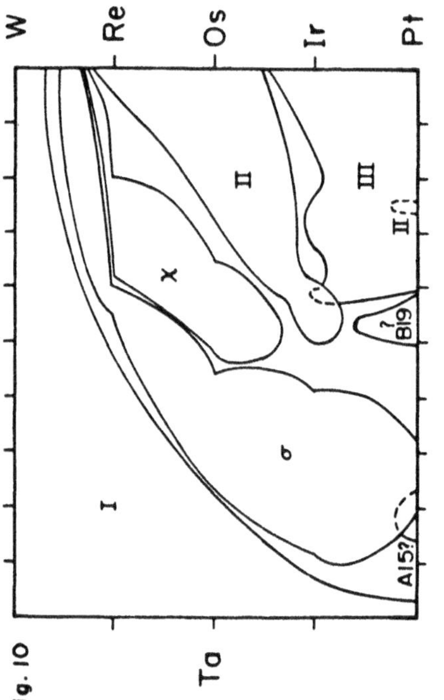

Fig. 11

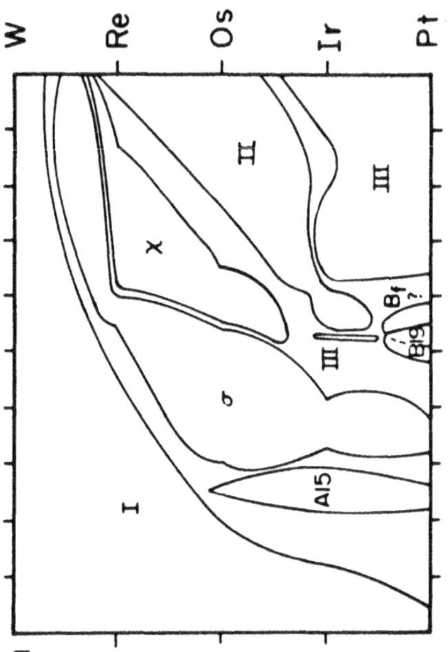

Fig. 12

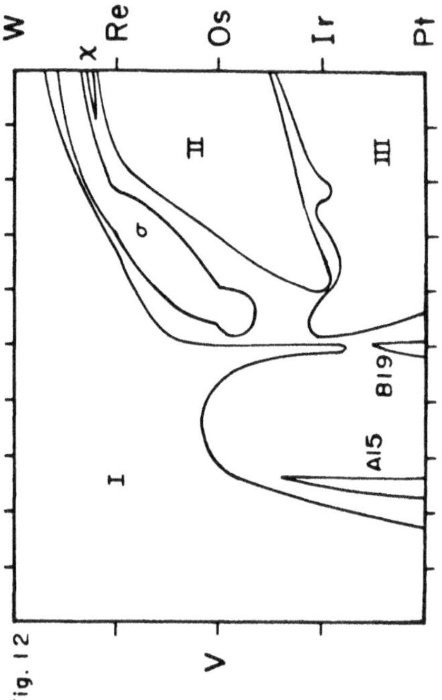

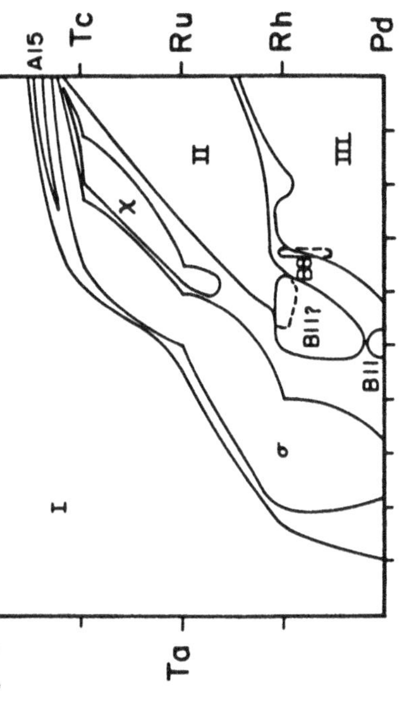

Fig. 13

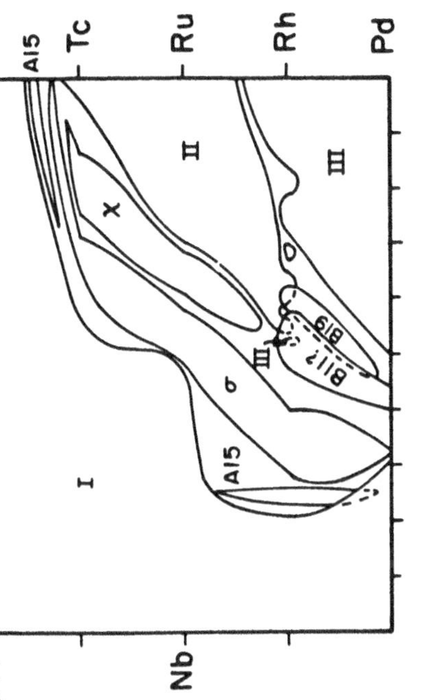

Fig. 14

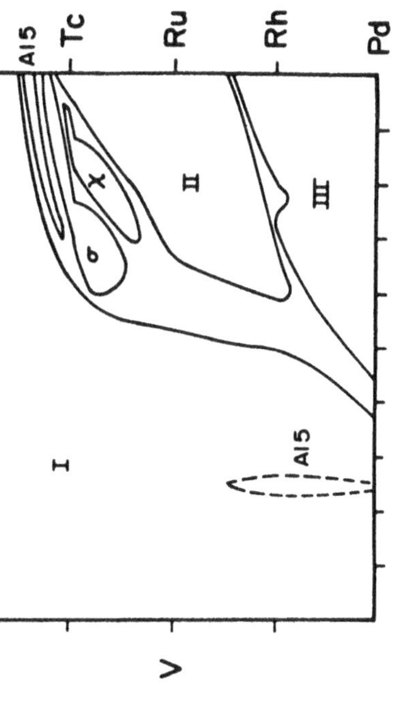

Fig. 15

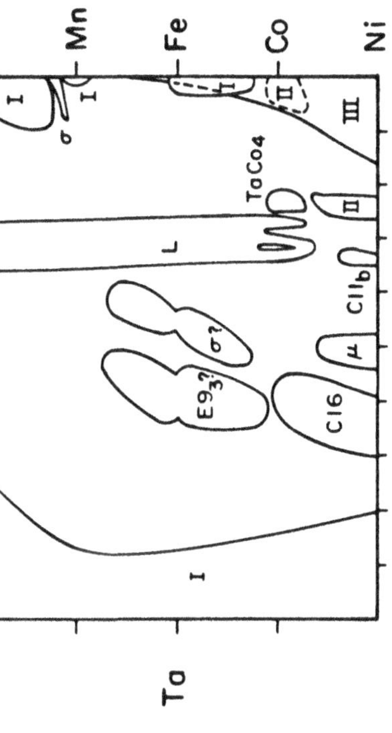

Fig. 16

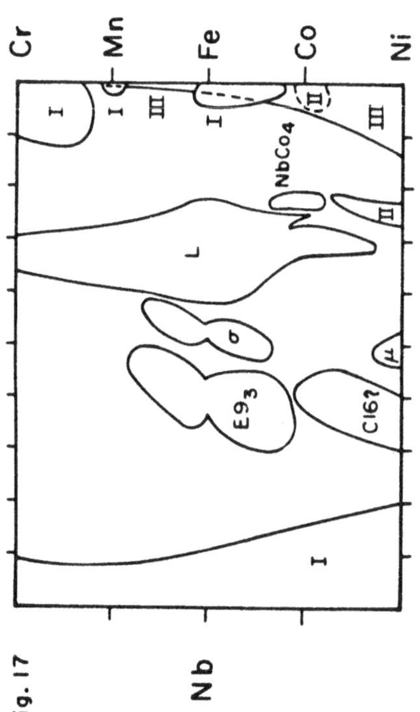

Fig. 17

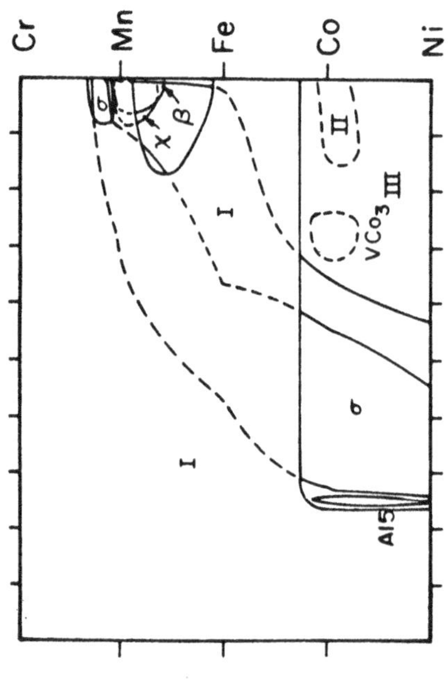

Fig. 18

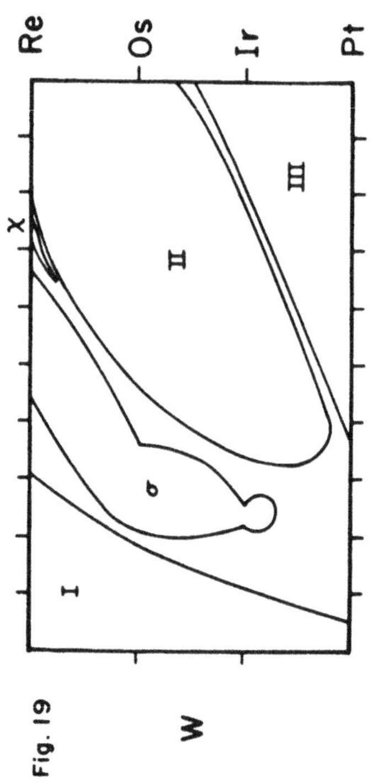

Fig. 19

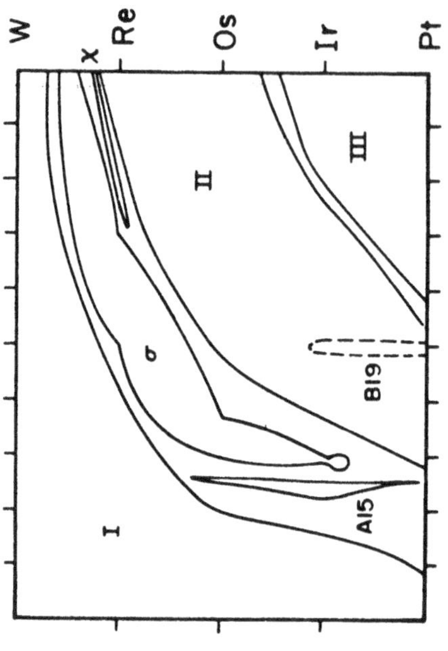

Fig. 20

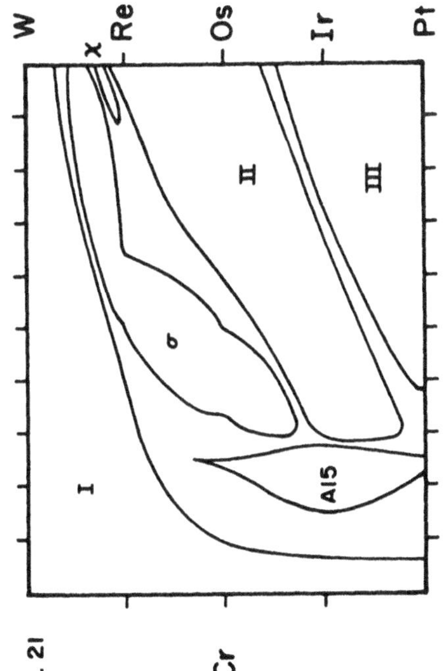

Fig. 21

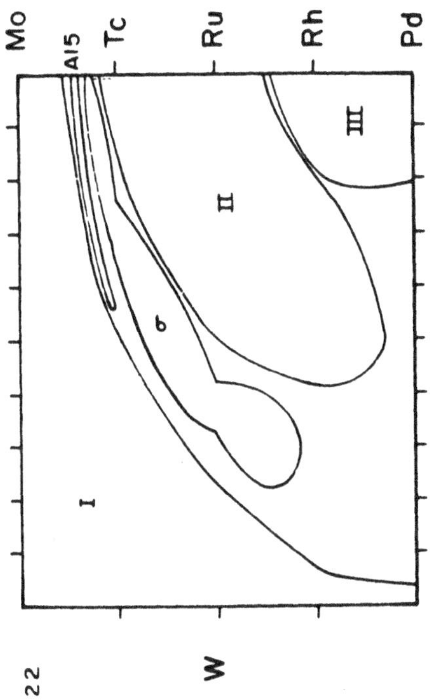

Fig. 22

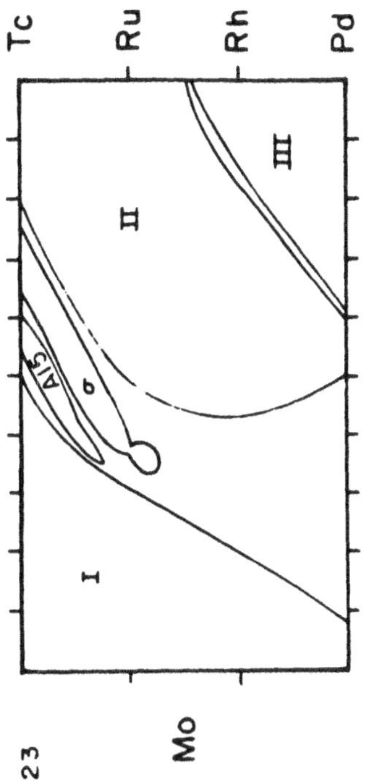

Fig. 23

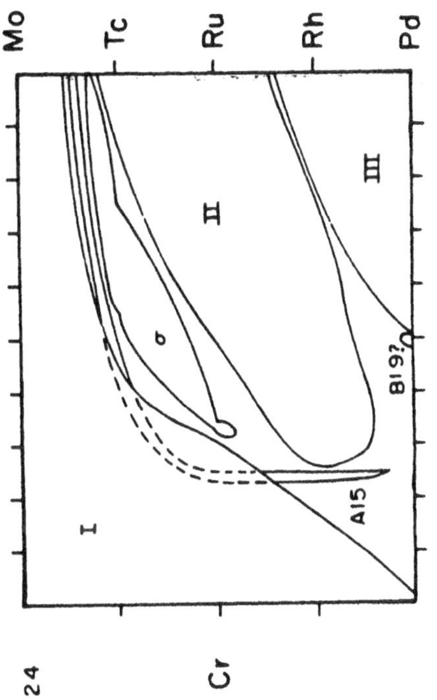

Fig. 24

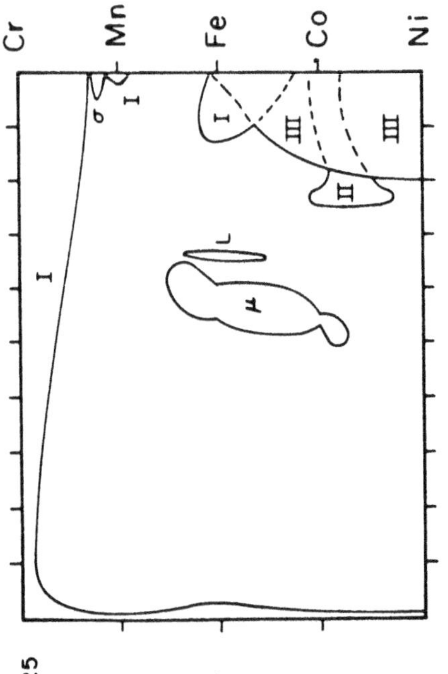

Fig. 25

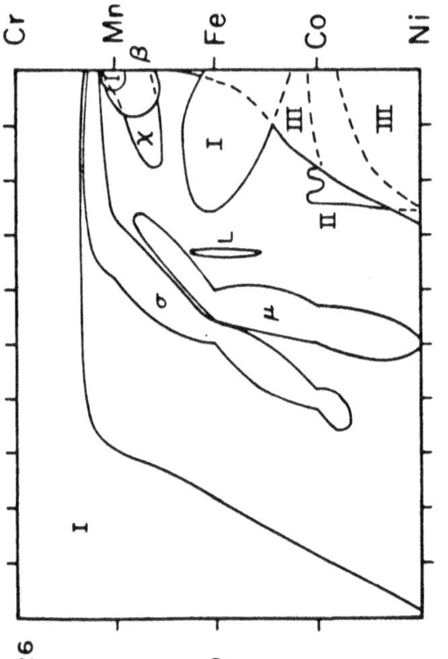

Fig. 26

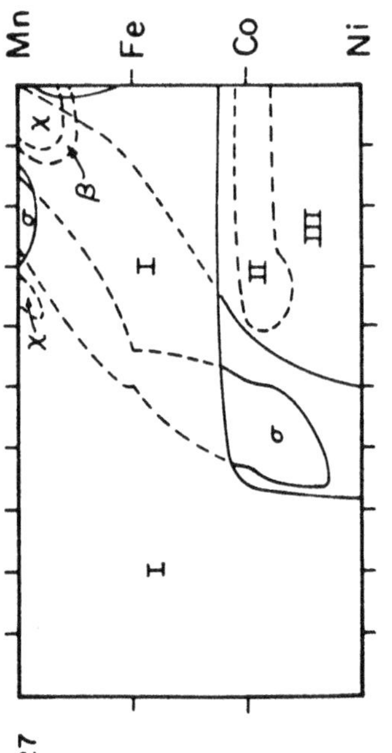

Fig. 27

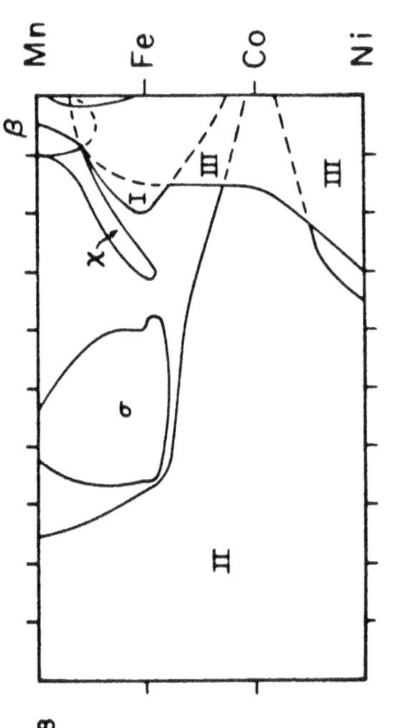

Fig. 28

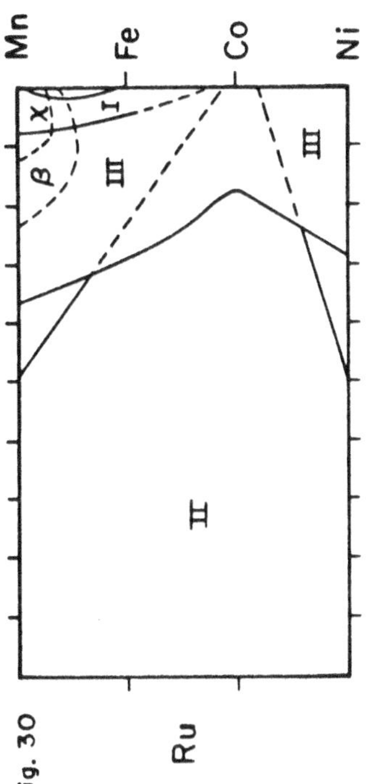

Fig. 30

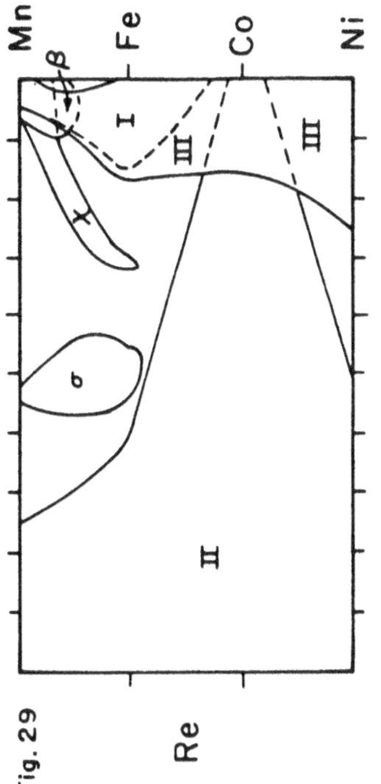

Fig. 29

Symbols Used in Figures to Designate Phase Regions

Figure Designations	Basic Crystal Type	Structure Designation	
I	body-centered cubic	A2 and B2 CsCl	
II	hexagonal close-packed	A3, DO_{19} $MgCd_3$, DO_{24} $TiNi_3$	
II	orthorhombic	$TiCu_3$ and α_1 TaRh	
III	face-centered cubic	A1 and $L1_2$ $AuCu_3$	
III	tetragonal	$TiAl_3$ DO_{22}, CuAu $L1_0$, and $MoNi_4$ $D1_a$	
L	cubic $MgCu_2$	C15	Laves phases
L	hexagonal $MgZn_2$	C14	Laves phases
L	hexagonal $MgNi_2$	C36	Laves phases
L	face-centered cubic UNi_5	$C15_b$	Laves phases
L	hexagonal $CaCu_5$	$D2_d$	Laves phases
A15	cubic Cr_3Si	A15	
σ	tetragonal β-U	$D8_b$	
ϕ	tetragonal HfRe	ϕ	
μ	rhombohedral W_6Fe_7	$D8_5$	
	and $Mo_7Fe_2Ni_5$	P	
μ	orthorhombic MoNi	δ	
	and Ti_2Mn_9	R	
χ	cubic α-Mn	A12	
β	cubic β-Mn	A13	
B_f	orthorhombic CrB and Zr_7Ni_{10}	B_f	

The face-centered cubic $E9_3$ Ti_2Ni; the hexagonal $B8_2$ Ni_2In; the tetragonal C11b $MoSi_2$, B11 TiCu, and C16 $CuAl_2$; and the orthorhombic B19 AuCd structures are designated by their Strukturbericht designations.

Fig. 31

Prediction of phase diagrams involving several elements is more complex than the prediction of phases for the pure elements because the atomic orbitals are further distorted by interactions. As a result, experimental data have been used extensively to derive the multicomponent phase diagrams. When there are quite complete data of unquestioned reliability for some systems then results are presented unaltered in the figures. In other instances, several conflicting observations have been reported in the literature and the Engel Correlation has been used to select from among the conflicting data. In other instances, single investigations have been reported which appear to be in error. When some internal inconsistencies are apparent that allow rejection of such data, this has been done. However, when there is no substantial basis for rejection of the data, they have been presented in the figures as reported. In very many instances, no data have been reported and the Engel Correlation has been used to predict existence of phases and their ranges, but this has been done in a conservative manner and extrapolations have not been carried too far from existing data, so that these figures can be used as representative of the existing experimental data.

Crystal Structures of the Elements[a]

Li I	Be II	B * *											C IV >=[f]	N =[b]	O =[b]	F —[c]
Na I	Mg II	Al III											Si IV	P ↖ *	S <[d]	Cl —
K I	Ca I III	Sc I II	Ti I II	V I	Cr I	Mn I III β χ	Fe I III I	Co III II	Ni III	Cu III	Zn II	Ga All	Ge IV	As ↖	Se <	Br —
Rb I	Sr I II III	Y I II	Zr I II	Nb I	Mo I	Tc II	Ru II	Rh III	Pd III	Ag III	Cd II	In (III)[g]	Sn A5 IV	Sb ↖	Te <	I —
Cs I	Ba I	La I III II	Hf I II	Ta I	W I	Re II	Os II	Ir III	Pt III	Au III	Hg A10	Tl I II	Pb III	Bi ↖		

Ce I III II	Pr I II	Nd I II	Pm	Sm I (II)[g]	Eu I	Gd I II	Tb I II	Dy I II	Ho I II	Er I II	Tm I II	Yb I III	Lu I II
Th I III	Pa *	U I σ A20	Np I **	Pu I * III ***	Am ? II	Cm ? II							

[a] I: Body centered cubic, A2. II: Hexagonal close packed, A3. III: Cubic close packed, Al. IV: Diamond, A4. Asterisk denotes complex structure. [b] Diatomic molecules with double or triple bonds. [c] Diatomic molecule with a single electron pair bond. [d] Atoms which form two single bonds per atom to form rings or infinite chains. [e] Three single bonds per atom, corresponding to a puckered planar structure. [f] The graphite structure where one resonance form consists of two single and one double bond per atom. [g] Parentheses indicate slight distortions. The structures are listed in order of temperature stability except for metastable diamond.

Fig. 32

Figures 1 through 30 present those diagrams where the changes in crystal structure that correspond with changes in the *s,p* electron concentration are most dramatic. As the *s,p* electron concentration increases from left to right and from top to bottom, the progression of crystal structures is: bcc, hcp, and ccp.

Using Fig. 1 as an example, the crystal structure and concentration of electrons for the solid elements proceeds as follows: Hf (bcc: $6s^15d^3$), W (bcc: $6s^15d^5$), Re (hcp: $6s^16p^15d^5$), Os (hcp: $6s^16p^15d^6$), Ir (ccp: $6s^16p^25d^6$), and Pt (ccp: $6s^16p^25d^7$).

For the compound $HfPt_3$, more bonding electrons are achieved by a sharing of the *d* orbital electrons than by remaining separate elemental phases. The $HfPt_3$ [ccp: Hf^{-3} ($6s^16p^25d^4$) Pt_3^{+1} ($6s^16p^25d^6$)] contains a total of seven bonding electron pairs per gram atom, which includes three *s,p* electrons to correlate with the ccp crystal structure. Hf (bcc: $6s^15d^3$) contains four bonding electrons and Pt (ccp: $6s^16p^25d^7$) contains six bonding electrons.

In a similar manner, the number of bonding electrons can be increased if Hf with its *d* orbitals, which are less than half-filled, is added to Os and Ir with their *d* orbitals more than half-filled. Note how the Os hcp phase and the Ir ccp phase extend to the left toward Hf.

In the case of the Hf–W and Hf–Re systems, the number of bonding electrons is not increased when compounds are formed because all of solid elements have *d* orbitals that are less than half-filled. For example, HfW (bcc: $6s^15d^4$) would have five bonding electrons, as does the average of Hf (bcc: $6s^15d^3$)—four bonding electrons—and of W (bcc: $6s^15d^5$)—six bonding electrons—taken by themselves. Since the total number of bonding electrons does not increase, HfW does not have an appreciably lower energy state than Hf and W as two immiscible solids and HfW does not form. Because the solubility parameters of Hf [105 $(cal/deg)^{1/2}$] and W (146) are significantly different, a large immiscibility gap separates them. A Laves phase appears because of a favorable size ratio at HfW_2.

Note the difference between Fig. 1 and Fig. 10. The solubility parameter of Ta [131 $(cal/deg)^{1/2}$] more closely matches the solubility parameters of W (146), Re (146), Os (148), Ir (136), and Pt (122) than does the solubility parameter of Hf (105). As a result, the ccp phases of Ir and Pt and the hcp phase of Os and Re extend further to the left toward Ta. Since the solubility parameters of Ta and W are closely matched, they are completely miscible even though no increase in the number of bonding electrons is achieved.

REFERENCES

1. Brewer, L., Prediction of High Temperature Metallic Phase Diagrams, in *High-Strength Materials*, Zackay, V. F., ed., Wiley, New York, 1965, pp. 12–103.

0018–1536/84/1700–0395$02.80

The Temperature–Composition Phase Diagram of the GeSe–GeTe System

HERIBERT WIEDEMEIER* AND PAUL A. SIEMERS†

Department of Chemistry, Rensselaer Polytechnic Institute, Troy, New York 12181

Received November 3, 1983; Accepted December 31, 1983

ABSTRACT

The temperature–composition phase diagram for the GeSe–GeTe system was determined between 25 and 675°C using high temperature X-ray powder diffraction techniques (*1*). The orthorhombic phase extends from 100 to 92 ± 3 mol% GeSe and is stable below 651 ± 5°C. For compositions between 86 ± 3 and 58 ± 3 mol% GeSe a hexagonal phase exists. Above 415 ± 5°C the hexagonal phase decomposes via a peritectoid reaction to form an orthorhombic and cubic phase. A rhombohedral phase was observed for compositions between GeTe and 52 ± 3 mol% GeSe. Above 375 ± 10°C the rhombohedral phase transforms into a cubic phase. A complete series of cubic GeSe–GeTe solid solutions exists above 651 ± 5°C.

Index Entries: Phase diagram, of the GeSe–GeTe system; GeSe–GeTe system, phase diagram of; SeGe–GeTe system, phase diagram of; TeGe–GeSe system, phase diagram of.

INTRODUCTION

The room temperature crystal structure of GeSe is orthorhombic with unit cell dimensions of a = 4.387 ± 0.008 Å, b = 3.837 ± 0.008 Å, and c = 10.83 ± 0.015 Å (*2*). High temperature X-ray diffraction studies of GeSe have shown that the orthorhombic structure transforms to a

*Author to whom all correspondence and reprint requests should be addressed.
†Present address: General Electric Company, R&D Center, Schenectady, New York 12345.

NaCl-type cubic structure at 651°C with a unit cell parameter of 5.730 ± 0.003 Å (*2*). The crystal structure of GeTe is rhombohedral at room temperature and transforms to a NaCl-type cubic structure at 375°C (*3*).

Previously, several investigations of the phase equilibria of the GeSe–GeTe system were performed. X-ray diffraction studies 25°C on GeSe–GeTe compositions annealed at elevated temperatures for various time periods yielded a single-phase rhombohedral region for compositions between GeTe and 63 mol% GeSe (*4*) and a single-phase orthorhombic region between GeSe and 95 mol% GeSe (*4*) separated by a two-phase region. These results are in partial agreement with those of Muir and Cashman (*5*). Based on X-ray diffraction measurements at 25°C of mixtures annealed at higher temperature, these authors (*5*) observed single-phase orthorhombic (GeSe–92 mol% GeSe), hexagonal (85–65 mol% GeSe) and rhombohedral (GeTe–53 mol% GeSe) regions. Employing dilatometry, differential thermal, and microstructural analyses (*6*), a two-phase region below 375°C and between 5 and 92 mol% GeSe and a single-phase region between about 300 and 600°C ranging from GeSe to 92 mol% GeSe were observed. Complete solid solution was reported for the high temperature forms of GeSe and GeTe above 626°C, without data concerning the crystal structure of the high temperature solid solution phase (*6*).

Based on high temperature X-ray diffraction and differential scanning calorimetry studies below 500°C of GeSe–GeTe mixtures prepared similarly to those described earlier (*5*), Muir and Beato (*7*) reported a single-phase rhombohedral region between GeTe and 55 mol% GeSe at temperatures below 370°C. Above 370°C, the rhombohedral phase transforms into a single-phase with cubic structure. For compositions between 65 and 85% GeSe, a single-phase with hexagonal structure stable up to 450°C was reported (*7*). Above 450°C, the hexagonal phase decomposes into orthorhombic and cubic GeSe–GeTe terminal solid solutions. Between 25 and 500°C, an orthorhombic single-phase was observed for GeSe–GeTe compositions between GeSe and 95 mol% GeSe (*7*).

In connection with crystal growth studies of GeSe–GeTe solid solutions at elevated temperature in our laboratory (*8*), accurate lattice parameters as a function of composition and structural information about the high temperature modifications of GeSe–GeTe solid solutions were required. For this reason and because of the lack of information on this system above 500°C, it was necessary to reinvestigate the temperature–composition phase equilibria of the GeSe–GeTe system.

EXPERIMENTAL METHODS

The starting materials for the synthesis of GeSe–GeTe solid solutions were the binary compounds GeSe and GeTe that were prepared from the elements in a manner described earlier (*2,3*). Two-gram mixtures of the binary constituents ranging from GeSe to GeTe in composition intervals of 10, 5, and 2.5 mol% GeSe were sealed in cleaned and outgassed (*9*) fused silica ampules at a pressure of 10^{-6} torr or less. All starting mixtures were subjected to the same heat treatment in a vertical, two-zone furnace. They were slowly heated to 770°C, which is about 50°C above the melting point of any GeSe–GeTe solid solution (*6*). After 24 h at 770°C, cooling to 630°C occurred over a 4-h period. After annealing at 630°C for 24 h, the samples were cooled at a rate of about 1°C/min to 400°C, at which temperature the samples were annealed for an additional 48 h. The final cooling to room temperature took about 24 h. During the above annealing procedures, the top portion of the ampules was maintained about 15°C hotter than the sample region to prevent sublimation of the samples. The above heat treatment converted the 2-g powder mixture of GeSe and GeTe into a small, well-defined ingot at the bottom of the ampule.

The room temperature phase equilibria for the annealed mixtures of GeSe and GeTe were determined using Debye-Scherrer powder X-ray diffraction techniques. Lattice parameters and unit cell volumes for each phase observed were calculated using the extrapolation method described by Cullity (*10*). From plots of unit cell volumes or lattice parameters versus the overall composition of the starting mixtures, the ranges of the single-phase solid solutions were identified.

For the investigation of the phase equilibria in this system above 25°C, a Unicam high temperature Debye-Scherrer camera as described earlier (*2*) was used. GeSe–GeTe solid solutions of different composition, annealed as discussed above, were heated in the high temperature camera to the desired temperature. The high temperature phase equilibria were established using a 1 h annealing time at a given temperature followed by a 2–3 h exposure using Ni-filtered CuKα radiation. Temperatures were varied in a nonsystematic way for each consecutive X-ray exposure. The calculation of lattice parameters of the high temperature phases from diffraction patterns has been discussed previously (*2,3*). Absolute temperature errors are estimated to be ± 5°C. The phase boundaries determined in these studies have estimated errors less than ± 3 mol%.

RESULTS AND DISCUSSION

Phase Equilibria in the GeSe–GeTe System at 25°C

Lattice parameters and unit cell volumes were determined for 26 annealed GeSe–GeTe samples with different overall compositions. A graphical representation of the unit cell volumes for each phase present at 25°C as a function of composition in Fig. 1 reveals the existence of solid solutions with three different crystal structures. A rhombohedral phase exists for GeSe–GeTe solid solutions of composition from 0 (GeTe) to 52

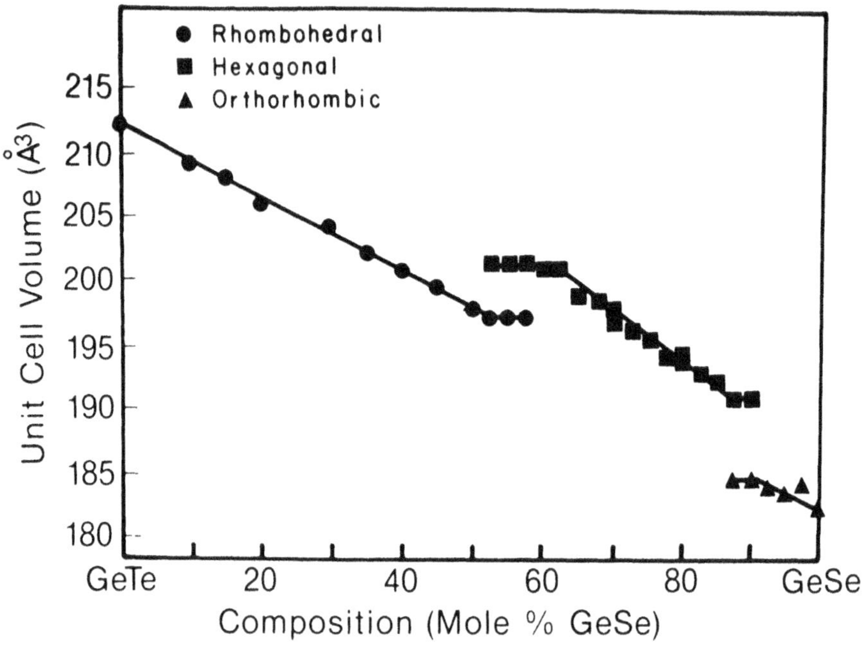

Fig. 1. The unit cell volumes of GeSe–GeTe solid solutions at 25°C as a function of composition.

mol% GeSe. For compositions from 58 to 86 mol% GeSe a single-phase with hexagonal structure is observed. Between GeSe and 92 mol% GeSe the orthorhombic phase exists. It is interesting to note that the width of the single-phase regions increases from the orthorhombic via the hexagonal to the rhombohedral phase. The covalent radii of tellurium and selenium are 1.35 Å (*11*) and 1.17 Å (*11*), respectively. Thus, with increasing GeTe content of the solid solution, the unit cell volume of the respective crystal structure increases. The rhombohedral structure can accommodate more of the smaller selenium atoms than the orthorhombic phase of the larger tellurium atoms. Similar trends with regard to the width of single-phase regions were observed in the PbS–SnS (*4*), PbTe–SnSe (*4*), and the SnTe–GeSe (*4*) systems. In none of the above systems (*4*) was a third phase, analogous to the hexagonal phase in the GeSe–GeTe system, reported.

The *a* parameter of the rhombohedral as well as the unit cell constants of the hexagonal phases are linear functions of composition, as shown in Figs. 2 and 3, respectively. Based on a linear least-squares treatment of the data shown in Fig. 2, the rhombohedral lattice parameter *a* can be expressed by the equation

$$a = 5.975 - 0.279\,N\ (\text{Å}) \tag{1}$$

for GeSe mole fractions, N, varying from 0 to 0.52. Extrapolation of Eq. (*1*) to $N = 1$ yields a "hypothetical" rhombohedral unit cell length of 5.696 Å for GeSe. This computed value is relatively close to the experimentally observed cubic lattice parameter of 5.730 ± 0.003 Å for GeSe at 656°C (2).

A linear least-squares treatment of the hexagonal lattice constants (Fig. 3) yields equations for the parameters

$$a = 3.997 - 0.249\,N\ (\text{Å}) \tag{2}$$

and

$$c = 16.19 - 0.815\,N\ (\text{Å}) \tag{3}$$

for GeSe mole fractions from 0.58 to 0.86. Based on these equations, the "hypothetical" hexagonal lattice parameters for GeSe (N = 1) are a = 3.75 Å and c = 15.37 Å. Accordingly, a hypothetical hexagonal modification of GeTe (N =0) would have lattice parameters of a = 3.997 Å and c = 16.19 Å. The c/a ratio for the single-phase hexagonal solid solutions is 4.09 ± 0.5% over the entire single-phase hexagonal region. This result

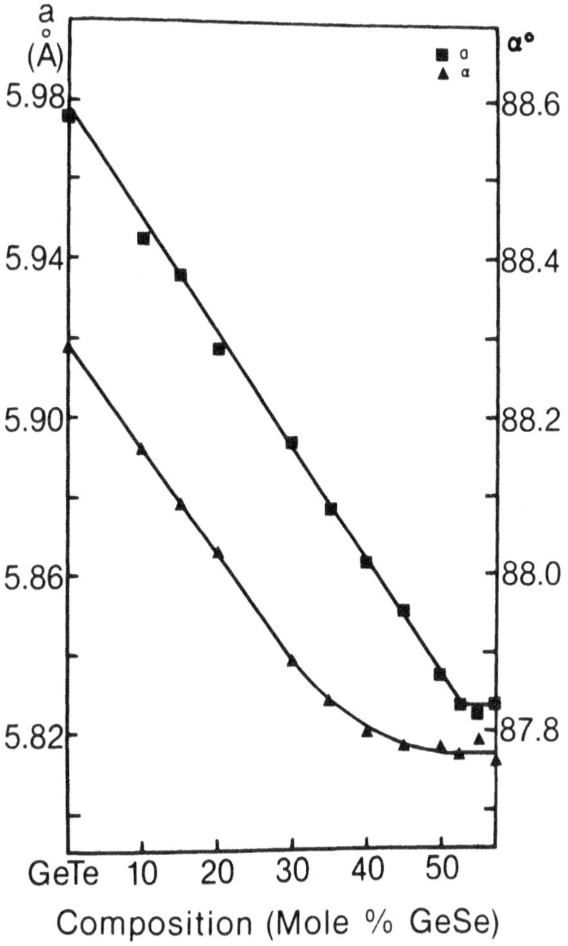

Fig. 2. The composition dependence of the lattice parameters for the rhombohedral phase of GeSe–GeTe solid solutions at 25°C. Closed square ■: lattice parameter a; closed triangle (▲): angle α.

indicates that the bonds in the hexagonal structure change at the same relative rates in all directions with composition of the GeSe–GeTe system.

Phase Equilibria in the GeSe–GeTe System above 25°C

The lattice parameters of GeSe–GeTe solid solutions as a function of composition and temperature determined from high temperature X-ray diffraction patterns are listed in Table 1. The uncertainties of the lattice constants for the orthorhombic phase are estimated to be less than ±0 .008 Å for the *a* and *b* parameter and less than ± 0.02 Å for the *c* parameter. For the rhombohedral phase the uncertainties are less than ± 0.006 Å and 0.1° for the *a* and α parameters, respectively. The cell constants for the hexagonal phase are accurate within ± 0.008 and ± 0.02 Å for the *a*- and *c*-axis, respectively. The uncertainties of the cubic lattice parameters are less than ± 0.006 Å.

Based on the data in Table 1, the temperature–composition phase diagram of the GeSe–GeTe system can be represented as a pseudobinary system, as shown in Fig. 4. There are two invariant points, namely, a eutectoid at 48 ± 3 mol% GeSe and 340 ± 10°C and a peritectoid at 415 ± 5°C and 56 ± 3 mol% GeSe. A two-phase region was not observed between the single-phase rhombohedral and cubic regions.

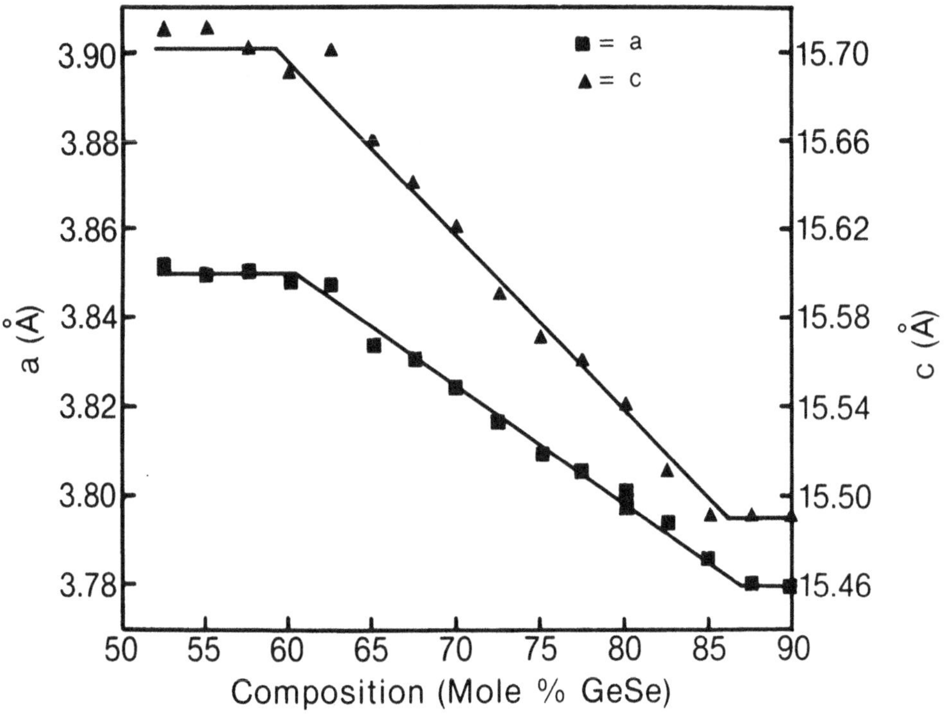

Fig. 3. The composition dependence of the lattice parameters for the hexagonal phase of GeSe–GeTe solid solutions at 25°C. Closed square (■): lattice parameter *a*; closed triangle (▲): lattice parameter *c*.

TABLE 1
Lattice Parameters of GeSe–GeTe Solid Solutions as a Function of Temperature and Composition

Composition, mol% GeSe	Temperature, °C	Phase[a]	a, Å	b, Å	c, Å	α degrees
20	26	R	5.916			87.97
20	315	R	5.953			89.09
20	351	R	5.956			89.33
20	366	R	5.959			89.45
20	376	C	5.956			
20	479	C	5.972			
20	523	C	5.980			
20	629	C	5.994			
35	23	R	5.889			87.74
35	255	R	5.908			88.61
35	358	R	5.919			89.28
35	376	C	5.918			
35	463	C	5.920			
35	554	C	5.920			
35	635	C	5.919			
45	26	R	5.845			87.81
45	300	R	5.881			88.80
45	355	R	5.885			89.35
45	359	R	5.887			89.30
45	372	C	5.877			
45	390	C	5.882			
45	436	C	5.884			
45	531	C	5.878			
45	619	C	5.879			
50	23	R	5.834			87.78
50	148	R	5.849			88.17
50	257	R	5.860			88.74
50	327	R	5.867			89.55
		H	3.872		15.85	
50	338	C	5.855			
		H	3.875		15.80	
50	380	C	5.869			
50	410	C	5.876			
50	514	C	5.884			
50	622	C	5.889			
55	23	R	5.816			87.77
		H	3.848		15.70	
55	200	R	5.837			88.60
		H	3.863		15.83	

(continued)

TABLE 1 (*continued*)

Composition, mol% GeSe	Temperature, °C	Phase[a]	a, Å	b, Å	c, Å	α degrees
55	315	R	5.844			89.10
		H	3.873		15.86	
55	329	R	5.845			89.33
		H	3.878		15.86	
55	362	C	5.850			
		H	3.884		15.82	
55	376	C	5.853			
		H	3.884		15.85	
55	410	C	5.854			
		H	3.880		15.86	
55	421	C	5.859			
55	520	C	5.867			
55	636	C	5.874			
60	23	H	3.848		15.69	
60	148	H	3.860		15.86	
60	315	H	3.870		15.91	
60	352	H	3.865		15.87	
60	382	H	3.875		15.90	
60	406	H	3.878		15.91	
60	413	H	3.868		15.87	
		C	5.850			
60	429	O	4.381	3.907	10.98	
		C	5.861			
60	432	C	5.844			
60	627	C	5.850			
65	23	H	3.834		15.73	
65	300	H	3.852		15.84	
65	360	H	3.856		15.85	
65	386	H	3.860		15.85	
65	394	H	3.861		15.84	
65	409	H	3.850		15.83	
		C	5.854			
65	419	C	5.855			
		O	4.393	3.910	10.98	
65	438	C	5.848			
		O	4.380	3.910	10.99	
65	461	C	5.816			
65	521	C	5.823			
65	558	C	5.829			
65	600	C	5.839			
70	23	H	3.824		15.62	
70	209	H	3.836		15.69	
70	315	H	3.843		15.72	

TABLE 1 (*continued*)

Composition, mol% GeSe	Temperature, °C	Phase[a]	*a*, Å	*b*, Å	*c*, Å	α degrees
70	382	H	3.848		15.74	
70	409	H	3.850		15.75	
70	419	O	4.390	3.912	10.96	
		C	5.850			
70	463	O	4.375	3.918	11.01	
		C	5.848			
70	482	C	5.830			
77.5	25	H	3.805		15.56	
77.5	201	H	3.818		15.63	
77.5	309	H	3.822		15.67	
77.5	320	H	3.826		15.69	
77.5	375	H	3.828		15.70	
77.5	402	H	3.823		15.71	
		O	4.395	3.905	10.99	
77.5	416	H	3.826		15.73	
		O	4.393	3.910	10.98	
77.5	432	O	4.381	3.907	10.99	
		C	5.856			
77.5	466	O	4.373	3.918	11.01	
		C	5.851			
77.5	491	O	4.365	3.908	11.00	
		C	5.825			
77.5	512	C	5.794			
77.5	603	C	5.796			
77.5	634	C	5.800			
82.5	23	H	3.792		15.62	
82.5	260	H	3.807		15.64	
82.5	333	H	3.811		15.66	
82.5	338	H	3.810		15.66	
82.5	352	H	3.813		15.67	
82.5	376	H	3.814		15.69	
82.5	384	H	3.814		15.72	
		O	4.397	3.906	10.96	
82.5	450	C	5.850			
		O	4.375	3.910	10.98	
82.5	507	C	5.786			
		O	4.362	3.910	10.98	
82.5	530	C	5.779			
		O	4. 360	3.920	10.99	
82.5	535	C	5.780			
82.5	550	C	5.782			
82.5	586	C	5.786			
82.5	604	C	5.790			

(*continued*)

TABLE 1 (*continued*)

Composition, mol% GeSe	Temperature, °C	Phase[a]	a, Å	b, Å	c, Å	α degrees
82.5	640	C	5.792			
87.5	23	O	4.397	3.860	10.89	
		H	3.780		15.49	
87.5	170	O	4.405	3.872	10.94	
		H	3.781		15.50	
87.5	300	O	4.400	3.895	10.98	
		H	3.790		15.54	
87.5	400	O	4.390	3.903	10.99	
		H	3.802		15.63	
90	23	O	4.405	3.860	10.89	
		H	3.774		15.49	
90	220	O	4.405	3.876	10.95	
		H	3.782		15.51	
90	310	O	4.398	3.892	10.98	
		H	3.788		15 .54	
90	382	O	4.392	3.893	10.98	
		H	3.796		15.60	
90	400	O	4.390	3.905	10.99	
		H	3.800		15.62	
90	470	O	4.370	3.920	11.00	
		C	5.795			
90	568	O	4.359	3.930	11.02	
		C	5.764			
90	592	C	5.752			
90	630	C	5.759			
95	23	O	4.398	3.845	10.86	
95	199	O	4.399	3.873	10.91	
95	303	O	4.393	3.885	10.92	
95	400	O	4.383	3.890	10.93	
95	452	O	4.391	3.907	10.95	
95	489	O	4.374	3.914	10.97	
95	507	O	4.362	3.913	10.98	
		C	5.783			
95	525	O	4.361	3.926	10.99	
		C	5.780			
95	563	O	4.345	3.944	11.01	
		C	5.765			
95	600	O	4.341	3.921	11.00	
		C	5.750			
95	613	C	5.736			
95	625	C	5.740			
95	645	C	5.736			

[a]R denotes rhombohedral structure; H denotes hexagonal structure; O denotes orthorhombic structure; C denotes cubic structure.

A multiparameter least-squares fit of the data in Table 1 reveals that the rhombohedral unit cell length, *a*, can be expressed as a function of temperature *T*(°C) and GeSe mole fraction *N* by the equation

$$a = 5.978 - 0.2898N + 1.094 \times 10^{-4}T \text{ (Å)} \quad (4)$$

An analogous least-squares treatment of the data in Table 1 yielded an equation for the *a*-parameter of the hexagonal solid solutions in terms of GeSe mole fraction *N* and temperature *T*(°C)

$$a = 4.003 - 0.2588N + 6.748 \times 10^{-5}T \text{ (Å)} \quad (5)$$

The *c*-parameter of the hexagonal phase can be expressed as a function of GeSe mole fraction *N* and temperature *T*(°C) by the equation

$$c = 16.37 - 1.058N + 2.760 \times 10^{-4}T \text{ (Å)} \quad (6)$$

The lattice parameters (Table 1) of the single-phase orthorhombic compounds (e.g., 95 mol% GeSe) change in a similar manner with increasing temperature as those of pure GeSe (2). The cubic phase exists up to the melting points of the solid solutions.

In Fig. 5, the lattice parameters, *a*, for the rhombohedral and cubic phases of GeTe and of different GeSe–GeTe solid solutions are plotted as a function of temperature. The graphical representation reveals anomalies in the change of the *a* parameter at the transformation temperature of about 370 ± 10°C and with increasing temperature. For GeTe and the composition with 20 mol% GeSe, the unit cell length decreases at the

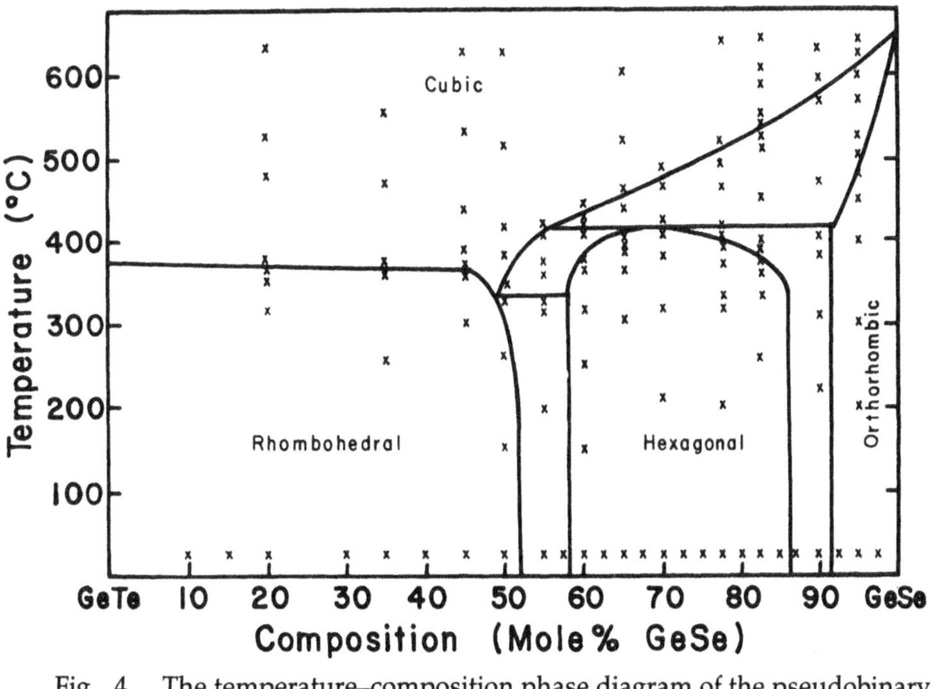

Fig. 4. The temperature–composition phase diagram of the pseudobinary GeSe–GeTe system. (The symbol × represents individual data points.)

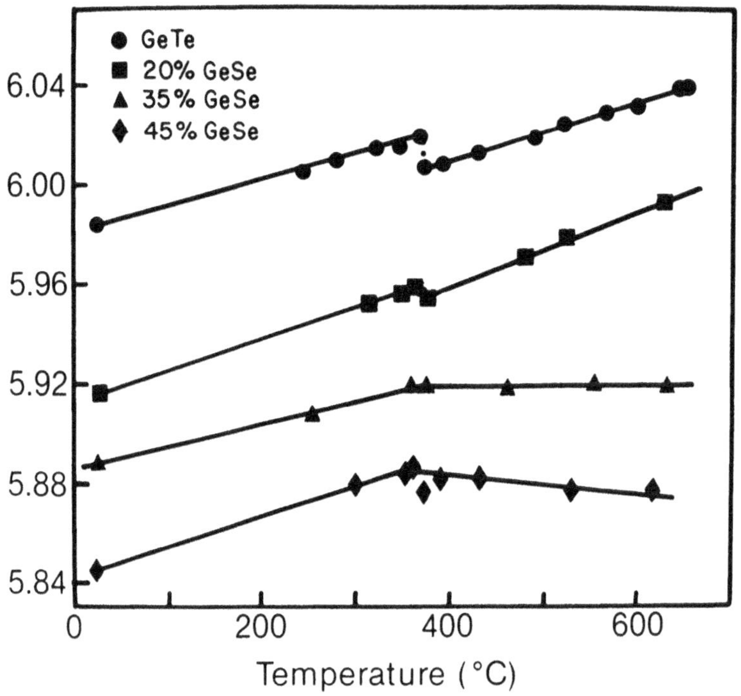

Fig. 5. The rhombohedral and cubic lattice parameter, *a*, as a function of temperature and composition for GeSe–GeTe solid solutions.

transformation from rhombohedral to cubic structure and subsequently increases with increasing temperature. However, for solid solutions with 35 and 45 mol% GeSe, the unit cell length remains unchanged at the transformation and the *a* parameter remains constant or decreases, above the transformation temperature, respectively. These results are in contrast to dilatometric measurements on GeSe–GeTe solid solutions (*6*), which indicated an increase in lattice dimensions at and above the rhombohedral-to-cubic phase transformation temperature. These discrepancies between X-ray diffraction and dilatometric observations could be resolved if the rhombohedral to cubic transformation were associated with a large increase in the vacancy concentration of GeSe–GeTe solid solutions. This effect could decrease the packing density by moving Ge atoms from the interior of a crystallite (forming vacancies) to its surface. In this case, a dilatometric measurement would reveal an increase in the size of a crystallite. X-ray diffraction measurements, however, might result in no change or in a decrease of the average size of a unit cell with an increase in vacancy concentration. The hypothesis of germanium vacancy formation during the rhombohedral-to-cubic transformation is consistent with the change in stoichiometry limits of GeTe at the transformation temperature (*12,13*). Below the transformation, the homogeneity limits of single-phase GeTe are 50.2 and 50.9 at% tellurium, and these limits change to 50.3 and 51.9 at% tellurium, respectively, above the transformation temperature. These data indicate that additional germanium is removed from the GeTe lattice for compositions near the germanium-rich homogeneity limit. The GeTe room temperature lattice parameter of 5.983 Å measured in this work indicates that the composition of GeTe used was very close to the germanium-rich homogeneity limit for GeTe (*12*).

SUMMARY AND LITERATURE CORRELATION

The results of the room temperature X-ray analysis of GeSe–GeTe solid solutions are in partial agreement with the results reported by Krebs and Langner (*4*) and are in excellent agreement with those of Muir and Cashman (*5*).

Krebs and Langner (*4*) reported a rhombohedral phase between GeTe and 63 mol% GeSe, whereas this work yielded a phase boundary of 52 ± 3 mol% GeSe. The room temperature limits of the orthorhombic phase reported earlier (*4*) (95 mol% GeSe) and observed in these studies (92 ± 3 mol% GeSe) are in good agreement. The hexagonal phase was not reported by Krebs and Langner (*4*). The results of the room temperature X-ray studies on GeSe–GeTe solid solutions reported by Muir and Cashman (*5*) are in excellent agreement with those of this work. The boundaries of the rhombohedral, hexagonal, and orthorhombic single-phases observed earlier (*5*) and in the present studies agree within limits of experimental error. Similarly good agreement is found for the unit cell constants of GeSe–GeTe hexagonal solid solutions at 25°C. Polytypes of the hexagonal phase reported previously (*5*) [usually caused by stacking faults in (0001) planes] were not observed in this work.

The temperature–composition phase diagram reported by Shelimova et al. (*6*) for GeSe–GeTe solid solutions bears almost no resemblance to the phase diagram determined by the X-ray studies of this work. It is possible that Shelimova et al. misinterpreted some of the dilatometric observations. At about 375 ± 10°C the phase diagram of Shelimova et al. indicates a phase transformation for compositions with GeSe mole fractions ranging from 0.92 to 0.10. In this work a rhombohedral to cubic transformation was observed at 375 ± 10°C for compositions ranging from GeTe to 46 ± 3 mol% GeSe. At the peritectoid temperature of 415 ± 5°C a tie-line exists in the GeSe–GeTe phase diagram for compositions ranging from 56 ± 3% to 92 ± 3% GeSe.

Muir and Beato (*7*) reported a phase diagram for the GeSe–GeTe system between room temperature and 500°C. Their work is in good agreement with our results except for the peritectoid decomposition temperature of the hexagonal phase. Muir and Beato (*7*) observed the decomposition at 450 ± 10 C, which differs from our value of 415 ± 5°C. The decomposition temperature of the hexagonal phase as determined by Muir and Beato (*7*) was obtained by differential scanning calorimetry. If the heating rate used was too high "for near-equilibrium" to be established in the calorimeter, superheating of the hexagonal phase could occur, resulting in transition temperature error. This is probably the reason for the discrepancy in the peritectoid transformation temperature since the rhombohedral–cubic transition temperature of 370 ± 10° reported by Muir and Beato (7) is in close agreement with that found in our work.

The results of the solid–solid phase equilibrium studies in the GeSe–GeTe system are in excellent agreement with those of the thermal expansion studies of GeSe (2) and GeTe (3). It has been demonstrated that the orthorhombic–cubic transformation of GeSe (2) and the rhombohedral–cubic transformation of GeTe (3) are consistent with the GeSe–GeTe phase diagram.

ACKNOWLEDGMENTS

The authors are pleased to acknowledge the support of this work by the National Aeronautics and Space Administration, Washington, DC, and through the George C. Marshall Space Flight Center, Huntsville, Alabama.

REFERENCES

1. Based on part of a thesis submitted by P. A. Siemers to the Graduate School of Rensselaer Polytechnic Institute in partial fulfillment of the requirements for the PhD degree.
2. Wiedemeier, H., and Siemers, P. A., *Z. Anorg. Allg. Chem.* **411,** 90 (1975).
3. Wiedemeier, H., and Siemers, P. A., *Z. Anorg. Allg. Chem.* **431,** 299 (1977).
4. Krebs, H., and Langner, D., *Z. Anorg. Allg. Chem.* **334,** 37 (1964).
5. Muir, J. A., and Cashman, R. J., *J. Phys. Chem. Solids* **28,** 1009 (1967).
6. Shelimova, L. E., Abrikosov, N. Kh., Zhdanova, V. V., and Sizov, V. V., *Izv. Akad. Nauk. SSSR, Neorg. Mater.* **2,** 2103 (1966).
7. Muir, J. A., and Beato, V., *J. Less Common Metals* **33,** 333 (1973).
8. Wiedemeier, H., and Siemers, P. A., to be published.
9. Wiedemeier, H., and Sigai, A. G., *J. Crystal Growth* **6,** 67 (1969).
10. Cullity, B. D., *Elements of X-Ray Diffraction,* Addison-Wesley, Reading, Mass. 1967, p. 335.
11. Sutton, L., ed., *Tables of Interatomic Distances and Configuration in Molecules and Ions,* Spec. Pubs. Nos. 11 and 18, The Chemical Society, London (1965).
12. Shelimova, L. E., Abrikosov, N. Kh., and Zhdanova, V. V., *Zh. Neorg. Khim.* **10,** 1200 (1965).
13. Dutta, S. N., and Jeffrey, G. A., *Inorg. Chem.* **4,** 1363 (1965).

0018–1536/84/1700–0409$02.00

Hydration and Dehydration of Calcium Oxide Powders Studied by Photoluminescence Spectroscopy

W. W. DULEY

Department of Physics, York University, Toronto, Ontario, Canada, M3J 1P3

Received September 23, 1983; Accepted December 31, 1983

ABSTRACT

Photoluminescence and photoexcitation spectroscopy have been used to characterize the surface properties of CaO powders containing low coordination OH^- species during outgassing at temperatures of up to 1200 K.

Index Entries: Hydration, of calcium oxide; dehydration, of calcium oxide; calcium oxide, photoluminescence studies of hydration and dehydration of; photoluminescence spectroscopy, of CaO hydration and dehydration.

INTRODUCTION

Photoluminescence spectroscopy has recently developed into a useful probe of the surface structure (*1,2*) and photocatalytic properties (*3,4*) of highly dispersed oxides. Much attention has been paid to the alkaline earth oxides in particular, with a view to the characterization of the properties of oxygen ions in low coordination states (*1,2,5,6*). Such centers are abundant on the surface of small particles of MgO, CaO, and SrO that have been activated by heating to temperatures in excess of 900 K. Reflectance spectroscopy (*7,8*) has shown that O^{2-} ions in low coordination states (O^{2-}_{LC} ions) are highly reactive, with the degree of reactivity with respect to a specific gaseous molecule being closely related to coordination.

Zecchina et al. (*7*) have studied the effect of H_2O adsorption on O^{2-}_{LC} ions on the surface of MgO and CaO powders using reflectance spectroscopy and find that the reaction of H_2O with the surface is progressive, with the first attack occurring at sites of threefold (O^{2-}_{3C}) or fourfold (O^{2-}_{4C}) coordination. Coluccia et al. (*9*) have used photoluminescence spectroscopy to study the hydration of activated MgO powder. Samples were excited with lines from a Hg lamp. In a recent set of experiments Duley has examined the thermal decomposition of $Mg(OH)_2$ using photoexcitation and photoluminescence spectroscopy (*10*). This work showed that the luminescence from O^{2-}_{LC} and OH^- ions on the surface of $Mg(OH)_2$–MgO powders was a sensitive function of outgassing temperature.

To date there has been no report of similar data on the CaO–H_2O system. In this paper I discuss the results of a preliminary investigation into the effect of H_2O adsorption on activated CaO powders obtained with the photoluminescence technique. Both photoexcitation and luminescence spectra have been obtained from CaO powders outgassed at 1200 K, exposed to H_2O vapor at 300 K, and then dehydroxylated by further heating to 1200 K.

EXPERIMENT

The starting CaO powder was prepared by the thermal decomposition of Johnson-Mathey "specpure" $CaCO_3$ in a spectrosil cell attached to a vacuum line pumped with a mercury diffusion pump. The base pressure in this system was 5×10^{-7} torr and was typically $\sim 10^{-5}$ torr during decomposition of the $CaCO_3$ powder. Water vapor was obtained from deionized liquid H_2O purified by several freeze–pump–thaw cycles in a separate vacuum line. Adsorption of H_2O was carried out in a sealed system with both the CaO sample and H_2O liquid at 300 K.

Photoluminescence and photoexcitation spectra were obtained by standard techniques (*1,2,6*). A 250 W xenon lamp was used as a light source. Luminescent emission was detected with a 9798B photomultiplier tube. All spectra were recorded with the sample at 77 K.

RESULTS AND DISCUSSION

Photoexcitation spectra recorded at various luminescent wavelengths, λ_L for clean CaO and for CaO with varying degrees of dehydroxlyation are shown in Fig. 1. The photoexcitation peaks near 4.4 eV (282 nm) in clean CaO have been previously identified with an electronic transition of surface O^{2-} ions in sites of fourfold coordination (*1,6*). The photoexcitation spectrum recorded for clean CaO with λ_L = 410 nm shows two peaks in this region (4.31 and 4.54 eV, Fig. 1b).

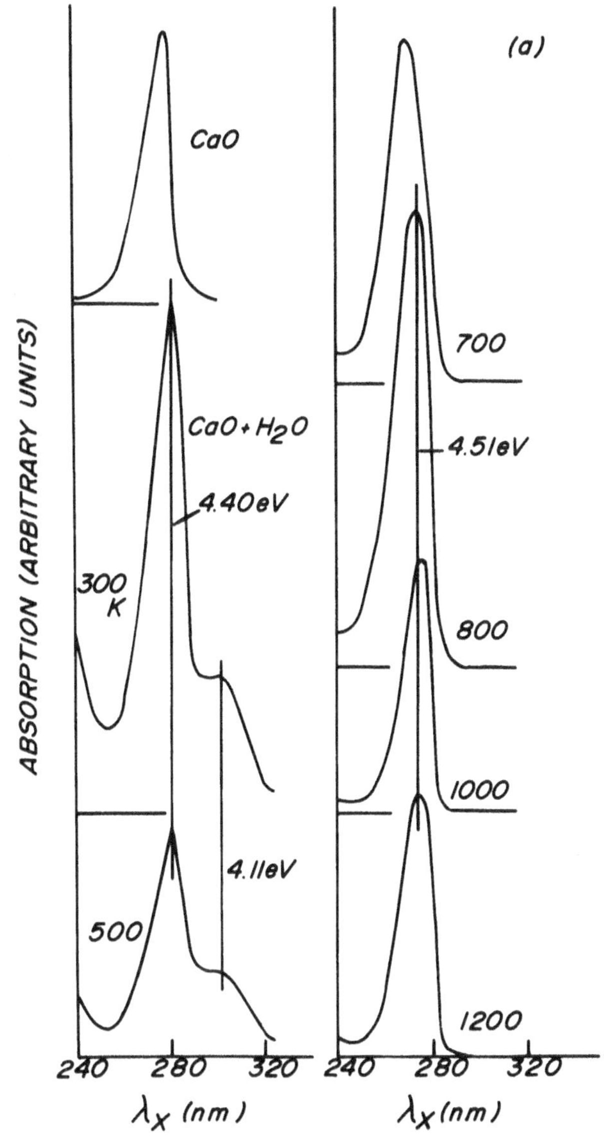

Fig. 1. Photoexcitation spectra of CaO powders recorded at 77 K with luminescence at (a) λ_L = 350 nm, (b) λ_L = 410 nm, and (c) λ_L = 480 nm. Upon exposure to H_2O vapor, samples were outgassed in vacuum by heating to the temperature shown before cooling to 77 K for recording of spectra.

These are associated with O^{2-}_{4C} ions in sites with differing Ca^{2+}_{LC} environments (*6*) and hence with different reactivities.

The adsorption of H_2O on activated CaO powders followed by subsequent outgassing at 1200 K was found to lead to sintering with an attendant reduction in the surface area of the powder. The effect of sintering can be seen in spectra of CaO + H_2O outgassed at 1200 K (Fig. 1b,c), which show that the 4.3 eV photoexcitation peak present in clean

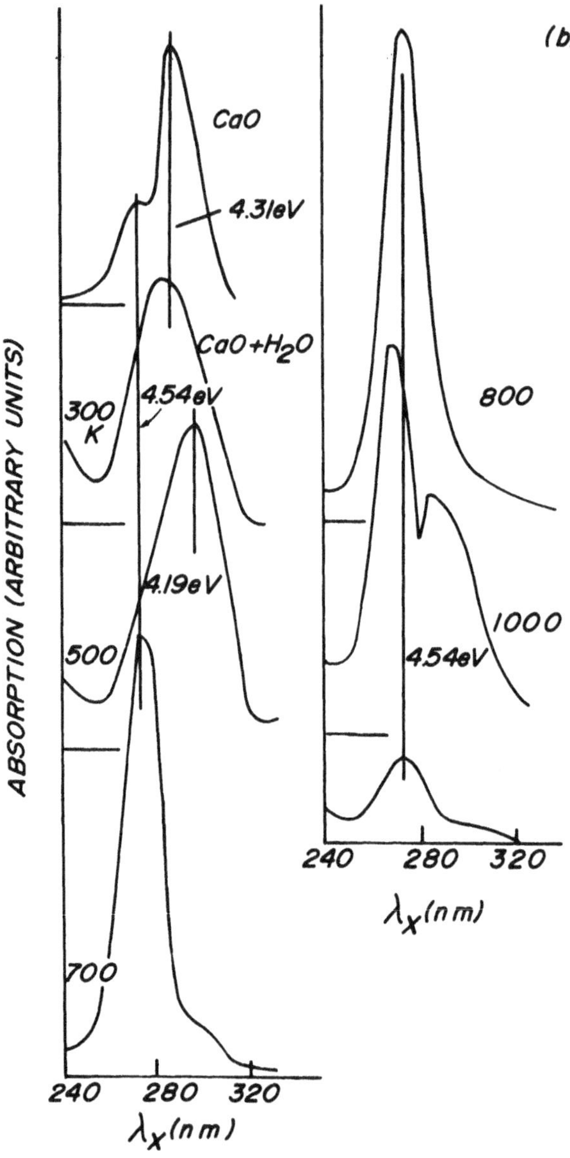

Fig. 1b.

CaO samples has been eliminated by H_2O adsorption followed by outgassing at 1200 K. The 4.6 eV peak evidently arises from a subset of O^{2-}_{LC} ions that are more resistant to sintering since it is little affected by the above treatment.

The addition of H_2O to clean CaO can be seen to produce a broad asymmetric photoexcitation peak at 4.54–4.57 eV in Figs. 1b and 1c. This is superimposed on a sloping background increasing to shorter wavelengths. This background is also apparent in scans with $\lambda_L = 350$ nm although here the 4.54–4.57 eV peak is not obvious. Instead a weaker feature occurs at 4.11 eV (Fig. 1a).

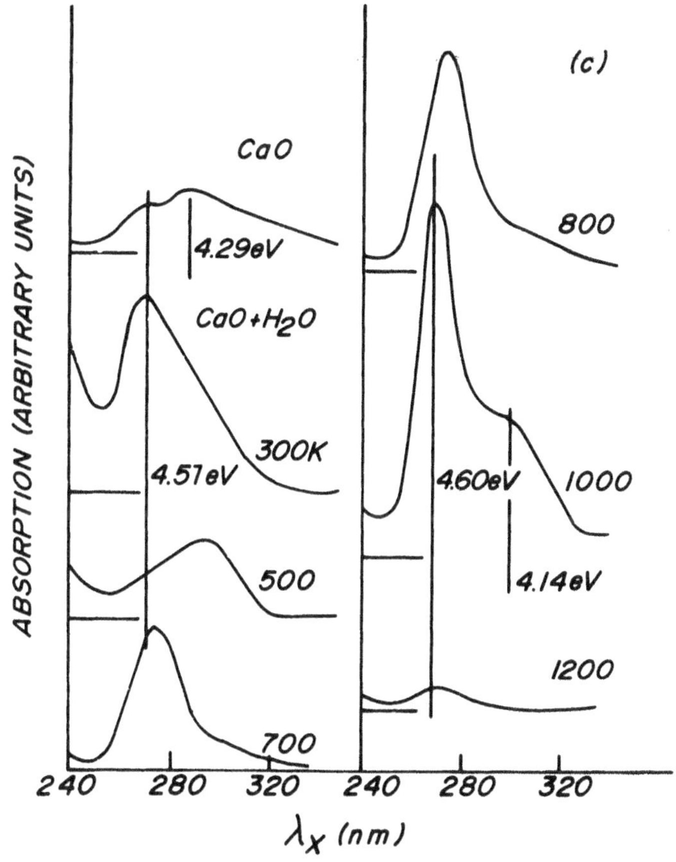

Fig. 1c.

By analogy with the spectra of $Mg(OH)_2$, the 4.54–4.57 eV photoexcitation peak can be associated with the presence of small amounts of $Ca(OH)_2$ created by exposure of CaO to H_2O vapor. Freund and Scharpen (*11*) have observed a transition at 4.6 eV in the X-ray stimulated photoemission spectrum of $Mg(OH)_2$ powders, which is also seen in photoexcitation spectra (*1*). Hence the peak at 4.54–4.57 eV in the present spectra likely arise from $Ca(OH)_2$ molecules in or on the CaO powders. The 4.1–4.2 eV peak appearing in some spectra would appear to be of similar origin. It likely arises from $Ca(OH)_2$ molecules in a different environment from that which yields the 4.54–4.57 eV feature.

Both the 4.1–4.2 and 4.54–4.57 eV photoexcitation features disappear as hydrated CaO powder is outgassed at high temperature. This presumably occurs when $Ca(OH)_2$ dissociates via the process $OH^-_{LC} \rightarrow O^-_{LC} + H$ followed by $2O^-_{LC} \rightarrow O^{2-}_{LC} + O$ as proposed by Martens et al. (*12*) for the $Mg(OH)_2$ system. The degree to which such a reaction has proceeded for a specific subset of low coordination OH^-_{LC} ions at a particular temperature will depend on temperature. It is expected that low outgassing temperature will release hydrogen from OH^-_{LC} sites of highest coordination, i.e., from OH^-_{6C} and OH^-_{5C} sites. Higher temperatures will be required to liberate hydrogen from the more reactive OH^-_{4C} and OH^-_{3C} low coordination sites.

Photoexcitation spectra obtained with $\lambda_L = 350$ nm have been shown (*1,6*) to sample absorption from O^{2-} ions in high coordination sites. As can be seen from Fig. 1a, outgassing to 700 K removes all trace of hydroxide absorption from the spectrum, in keeping with the thermal release of hydrogen from high coordination sites. On the other hand,

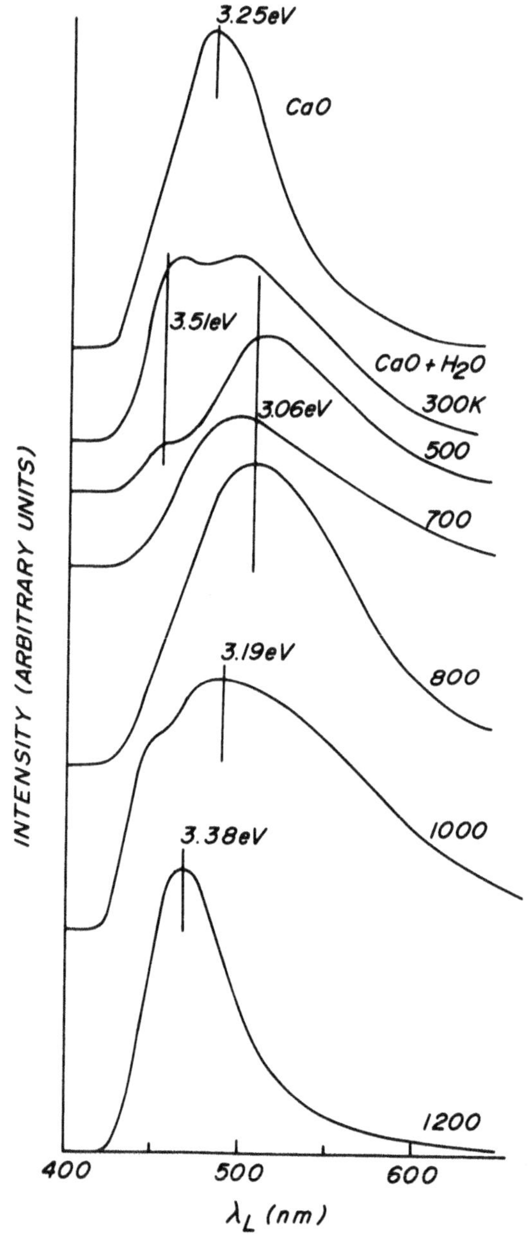

Fig. 2. Luminescence spectra of hydrated CaO powders outgassed at various temperatures. Spectra recorded at 77 K with $\lambda_X = 280$ nm.

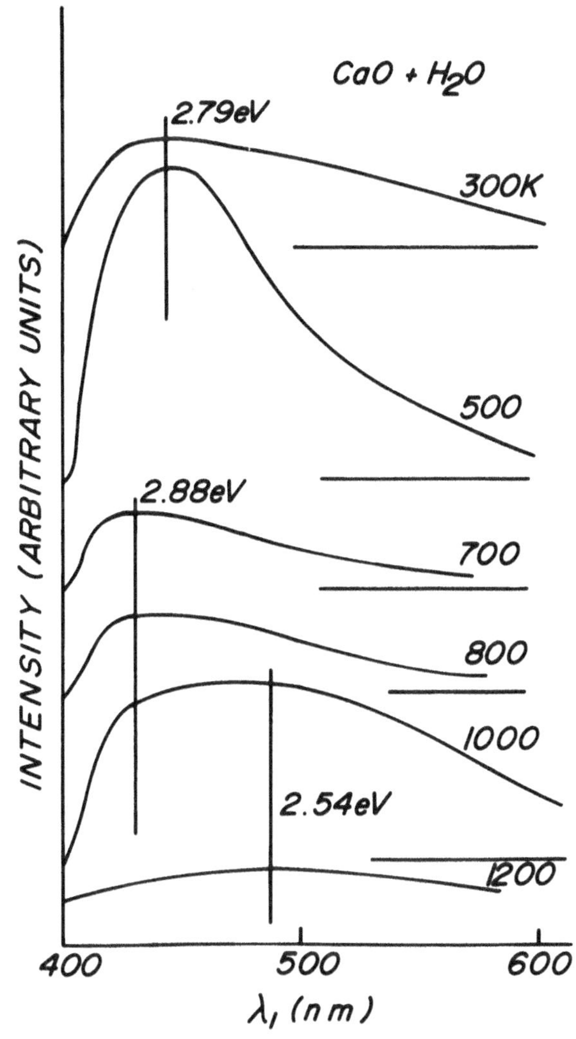

Fig. 3. As Fig. 2, but with λ_X = 350 nm.

spectra obtained with λ_L = 410 or 480 nm sample sites of lower coordination to which hydrogen is more strongly bound (*3,6*). As expected, the effect of hydroxide ion absorption can be seen to higher temperatures in these spectra.

Some evidence for surface reconstruction, driven perhaps by reaction of lattice hydroxide with low coordination O^-_{LC} sites can be seen in the 1000 K spectra of Fig. 1c. This spectrum is almost identical with that shown in the 300 K curve of Fig. 1a. It suggests that water vapor or hydroxide released during the final stages of outgassing may re-react with certain surface sites to reform trace amounts of $Ca(OH)_2$.

Luminescence spectra of CaO, $Ca(OH)_2$ powders are shown in Figs. 2 and 3. Clean CaO shows a broad single peak centered at 3.25 eV when excited with λ_X = 280 nm radiation. The addition of H_2O followed by outgassing at 300 K eliminates this peak and generates features at 3.51 and 3.06 eV. Both peaks are likely related to the presence of OH^-, but it is uncertain whether either arises from an intrinsic excitation of this ion. By analogy with similar spectra for rehydrated MgO (*10*), the 3.06 eV may be identified with emission from O^{2-}_{4C} ions perturbed by neighboring OH^- groups. Outgassing of rehydrated samples to 1200 K leads to a shift of the emission peak to 3.38 eV, providing further evidence for sintering and a reduction in the proportion of surface O^{2-}_{3C} ions.

Clean CaO powder excited with 350 nm radiation shows only a weak luminescence peak at ~2.6 eV (*6*). Hydration followed by outgassing at 300 K yields the luminescence spectrum shown in Fig. 3 on excitation with 350 nm light. An intense emission centered at ~2.8 eV appears in this spectrum and persists on outgassing to ~1000 K. Heating above 1000 K eliminates this peak and reconstitutes the 2.6 eV feature characteristic of clean CaO. It is possible that the 2.8 eV peak arises from OH^- emission at sites of low coordination although luminescent emission from the hydroxyl radical should also occur in this energy range. Further information on this subject could be obtained from combined photoluminescence-infrared absorption experiments.

ACKNOWLEDGMENT

This research was supported in part by a grant from the NSERCC.

REFERENCES

1. Coluccia, S., Deane, A. M., and Tench, A. J., *J. Chem. Soc. Faraday 1* **74,** 2913 (1978).
2. Coluccia, S., Tench, A. J., and Segal, R. L., *J. Chem. Soc. Faraday II* **75,** 1769 (1979).
3. Coluccia, S., and Tench, A. J., *Proc. Int. Congr. Catal.*, 7th B35 (1980).
4. Che, M., and Tench, A. J., in *Advances in Catalysis,* Eley, D. D., Pines, H., and Weisz, P. R., eds., Academic Press, New York, vol. 31, 1982, p. 78.
5. MacLean, S., and Duley, W. W., *J. Phys. Chem. Sol.* **45,** 227 (1984).
6. Duley, W. W., *Phil. Mag. B,* **49,** 159 (1984).
7. Zecchina, A., Lofthouse, M. G., and Stone, F. S., *J. Chem. Soc. Faraday I* **71,** 1476 (1975).
8. Zecchina, A., and Stone, F. S., *J. Chem. Soc. Faraday I* **72,** 2364 (1976).
9. Coluccia, S., Deane, M., and Tench, A. J., *Proc. 6th Int. Congr. Catalysis* **1,** 171 (1976).
10. Duley, W. W., *J. Chem. Soc. Faraday I,* in press (1984).
11. Freund, F., and Scharpen, L. H., *J. Electron. Spectros. Rel. Phenom.* **3,** 305 (1974).
12. Martens, R., Gentsch, H., and Freund, F., *J. Catalysis* **44,** 366 (1976).

0018–1536/84/1700–0417$02.00

Reactions of Transition Metal Atoms with 2-Nitropropane

A. B. M. ANWARUL BASHAR AND PETER L. TIMMS*

School of Chemistry, University of Bristol, BS8 1TS, UK

Received December 6, 1983; Accepted December 31, 1983

ABSTRACT

Atoms of Cr, Mn, Fe, Co, Ni, and Cu formed by vacuum vaporization of the metals react with 2-nitropropane at −196°C. The products isolated at 0°C are new propane-2-nitronato complexes of the metals, different from those made by conventional metathesis reactions. Cr, Mn, and Fe form basic complexes $[M\{O_2NC(CH_3)_2\}_2OH]$ with additional H_2O in the lattice, Co gives $[Co\{O_2NC(CH_3)_2\}_2]$, and Cu thermally and air-sensitive $[Cu_4O\{O_2NC(CH_3)_2\}_2]$. Ni gives two propane-2-nitronato complexes, one of which contains coordinated NO. Reduction and rearrangements occur in the reactions of Mo and Ru atoms with 2-nitropropane. The results suggest that using high temperatures to form metal atoms may open a new chapter in "classical" coordination chemistry.

Index Entries: Transition metal atoms, reactions with 2-nitropropane; 2-nitropropane, reactions with transition metal atoms; metals, reactions with 2-nitropropane; chromium, reaction with 2-nitropropane; manganese, reaction with 2-nitropropane; iron, reaction with 2-nitropropane; cobalt, reaction with 2-nitropropane; nickel, reaction with 2-nitropropane; copper, reaction with 2-nitropropane.

INTRODUCTION

The technique of carrying out reactions at low temperatures on gaseous chemical species formed at high temperatures has had an influence

*Author to whom all correspondence and reprint request should be addressed.

on many areas of chemistry over the last 20 years (*1,2*). Gaseous transition metal atoms have been condensed at low temperatures with a wide range of ligands that stabilize the metals in low oxidation states (*2,3*). Less work has been done on the reactions of transition metal atoms with compounds that can form "classical" coordination complexes in which the metal atom has been both oxidized and complexed. One study showed that reactions of metal atoms with acetylacetone gave known metal acetylacetonates (*4*). We have been interested in the reactions of transition metal atoms with 2-nitropropane. This compound has the potential to oxidize the metal atoms either through release of oxygen from the nitrogroup or by loss of its α-hydrogen atom and complexation of the metal by the resulting propane-2-nitronato group. Studies by Gladysz on the reaction of chromium atoms with nitrobenzene showed that the metal abstracted oxygen to give nitrosobenzene and chromium oxides (*6*).

A range of transition metal compounds of 2-nitropropane is known; these have been prepared by metathesis reactions using the alkali metal propane-2-nitronates and transition metal salts (*7*). Few of these compounds have been obtained in a crystalline state, but the structure of copper(II) propane-2-nitronate has been reported (*8*).

EXPERIMENTAL

Each of the metals Cr, Mn, Fe, Co, Ni, and Cu was evaporated from a resistively heated alumina crucible inside an evacuated 250-mm-diameter stainless-steel reaction vessel with liquid nitrogen-cooled walls. The apparatus has been described in detail elsewhere (*9*). During each experiment lasting about an hour, 30 mmol of the vapor of the metal was condensed on to the cold walls with the vapor from 300 mmol of dry 2-nitropropane (containing 1% 1-nitropropane). The pressure in the reactor was kept below 3×10^{-5} torr during the codeposition. A similar procedure was used for experiments with Mo and Ru except that the metal was evaporated by electron bombardment heating using an electrostatically focused gun (G.V. Planer). At the end of a run, the reaction vessel was filled with gaseous nitrogen and warmed to 0°C. A solution of the metal/2-nitropropane reaction product dissolved in excess 2-nitropropane was then sucked out of the reactor, filtered, and concentrated by vacuum evaporation of part of the solvent. In the majority of cases, the metal-containing product was precipitated from the 2-nitropropane solution by addition of hexane and was then filtered off, washed, and dried under vacuum. The products thus obtained were all amorphous powders and attempts to crystallize them using mixed solvent, solvents diffusion, and slow cooling all failed. The special problems with the Co and Ni products are described in the Results section. Yields of the products isolated were in the range 60–80% of theoretical based on the metal evaporated.

The products were routinely handled under nitrogen and treated only with dry, degassed solvents and reagents; work on the Cu product was carried out at 0°C to avoid its decomposition. Analytical data for the products is summarized in Table 1. Analyses for C, H, and N were performed by our Microanalytical Laboratory and metals were analyzed by standard gravimetric or volumetric procedures. Infrared spectra were taken in Nujol mulls or as thin films evaporated from 2-nitropropane solution, using a Perkin-Elmer 457 spectrometer. Ultraviolet/visible spectra were taken by solid reflectance using a Beckmann DK-2A spectrophotometer. Molecular weights were determined by vapor pressure depression in $CHCl_3$ solution. Magnetic measurements were made on the solids at room temperature using a simple Guoy balance or on 2-nitropropane solutions of the complexes by Evan's method (*10*) with a JEOL FX200 spectrometer at temperatures down to −80°C.

The Mössbauer spectrum on the Fe product was measured in the Department of Chemistry, Birkbeck College, London by I. Sayer and B.W. Fitzsimmons. The product was cooled to −196°C and irradiated from a ^{57}Co source. The spectrum showed two equal peaks from which the following data were determined: isomer shift +0.4841 ± 0.001 mm/s; quadrupole splitting 0.9691 ±0.002 mm/s; line width 0.57 ±0.01 mm/s.

RESULTS AND DISCUSSION

Atoms of Cr, Mn, Fe, Co, Ni, and Cu were condensed with a large excess of the vapor of 2-nitropropane on a surface cooled by liquid nitrogen. The condensates were allowed to warm to 0°C and soluble products were extracted in 2-nitropropane to give colored solutions. With the exception of the cobalt product, the compounds isolated from the solutions were noncrystalline.

The products were characterized by elemental analysis, infrared and ultraviolet/visible spectroscopy, and magnetic measurements (see Table 1). The compounds isolated were found to be propane-2-nitronato complexes of the metals; some also contained hydroxo or oxo ligands (i.e., they were basic complexes) and coordinated or lattice water. Treatment of the compounds with acetylacetone at room temperature or above gave metal acetylacetonates and liberated 2-nitropropane; water was also liberated from the basic compounds. The identity of the compounds as propane-2-nitronates was shown most conclusively by their infrared spectra in which strong bands at 1650–1600 cm^{-1} (ν C=N), 1170–1100 cm^{-1} (ν_{asym} NO_2), and 950–940 cm^{-1} (ν_{sym} NO_2) were diagnostic for the propane-2-nitronato group (*7*). Many of the products also showed strong, broad νOH bands and bands in the 600–300 cm^{-1} region associated with νM–O modes.

Chromium and iron atoms yielded complexes of Cr(III) (grey-green powder) and Fe(III) (red-brown powder) of formula $[M\{O_2NC(CH_3)_2\}_2OH]$ with some additional lattice water. The magnetic

TABLE 1
Analytical Data on Metal Atom/2-Nitropropane Reaction Products

Metal		Product composition, %				Product formula [E = $O_2NC(CH_3)_2$]	Infrared bands (cm^{-1})				Uv/visible bands, nm	U_{eff}, BM	Molecular weight
		Metal	C	H	N		$\nu C{=}N$	$\nu_{asym}NO_2$	$\nu_{sym}NO_2$	$\nu O{-}H$			
Cr	Found	20.95	28.09	5.70	11.25	$CrE_2OH \cdot \frac{1}{2}H_2O$	1646s	1159s	950m	3880s	578, 400, 340	3.70[a]	~2300
	Calc.	20.46	28.33	5.55	11.02		1626sh	1124s			240	3.89[b], 3.87[c]	
Fe	Found	22.25	28.16	5.32	11.36	$FeE_2(OH)$	1630s	1160s	945m	3315m	924, 470, 430	3.81[a]	—
	Calc.	22.43	28.92	5.26	11.24		1622sh	1133s			400, 240		
Mn	Found	21.95	27.12	5.39	10.92	$MnE_2OH \cdot \frac{1}{2}H_2O$	1641sh	1159s	945m	3300s	830, 500, 470	5.20[a]	~1800
	Calc.	21.37	28.01	5.49	10.89		1635s	1130s			360, 300	5.91[b], 5.72[c]	
Co	Found	24.95	30.55	5.31	11.58	CoE_2	1641vs	1146s	945m	—	1320, 840, 588	4.90[a]	—
	Calc.	25.10	30.65	5.11	11.92			1110s			552, 492, 300		—
Ni	Found	32.96	19.35	4.80	7.76	$NiE(OH) \cdot H_2O$	1638sh	1158s	940m	3300s	1096, 646, 448	3.42[a]	—
	Calc.	32.31	19.81	4.95	7.70		1628s	1140sh			340		
	Found	24.70	29.41	5.67	12.55	$(NiE_2)_5NO \cdot H_2O$	1638sh	1153s	938m	3300m	—	—	—
	Calc.	23.86	29.46	5.12	12.51		1628s	1125s					
Cu	Found	55.03	15.78	2.95	6.20	Cu_4OE_2	1608s	1153s	937s	—	~250(broad)	0.45[b]	—
	Calc.	56.95	16.13	2.69	6.27			1114s				1.05[a, d]	—
	Found	36.20	21.71	4.16	9.44	$CuE(OH)$	1650sh	1155s	937s	3160m	650, 397, 270	1.20[a]	~4500
	Calc.	37.60	21.37	4.15	8.31		1643s	1116s				1.30[b], 1.17[c]	

[a]Magnetic moment measured with a Gouy balance at 18°C.
[b]Magnetic moment measured by Evan's method at 29°C.
[c]Magnetic moment measured by Evan's method at −80°C.
[d]Magnetic moment taken after auto-oxidation from Cu(I) to Cu(II).

moment of the Fe complex at 18°C was 3.81 BM, much lower than the spin-only value for a high-spin d^5 ion, but the +3 oxidation state was indicated by the ultraviolet/visible spectrum and by the Mössbauer spectrum, which showed an isomer shift and quadrupole splitting typical of Fe(III). The Mn atom/2-nitropropane reaction was more complex. The character of the product varied from run to run unpredictably. From some runs the product was deep red-brown in color and reacted with acetylacetone to give $Mn(acac)_3$; from other runs the product was brown and reacted with acetylacetone to give a little $Mn(acac)_3$ but more $Mn(acac)_2$. Analytical data for the two types of products was similar and it would appear that the product was a mixture of varying amounts of $[Mn(II)\{O_2NC(CH_3)_2\}_2]$ and $[Mn(III)\{O_2NC(CH_3)_2\}_2OH]$. The reason for the variability is not known, but oxidation of Mn(II) to Mn(III) may have occurred on the cold surface because of (variable) small amounts of nitrous acid vapor formed from thermal decomposition of 2-nitropropane (*11*).

In the case of cobalt, the 2-nitropropane solution taken from the reactor, after concentration by evaporation, deposited a few blue crystals on standing at 0°C for 12 h. These proved to be insoluble in 2-nitropropane and to be of rather variable composition. The main product, $[Co\{O_2NC(CH_3)_2\}_2]$ was precipitated from the mother liquor by addition of diethyl ether; it formed blue microcrystals and it too was insoluble in 2-nitropropane or in other inert solvents, and could not be recrystallized. Its infrared spectrum was very sharp compared to the noncrystalline products from other metals; its ultraviolet/visible spectrum showed that it contained octahedrally coordinate Co(II).

The nickel atom/2-nitropropane product was a blue-green solid soluble in organic solvents. Its infrared spectrum showed that it contained propane-2-nitronato groups, but there was an additional band at 1756 cm^{-1} typical of νNO of nitrosyl complexes. Analysis indicated a $Ni:O_2NC(CH_3)_2$ ratio of about 1:2, but with a high nitrogen content. The product was revealed as a mixture by treatment with water and extraction with toluene. A green aqueous layer formed, yielding a green solid $[Ni\{O_2NC(CH_3)_2\}OH]$, presumably formed by partial hydrolysis of $[Ni\{O_2NC(CH_3)_2\}_2]$. A blue solid was obtained from the toluene extract and this showed a strong νNO band at 1756 cm^{-1}; analysis indicated a composition close to $[Ni\{O_2NC(CH_3)_2\}_5(NO)(H_2O)]$

The Cu atom/2-nitropropane reaction product was thermally unstable. The product isolated at 0°C was a basic Cu(I) propane-2-nitronate of composition $[Cu_4O\{O_2NC(CH_3)_2\}_2]$, but at room temperature this changed rapidly to a Cu(II) propane-2-oximate without change in composition, i.e., there was formally a shift of an oxygen atom.

$$Cu\{O_2NC(CH_3)_2\} \xrightarrow{>0°C} —O—Cu\{ONC(CH_3)_2\}$$

The change was seen in the infrared spectrum by replacement of a sharp band at 1608 cm^{-1} [νC=N for a Cu(I) propane-2-nitronate] by a broad band centred at 1620 cm^{-1} [νC=N for a Cu(II) propane-2-oximate] and by corresponding changes in the 1200–1000 cm^{-1} region, especially the growth of a band at 1080 cm^{-1} caused by the oximato-complex. The Cu(I) propane-2-nitronate was also extremely oxygen-sensitive and reaction with dry O_2 yielded a basic Cu(II) propane-2-nitronate with a strong νC=N band at 1643 cm^{-1}. A freshly prepared sample of the Cu(I) product was virtually diamagnetic, but the Cu(II) propane-2-oximato and propane-2-nitronato complexes had magnetic moments of 1.05 and 1.30 BM, respectively, appreciably lower than the spin-only value for a d^9 ion.

2-Nitropropane was reacted with atoms of two Second Row transition metals, Mo and Ru. The reactions appeared less simple than those with the First Row transition metal atoms. With molybdenum, the main product appeared to be a propane-2-oximato-oxo complex of Mo(V), but the C:N ratio was less than 3:1 showing the presence of additional nitrogen (possibly as NH groups judging by the complexity of the infrared spectrum in the 3500–3000 cm^{-1} region). With ruthenium, the infrared spectrum of the product was dominated by bands in the region 1870–1820 cm^{-1}, presumably associated with νNO for a nitrosyl and there were also bands around 2000 cm^{-1} (νRu-H ?), but the fate of the organic moiety from 2-nitropropane is not yet known.

Assignment of structures to the reaction products is very difficult in the absence of X-ray diffraction data. In the known structure of $[Cu\{O_2NC(CH_3)_2\}_2]$ (*8*), the propane-2-nitronato groups act as both chelating and bridging ligands. This mixed bonding mode gives rise to splitting of the ν_{asym} NO_2 infrared bands in the 1160–1100 cm^{-1} region, whereas the alkali metal propane-2-nitronates, with chelating ligands, show only one ν_{asym} NO_2 band (*12*). All the propane-2-nitronates we have made show two bands in the 1160–1100 cm^{-1} region, implying that they all contain both chelating and bridging propane-2-nitronato ligands or propane-2-nitronato groups acting simultaneously as chelating and bridging ligands, i.e.

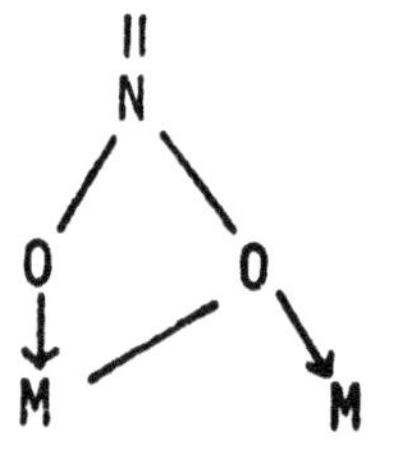

This last mode of bonding occurs to a small extent in $[Cu\{O_2NC(CH_3)_2\}_2]$ in which there is a longer range interaction giving the copper distorted octahedral coordination.

The other problem in assigning structures is the difficulty in distinguishing between OH groups acting as terminal or bridging ligands and a mixture of bridging oxo groups and coordinated water. The δ OH band of water at 1600–1650 cm^{-1}, is obscured in most cases by the strong νC=N band of the propane-2-nitronato groups, whereas in all cases the νOH region shows only broad bands.

In cases for which measurements have been possible, the products have been shown to have high molecular weights (2000–5000). This fact, coupled with their lack of crystallinity suggests chain structures with random chain lengths. Thus, the chromium product may have a structure like that in Fig. 1 with an additional 1 mol H_2O/2 Cr atoms, hydrogen-bonded within the lattice.

The structure of the anhydrous cobalt product may be more regular with all the propane-2-nitronato groups in the same environment, as shown in Fig. 2. The νC=N band in the infrared spectrum was much sharper than any of the νC=N bands for other products and showed no shoulders, whereas the ν_{asym} NO_2 bands were not much sharper than the corresponding bands for other products. This supports the structure shown. Infrared spectroscopy suggests that the structure of the other propane-2-nitronato complexes are similar, but not identical to those shown in the figures with slightly different bonding arrangements for the propane-2-nitronato ligands and any hydroxo or oxo ligands.

In the vapor phase, 2-nitropropane occurs exclusively in the nitro form, but as a liquid it is in equilibrium with a low concentration of the nitronic acid form $(CH_3)_2C{=}N(O)OH$. It is probable that the initial reac-

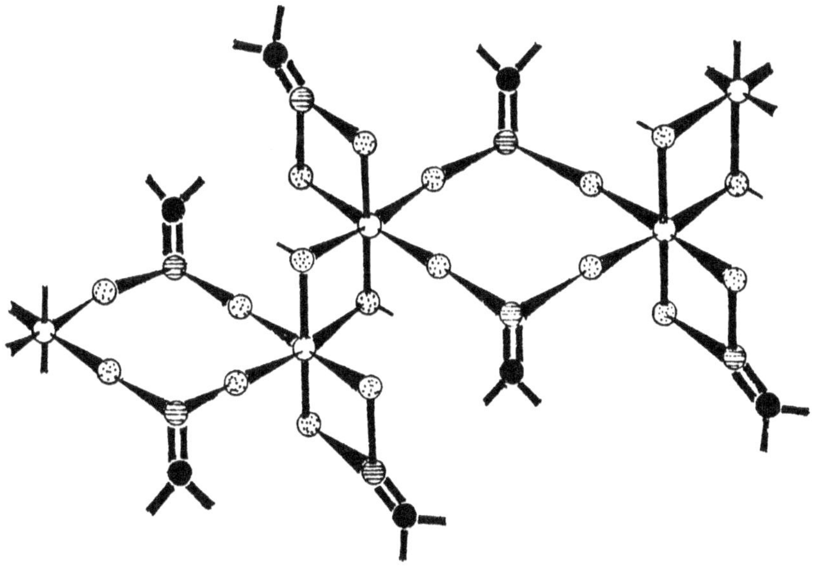

○ = chromium, ⊙ = oxygen, ⊜ = nitrogen, ● = carbon, ⊙ = hydroxo

Fig. 1. Proposed structure of $[Cr\{O_2NC(CH_3)_2\}(OH)] \cdot ½H_2O$ (lattice water omitted).

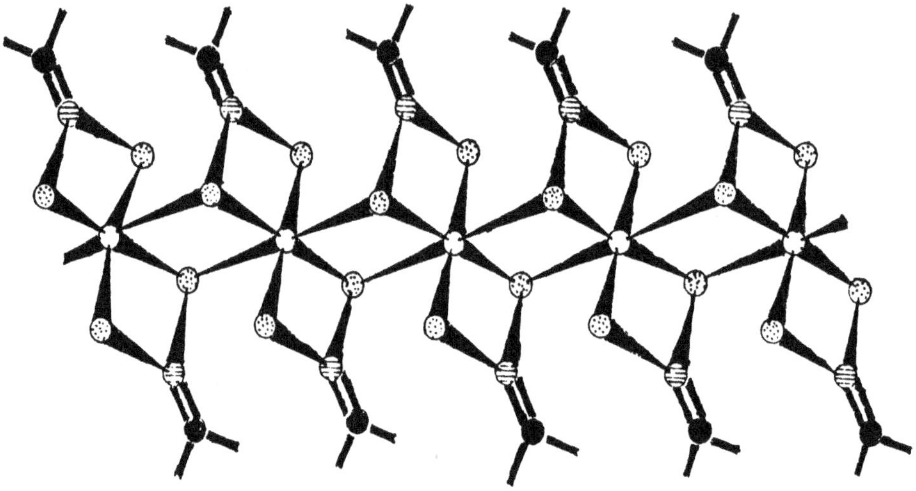

○ = cobalt, ⊙ = oxygen, ⊜ = nitrogen, ● = carbon

Fig. 2. Proposed structure of $[Co\{O_2NC(CH_3)_2\}_2]$.

tion of a transition metal atom with excess 2-nitropropane at low temperatures is to form a weak complex of type

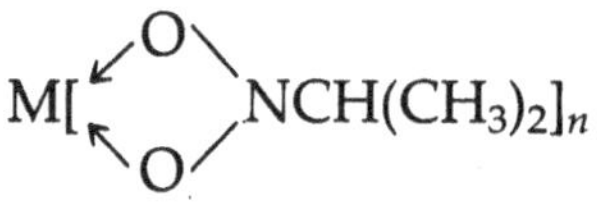

The conversion of this to the observed (basic) propane-2-nitronates may occur on warm-up by reaction with propane-2-nitronic acid in the excess 2-nitropropane. No hydrogen evolution has been observed and there is clearly a series of redox reactions leading to the final products. Schorpp and Beck proposed a quite different mechanism for the reaction of 2-nitropropane with $Pt(PPh_3)_4$ (*13*), involving an intermediate containing the moiety

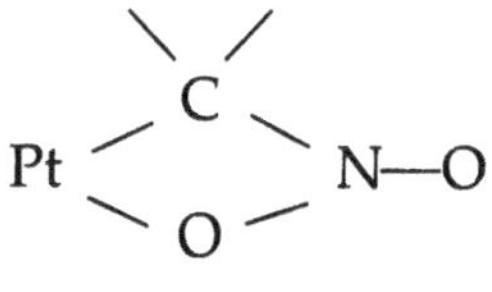

that lost propene to give their final product $[cis\text{-}Pt(NO_2)_2(PPh_3)_2]$. This type of mechanism may be involved in the nickel atom/2-nitropropane reaction in which a complex containing a Ni—NO group was one of the products.

Our products all appear to be slightly different in molecular formula and structure from the propane-2-nitronates prepared by metathesis from transition metal salts and alkali metal propane-2-nitronates. We believe that this will emerge as the pattern for reactions of transition metal atoms with many other ligands that, like 2-nitropropane, are capable of oxidizing and complexing the metal atoms. If this proves true, high temperature chemistry will soon be seen as having a profound influence on coordination chemistry. There is also the prospect that where known complexes have practical applications, the related but different complexes accessible via metal atoms will be more useful in some cases.

ACKNOWLEDGMENTS

We wish to acknowledge helpful discussions with Dr. N. Winterton and to thank Imperial Chemical Industries for research support.

REFERENCES

1. Timms, P. L., *Advances in Inorg. Chem. and Radiochem.* **14,** 121 (1972).
2. Klabunde, K. J., *Chemistry of Free Atoms and Particles,* Academic Press, New York, 1980.
3. Blackborow, J. R., and Young, D., *Metal Vapour Synthesis in Organometallic Chemistry,* Springer-Verlag, 1979.
4. Blackborow, J. R., Eady, C. R., Koerner von Gustorf, E. A., Scrivanti, A., and Wolfbeiss, O., *J. Organomet. Chem.* **108,** C32 (1976).
5. Bashar, A. B. M. A., PhD Thesis, University of Bristol, 1983.
6. Gladysz, J. A., Fulcher, J. G., and Tagashi, S. *Tetrahedron Lett.* 521 (1977).
7. Cook, J. A., Osborne, M. J., and Rice, D. A., *J. Inorg. Nucl. Chem.* **38,** 711 (1976).
8. Simonsen, O., *Acta Crystallogr.* **B29,** 2600 (1973).
9. Timms, P. L., *J. Chem. Soc. A.* 2526 (1970); also see ref. (3).
10. Evans, D. F., *J. Chem. Soc.,* 2003 (1959).
11. Spokes, G. N., and Benson, S. W., *J. Am. Chem. Soc.* **89,** 6030 (1967).
12. Brooks, M. J., and Jonathan, N., *Spectrochim. Acta* **25A,** 187 (1969).
13. Schorpp, K., and Beck, W., *Chem. Ber.* **107,** 1371 (1974).

0018–1536/84/1700–0427$03.40

A Predictive Procedure for Vapor Pressure

GOLLAKOTA R. SOMAYAJULU

Thermodynamics Research Center, Texas A&M University, College Station, Texas, 77843

Received September 23, 1983; Accepted December 31, 1983

ABSTRACT

A vapor pressure equation containing only three parameters has been proposed to calculate vapor pressures from the triple point to the normal boiling point of both organic and inorganic liquids. One of the three parameters is the normal boiling point. The other two parameters are obtained from only two pieces of experimental data, namely, the enthalpy of vaporization at the normal boiling point, and the enthalpy of vaporization at 298.15 K. There are several empirical procedures for the calculation of the boiling points of most liquids. The enthalpy of vaporization at the boiling point is obtainable from the Trouton's constant. The enthalpy of vaporization at 298.15 K is obtainable from group contributions. Thus one obtains a simple predictive procedure for the vapor pressure. We call our vapor pressure equation the quadratic equation. By the addition of one more constant we may make the quadratic equation a cubic equation.

Comparison of the quadratic equation with the Antoine equation has shown that the new equation is much more accurate than the latter in the prediction of vapor pressures and enthalpies of vaporization, partircularly when large extrapolations to low temperatures are necessary. The Cox equation, which employs four parameters, is closely related to the quadratic equation. It is, however, inferior to the quadratic equation when it comes to compounds of high molecular weight, which usually have very low pressures at 298.15 K. The cubic equation, which contains one parameter more than the quadratic equation, is equally defective when it comes to compounds of high molecular weight. The attractive features of the three-constant quadratic equation are that (1) its constants have some simple physical significance, (2) its constants can be determined from the molecular structure, and (3) it is the most accurate equation for the representation of vapor pressure data from the triple point to the normal boiling point.

Index Entries: Vapor pressures; enthalpies of vaporization; Trouton's constant; Antoine equation; Cox equation; vapor pressure, of alkanes; correlation, of enthalpies of vaporization and boiling point with molecular structures; molecular structure, correlation with boiling point and enthalpies of vaporization.

INTRODUCTION

Vapor pressure equations are legion. Several of them have been reviewed by Partington (*1*), Reid et al. (*2*), and Ambrose (*3*). The most recent ones are those of Wagner (*4*), and Scott and Osborn (*5*). Some of these equations are quite accurate in describing vapor pressures from the triple point to the critical point. Nonetheless, the equation that has been widely used is the Antoine equation (*6*). The reason for this is that it is perhaps the only equation with constants that can be correlated with molecular structure. Such a facility makes the Antoine equation unique for the prediction of vapor pressures of compounds for which the vapor pressure data are either sparse or unknown. Wilingham et al. (*7*) have recommended the use of the Antoine equation for pressures in the range 1–200 kPa. Several authors (*8–11*) have presented graphical procedures for the determination of the Antoine constants. Some authors (*12–14*) have developed emipirical procedures for the calculation of the Antoine constants. Currently there are three compilations (*15–18*) of vapor pressure data based on the Antoine equation. The usefulness of this data has been recently demonstrated by Smith et al. (*19*).

Serious objections have also been raised by several authors to the use of the Antoine equation. Scott and Osborn (*5*) have pointed out that the Antoine equation and several other three-constant equations give very poor extrapolations, both to lower and higher temperatures, and obviously are not appropriate for representing accurate vapor pressure data. For high molecular weight compounds that have low vapor pressures at room temperature, the vapor pressure is often determined at a convenient range of higher temperatures. The Antoine equation selected to represent such data usually fails to extrapolate to lower temperatures, resulting in inaccurate values for the enthalpies of vaporization at 298.15 K. The Cox equation, according to Scott and Osborn (*5*), provides the best extrapolation to lower temperatures and yields reliable values for the enthalpies of vaporization at lower temperatures. Unfortunately no one has found any suitable procedures for the calculation of the Cox constants on the basis of the molecular structure. Scott and Osborn do not expect any of the simple equations with three or four constants to extrapolate well to higher temperatures up to the critical temperature and one may have to use as many as 11 constants for the entire liquid range.

We wish to point out that even in the case of the Antoine equation the procedures for the estimation of the Antoine constants from molecular structure are neither simple nor accurate. We have therefore looked for a vapor pressure equation that is reasonably accurate at pressures below one atmosphere, but has constants that can be correlated with molecular structure and at the same have some simple physical significance. We have found the following three-parameter equation to meet our requirements.

$$\ln P = A(\phi - 1) + B(\phi - 1)^2 \tag{1}$$

where $\phi = C/T$. When the pressure is expressed in atmospheres $C = \theta$, the normal boiling point. Since our equation is restricted to low pressures, we assume the Clausius-Clayperon equation

$$\Lambda = -R\,[d \ln P/d(1/T)] \tag{2}$$

where Λ is the enthalpy of vaporization at temperature T to hold and derive from Eq. (1) the following:

$$\Lambda = -RC[A + 2B(\phi - 1)] \tag{3}$$

We now introduce the tems Λ_1 for Λ at $T = \theta$ and Λ_2 for Λ at $T =$ 298.15 K. Then

$$\Lambda_1 = -RA\theta \tag{4}$$

and

$$-A = (\Lambda_1/\theta)/R = \Delta S_V/R \tag{5}$$

where ΔS_V is the entropy of vaporization at the boiling point and is usually referred to as the Trouton's constant. A is a dimensionless quantity with a value approximately equal to -11.5 for most liquids. Strictly speaking, the Trouton's constant is no constant and has been shown to vary from compound to compound.

From Eqs. (3) and (4) one can now write the following expression for B.

$$B = (\Lambda_1 - \Lambda_2)/2RC\{(C/298.15) - 1)\} \tag{6}$$

The value of B is characteristic of each molecule and can be determined from molecular structure. The constants of Eq. (1) can thus be determined from three pieces of experimental data, namely, Λ_1, Λ_2, and θ. Even when these three pieces of data are not available, one can calculate them from molecular structure for most compounds and obtain a vapor pressure equation for the compound in question.

For still greater accuracy in the representation of the vapor pressure data, one may use a four-parameter equation, in which case, the proposed equation is

$$\ln P = A(\phi - 1) + B(\phi - 1)^2 + D(\phi - 1)^3 \quad (7)$$

Then

$$\Lambda = -RC[A + 2B(\phi - 1) + 3D(\phi - 1)^2] \quad (8)$$

and

$$\Lambda_1 = -RAC = -RA\theta \quad (9)$$

The Eq. (7) lacks the simplicity of the three-parameter Eq. (1) and we shall consider it here only for academic reasons.

METHOD OF APPLICATION

Equation (1) is a quadratic equation and can be applied as such by using a nonlinear regression procedure for the determination of its constants. It can also be made amenable to linear least-squares regression by rewriting it in the following way:

$$T^2 \ln P = BC^2 + T(AC - 2BC) + T^2(B - A) \quad (10)$$

When P is expressed in atmospheres $\ln P = 0$ when $P = 1$ atmosphere and

$$BC^2 + T(AC - 2BC) + T^2(B - A) = 0 \quad (11)$$

By using $T^2 \ln P$ as the dependent variable and T and T^2 as the independent variables, one can use Eq. (10) for linear least-squares regression and derive the constants A, B, and C by the solution of the quadratic equation (11).

Equation (7) may also be made amenable to linear least squares regression by rewriting it in the following way:

$$T^3 \ln P = (B - A - D)T^3 + T^2(AC - 2BC + 3CD) + T(BC^2 - 3C^2D) + DC^3 \quad (12)$$

When the pressure is expressed in atmospheres $\ln P = 0$ when $P = 1$ atmosphere and

$$DC^3 + (BC^2 - 3C^2D)T + (AC - 2BC + 3CD)T^2 + (B - A - D)T^3 = 0 \quad (13)$$

By making $T^3 \ln P$ the dependent variable, and T, T^2, and T^3 the independent variables, one can use Eq. (12) for least-squares regression and determine the constants A, B, C, and D by the solution of the cubic Eq. (13).

For the sake of comparison we now introduce the Antoine and the Cox equations. The Antoine equation is usually written in the following forms:

$$\ln P = A - [B/(T + C)] \tag{14}$$

or as

$$T \ln P = (AC - B) + AT - C \ln P \tag{15}$$

The quantities Λ, Λ_1, and θ corresponding to the Antoine equation are given by the following expressions:

$$\Lambda = RBT^2/(T + C)^2 \tag{16}$$

$$\Lambda_1 = R\theta^2 A^2/B \tag{17}$$

$$\theta = (B - AC)/A \tag{18}$$

The Cox equation is written as follows:

$$\ln P = A_1[1 - (D/T)] \tag{19}$$

where

$$A_1 = \exp (A + BT + CT^2) \tag{20}$$

When the pressure is expressed in atmospheres

$$\theta = D \tag{21}$$

$$\Lambda = RA_1\{(D + [1 - (D/T)]\ (BT^2 + 2CT^3)\} \tag{22}$$

and

$$\Lambda_1 = RA_1D = RA_1\theta \tag{23}$$

The quadratic, cubic, and Antoine equations may be used in their linear forms for least-squares regressions, but in the case of the Cox equation, one has to resort to a nonlinear least-squares procedure. The standard deviation σ is calculated as follows:

$$\sigma = \{[\Sigma(P_1 - P_2)^2/(N - k)]\}^{1/2} \tag{24}$$

where P_1 and P_2 are the observed and the calculated pressures in kPa and k is the number of parameters. All the regressions are, however, carried out with pressures expressed in atmospheres and temperatures in K. All temperatures , if necessary, are converted to the IPTS-68. Except for certain bad points, all observations are weighted equally for all the regressions. For the sake of comparison of the various equations, we have tested a special weighting scheme, but found no significant changes in the results. This is presumably caused by the range of pressures under consideration being very narrow.

RESULTS AND DISCUSSION

We have applied the quadratic equation to various classes of organic compounds from temperatures close to the triple point to temperatures up to the normal boiling point and found it to be quite satisfactory. In this paper we have restricted our attention mainly to 60 alkanes in the range C_1 to C_{18}. The vapor pressure data for these compounds are from reliable sources (*7, 20–38*). We have compared in Table 1 the results obtained on the basis of the quadratic equation for 3,3-diethylpentane with the results obtained on the basis of the other three equations, namely, the cubic equation, the Cox equation, and the Antoine equation. The vapor pressure data are those determined by Forziati et al. (*32*) combined with the low-temperature data obtained by Osborn and Scott (*35*). The constants of the various vapor pressure equations are recorded in Table 2. The results indicate that the standard deviation is the lowest for the quadratic equation and highest for the Antoine equation, whereas those of the other equations fall in between. This has been the pattern with a substantial number of compounds. For several compounds the pressures extended only from 6 to 101 kPa. In such cases when the pressure range is narrow, the Antoine equation proved to be the best of all. For some compounds, for example, 2-methylheptane and 3-methylheptane, where the pressures extended from 0.2 to 101 kPa, the cubic and the Cox equations proved to be better. In cases such as methane, normal pentane, and neopentane, the equations are satisfactory even up to pressures of 270 kPa. It is, however, not our intention to apply the quadratic equation well beyond the normal boiling point. The data necessary for testing the validity of all the four equations up to the triple point are either of poor quality or missing in most cases. An equation may fit the data more accurately than the other equations, but that does not necessarily qualify the equation as the best equation.

According to Scott and Osborn, the merit of an equation depends largely on how well it extrapolates to lower and higher temperatures and also gives rise to an enthalpy of vaporization close to the calorimetric value. For this reason we have compared in Table 3 the Λ_2 values obtained at 298.15 K with the ΔH°_V values determined calorimetrically (*24, 25, 39, 40, 41*). Unfortunately, calorimetric values are available only for 38 alkanes and the values recorded as observed for the rest of the compounds are either from TRC tables (*15*) or from a generalized correlation (*42*). Methane is well above its critical temperature at 298.15 K and hence its enthalpy of vaporization is zero at 298.15 K. The Λ_2 values for the first few members from ethane to neopentane deviate appreciably for all the equations since the pressures for these compounds at 298.15 K are quite high, necessitating corrections for nonideality. The Λ_2 values for all the alkanes from normal hexane to normal decane, with the exception of 2,2,3,3-tetramethylbutane, determined from all the four vapor pressure equations compare favorably with the calorimetric values. In the case of 2,2,3,3-tetramethylbutane, the quadratic equation gives a value that devi-

TABLE 1
Comparison of Quadratic Equation with Other Equations with Respect to 3,3-Diethylpentane

Observed		Quadratic Equation		Cubic Equation		Cox Equation		Antoine Equation	
T/K	P_1/kPa	P_2/kPa	Δ/kPa	P_2/kPa	Δ/kPa	P_2/kPa	Δ/kPa	P_2/kPa	Δ/kPa
273.150	0.199	0.197	0.002	0.198	0.001	0.198	0.001	0.197	0.002
288.144	0.532	0.535	-0.003	0.535	-0.003	0.535	-0.003	0.534	-0.002
293.143	0.724	0.726	-0.002	0.726	-0.002	0.725	-0.001	0.726	-0.002
298.142	0.969	0.973	-0.004	0.972	-0.003	0.971	-0.002	0.973	-0.004
303.141	1.288	1.288	0.000	1.288	0.000	1.287	0.001	1.289	-0.001
308.140	1.689	1.688	0.001	1.687	0.002	1.686	0.003	1.690	-0.001
313.140	2.188	2.189	-0.001	2.187	0.001	2.186	0.002	2.191	-0.003
318.140	2.810	2.810	0.000	2.809	0.001	2.807	0.003	2.813	-0.003
323.140	3.575	3.576	-0.001	3.574	0.001	3.573	0.002	3.578	-0.003
328.140	4.509	4.509	0.000	4.508	0.001	4.507	0.002	4.511	-0.002
333.141	5.640	5.640	0.000	5.639	0.001	5.638	0.002	5.641	-0.001
338.141	7.000	6.999	0.001	6.998	0.002	6.999	0.001	6.999	0.001
336.023	6.402	6.393	0.009	6.392	0.010	6.392	0.010	6.395	0.007
340.381	7.694	7.691	0.003	7.690	0.004	7.691	0.003	7.690	0.004
344.168	8.994	8.989	0.005	8.989	0.005	8.991	0.003	8.988	0.006
347.627	10.332	10.329	0.003	10.330	0.002	10.333	-0.001	10.326	0.006
350.843	11.723	11.719	0.004	11.721	0.002	11.725	-0.002	11.714	0.009
355.188	13.844	13.840	0.004	13.843	0.001	13.848	-0.004	13.832	0.013
360.146	16.637	16.635	0.002	16.641	-0.004	16.647	-0.010	16.623	0.014
365.196	19.948	19.942	0.006	19.950	-0.002	19.958	-0.010	19.925	0.022
369.886	23.478	23.476	0.002	23.487	-0.009	23.495	-0.017	23.455	0.023
376.153	28.982	28.977	0.004	28.992	-0.010	29.000	-0.018	28.950	0.032
388.883	43.364	43.362	0.002	43.383	-0.018	43.386	-0.021	43.326	0.038
396.061	53.702	53.704	-0.002	53.725	-0.023	53.723	-0.021	53.670	0.032
403.720	66.800	66.813	-0.013	66.829	-0.029	66.820	-0.021	66.789	0.011
412.028	83.754	83.768	-0.014	83.770	-0.015	83.758	-0.003	83.776	-0.021
417.893	97.651	97.653	-0.002	97.635	0.016	97.625	0.025	97.700	-0.049
418.524	99.240	99.248	-0.008	99.228	0.012	99.219	0.021	99.300	-0.060
419.110	100.733	100.748	-0.015	100.725	0.008	100.717	0.016	100.804	-0.071
419.775	102.468	102.469	-0.001	102.443	0.025	102.436	0.032	102.532	-0.065
420.363	104.005	104.012	-0.007	103.983	0.022	103.977	0.028	104.080	-0.075

TABLE 2
Constants of the Vapor Pressure Equations for 3,3-Diethylpentane

Constant	Quadratic Equation	Cubic Equation	Cox Equation	Antoine Equation
A	-10.69805	-10.68168	2.78096	9.33414
B	-1.80053	-1.90129	-1.58336†	3403.63337
C	419.33415	419.34338	1.43421‡	-54.66405
D		0.14823	419.34633	
θ/K	419.33415	419.34338	419.34633	419.30765
Λ_1/kJ	37.29893	37.24268	37.26756	37.41990
Λ_2/kJ	42.40200	42.37569	42.35801	42.43225
σ	±0.00583	±0.01200	±0.01416	±0.03082
Corr. Coef.	1.00000	1.00000	1.00000	1.00000

† This value should be divided by 1000.
‡ This value should be divided by 1000000.

ates only by 2.4 kJ mol^{-1}, while the value given by the Antoine equation deviates by 4.6 kJ mol^{-1}. The four-parameter equations fit the vapor pressure data quite favorably, but fail to yield a proper value for the enthalpy of vaporization of the undercooled liquid at 298.15 K, although the extrapolation is from 374 K to 298.15 K only. A somewhat large deviation found for 2,2,4,4-tetramethylpentane necessitates a revision of the literature value.

From normal decane onwards up to normal octadecane, the Λ_2 value calculated on the basis of the Antoine equation deviates appreciably from the literature value, the discrepancy becoming more pronounced with increasing carbon number. A similar trend is also shown by the quadratic equation, but the deviations are not appreciable. The Cox equation also fails to produce results close to the literature values for these high molecular weight alkanes. The performance of the cubic equation is extremely satisfactory in the case of tetradecane, pentadecane, and hexadecane. The data for heptadecane and octadecane are of poor quality and cover a very narrow range of pressures on the low temperature side; then a wide gap exists between these temperatures and the normal boiling point. Also a large extrapolation is needed to calculate the Λ_2 value at 298.15 K. Even in this case the performance of the quadratic equation is quite remarkable. In the case of normal octadecane the quadratic equation gives a value of 91 kJ mol^{-1} for Λ_2 at 298.15 K of the undercooled liquid, while the literature value is 90.8 kJ mol^{-1}. This feature of the quadratic equation is noteworthy.

It is pertinent to point out here that all the four equations yield nearly the same value for the Λ_1. This shows that all the four equations yield nearly the same Trouton's constant. This enables us to write the following equalities:

$$A(\text{quadratic Eq.}) = A(\text{cubic Eq.}) \tag{25}$$

$$A(\text{quadratic Eq.}) = -\theta A^2/B \text{ (Antoine Eq.)} \tag{26}$$

$$A(\text{quadratic Eq.}) = -\exp(A + BD + CD^2) \text{ (Cox Eq.)} \tag{27}$$

There is yet one more criterion to judge the merits of these vapor pressure equations. It is attached to the physical significance of the constants employed in these equations and also on their dependence on the molecular structure. Fortunately, in the case of the quadratic equation, all the three constants have some simple physical significance. We now wish to show that all these three constants can be determined from the molecular structure and this facility makes the quadratic equation unique among these four vapor pressure equations. Following our generalized treatment of the alkanes (42), we now propose the following group additivity equation for the correlation of the quadratic constant A.

$$A = n^{2/3}\,\alpha + n_1G_1 + n_2G_2 + n_3G_3 + n_4G_4 \qquad (28)$$

where n = the number of carbon atoms in the molecule
n_1 = the number of primary carbon atoms
n_2 = the number of secondary carbon atoms
n_3 = the number of tertiary carbon atoms
n_4 = the number of quaternary carbon atoms

and α, G_1, G_2, G_3, and G_4 are their respective contributions. It is possible to use methane also in this treatment by making the group additivity equivalent to the trigonal additivity as shown below:

$$G_1 = T_{111} + 3T_{112} \qquad (29)$$

$$G_2 = 2T_{112} + 2T_{122} \qquad (30)$$

$$G_3 = 3T_{122} + T_{222} \qquad (31)$$

$$G_4 = 4T_{222} \qquad (32)$$

where T_{111}, T_{112}, T_{122}, and T_{222} are the contributions of the nonbonded trios HHH, HHC, HCC, and CCC, respectively. Methane has four nonbonded HHH trios and hence we may write the following relationship:

$$4T_{111} = 4G_1 - 6G_2 + 4G_3 - G_4 \qquad (33)$$

In view of the above relationship one can set $n_1 = 4$, $n_2 = -6$, $n_3 = 4$, and $n_4 = -1$ for methane and use it also as one of the compounds in this special group additivity procedure.

The quadratic constant B is given by the following relation:

$$B = n^{2/3}\,\alpha + n_1G_1 + n_2G_2 + n_3G_3 + n_4G_4 + p\beta \qquad (34)$$

where the additional parameter p is the familiar Platt number and represents the number of C—C—C—C structural features in the molecule and β is the contribution of one such structural feature.

The third constant C of the quadratic equation is the normal boiling point itself. A number of procedures exist for the calculation of the normal boiling point and have been reviewed by us elsewhere (43). Practically all the procedures proposed for the calculation of the normal boiling point overlook one important factor that the critical pressure of a normal

TABLE 3
Comparison of the Calorimetric Enthalpies of Vaporization with the Λ_2 Values Obtained from the Vapor Pressure Equations in kJ mol^{-1}

Compound	obsd†	Ref.	Quadratic Equation		Cubic Equation		Cox Equation		Antoine Equation	
	ΔH_v°/kJ		Λ_2/kJ	Δ/kJ	Λ_2/kJ	Δ/kJ	Λ_2/kJ	Δ/kJ	Λ_2/kJ	Δ/kJ
1	0.00									
2	9.47	15	14.11	-4.64	14.72	-5.25	19.71	-10.23	14.19	-4.72
3	16.59	15	18.62	-2.03	18.01	-1.42	17.32	-0.73	18.75	-2.17
4	21.72	24	22.98	-1.26	22.59	-0.87	22.32	-0.60	23.07	-1.35
2m3	20.08	25	21.48	-1.40	21.25	-1.17	21.42	-1.34	21.58	-1.50
5	26.76	39	27.46	-0.70	27.47	-0.71	27.46	-0.70	27.46	-0.70
2m4	25.23	40	26.12	-0.89	26.06	-0.83	26.10	-0.87	26.15	-0.92
22mm3	22.36	15	23.58	-1.22	23.58	-1.22	23.58	-1.22	23.58	-1.22
6	31.61	39	32.05	-0.44	32.05	-0.44	32.05	-0.44	32.05	-0.44
2m5	29.96	39	30.50	-0.54	30.50	-0.54	30.50	-0.54	30.50	-0.54
3m5	30.35	39	30.83	-0.47	30.82	-0.47	30.82	-0.47	30.83	-0.47
22mm4	27.83	39	28.43	-0.60	28.43	-0.60	28.42	-0.59	28.42	-0.60
23mm4	29.23	39	29.75	-0.53	29.75	-0.53	29.75	-0.53	29.75	-0.52
7	36.56	39	36.86	-0.29	36.95	-0.39	36.94	-0.37	36.91	-0.35
2m6	34.83	15	35.13	-0.30	35.12	-0.29	35.12	-0.28	35.15	-0.32
3m6	35.11	15	35.35	-0.24	35.35	-0.24	35.34	-0.23	35.38	-0.27
3e5	35.25	39	35.49	-0.24	35.50	-0.25	35.49	-0.24	35.52	-0.27
22mm5	32.48	39	32.84	-0.36	32.85	-0.36	32.85	-0.37	32.85	-0.37
23mm5	34.27	39	34.51	-0.24	34.56	-0.29	34.57	-0.30	34.47	-0.20
24mm5	32.94	39	33.33	-0.39	33.32	-0.39	33.32	-0.39	33.34	-0.40
33mm5	33.01	39	33.36	-0.35	33.36	-0.35	33.36	-0.35	33.37	-0.36
223mmm4	32.09	39	32.41	-0.32	32.41	-0.32	32.41	-0.32	32.41	-0.33
8	41.48	39	41.73	-0.24	41.93	-0.44	41.74	-0.25	41.97	-0.49
2m7	39.68	39	39.79	-0.11	39.83	-0.15	39.85	-0.16	39.74	-0.06
3m7	39.84	39	39.96	-0.12	39.98	-0.14	40.00	-0.16	39.92	-0.08
4m7	39.68	39	39.90	-0.22	39.94	-0.27	39.86	-0.18	40.05	-0.37
3e6	39.65	39	39.85	-0.21	39.88	-0.23	39.78	-0.13	40.00	-0.35
22mm6	37.30	39	37.52	-0.22	37.58	-0.28	37.53	-0.23	37.61	-0.31
23mm6	38.79	39	39.02	-0.22	39.13	-0.33	39.03	-0.23	39.14	-0.35

24mm6	37.78	39	38.02	-0.24	38.08	-0.30	38.01	-0.23	38.11	-0.33
25mm6	37.87	39	38.12	-0.25	38.17	-0.30	38.11	-0.24	38.21	-0.34
33mm6	37.54	39	37.76	-0.22	37.80	-0.26	37.74	-0.20	37.85	-0.31
34mm6	38.98	39	39.19	-0.21	39.19	-0.21	39.10	-0.13	39.31	-0.33
23me5	38.53	39	38.71	-0.18	38.76	-0.23	38.69	-0.16	38.83	-0.30
33me5	37.99	39	38.18	-0.19	38.23	-0.24	38.15	-0.16	38.28	-0.29
223mmm5	36.93	39	37.09	-0.17	37.09	-0.16	37.04	-0.11	37.17	-0.24
224mmm5	35.15	39	35.46	-0.30	35.46	-0.31	35.46	-0.30	35.39	-0.24
233mmm5	37.23	39	37.45	-0.23	37.54	-0.32	37.47	-0.24	37.54	-0.32
234mmm5	37.71	39	37.79	-0.08	37.88	-0.16	37.90	-0.18	37.74	-0.02
2233mmmm4	35.19	15‡	32.75	2.44					39.76	-4.57
9	46.45	39	46.76	-0.31	47.17	-0.73	46.66	-0.22	47.31	-0.87
223mmm6	41.30	42	41.25	0.05	41.51	-0.21	41.53	-0.23	41.18	0.12
224mmm6	39.66	42	39.79	-0.13	39.84	-0.19	39.83	-0.18	39.74	-0.08
225mmm6	40.08	42	40.23	-0.14	40.28	-0.19	40.30	-0.21	40.14	-0.05
244mmm6	40.25	42	40.50	-0.25	40.32	-0.07	40.24	0.01	40.68	-0.43
33ee5	42.57	42	42.40	0.17	42.38	0.19	42.36	0.21	42.43	0.14
2233mmmm5	41.17	15	41.46	-0.29	41.61	-0.44	41.40	-0.23	41.71	-0.54
2234mmmm5	40.84	42	40.44	0.40	40.41	0.43	40.28	0.56	40.63	0.21
2244mmmm5	37.74	42	38.63	-0.89	38.60	-0.86	38.52	-0.78	38.76	-1.02
2334mmmm5	41.63	42	41.67	-0.04	41.56	0.07	41.41	0.22	41.90	-0.27
10	51.36	39	51.81	-0.45	51.67	-0.31	50.98	0.39	52.89	-1.53
11	56.35	42	57.11	-0.75	58.93	-2.58	56.83	-0.48	58.95	-2.59
2m10	54.74	42	54.86	-0.12	54.61	0.13	54.50	0.24	55.03	-0.30
12	61.30	41	62.65	-1.36	66.46	-5.16	62.65	-1.35	65.62	-4.33
13	66.26	42	67.80	-1.55	68.37	-2.11	65.42	0.84	71.85	-5.59
14	71.22	42	73.17	-1.95	70.67	0.55	68.48	2.74	78.77	-7.55
15	76.18	42	78.90	-2.72	77.11	-0.93	74.12	2.06	86.69	-10.51
16	81.38	41	84.76	-3.38	82.33	-0.95	78.04	3.34	95.61	-14.24
17	86.10	42	91.38	-5.28	74.65	11.44	76.81	9.28	99.26	-13.17
18	90.79	41‡	91.01	-0.22	74.08	16.71	83.92	6.87	101.37	-10.57

† The word 'obsd' has to be qualified with certain compounds for which only the best correlated values exist. Also some of the values for which the reference is 15 are not calorimetrically determined. Methane is not a liquid at 298.15 and hence its enthalpy of vaporization is taken as equal to 0.

‡ The values are for the undercooled liquid at 298.15 K.

TABLE 4
Comparison of the Constants of the Quadratic Equation with the Values Calculated on the Basis of Group Additivity Eqs. (33), (34), and (36)

Compound	Constant A			Constant B			Constant C			References for P
	obsd	calcd	devn	obsd	calcd	devn	obsd	calcd	devn	
1	-9.235	-9.327	0.092	-0.561	-0.531	-0.030	111.644	111.931	-0.287	20,21
2	-9.956	-9.862	-0.094	-0.994	-0.970	-0.024	184.506	183.846	0.660	22
3	-10.226	-10.124	-0.102	-1.180	-1.240	0.060	231.055	230.604	0.451	23
4	-10.379	-10.351	-0.028	-1.409	-1.462	0.053	272.638	272.532	0.106	24
2m3	-10.229	-10.212	-0.017	-1.415	-1.394	-0.021	261.420	261.860	-0.440	25
5	-10.562	-10.554	-0.008	-1.595	-1.668	0.073	309.208	309.163	0.045	7,26,27
2m4	-10.409	-10.415	0.006	-1.532	-1.575	0.043	300.965	300.719	0.246	7,28
22mm3	-10.194	-10.205	0.011	-1.530	-1.472	-0.058	282.643	283.626	-0.983	27,29,30
6	-10.731	-10.739	0.008	-1.857	-1.861	0.004	341.882	341.886	-0.004	7
2m5	-10.599	-10.600	0.001	-1.700	-1.769	0.069	333.411	333.442	-0.031	7
3m5	-10.586	-10.600	0.014	-1.697	-1.745	0.048	336.422	335.669	0.753	7
22mm4	-10.312	-10.390	0.078	-1.672	-1.617	-0.055	322.883	323.038	-0.155	7,31
23mm4	-10.441	-10.461	0.020	-1.649	-1.652	0.003	331.128	330.747	0.381	7
7	-10.899	-10.910	0.011	-2.093	-2.046	-0.047	371.576	371.553	0.023	7
2m6	-10.776	-10.771	-0.005	-1.968	-1.953	-0.015	363.200	363.109	0.091	32
3m6	-10.761	-10.771	0.010	-1.978	-1.929	-0.049	364.998	365.336	-0.338	32
3e5	-10.750	-10.771	0.021	-1.945	-1.905	-0.040	366.626	365.129	1.497	32
22mm5	-10.540	-10.561	0.021	-1.846	-1.802	-0.044	352.347	352.705	-0.358	7
23mm5	-10.668	-10.632	-0.036	-1.770	-1.812	0.042	362.916	362.641	0.275	27,32
24mm5	-10.627	-10.632	0.005	-1.904	-1.861	-0.043	353.644	354.664	-1.020	32
33mm5	-10.472	-10.561	0.089	-1.703	-1.753	0.050	359.210	359.394	-0.184	32
223mmm4	-10.381	-10.422	0.041	-1.678	-1.661	-0.017	354.026	353.948	0.078	32
8	-11.062	-11.070	0.008	-2.253	-2.222	-0.031	398.827	398.737	0.090	7
2m7	-10.966	-10.931	-0.035	-2.058	-2.130	0.072	390.794	390.293	0.501	7,27
3m7	-10.957	-10.931	-0.026	-2.063	-2.106	0.043	392.062	392.520	-0.458	7,27
4m7	-10.941	-10.931	-0.010	-2.148	-2.106	-0.042	390.865	392.520	-1.655	7
3e6	-10.910	-10.931	0.021	-2.115	-2.081	-0.034	391.692	392.313	-0.621	7
22mm6	-10.739	-10.721	-0.018	-2.071	-1.978	-0.093	379.995	379.889	0.106	7
23mm6	-10.814	-10.792	-0.022	-2.066	-1.989	-0.077	388.765	388.758	0.007	7

24mm6	-10.779	-10.792	0.013	-2.070	-2.013	-0.057	382.584	384.076	-1.492	7
25mm6	-10.812	-10.792	-0.020	-2.093	-2.037	-0.056	382.259	381.848	0.411	7
33mm6	-10.677	-10.721	0.044	-1.915	-1.930	0.015	385.125	386.578	-1.453	7
34mm6	-10.811	-10.792	-0.019	-2.005	-1.965	-0.040	390.883	390.985	-0.102	7
23me5	-10.761	-10.792	0.031	-1.997	-1.965	-0.032	388.808	388.550	0.258	7
33me5	-10.620	-10.721	0.101	-1.777	-1.882	0.105	391.418	390.832	0.586	7
223mmm5	-10.587	-10.582	-0.005	-1.866	-1.813	-0.053	382.996	383.360	-0.364	7
224mmm5	-10.519	-10.582	0.063	-1.872	-1.886	0.014	372.388	371.444	0.944	7,33
233mmm5	-10.549	-10.582	0.033	-1.766	-1.789	0.023	387.918	387.821	0.097	7
234mmm5	-10.707	-10.653	-0.054	-1.770	-1.873	0.103	386.599	387.130	-0.531	7,27
2233mmmm4†	-10.499	-10.372	-0.127	-0.066	-1.638	1.572	379.360	380.502	-1.142	34
9	-11.214	-11.220	0.006	-2.430	-2.392	-0.038	423.971	423.845	0.126	32
223mmm6	-10.785	-10.732	-0.053	-1.915	-1.984	0.069	407.159	408.468	-1.309	27
224mmm6	-10.647	-10.732	0.085	-1.917	-2.032	0.115	400.160	398.780	1.380	27
225mmm6	-10.817	-10.732	-0.085	-2.050	-2.056	0.006	397.221	396.553	0.668	27,32
244mmm6	-10.703	-10.732	0.029	-1.917	-2.008	0.091	403.809	403.242	0.567	32
33ee5	-10.698	-10.871	0.173	-1.800	-2.004	0.204	419.334	420.194	-0.860	7,35
2233mmmm5	-10.605	-10.522	-0.083	-1.883	-1.760	-0.123	413.442	412.300	1.142	32
2234mmmm5	-10.614	-10.593	-0.021	-1.876	-1.867	-0.009	406.180	405.773	0.407	32
2244mmmm5	-10.528	-10.522	-0.006	-1.871	-1.904	0.033	395.443	395.443	0.000	32
2334mmmm5	-10.656	-10.593	-0.063	-1.827	-1.819	-0.008	414.718	414.173	0.545	32
10	-11.368	-11.363	-0.005	-2.562	-2.557	-0.005	447.305	447.181	0.124	7
11	-11.504	-11.498	-0.006	-2.737	-2.717	-0.020	469.084	468.978	0.106	36
2m10	-11.427	-11.359	-0.068	-2.582	-2.625	0.043	462.332	460.534	1.798	27
12	-11.629	-11.627	-0.002	-2.934	-2.873	-0.061	489.483	489.421	0.062	7
13	-11.769	-11.751	-0.018	-3.020	-3.026	0.006	508.643	508.657	-0.014	36
14	-11.889	-11.870	-0.019	-3.142	-3.175	0.033	526.730	526.809	-0.079	36
15	-11.999	-11.984	-0.015	-3.307	-3.320	0.013	543.834	543.980	-0.146	36
16	-12.109	-12.095	-0.014	-3.470	-3.464	-0.006	560.018	560.256	-0.238	36
17‡	-12.133	-12.201	0.068	-3.755	-3.604	-0.151	575.113	575.711	-0.598	37,38
18‡	-12.255	-12.304	0.049	-3.211	-3.743	0.532	590.217	590.408	-0.181	37

† This compound is excluded from the correlation of B.
‡ This compound is excluded from the regression.

alkane drops below one atmosphere around C_{70} and a normal boiling point cannot occur above that carbon number. How far they are successful below the carbon number 70 is also open to question. With respect to the branched alkanes the most successful of all the procedures is the procedure developed by Greenshields and Rossini (44). This is an isomeric variation procedure and strictly meant for the calculation of the boiling point of an isomeric alkane, given the boiling point of the normal alkane. Our study has revealed that a number of factors affect the boiling point. They are (1) the chain length, (2) the branching, (3) the steric interactions between hydrogen atoms attached to the 1 and 4 carbon atoms, (4) the length of the branches, and (5) the crowding or buttressing effects. Both isobutane and neopentane are not affected by the factors 3–5 and may be treated along with the normal alkanes. We propose for such a limited set of molecules, which include all the normal alkanes in the range C_1 to C_{18}, as well as isobutane and neopentane, the following equation:

$$C = n^{2/3}\,\alpha + n_1 G_1 + n_2 G_2 + n_3 G_3 + n_4 G_4 + p\beta \qquad (35)$$

For all the alkanes in the range C_1 to C_{18} we propose the following 13 parameter equation:

$$C = n^{2/3}\,\alpha + n_1 G_1 + n_2 G_2 + n_3 G_3 + n_4 G_4 + p\beta + n_{23}\,\gamma_{23} + n_{24}\,\gamma_{24} + n_{33}\,\gamma_{33} + n_{34}\,\gamma_{34} + n_{44}\,\gamma_{44} + m_1\delta_1 + m_2\delta_2 \qquad (36)$$

TABLE 5

Comparison of the Observed Normal Boiling Points with Calculated Ones for the Alkanes Without the Steric Interactions

Compound	Normal Boiling Point/ K		
	Observed	Calculated	Deviation
1	111.644	111.657	-0.013
2	184.506	184.453	0.053
3	231.055	230.888	0.167
4	272.638	272.765	-0.127
2m3	261.420	261.367	0.053
5	309.208	309.352	-0.144
22mm3	282.643	282.656	-0.013
6	341.882	342.035	-0.153
7	371.576	371.666	-0.090
8	398.827	398.817	0.010
9	423.971	423.894	0.077
10	447.305	447.202	0.103
11	469.084	468.972	0.112
12	489.483	489.389	0.094
13	508.643	508.602	0.041
14	526.730	526.732	-0.002
15	543.834	543.882	-0.048
16	560.018	560.138	-0.120
17†	575.113	575.574	-0.461
18†	590.217	590.253	-0.036

† This compound is not included in the regression.

TABLE 6
Constants of the Group Additivity Eqs. (33), (34), (35), and (36)

Constant	A	B	C	C
a	-0.67661	-0.45676	148.64915	148.82868
G_1	-4.39375	-0.12239	-25.75646	-26.20250
G_2	0.07065	-0.04519	-26.80130	-26.56692
G_3	4.67417	0.12438	-35.93611	-34.55774
G_4	9.34854	0.35321	-48.97062	-46.74204
β		0.02415	3.30813	3.04575
γ_{23}				-0.81796
γ_{24}				0.59796
γ_{33}				1.06772
γ_{34}				3.37599
γ_{44}				8.47469
δ_1				9.29427
δ_2				-2.43553
σ	±0.05346	±0.05509	±0.11039	±0.76934

where n_{23}, n_{24}, n_{33}, n_{34}, n_{44} are the numbers of the C_2—C_3, C_2—C_4, C_3—C_3, C_3—C_4, and C_4—C_4 bonds and γ_{23}, γ_{24}, γ_{33}, γ_{34}, and γ_{44} are the corresponding contributions. These five parameters take care of five kinds of 1–4 H . . . H interactions. The quantity m_2 is the number of branches higher than the methyl and δ_2 is the contribution of one such branch; m_1 is the number of structural features such as C_4—C_2—C_3, C_4—C_3—C_4, C_4—C_4—C_3 and δ_1 is the contribution of one such feature. The values of *A*, *B*, and *C* calculated on the basis of the Eqs. (33), (34), and (36) are compared with those determined from the vapor pressure equations in Table 4 for 60 alkanes. The normal boiling points calculated on the basis of Eq. (35) are compared with those determined from vapor pressures in Table 5. The constants of the equations are given in Table 6. The standard deviations for all the regressions are satisfactory.

Thus judging from all these criteria we find the quadratic equation to be quite unique for the representation of the vapor pressure from temperatures close to the triple point to the normal boiling point. Extension of this equation to various other classes of organic compounds would enable us to generate the required group additivity contributions for the calculation of the quadratic constants for other compounds.

ACKNOWLEDGMENTS

My thanks are due to Dr. K. R. Hall and Dr. R. C. Wilhoit for their kind interest in this work. My special thanks are due to Mrs. Carol Chen for her invaluable help with the computer programming for this work. This work is financially supported by the Thermodynamics Research Center.

REFERENCES

1. Partington, J. R., *An Advanced Treatise on Physical Chemistry,* Vol. II, Longmans Green, London, 1955.
2. Reid, R. C., Prausnitz, J. M., and Sherwood, T. K., *The Properties of Gases and Liquids,* 3rd edn., McGraw-Hill, New York, 1977.
3. Ambrose, D., *Chemical Thermodynamics,* Vol. I., The Chemical Society, London, 1973.
4. Wagner, W., *Cryogenics* **12,** 214 (1972).
5. Scott, D. W., and Osborn, A. G., *J. Phys. Chem.* **83,** 2714 (1979).
6. Antoine, C., *Compt. Rend.* **107,** 681 (1888).
7. Willingham, C. B., Taylor, W. J., Pignocco, J. M., and Rossini, F. D., *J. Res. Natl. Bur. Stand.* **35,** 219 (1945).
8. Thomson, G. W., *Chem. Rev.* **38,** 1 (1946).
9. Miller, D. G., *Ind. Eng. Chem. Fund.* **2,** 68 (1963).
10. Lu, B. C. Y., *Can. J. Chem. Eng.* **38,** 33 (1960).
11. Rehberg, C. E., *Ind. Eng. Chem.* **42,** 829 (1950).
12. Fishtine, S. H., *Ind. Eng. Chem.* **55,** 20, 49 (1963).
13. Kudchadker, A. P., and Zwolinski, B. J., *J. Chem. Eng. Data* **11,** 253 (1966).
14. Kudchadker, A. P., Holcomb, W. D., and Zwolinski, B. J., *J. Chem. Eng. Data* **13,** 182 (1968).
15. "Selected Values of Properties of Hydrocarbons and Related Compounds", Thermodynamics Research Center Hydrocarbon Project, Thermodynamics Research Center, Texas A&M University, College Station, Texas (loose-leaf data sheets, extant, 1983).
16. "Selected Values of Properties of Chemical Compounds", Thermodynamics Research Center Data Project, Thermodynamics Research Center, Texas A&M University, College Station, Texas (loose-leaf data sheets, extant, 1983).
17. Boublik, T., Fried, V., and Hala, E., *Vapor Pressure of Pure Substances,* Elsevier, New York, 1973.
18. Ohe, S., *Computer Aided Data Book of Vapor Pressure,* Data Book Publishing Co., Tokyo, 1976.
19. Smith, G., Winnick, J., Abrams, D. S., and Prausnitz, J. M., *Can. J. Chem. Eng.* **54,** 337 (1976).
20. Stock, A., Henning, F., and Kuss, E., *Ber.* **54,** 1119 (1921).
21. Hestermans, P., and White, D., *J. Phys. Chem.* **65,** 362 (1961).
22. Loomis, A. G., and Walters, J. E., *J. Am. Chem. Soc.* **48,** 2051 (1926).
23. Kemp, J. D., and Egan, C. J., *J. Am. Chem. Soc.* **60,** 1521 (1938).
24. Aston, J. G., and Messerly, G. H., *J. Am. Chem. Soc.* **62,** 1917 (1940).
25. Aston, J. G., Kennedy, R. M., and Schumann, S. C., *J. Am. Chem. Soc.* **62,** 2059 (1940).
26. Messerly, G. H., and Kennedy, R. M., *J. Am. Chem. Soc.* **62,** 2988 (1940).
27. Osborn, A. G., and Douslin, D. R., *J. Chem. Eng. Data* **19,** 114 (1974).
28. Schumann, S. C., Aston, J. G., and Sagenkahn, M., *J. Am. Chem. Soc.* **64,** 1039 (1942).
29. Aston, J. G., and Messerly, G. H., *J. Am. Chem. Soc.* **58,** 2354 (1936).
30. Whitmore, F. C., and Fleming, G. H., *J. Am. Chem. Soc.* **55,** 3803 (1933).
31. Kilpatrick, J. E., and Pitzer, K. S., *J. Am. Chem. Soc.* **68,** 1066 (1946).
32. Forziati, A. F., Norris, W. R., and Rossini, F. D., *J. Res. Natl. Bur. Stands.* **43,** 555 (1949).
33. Milazo, G., *Ann. Chim.* **46,** 1105 (1956).

34. Caligaert, G., Soroos, H., Hnizda, V., and Shapiro, H., *J. Am. Chem. Soc.* **66,** 1389 (1944).
35. Osborn, A. G., and Scott, D. W., *J. Chem. Thermodyn.* **12,** 429 (1980).
36. Camin, D. L., and Rossini, F. D., *J. Phys. Chem.* **59,** 1173 (1955).
37. Kraft, F., *Ber.* **15,** 1687 (1882).
38. American Petroleum Institute Research Project 42, Pennsylvania State University, University Park, Pennsylvania (1947).
39. Osborne, N. S., and Ginnings, D. C., *J. Res. Natl. Bur. Stands.* **39,** 453 (1947).
40. Scott, D. W., McCullough, J. P., Williamson, K. D., and Waddington, G., *J. Am. Chem. Soc.* **73,** 1707 (1951).
41. Morawetz, E., *J. Chem. Thermodyn.* **4,** 139, 145 (1972).
42. Somayajulu, G. R., and Zwolinski, B. J., *J. Chem. Soc. Farady II* **72,** 2213 (1976).
43. Somayajulu, G. R., Kudchadker, A. P., and Zwolinski, B. J., *Ann. Rev. Phys. Chem.* **16,** 213 (1965).
44. Greenshields, J. B., and Rossini, F. D., *J. Phys. Chem.* **62,** 271 (1958).

0018–1536/84/1700–0445$02.40

Empirical Calculations of Molecular Properties

CHIN-AN CHANG

IBM T. J. Watson Research Center, Yorktown Heights, New York 10598

Received September 23, 1983; Accepted December 31, 1983

ABSTRACT

Empirical equations for the stretching and bending force constants, bond energies, and bond lengths are reviewed. Applications to gaseous and matrix-isolated molecules, and to chemisorptive bonds on surfaces are illustrated.

Index Entries: Empirical calculations, for bond energies, lengths, and force constants; bond energies, empirical calculations for; bond lengths, empirical calculations for; force constants, empirical calculations for; matrix-isolated molecules, empirical calculations of properties for; chemisorptive bonds, empirical calculations for.

I. INTRODUCTION

The nature and properties of chemical bonds and structures have long been of central interest in research throughout both experimental and theoretical chemistry. Various empirical relations have been proposed to estimate molecular properties from available experimental data, and have considerably helped the studies and understandings of these molecular properties.

In this paper we review the empirical equations proposed for the force constants, both stretching and bending, for bond energies, and for bond lengths. These equations have been applied mainly to free molecules and in some cases to those trapped in matrices. Recent work has extended such calculations to molecules adsorbed on surfaces and to the chemisorptive bonds involved. These will be described and comparison with experiment discussed.

II. EMPIRICAL EQUATIONS

In this Section we review the empirical equations for the stretching and bending force constants, bond energies, and bond lengths.

A. Stretching Force Constant

The stretching force constant, F, and vibrational frequencies, ν, for both diatomic and polyatomic molecules have been calculated by Badger (*1*), Clark (*2*), Allen and Longair (*3*), Gordy (*4*), Guggenheimer (*5*), Somayajulu (*6*), Varshni (*7*), Herschbach and Laurie (*8*), and Chang (*9*).

Badger (*1*) proposed the equation

$$F(r_e - d_{ij})^3 = 1.86 \times 10^5 \tag{1}$$

where r_e is the internuclear distance, and d_{ij} is a constant characteristic of a diatomic molecule with the two elements from the ith and jth rows of the periodic table. Values of d_{ij} are given by Badger for a number of molecules.

A modified relation was also proposed by Badger (*1*) as

$$r_e = (C_{ij}/F)^{1/2} + d_{ij} \tag{2}$$

This is compared with two similar relations by Clark (*2*)

$$r_e^2 = C_{ij}' \ (\mu/nF)^{1/2} \tag{3}$$

and by Allen and Longair (*3*)

$$r_e^2 = C_{ij}''/F^{1/2} \tag{4}$$

C_{ij}, C_{ij}', and C_{ij}'' are constants determined empirically, μ is reduced mass, n is the "group number," or the sum of valence electrons of the molecule. In Eq. (2) C_{ij} and d_{ij} have different values for different groups of molecules. Applications to both diatomic and polyatomic molecules showed good agreement with experiment.

Gordy (*4*) considered the relations of bond order, N, bond length and electronegativity, x, and proposed the relation

$$F = aN(x_A \ x_B/r^2)^{3/4} + b \tag{5}$$

With F in mdyn/cm $\times 10^{-5}$ and r in Å, different sets of values are assigned to a and b for different groups of molecules A—B. For example, $a = 1.67$ and $b = 0.30$ for molecules exhibiting their normal covalencies, except those in which both bonded atoms have only one valence electron; for diatomic alkalis, $a = 1.180$ and $b = -0.013$ are assigned, etc. Equation (5) was applied to many molecules to calculate one of F, N, x, and r from the available data for the others. Good agreement with experiment was shown for both diatomic and polyatomic molecules.

Guggenheimer (5) proposed several equations for different types of bondings

$$\text{Single bond: } \nu = 2.156 \times 10^3 (Z_1Z_2)^{1/4} \mu^{-1/2} r_e^{-1.23} \ (\text{cm}^{-1}) \quad (6)$$

$$\text{Multiple bond: } \nu = 2.976 \times 10^3 (Z_1Z_2)^{1/4} \mu^{-1/2} r_e^{-1.23} \ (\text{cm}^{-1}) \quad (7)$$

$$\text{Polar bond: } \nu = 2.065 \times 10^3 (Z_1Z_2)^{1/4} \mu^{-1/2} r_e^{-0.92} \ (\text{cm}^{-1}) \quad (8)$$

Z_1 and Z_2 are the number of electrons in the outer shells of the two atoms, with practically equal contributions of *s*- and *p*-electrons to the force constant and vibrational frequency. The relations are shown to have quantum mechanical foundations and have been successfully applied to many diatomic and polyatomic molecules.

Somayajula (6) suggested the relation

$$F_{AB} = (F_{AA} \times F_{BB})^{1/2} + \Delta \quad (9)$$

F_{AA} and F_{BB} are the force constants of homonuclear A—A and B—B molecules, respectively, Δ is an ionic contribution term. It is found that for single bonds, $\Delta^{1/2}$ obtained from the Δ of Eq.(9) corresponds to the difference in electronegativities of the bonded atoms in their valence states; for multiple bonds, $\Delta^{1/2}$ gives the correct order for the percentage ionic character of the bonds. The relation is shown to apply to both diatomic and polyatomic molecules.

Varshni (7) proposed the relation

$$F(r_e - d)^P = A \quad (10)$$

d and A are constants, having different values for different groups of molecules. A value of $P = 2.0$ was found to give the best fit to many diatomic molecules.

Herschbach and Laurie (8) calculated quadratic, cubic, and quartic force constants of diatomic molecules and ions from spectroscopic data, using the relations

$$F_2 = 5.889 \times 10^{-7} \mu\nu^2$$
$$F_3 = -a_1(F_2 / r_e)$$
$$F_4 = a_2(F_2 / r_e^2)$$

with

$$-a_1 = 1 + (\alpha_e \omega_e / 6B_e^2)$$
$$a_2 = (5/4)\, a_1^2 - (2/3)(\omega_e x_e / B_e)$$

where α and β are rotational constants, and ω_e and $\omega_e x_e$ are vibrational constants. They found that the experimental data fit well with the general formula

$$(-1)^n F_n = 10^{-(r_e - a_{ij})/b_{ij}} \quad n = 2, 3, 4 \quad (11)$$

where a_{ij} and b_{ij} are constants characteristic of the groups of molecules with atoms from the ith and jth families of the periodic table.

Chang (9) proposed the equations

$$F_{AX} \cong F_{BX} \, (F_{AA} / F_{BB})^{1/2} \qquad (12)$$

and

$$F_{AX} \cong F_{BX} \, (F_{AY} / F_{BY}) \qquad (13)$$

Here atoms A and B belong to one family, and X and Y another family or same as that of A and B. The equations applied quite well to diatomic alkalis, halogens, and alkali halides. Figure 1 shows the ratios of force constants of chloride to bromide for a number of diatomic molecules using values from Jones (10). The ratio of F_{XCl} / F_{XBr} is seen to agree quite well with the value of 1.16 for $(F_{Cl-Cl} / F_{Br-Br})^{1/2}$ as suggested in

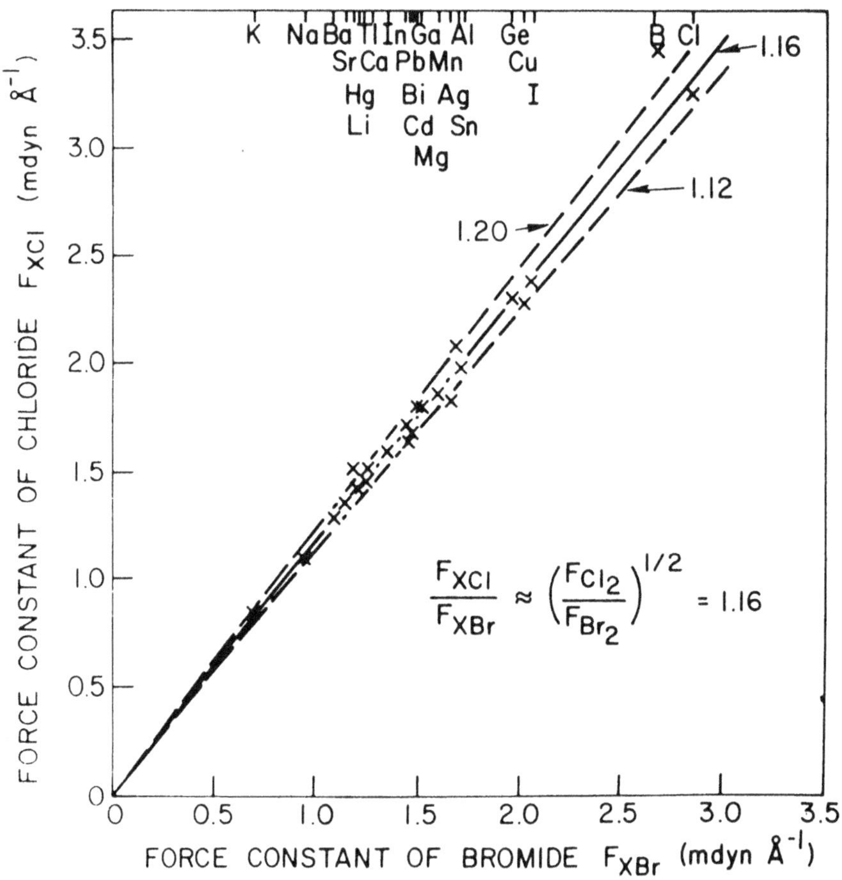

Fig. 1. Force constant ratios of diatomic chlorides and bromides, F_{XCl} F_{XBr}. Data points are labeled by the elements X. Solid line is for a ratio of 1.16, as calculated from the inserted equation; upper and lower dashed lines are for ratios of 1.20 and 1.12, respectively.

Eq. (12). As will be seen later, Eq. (12) also applies to chemisorptive bonds on surfaces.

B. Bending Force Constant

Unlike stretching force constants, much less effort has been devoted to calculating bending force constants empirically. Several semiempirical methods have been used to calculate this quantity (*11, 12*). Recently, an empirical equation was derived by Chang (*13*), assuming a Coulomb potential for the interaction of the two bonds involved in the bending motion. It led to the relation for the bending force constant F_ϕ

$$F_\phi / r_1 r_2 = c z_1 z_2 R^{-5}(3 r_1 r_2 \sin^2\phi - R^2 \cos\phi) \tag{14}$$

where z_1 and z_2 are the charges of the two bonds involved, ϕ is the bond angle, r_1 and r_2 are the bond lengths, R is the distance between the two end atoms, and c is a proportionality constant. The assumption is made that the product $c z_1 z_2$ should be nearly constant when molecules of the same type, such as CO_2, CS_2, and OCS, or NH_3, PH_3, and AsH_3, are compared. It is shown that for many different types of triatomic and larger polyatomic molecules, $c z_1 z_2$ indeed remains nearly constant for each group of molecules. Equation (14) can thus be used to calculate F_ϕ for a molecule from data available for molecules of the same type. Furthermore, for two linear symmetrical triatomic molecules A and B of the same type, Eq. (14) gives

$$(F_\phi)_A \,/\, (F_\phi)_B = r_B \,/\, r_A$$

which provides a quantitative relation for the inverse dependence of F_ϕ on r often observed in spectroscopic studies.

C. Bond Energies

Empirical methods for bond energies, D, have been proposed by Pauling (*14*), Evans and Huheey (*15*), Van Hooydonk (*16*), Chang (*9*), Sanderson (*17*), and Matcha (*18*).

Pauling (*14*) proposed the equations

$$D_{AB} = (D_{AA} \times D_{BB})^{1/2} + 30\,(x_A - x_B)^2 \tag{15}$$

and

$$D_{AB} = 1/2\,(D_{AA} + D_{BB}) + 23\,(x_A - x_B)^2 \tag{16}$$

x_A is the electronegativity of the element A, etc. Bond energies obtained by these two equations generally agree quite well, except for molecules like alkali hydrides, MH, where the bond energies D_{HH} and D_{MM} differ significantly.

Evens and Huheey (*15*) considered the contributions of covalent energy, electronegativity energy, and electrostatic attraction energy in calculating the bond energy of a heteropolar bond. The total energy function of these three energy terms is then minimized with respect to charge to find the minimum total energy. Readers are referred to the original papers of Evans and Huheey for detailed definition of the terms used.

Van Hooydonk (*16*) considered electronegativity equalization and proposed the equations

$$D_{AB} = (-1/2)\, e\, (x_A + x_B)\{1 + [(x_B - x_A)/(x_B + x_A)]^2\} \quad (17)$$

and

$$D_{AB} = (1/2)(D_{AA} + D_{BB})\{1 + [(D_{BB} - D_{AA})/(D_{BB} + D_{AA})]^2\} \quad (18)$$

where the x values are the electronegativities and D values are the bond dissociation energies. The values obtained by this method agree well with those of Evans and Huheey (*15*).

Chang (*9*) suggested equations for bond energy similar to those for the stretching force constant

$$D_{AX} \cong D_{BX}(D_{AA}/D_{BB})^{1/2} \quad (19)$$

and

$$D_{AX} \cong D_{Bx}(D_{AY}/D_{BY}) \quad (20)$$

Equations (19) and (20) apply quite well to diatomic alkalis, halogens, and alkali halides. Figure 2 shows the ratio of bond energies of chlorides to bromides for a number of molecules, using Gaydon's compilation (*19*). It is seen that the experimental (D_{XCl}/D_{XBr}) agree very well with the value of 1.12 of $(D_{Cl-Cl}/D_{Br-Br})^{1/2}$ using Eq. (19). Further applications to surface chemisorptive bonds are described later.

Sanderson (*17*) considered equalization of electronegativities in his calculations of bond energies. Assuming a geometric mean for the final value of electronegativity, e.g., $x_{AB} = (x_A x_B)^{1/2}$ for a diatomic AB, the change of each of x_A and x_B from x_{AB}, i.e., $\Delta x_A = x_A - x_{AB}$ for example, is then divided by that expected if the bond were completely ionic, which is defined as Δx_i. The ratio $\Delta x_A/\Delta x_i$ is taken as the partial electrical charge on atom A, and also the fraction of the bond energy that is ionic. The bond energy thus becomes

$$E_{AB} = (1 - \Delta x_A/\Delta x_i)\, E_c + (\Delta x_A/\Delta x_i)E_i \quad (21)$$

where E_c is the energy of a completely covalent bond and E_i that of a completely ionic bond. Sanderson found E_c and E_i as

$$E_c = R_c(E_{AA} \times E_{BB})^{1/2}/R_o$$

$$E_i = 33{,}200/R_o$$

where R_c is the nonpolar covalent bond length, R_o is the experimental bond length, E_{AA} and E_{BB} are one-half the bond energies of molecules

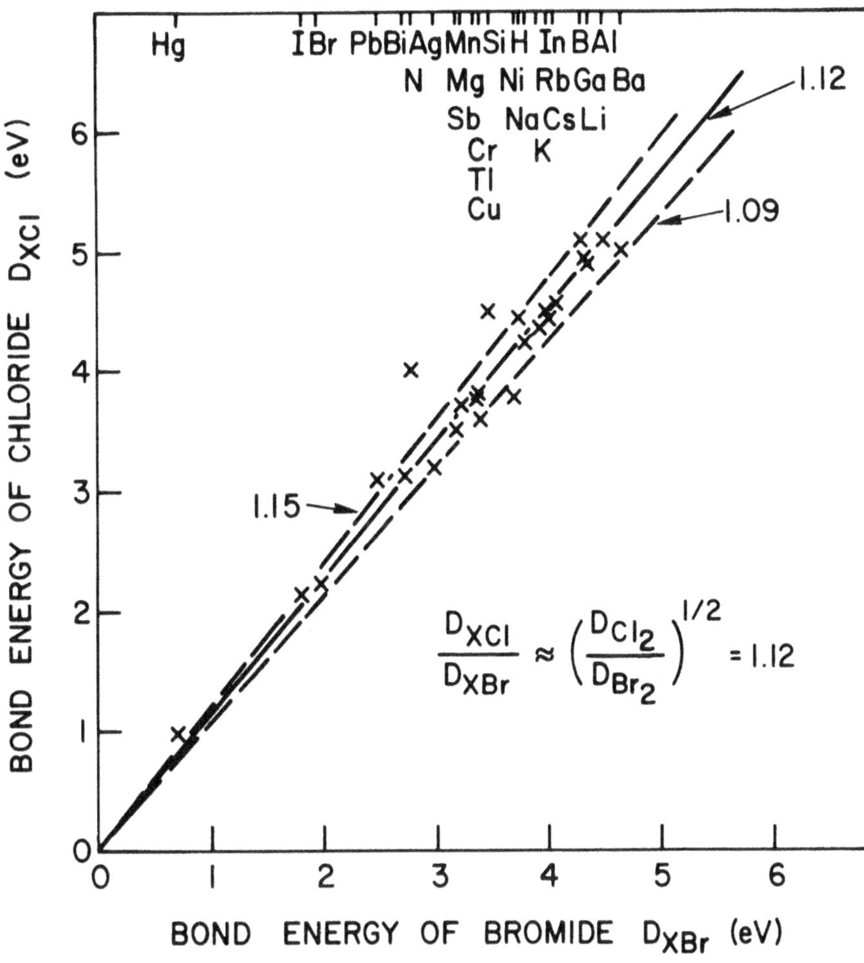

Fig. 2. Bond energy ratios of diatomic chlorides and bromides, D_{XCl}/D_{XBr}. Data points are labeled by the elements X. Solid line is for a ratio of 1.12 as calculated from the inserted equation; upper and lower dashed lines are for a ratio of 1.15 and 1.09, respectively.

A—A and B—B, respectively. The bond energies and atomization energies have been calculated for many molecules and some solids. The results are in excellent agreement with experiment.

Matcha (*18*) proposed the relation

$$D_{AB} = (1 - f)(D_{AA} + D_{BB})/2 + 252(f/R_e) \tag{22}$$

with

$$f/R_e \cong (2/3)\,[1 - \exp(-\Delta x^2/4)]$$

where f is the bond polarity and is related to Pauling's ionicity as shown, and R_e is the equilibrium bond length. Eq. (22) is shown to give good agreement with experiment for both ionic and covalent bonds.

D. Bond Length

The empirical equations described above for the stretching force constant involve bond length in several cases, and have been used to calculate the latter using observed vibrational frequencies (*1,4,7*). Other empirical equations for bond length have been reported by Schomaker and Stevenson (*20*), and Chang (*9*).

Schomaker and Stevenson (*20*) considered the contributions of electronegativity and proposed the equation

$$r_{AB} = r_A + r_B - 0.09(x_A - x_B) \qquad (23)$$

where r_A and r_B are the single-bond covalent radii of atoms *A* and *B*. The calculated bond lengths are in general better than those using additivity rule when the ionic character is significant.

Chang (*9*) suggested that for two diatomic molecules *AX* and *BX*, with *A* and *B* belonging to the same family of the periodic table, the bond lengths can be approximated by

$$R_{AX} \cong R_{BX} + \tfrac{1}{2}(R_{AA} - R_{BB}) \qquad (24)$$

and

$$R_{AX} \cong R_{BX} + (R_{AY} - R_{BY}) \qquad (25)$$

with *Y* being the same family as *X*. Applications to diatomic molecules show good agreement with experiment.

III. APPLICATIONS

All the empirical methods reviewed above have been applied mainly to gaseous molecules. It is interesting and useful to see whether the various equations proposed can also be applied to other systems. In this Section we illustrate the use of some empirical equations to the molecules trapped in matrices, and to those adsorbed on surfaces and to the chemisorptive bonds formed.

Using matrix isolation technique for transition metal oxides and sulfides, Hauge and Margrave (*21*) found a relation of force constants and bond energies as

$$(D_{MO}/D_{MS})/(F_{MO}/F_{MS}) \sim \text{constant}$$

where *MO* and *MS* stand for oxides and sulfides, respectively. The ratios have values between 0.61 and 0.81. From Eqs. (12) and (19) the above ratio becomes

$$(D_{MO}/D_{MS})/(F_{MO}/F_{MS}) \cong (D_{O\text{–}O}/D_{S\text{–}S})^{1/2}\ (F_{S\text{–}S}/F_{O\text{–}O})^{1/2} = 0.70 \qquad (26)$$

using data on O_2 and S_2 (*10,19*). The value obtained from Eq. (26) agrees quite well with the observations of Hauge and Margrave, and provides some physical meaning to these observations.

Similar matrix studies on the first-period transition metal sulfides have been reported by DeVore and Franzen (*22*). An averaged value of 0.769 ± 0.041 is found for $D_{MO}K_{MS}/D_{MS}K_{MO}$; this is compared with the value of 0.70 from Eq. (26). DeVore and Gallaher (*23*) also reported vibrational frequencies of gaseous group IV transition metal nitrides and found the value of $(D_{MN}/D_{MO})/(K_{MO})$ to be 0.87~0.90, MN standing for metal nitrides. This is compared with the value of 1.0 using Eq. (26) by substituting N for S. It seems that these experimental observations can be correlated with the empirical equations described here, which can be used for estimating the molecular properties of other molecules.

We now describe several examples in which the empirical equations originally proposed for gaseous molecules can be applied to molecules adsorbed on surfaces and to the chemisorptive bonds formed. The successful application to surface properties is encouraging for the various empirical methods proposed which should provide useful information to the surface studies.

The binding energies and vibrational frequencies of halogens adsorbed on metal surfaces have been calculated using Eqs. (12) and (19) (*24*). The observed binding energies of F, Cl, and Br on Mo surfaces are 4.6–4.65, 4.11–4.15, and 3.65–3.70 eV, respectively (*25*). Similar binding energies have also been observed for Nb and W surfaces (*26*). Using Eq. (19) in the form

$$E_{A\text{-}M} \cong E_{B\text{-}M}\,(E_{AA}/E_{BB})^{1/2} \tag{27}$$

with M being the metal and A and B the halogens, the calculated binding energies are 4.81–4.86 eV for F–Mo, 3.67–3.71 eV for Br–Mo, and 3.24–3.27 eV for I–Mo; the observed value for Cl–Mo is used as reference in these calculations. The agreement with experiment is excellent for Br–Mo and fairly good for F–Mo.

The vibrational frequencies of halogens adsorbed on Ag surfaces are 240 cm^{-1} for Cl–Ag, 170 cm^{-1} for Br–Ag, and 110 cm^{-1} for I–Ag (*27*). Using Eq. (12) and the relation $2\pi\nu = (F/\mu)^{1/2}$, ν being the vibrational frequency, F the force constant, and μ the reduced mass, the following relation is obtained

$$\nu_{A\text{-}M}{}^{2} \cong \nu_{B\text{-}M}{}^{2}\,(m_B/m_A)\,(F_{AA}/F_{BB})^{1/2} \tag{28}$$

Here, for the "diatomic" A-M, where M is the metal, μ becomes essentially m_A, the mass of atom A. The calculated vibrational frequencies are 424 cm^{-1} for F–Ag, 149 cm^{-1} for Br–Ag, and 108 cm^{-1} for I–Ag; the observed Cl–Ag value is used as reference. The calculated frequencies are excellent for I–Ag, but show a deviation of 21 cm^{-1} for Br–Ag. It is interesting to notice that similar results have been obtained from a theoretical calculation that reported 437 cm^{-1} for F–Ag, 241 cm^{-1} for Cl–Ag, 150 cm^{-1} for Br–Ag, and 113 cm^{-1} for I–Ag, with a similarly large deviation for Br–Ag (*28*). These results indicate that Eq.(28) can be useful to calculating vibrational frequencies for the surface chemisorptive bonds.

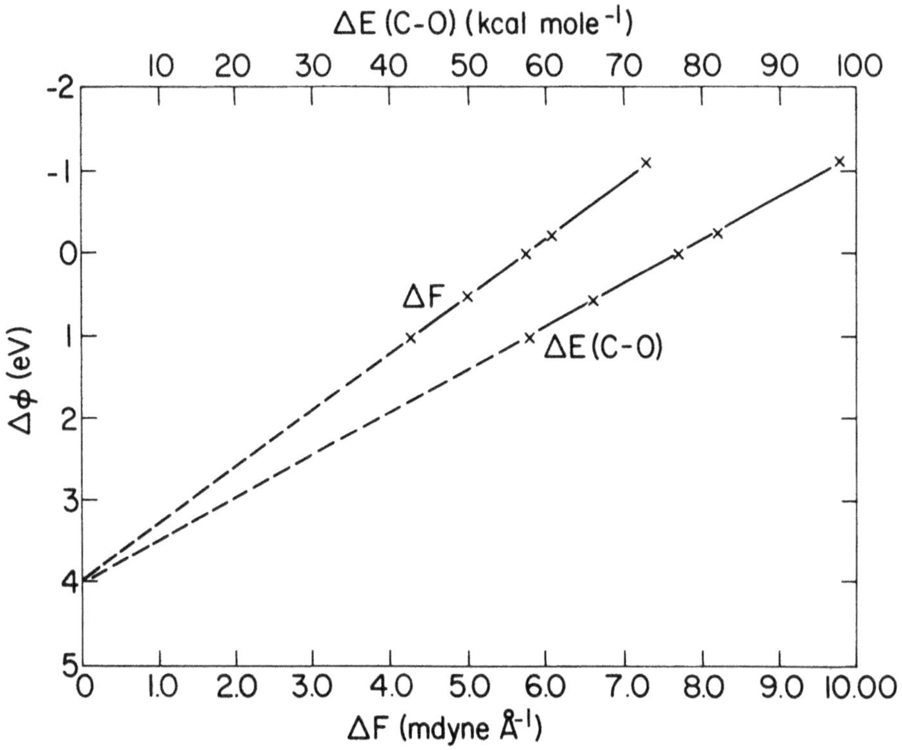

Fig. 3. Calculated force constant changes of chemisorbed CO, ΔF, and weakenings in C—O bond energy, ΔE(C—O), at various work function changes of Ni(*111*), $\Delta\phi$, for the CO—Ni(*111*) system.

Another example of calculating surface bonding properties is the chemisorptive CO—Ni (*111*) system (*29*). The stretching frequency of the chemisorbed C—O decreases with decreasing coverage, Θ, of CO from $\Theta \sim 0.41$ to 0, and further decreases with coadsorption of ethylene and benzene (*30*). These are accompanied by a decrease in work function of Ni (*30*). The vibrational frequencies observed are less than the gas-phase value of CO, an indication of weakening of the C—O bond upon chemisorption. The observed vibrational frequencies were used to calculate such weakenings in the force constant of C—O and hence those in the C—O bond energy. For example, the calculated force constant of C—O is 14.75 mdyn $Å^{-1}$ for $\nu = 1911$ cm^{-1} at $\Theta = 0.41$. This compares excellently with the theoretical value of 14.9 mdyn $Å^{-1}$ of Richardson and Brashaw, who used $\nu = 1932$ cm^{-1} in their cluster calculations (*31*). The empirically calculated changes in both force constant, ΔF, and bond energy of C—O, ΔE(C—O), are correlated with the observed changes in work function of Ni, $\Delta\phi$, and shown in Fig. 3. The extrapolated value of $\Delta\phi = 4.0$ eV for zero changes of ΔF and ΔE (C—O) means that the weakening in C—O bond for the adsorbed CO molecule should vanish at this $\Delta\phi$. In other words, CO would cease to adsorb on Ni(*111*) beyond this threshold $\Delta\phi$, which is interesting and can be technologically significant for surface catalysis. Further analysis gives a C—Ni bond energy of 34.5 kcal mol^{-1} for the O—C—Ni(*111*) system at $\Theta = 0.41$. This agrees very well with the Ni—CO bond energy of 35.1 kcal mol^{-1} for the $Ni(CO)_4$ molecule (*32*).

These examples demonstrate the applicability of empirical methods to the surface bonding properties. Extension to other surface systems using the empirical equations reviewed here should allow more understanding of the molecular properties of surfaces. Combining empirical and *ab initio* results can lead to further insights into the various molecular properties. For example, Ferrante et al. have shown a single binding energy relation for diatomic molecules, metallic adhesion, cohesion, and chemisorption (*33*).

IV. SUMMARY

The various empirical methods for calculating force constants, bond energies, and bond lengths have been briefly reviewed. The equations proposed allow calculations of one of these properties by relating it either to the other two, to such properties as electronegativity, or to those of similar molecules. Using available data one is able to estimate these molecular properties for different types of bonding.

It is interesting and encouraging that the empirical equations, originally proposed for gaseous molecules, also apply to molecules trapped in matrices, adsorbed on surfaces, and to the chemisorptive bonds involved. Extension to other chemical systems and to different molecular properties should lead to further applications of the many empirical equations proposed, and to better understandings of the various systems studied.

REFERENCES

1. Badger, R. M., *J. Chem. Phys.* **2,** 128 (1934); **3,** 710 (1935).
2. Clark, C. H. D., *Phil. Mag.* **18,** 459 (1934).
3. Allen, H. S., and Longair, A. K., *Nature* **135,** 764 (1935).
4. Gordy, W., *J. Chem. Phys.* **14,** 305 (1946).
5. Guggenheimer, K. M., *Proc. Phys. Soc. (London)* **58,** 456 (1946).
6. Somayajulu, G. R., *J. Chem. Phys.* **28,** 814 (1958).
7. Varshni, Y. P., *J. Chem. Phys.* **28,** 1078, 1081 (1958).
8. Herschbach, D. R., and Laurie, V. W., *J. Chem. Phys.* **35,** 458 (1961).
9. Chang, C-A., *High Temp. Sci.* **6,** 276 (1974).
10. Jones, L. H., *Inorganic Vibrational Spectroscopy,* Vol. 1, Dekker, NY 1971.
11. Bartlett, R. J., and Parr, G. G., *J. Chem. Phys.* **67,** 5828 (1977).
12. Ohwada, K., *J. Chem. Phys.* **72,** 1 (1980), and references therein.
13. Chang, C-A., *J. Phys. Chem.* **87,** 1694 (1983).
14. Pauling, L., *Nature of the Chemical Bond,* 3rd Ed., Cornell Press, New York, 1960.
15. Evans, R. S., and Huheey, J. E., *J. Inorg. Nucl. Chem.* **32,** 373, 383, 777 (1970).

16. Van Hooydonk, G., *Theoret. Chim. Acta* **22,** 157 (1971).
17. Sanderson, R. T., *Chemical Bonds and Bond Energy,* Academic Press, New York, 1976.
18. Matcha, R. L., *J. Am. Chem. Soc.* **105,** 4859 (1983).
19. Gaydon, A. G., *Dissociation Energies and Spectra of Diatomic Molecules,* 3rd Ed., Chapman and Hall, London, 1968.
20. Schomaker, V., and Stevenson, D. P., *J. Amer. Chem. Soc.* **63,** 37 (1941).
21. Hauge, R. H., and Margrave, J. L., *High Temp. Sci.* **4,** 170 (1972).
22. DeVore, T. C., and Franzen, H. F., *High Temp. Sci.* **7,** 220 (1975).
23. DeVore, T. C., and Gallaher, T. N., *J. Chem. Phys.* **70,** 3497 (1979).
24. Chang, C-A. *Chem. Phys. Lett.* **91,** 277 (1982).
25. Bulbach, G., and Balais, J. C., *Surf. Sci.* **111,** 575 (1981); Scheer, M. D., Klein, R., and McKinley, J., D., *Surf. Sci.* **30,** 251 (1972).
26. Fanlian, K., and Bauer, E., *Phys. Lett.* **54A,** 313 (1975); Biedermann, B., and Wasmuth, H. W., *Proceedings of the 7th International Vacuum Congress* and *3rd International Conference on Solid Surfaces,* Vienna, 1979, p. 1097; Price, G. G., Rawlings, K. L., and Hopkins, B. J., *Surf. Sci.* **85,** 379 (1979); Bonczek, F., Engel, T., and Bauer, E., *Surf. Sci.* **97,** 595 (1980).
27. Creghton, J. A., Albrecht, G. M., Hester, R. E., and Matthew, J. A. D., *Chem. Phys. Lett.* **55,** 55 (1978); Pemble, M. M. E., as quoted in ref. 28.
28. Nicols, H., and Hexter, R. M., *J. Chem. Phys.* **24,** 2059 (1981).
29. Chang, C-A., *Surf. Sci.* **95,** L239 (1980).
30. Bertolini, J. C., Dalmai-Imelik, G., and Rousseau, J., *Surf. Sci.* **68,** 539 (1977).
31. Richardson, N. V., and Bradshaw, A. M., *Surf. Sci.* **88,** 255 (1979).
32. Behrens, R. G., *J. Less-Common Metals* **58,** 47 (1978).
33. Ferrante, J., Smith, J. R., and Rose, J. H., *Phys. Rev. Lett.* **50,** 1385 (1983).

Author Index

Subject Index

L

M

N

P

GPSR Compliance

The European Union's (EU) General Product Safety Regulation (GPSR) is a set of rules that requires consumer products to be safe and our obligations to ensure this.

If you have any concerns about our products, you can contact us on ProductSafety@springernature.com

In case Publisher is established outside the EU, the EU authorized representative is:

Springer Nature Customer Service Center GmbH
Europaplatz 3
69115 Heidelberg, Germany

Batch number: 10837784

Printed by Printforce, the Netherlands